# Advanced Electric Drive Vehicles

# Energy, Power Electronics, and Machines

Series Editor
*Ali Emadi*

# Advanced Electric Drive Vehicles

## Ali Emadi

McMaster University, Hamilton, Ontario, Canada

CRC Press
Taylor & Francis Group
Boca Raton   London   New York

CRC Press is an imprint of the
Taylor & Francis Group, an **informa** business

CRC Press
Taylor & Francis Group
6000 Broken Sound Parkway NW, Suite 300
Boca Raton, FL 33487-2742

First issued in paperback 2017

© 2015 by Taylor & Francis Group, LLC
CRC Press is an imprint of Taylor & Francis Group, an Informa business

No claim to original U.S. Government works

Version Date: 20140825

ISBN 13: 978-1-138-07285-5 (pbk)
ISBN 13: 978-1-4665-9769-3 (hbk)

**Visit the Taylor & Francis Web site at**
**http://www.taylorandfrancis.com**

**and the CRC Press Web site at**
**http://www.crcpress.com**

*To my family*

**Ali Emadi**

# Contents

# Preface

Electrification is an evolving paradigm shift in the transportation industry toward more efficient, higher performance, safer, smarter, and more reliable vehicles. There is in fact a clear trend to shift from internal combustion engines (ICEs) to more integrated electrified powertrains. Nonpropulsion loads, such as power steering and air-conditioning systems, are also being electrified. Electrified vehicles include more electric vehicles (MEVs), hybrid electric vehicles (HEVs), plug-in hybrid electric vehicles (PHEVs), range-extended electric vehicles (REEVs), and all electric vehicles (EVs) including battery electric vehicles (BEVs) and fuel cell vehicles (FCVs).

This book begins with an introduction to the automotive industry and explains the need for electrification in Chapter 1. Parallels with other industries such as the telecommunications industry are highlighted. Chapter 1 also explains how the paradigm shift began with MEVs, was established by HEVs, is gaining momentum by PHEVs and REEVs, and will be completed by EVs.

Chapters 2 and 3 present the fundamentals of conventional vehicles and ICEs, respectively. Chapters 4 through 7 focus on the major components of electrified vehicles including power electronic converters, electric machines, electric motor controllers, and energy storage systems. Chapter 8 introduces hybrid battery/ultra-capacitor energy storage systems with applications in advanced electric drive vehicles.

Chapter 9 presents the electrification technologies applied to nonpropulsion loads with low-voltage electrical systems. 48 V electrification and belt-driven starter generator systems are explained in Chapter 10, while Chapters 11 and 12 introduce the fundamentals of hybrid powertrains and HEVs, respectively. Chapter 13 is then focused on chargers needed for plug-in vehicles. PHEVs are studied in Chapter 14. EVs and REEVs are described in Chapter 15. In addition, vehicle-to-grid (V2G) interface and electrical infrastructure issues are presented in Chapter 16. Finally, Chapter 17 deals with energy management and optimization in advanced electric drive vehicles.

This book is planned as a comprehensive textbook covering major aspects of advanced electric drive vehicles for a graduate or senior-level undergraduate course in engineering. Each chapter includes various illustrations, practical examples, and case studies. This book is also an easy-to-follow reference on electrified vehicles for engineers, managers, students, researchers, and other professionals who are interested in transportation electrification.

I would like to acknowledge the efforts and assistance of the staff of Taylor & Francis/CRC Press, especially Ms. Nora Konopka, Ms. Jessica Vakili, and Ms. Michele Smith. I would also like to thank Mr. Weisheng Jiang for his kind efforts in preparing many of the illustrations in this book.

**Ali Emadi**
*November 2014*

# Editor

**Ali Emadi** (IEEE S'98-M'00-SM'03-F'13) received BS (1995) and MS (1997) in electrical engineering with the highest distinction from Sharif University of Technology, Tehran, Iran. He received his PhD (2000) in electrical engineering from Texas A&M University, College Station, Texas. He is currently the Canada Excellence Research Chair (CERC) in Hybrid Powertrain and the director of McMaster Institute for Automotive Research and Technology (MacAUTO) at McMaster University in Hamilton, Ontario, Canada. Before joining McMaster University, Dr. Emadi was the Harris Perlstein endowed chair professor of engineering and the director of the Electric Power and Power Electronics Center and Grainger Laboratories at Illinois Institute of Technology (IIT) in Chicago, where he established research and teaching facilities as well as courses in power electronics, motor drives, and vehicular power systems. In addition, Dr. Emadi was the founder, chairman, and president of Hybrid Electric Vehicle Technologies, Inc. (HEVT)—a university spin-off company of IIT.

Dr. Emadi is the recipient of numerous awards and recognitions. He was named a Chicago Matters Global Visionary in 2009. He was named the Eta Kappa Nu Outstanding Young Electrical Engineer of the Year 2003 (a single international award) by virtue of his outstanding contributions to hybrid electric vehicles by the Electrical Engineering Honor Society. He also received the 2005 Richard M. Bass Outstanding Young Power Electronics Engineer Award from the IEEE Power Electronics Society. In 2005, he was selected as the best professor of the year by the students of IIT. Dr. Emadi is the recipient of the 2002 University Excellence in Teaching Award from IIT as well as the 2004 Sigma Xi/IIT Award for Excellence in University Research. He directed a team of students to design and build a novel motor drive, which won the First Place Overall Award of the 2003 IEEE International Future Energy Challenge for Motor Competition. In addition, he was the advisor of the Formula Hybrid Teams at IIT and McMaster University, which won the *GM Best Engineered Hybrid Systems Award* at the 2010 and 2013 Formula Hybrid Competitions, respectively.

Dr. Emadi is the principal author/coauthor of over 300 journal and conference papers as well as several books including *Vehicular Electric Power Systems: Land, Sea, Air, and Space Vehicles* (Marcel Dekker, 2003), *Energy Efficient Electric Motors* (Marcel Dekker, 2004), *Uninterruptible Power Supplies and Active Filters* (CRC Press, 2004), *Modern Electric, Hybrid Electric, and Fuel Cell Vehicles: Fundamentals, Theory, and Design,* Second Edition (CRC Press, 2009), and *Integrated Power Electronic Converters and Digital Control* (CRC Press, 2009). Dr. Emadi is also the editor of the *Handbook of Automotive Power Electronics and Motor Drives* (CRC Press, 2005).

Dr. Emadi was the general chair of the 1st IEEE Vehicle Power and Propulsion Conference (VPPC'05), which was co-located under his chairmanship with the Society of Automotive Engineers (SAE) International Future Transportation Technology Conference. He was also the general chair of the 2011 IEEE VPPC. He is currently the steering committee chair of the IEEE Transportation Electrification Conference and Expo (ITEC). In addition, Dr. Emadi was the inaugural general chair of the 2012 ITEC. He has served as the chair of the IEEE Vehicle Power and Propulsion Steering Committee, the chair of the Technical Committee on Vehicle and Transportation Systems of the IEEE Power Electronics Society, and the chair of the Power Electronics Technical Committee of the IEEE Industrial Electronics Society. He has also served as the chair of the 2007 IEEE International Future Energy Challenge.

Dr. Emadi is the editor-in-chief of the *IEEE Transactions on Transportation Electrification*. He is also the editor (North America) of the *International Journal of Electric and Hybrid Vehicles*. He was the guest editor-in-chief of the special issue on *Transportation Electrification and Vehicle Systems, IEEE Transactions on Power Electronics*. He has also served as the guest editor-in-chief of the special issue on *Transportation Electrification and Vehicle-to-Grid Applications*, *IEEE Transactions on Smart Grid*, and the guest editor-in-chief of the special issue on *Automotive Power Electronics and Motor Drives*, *IEEE Transactions on Power Electronics*. He has also been the guest editor of the special section on *Hybrid Electric and Fuel Cell Vehicles*, *IEEE Transactions on Vehicular Technology*, and the guest editor of the special section on *Automotive Electronics and Electrical Drives*, *IEEE Transactions on Industrial Electronics*.

# Contributors

**Florence Berthold**
Department of Electrical and Computer
  Engineering
Concordia University
Montreal, Quebec, Canada

**Berker Bilgin**
McMaster Institute for Automotive Research
  and Technology (MacAUTO)
McMaster University
Hamilton, Ontario, Canada

**Giampaolo Carli**
Department of Electrical and Computer
  Engineering
Concordia University
Montreal, Quebec, Canada

**Ilse Cervantes**
Institute for Scientific and Technological
  Research of San Luis Potosi (IPICyT)
San Luis Potosi, Mexico

**Ali Emadi**
McMaster Institute for Automotive Research
  and Technology (MacAUTO)
McMaster University
Hamilton, Ontario, Canada

**Lucia Gauchia**
Michigan Technological University
Houghton, Michigan

**Oliver Gross**
Oxford, Michigan

**Ruoyu Hou**
McMaster Institute for Automotive Research
  and Technology (MacAUTO)
McMaster University
Hamilton, Ontario, Canada

**Weisheng Jiang**
McMaster Institute for Automotive Research
  and Technology (MacAUTO)
McMaster University
Hamilton, Ontario, Canada

**Alireza Khaligh**
Electrical and Computer Engineering
  Department
University of Maryland
College Park, Maryland

**Mariam Khan**
McMaster Institute for Automotive Research
  and Technology (MacAUTO)
McMaster University
Hamilton, Ontario, Canada

**Mahesh Krishnamurthy**
Electric Drives and Energy Conversion
  Laboratory
Illinois Institute of Technology
Chicago, Illinois

**William Long**
McMaster Institute for Automotive Research
  and Technology (MacAUTO)
McMaster University
Hamilton, Ontario, Canada

**Pierre Magne**
McMaster Institute for Automotive Research
  and Technology (MacAUTO)
McMaster University
Hamilton, Ontario, Canada

**Pawel P. Malysz**
McMaster Institute for Automotive Research
  and Technology (MacAUTO)
McMaster University
Hamilton, Ontario, Canada

**Fariborz Musavi**
Novum Advanced Power, CUI, Inc.
Portland, Oregon

**Nicholas J. Nagel**
Research and Development, Triumph
  Aerospace Systems
Seattle, Washington

**Omer C. Onar**
National Transportation Research Center
Oak Ridge National Laboratory
Knoxville, Tennessee

**Josipa G. Petrunić**
McMaster Institute for Automotive Research
    and Technology (MacAUTO)
McMaster University
Hamilton, Ontario, Canada

**Anand Sathyan**
Electrical and Computer Engineering
    Department
McMaster University
Hamilton, Ontario, Canada

**Arash Shafiei**
Department of Electrical and Computer
    Engineering
Concordia University
Montreal, Quebec, Canada

**Xiaodong Shi**
Electric Drives and Energy Conversion
    Laboratory
Illinois Institute of Technology
Chicago, Illinois

**Piranavan Suntharalingam**
McMaster Institute for Automotive Research
    and Technology (MacAUTO)
McMaster University
Hamilton, Ontario, Canada

**Sheldon S. Williamson**
Department of Electrical and Computer
    Engineering
Concordia University
Montreal, Quebec, Canada

**Sanjaka G. Wirasingha**
Electrical and Computer Engineering
    Department
McMaster University
Hamilton, Ontario, Canada

**Fengjun Yan**
Department of Mechanical Engineering
McMaster University
Hamilton, Ontario, Canada

**Hong H. Yang**
Electrical and Computer Engineering
    Department
McMaster University
Hamilton, Ontario, Canada

**Yinye Yang**
McMaster Institute for Automotive Research
    and Technology (MacAUTO)
McMaster University
Hamilton, Ontario, Canada

**Mengyang Zhang**
Chrysler Group LLC
Auburn Hills, Michigan

# 1 Automotive Industry and Electrification

*Ali Emadi and Josipa G. Petrunić*

## CONTENTS

Vehicles constitute the crucial components of modern industrial life. The ubiquity of cars, vans, sports utility vehicles (SUVs), and trucks for personal and industrial transportation today is the result of an industrial revolution that began in Europe and North America in the nineteenth century—a revolution that privileged the use of internal combustion engines (ICEs) as the primary means of motive power. Yet, the twentieth century witnessed the rise of serious environmental, economic, and social concerns related to greenhouse gases (GHGs) from ICE-powered transportation. These concerns reshaped the manufacturing landscape, forcing automotive manufacturers to rethink the way they designed automobiles. They also recalibrated consumer and public expectations of efficiency and sustainability in transportation overall.

A new "sustainability" mantra is emerging as the pre-eminent metric of transportation technologies. This chapter has been prepared with one overarching principle in mind—the electrification of transportation is the primary means by which we can ensure the automotive industry becomes "sustainable" over the next half-century.

This globally sustainable transportation system, which is electrified, is what we call Transportation 2.0 (Emadi 2011). Transportation 2.0 describes a world in which profound technological shifts in advanced electric drive vehicles and transportation networks reshape the automotive industry, focusing its attention on the most efficient source of motive power—namely, electricity. Companies will seek to ensure that these new technologies are scalable, marketable, and profitable, while consumers will increasingly demand electric drive vehicles, which are low-cost, low maintenance, safe, secure, reliable, rugged, and eco-friendly.

Achieving these varied outcomes will require industrial and academic investment in hybrid, plug-in hybrid, and fully electric power trains, including the development of superior electric motors, power electronics and controllers, embedded software, batteries and energy-storage devices, and micro-and smart grid interface systems. It will also require highly skilled workers who understand the theoretical and practical aspects of these advanced electric drive vehicles to guide automotive companies over the next few decades of innovation.

This introductory chapter presents readers with an overview of the innovations that have informed the development of advanced electric drive vehicles over the past century and it points to innovative pathways for the future. First, we begin with a brief historical review of the automotive technologies that formed the Transportation 1.0 paradigm throughout the twentieth century. We present the concept of "sustainability" as a core concept embodied in the Transportation 2.0 paradigm—a worldview that incorporates a radical shift toward advanced electric drive vehicles and electrified transportation. Then, we explore disruptive technologies already shaping industrial trends in transportation electrification, including the continuum of more electric vehicles (MEVs), hybrid electric vehicles (HEVs), plug-in hybrid electric vehicles (PHEVs), and electric vehicles (EVs). Finally, we explore future technologies associated with power electronics and controllers, electric machines, batteries and ultra-capacitors, and electrical grid innovations—all of which will enable the development of improved electric drive vehicles throughout the twenty-first century.

## 1.1  FROM THE FIRST EVs TO A TRANSPORTATION 2.0 FUTURE

Electrified transportation has a rich history extending back more than 150 years. Electric motors were first developed in the early nineteenth century based on experiments with magnets. But, electric motors were not used for transportation purposes until after the 1850s, when electric machines were experimentally incorporated into trains, ships, and personal carriages (or "cars"). These experiments resulted in a series of novel electric drive vehicles for personal transportation—many of which dominated the car market until World War One.

Thomas Davenport is usually credited with building the first practical one-person "electric vehicle" in 1834, followed by a two-person version in 1847. The first EV to emerge, which resembled what might be called today a "car," appeared in 1851 and traveled at a pace of approximately 20 mph (32 km/h).

Decades later, the first mass-manufactured electric cars hit the marketplace with the development of the Edison Cell, a nickel–iron battery that enabled what was an already-burgeoning marketplace. The Edison Cell had greater storage capacity than the batteries used in early EVs and prototypes. They were also rechargeable, allowing car designers to produce vehicles that were convenient for middle-class consumers who could recharge their cars at public or private charging stations (many of which were installed along city streets or at homes with preferential electricity rates). By the 1900s, EVs had captured a notable share of the leisure car market. Among the 4200 automobiles sold in the United States in 1900, 38% were electric and only 22% were gasoline, while another 40% were still steam driven (Electric Vehicles, 2008).

Despite the popularity of electric cars, however, they were expensive and their driving ranges were limited. In contrast, gasoline-and diesel-powered vehicles emerged by World War One as cheaper, more powerful modes of transportation. Between 1890 and 1914, German and American engineers and manufacturers shifted interest *away* from EVs and steam engines toward ICEs by designing heat engines that could power ships, trains, and vehicles, as well as many other industrial and manufacturing applications.

In 1872, the American inventor, George Brayton, manufactured "Brayton's Ready Motor"; he used constant pressure combustion and liquid fuel to produce what is often tagged as the first "internal combustion engine" (ICE). Gottlieb Daimler and his partner, Wilhelm Maybach, issued the world's first combustion engine *vehicle* in the 1880s when they patented a gasoline-powered car that used a version of Nicolaus Otto's "Atmospheric Engine." In the 1890s, Rudolf Diesel recognized the potential for a combustion engine that did not require an external ignition system. He obtained a patent for the design of a compression–ignition engine in which fuel was injected at the end of the compression cycle. The fuel was ignited by the high temperatures resulting from compression. Diesel's engine was marketed as a high-pressure, high-efficiency engine durable enough for railroad locomotives, large trucks, ships, and automobiles. The German manufacturer, Karl Benz, then developed innovations such as the carburetor, an electrical ignition system, water-cooling systems, and improved steering, and by the early 1900s, Benz had marketed ICEs widely as "reliable, fast, durable, elegant."

It was in the context of these emergent ICE technologies that the American business guru and inventor, Henry Ford, launched an assembly line-manufacturing process to mass produce Model T Fords (also known as "Tin Lizzies"), which used ICEs for motive power. In 1908, these low-cost cars were sold for less than a quarter of the price of contemporary EVs. The Model T thus made personal vehicle ownership a viable option for working-class families. Throughout the 1910s, petroleum companies also lobbied municipalities and regional governments to switch to ICE-powered public transit systems, since many early transit systems had been originally electric.

The rising prominence of ICEs throughout the first decades of the twentieth century created what we call the Transportation 1.0 paradigm. The transportation systems at the core of this paradigm operated on the basis of unhindered, cheap, and constant supplies of fossil fuels as sources for motive power. In the Transportation 1.0 world, EVs were relegated to niche markets, useful for lightweight delivery trucks or milk vans, but they were not marketed for mass consumption or designed for daily commutes.

The Transportation 1.0 paradigm also shaped the public's perception of automobiles. The inefficiency of ICEs became an acceptable norm for manufacturers, while cutting-edge automotive innovations in power-train efficiency stagnated. Indeed, the most significant automotive innovations in the post-World War Two era were those advancing greater horsepower, bigger engines, and stronger materials.

The concept of highly efficient, clean, smart, and interconnected transportation networks remained largely unexplored until the 1990s. By that decade, a combination of environmental, geo-political, and social concerns impelled governments across North America and Europe to change the way they understood and manufactured automobiles. Combustion-powered transportation became the source of a never-ending stream of problems. It was dirty. It was insecure. And, it promised to cause astronomical increases in health-care costs due to urban smog and environmental degradation. By 2000, the belief in low-priced, easily available petroleum had also dissipated in many regions of the world, and especially in North America and Europe where governments had experienced serious price shocks due to repeated oil crises throughout the 1970s. Gone, too, was the belief that humans had little to do with environmental decay. Rapidly rising concentrations of GHGs in the Earth's atmosphere, combined with predictive models demonstrating the catastrophic outcomes associated with global climate change (including unpredictable weather patterns, damaged food crops, and rising ocean levels), motivated the United States in particular to develop more rigorous and stringent Corporate Average Fuel Economy (CAFE) standards, which today constitute some of the world's most stringent emission standards. The CAFE standards have, in fact, served as a prime motivating factor over the past decade informing the redesign and manufacture of newly efficient and highly electrified vehicles.

In sum, the recent combination of environmental, geo-political, economic, and health concerns related to ICE-powered transportation has created fertile ground for renewed interest and investment in electric drive vehicles and electrified transportation.

Today, there are more than 1 billion registered vehicles in the world—approximately one for every seven people on the Earth. That number is set to drastically rise over the next few decades with the increasing wealth and purchasing power among citizens in rapidly developing economies in Asia and South America. Already, more than 80 million vehicles are produced worldwide on an annual basis. There is evident need to make these cars cleaner, more efficient, and less socially harmful so as to avoid both environmental and economic disaster over the next century.

The Transportation 2.0 paradigm envisions a world that travels using clean, efficient, safe, reliable, powerful, and intelligent mobility options. It is based on the premise that "transportation electrification" will involve a series of enabling technologies that allow consumers, manufacturers, and governments to achieve sustainable mobility over the next century. The profundity of a shift to a Transportation 2.0 world is best captured by an analogy to a previous period of irrevocably profound revolution in transportation networks—namely, the age of steam.

The development of the steam engine in the late-1700s and its incorporation into transcontinental travel throughout the 1800s radically transformed global society and domestic economies. The Scottish inventor, James Watt, propelled the earliest innovations in steam-powered mobility by patenting an efficient steam engine in 1769. By the 1780s, the American inventor, John Fitch, used Watt's engine to power a boat, and by the 1810s, the business leader and inventor, Robert Fulton, turned steam-driven boats and ships into major commercial successes, using them to ferry goods and people between Europe and North America. Over the span of 40 years, steam mobility sparked an industrial revolution and enabled the relatively rapid shipment of raw products (such as cotton) and manufactured goods (such as linens) worldwide, creating new international markets and feeding the rising powers of Great Britain and the United States. Indeed, steam-powered mobility enabled the establishment and expansion of the British Empire as the world's most dominant political and economic power from 1790 to 1914—technologically fostering the colonization of huge swathes of the world from South America to Africa across to Asia.

Transportation 2.0 implies a similar level of profound social, economic, political, and technological transformation. This new paradigm may well be emblematic of the twenty-first century. It will alter the way people move, the way they transit from place to place, and the way they think about transportation in their daily lives. It will alter global trade networks by making modes of goods transportation more efficient, less costly, and more reliable. The cost savings associated with a shift to electrified mobility hold the potential to grow economies in the developed and developing worlds alike.

The rise of advanced electric drive vehicles will also provide individual consumers and travelers with the power to determine the price of their transport, by choosing when and how to charge their vehicles and when to connect with a grid system to sell electricity back to grid operators. It will liberate them from the stranglehold of limited, nonrenewable petro-fuel supplies (since electric drive vehicles can be powered by batteries charged through micro- and off-grid systems using solar panels, wind turbines, and local energy-storage devices). Ubiquitous electrified transportation systems will also liberate nation–states from uncertain foreign supplies of petroleum by allowing them to rely on domestically produced electricity as a primary source of fuel for mobility.

Meanwhile, the incorporation of "smart" technologies will liberate time associated with vehicle maintenance by freeing drivers from dreary duties at pumping stations. The integration of "autonomous" vehicle technologies—and intelligent driving mechanisms (as with "autonomous vehicles")—holds out the possibility of creating urban environments free from traffic-congested highways and roadways, where efficient horizontal transit systems are optimized to transport people and goods using ideal routes.

In some ways, the concept of Transportation 2.0 implies a world of infinite possibilities. Just as eighteenth-century traders who relied on sail power could not have foreseen the highly industrialized steam-powered revolution that altered the world a century later, we believe many consumers, manufacturers, and policy makers today will struggle to envision what the Transportation 2.0 *could* look like, precisely because it be so fundamentally different from the Transportation 1.0 world we currently live in. But, what is known is that this new world of electrified transportation will be one in which vehicles and grid infrastructures will be seamlessly interwoven into a network of efficient, low-cost, safe, reliable, smart, high-performance, and ultimately "sustainable" mobility options. When it arrives, Transportation 2.0 will seem entirely natural and obvious to the transit user or car driver. Vehicle operators will wonder how they ever lived in a world dominated by ICEs in the first place.

Just as steam power changed the way people traveled, the way enterprises interacted, and the way societies were networked, the rise of ubiquitous electric drive vehicles will also change the way humanity operates on a daily basis, including daily commutes, leisure travel, business trade, or nation–state interaction.

## 1.2   THE CONTINUUM OF AUTOMOTIVE ELECTRIFICATION

For over a century, vehicle performance has dominated the consumer dialogue about automobiles. Greater power, greater range, more space, and increasing luxuriousness have been norms used to judge the value of automotive innovation. Starting in the 1990s, contemporary EVs suffered from a branding crisis. They were perceived to be weak, slow, and short ranged compared to their powerful ICE cousins. Twenty years later, however, consumers and industrial manufacturers have reassessed these initial assumptions. Today, the powerful performance characteristics of electrically driven vehicles are evident. The efficiency of electric motors combined with their instantaneous torque has helped EVs enter the contemporary automotive marketplace as clean *and* powerful options. Meanwhile, performance-minded consumers are lured by the fact that some of the world's fastest vehicles now use electric drives. Since 2010, a fully electric car designed by Venturi Automobiles (in collaboration with Ohio State University) has held the Fédération Internationale de l'Automobile (FIA) record for the fastest electric land vehicle, recording a top speed of just over 495 km/h. Venturi's new model, named "JamaisContente," is aiming for a top speed that will break through the 700 km/h barrier by 2014 (Prince Albert of Monaco unveils for the first time the Venturi VBB-3 the world's most powerful electric car, 2013). These feats of performance dissipate lingering doubts about the power and coolness factors associated with electric drives.

As with most luxury, high-performance vehicles, the vast majority of car owners will not be able to afford a Venturi or even the much less-expensive (but still luxurious) electric Tesla Model S. There are, however, numerous other competitively priced electrified cars in the marketplace today, offering a range of electrification levels from MEVs, HEVs, PHEVs, and fully EVs. The continuum of price-convenient electrified vehicles is going to fundamentally shape the mass transition to a Transportation 2.0 world.

To achieve a sustainable transportation future that produces minimal emissions and maximum efficiencies, the electrification level of all vehicles will need to increase significantly over the next 50 years. Since electrification in automobiles can occur in both propulsion and nonpropulsion systems, the "electrification level" of any given automobile is defined as the percentage of a vehicle's electric power to its total power. Automobiles vary in their electrification level from 0% (where the vehicle contains no electrical system) to 100% (where the vehicle contains only electrical systems). Most conventional cars today hover around 5%–10% in their respective electrification levels. Below, we explore the increasing levels of electrification associated with MEVs, HEVs, PHEVs, and EVs.

### 1.2.1   MORE ELECTRIC VEHICLES

MEVs include those vehicles that incorporate ever-increasing levels of electrification for propulsion and nonpropulsion loads. In general, four different power transfer systems are used in vehicles: mechanical, hydraulic, pneumatic, and electrical systems. Electrical systems are normally much more efficient, faster, and can be controlled more easily. The MEV concept is based on mechanical, hydraulic, and pneumatic systems with with electrical systems.

Demands for higher fuel economy and lower emissions to meet increasingly more stringent government standards and consumer expectations are pushing the automotive industry to seek electrification of ancillary and nonpropulsion loads in addition to the power trains. Nonpropulsion loads include electrically assisted power steering, electrically driven air conditioning, interior and exterior lighting, seat heaters, power windows, power mirrors, pumps, fans, throttle actuation, antilock brak-ing, electrically heated catalytic converters, and so on. Because of the high efficiency of electricity, these auxiliary systems operate most efficiently when serviced by electrical power. Indeed, most vehicles on roads in North America and Europe today are already MEVs, because they incorporate electrified auxiliary systems. The shift toward electrifying an ever-increasing number of auxiliary systems within ICE-powered vehicles has helped to shift conventional automobiles toward ever-higher levels of efficiency overall.

Electrifying nonpropulsion loads raises the electrification factor by modest increments up to 15%–20%. Electrifying a vehicle's mode of propulsion produces a much greater electrification fac-tor, reaching as high as 50%–70% in hybrid electric power trains and nearly 100% in all EVs.

### 1.2.2 HYBRID ELECTRIC VEHICLES

Without doubt, significantly increasing the efficiency of ICE-powered vehicles requires power-train hybridization that goes beyond the electrification of nonpropulsion loads. Radical increases in propulsion efficiency and concomitant decreases in emissions require the introduction of elec-tric machines into a vehicle's drivetrain for the purposes of supplying motive power. HEVs are in fact dual-powered vehicles. They contain both an electric motor and a combustion engine. There are multiple hybrid power-train topologies designed to deliver motive power in these types of vehicles.

In HEVs, a battery powers the electric motor(s). The battery is capable of being charged through regenerative braking, which is a practical system to adopt in city environments where cars are starting and stopping frequently. Regenerative braking takes advantage of the fact that an electric machine can also operate as a generator. Whenever a driver applies the brakes in this scenario, the electric motor turns into a generator to convert the vehicle's forward momentum into electricity (HEVs also use regular friction brakes in combination with regenerative braking).

The battery in an HEV can also be charged by the engine. The heat engine, which is powered by gasoline, diesel, compressed natural gas (CNG), or biofuel, feeds a generator, which then feeds electricity into the battery. Here, the motor is used as a generator to recharge the battery when bat-tery charge runs low. Hybrid vehicles normally operate in charge-sustaining (CS) mode. This means the batteries onboard, in CS mode, never fall below a certain level of charge, because they are con-stantly recharged by virtue of regenerative braking or the use of the electric machine as a generator.

Depending on the make and model of the car, hybrid vehicles can travel part time on electric power alone (with the heat engine shutdown until battery depletion reaches a critical level) or they can be jointly propelled by the electric motor and heat engine working complementarily.

Hybrid vehicles are divided into microhybrids, mild hybrids, power (full) hybrids, and energy hybrids based on the relative size of the electric propulsion system with respect to the ICE and the role and functions performed by the electrical and mechanical propulsion systems. Hybridization factor, in its simplest form, is defined as the ratio between the peak electrical propulsion power and the peak total electrical and mechanical propulsion power. Microhybrids usually have a hybridiza-tion factor in the range of 5%–10% and benefit from the start/stop technology. The hybridization factor of mild hybrids is usually in the range of 10%–25%. Higher hybridization factors are usually associated with power (full) hybrids. An energy hybrid has an energy-storage system (ESS) larger than power hybrids. PHEVs in CS mode are also referred to as energy hybrids because of their larger ESS.

Hybrids can generally come in three topological power-train forms: parallel, series, and series–parallel. In parallel systems, the power train contains an electric motor and an ICE, which are coupled together in parallel mechanically and can power the vehicle in combination or individually. In other words, there are parallel propulsion systems at play. Typically, there is only one electric machine installed in a parallel hybrid power train. The ICE, the electric motor, and the gearbox are coupled using automatic clutches.

In a series hybrid, traction power is delivered by the electric motor(s) while the ICE drives an electric generator that produces power to charge the batteries and drive the electric motor. In series hybrid architecture, the ICE is mechanically decoupled from the wheels. The electric motor(s) is attached to the transmission or directly to the differential/wheels.

The Toyota Prius is an interesting case study to consider when thinking about parallel and series hybrid designs, because the Prius uses a power-split device that allows it to operate either as a series or a parallel hybrid. It is in fact a series–parallel hybrid power train. The power-split device includes a gearbox that links the gasoline engine, the generator, and the electric motor. This device allows the car to operate like a parallel hybrid so that the electric motor can power the car on its own, or the gasoline engine can power the car on its own or in combination with the electric motor. But, the power-split device also enables the car to operate like a series hybrid, where the gasoline engine powers a generator to charge the onboard battery. The Prius's power-split device also allows the generator to start the engine, as with other start/stop systems.

In a series–parallel hybrid power train, two electric machines are used to provide both parallel and series paths for the power. These types of advanced electromechanical hybrid power trains constitute a major technological trend in the industry today.

## 1.2.3  Plug-In Hybrid Electric Vehicles

Plug-in hybrids have an ESS of high-energy density that can be externally charged. They can solely run on electric power for a range longer than regular hybrids, resulting in better fuel economy. This is because the battery packs in PHEVs are typically much larger than those found in HEVs; therefore, their fully electric ranges are also much longer. Plug-in hybrids combine the superior efficiency of powerful electric machines with the security of long-distance range by virtue of onboard engines.

The batteries in PHEVs are designed to deplete so as to maximize the use of efficient electric power. This is called charge-depleting (CD) mode. In a typical CD mode, the engine is off and the onboard batteries are depleted to a predefined low state of charge (SOC); then, the engine turns on and the vehicle operates in a CS mode in which the SOC is usually sustained in a predefined range.

When the vehicle is stationary, recharging occurs by virtue of the plug-in mechanism, which draws power from the main electrical grid system. Deep discharge cycles mean that PHEVs maximize the use of electricity that originates from the electrical grid system. Since the grids that PHEVs can plug into can be supplied with electricity from renewable sources, such as wind and solar, the source of electricity used by PHEVs can be significantly cleaner than the source of electricity currently used by conventional hybrid vehicles. Finally, because PHEVs contain heat engines (of varying sizes depending on the make and model of the vehicle), they liberate drivers from the "range anxiety" often associated with the limited-range barrier tagged to pure battery electric vehicles (BEVs).

PHEVs serve as a crucial bridging technology toward low emissions, and sustainable transportation networks in the future. This newly emerging line of PHEV models emanating from major manufacturing companies ensures that consumers can dramatically increase the electrification and efficiency levels of their personal vehicles. Current examples of PHEVs include the Toyota Prius plug-in hybrid and the Ford C-MAX Energi plug-in hybrid.

Similar to hybrids, PHEVs can also be built using parallel, series, or series–parallel power-train architectures. An external electrical system can charge the onboard battery directly from the grid, while onboard generators powered by heat engines and regenerative braking can still be used to recharge batteries as the vehicle is in use and after the battery has depleted to below a critical SOC level.

In the case of range-extended EV designs, as with the Chevrolet Volt, the topology mostly looks like a series design, with an ICE powering a generator, which powers a motor. In this case, the battery is charged via the plug-in mechanism, which connects the vehicle to the grid system overnight or during other daytime-charging periods. The Chevrolet Volt also allows the heat engine to power the vehicle through a coupling mechanism. The Volt is therefore neither a pure series nor a pure parallel PHEV design; rather, it is referred to as a "range-extended" EV.

Over the past two decades, the emergence of highly effective power electronics has further enabled the development of marketable PHEVs, because improvements to power electronic convert- ers have allowed for innovations in vehicle-interfacing modes, including vehicle-to-home (V2H), vehicle-to-grid (V2G), and vehicle-to-building (V2B) systems. Bidirectional alternating/direct cur- rent (AC/DC) converters, for example, provide an interface between the vehicle and the grid. These converters allow a given vehicle to draw AC power from the grid at a high-power factor and low total harmonic distortion (THD); they then allow the vehicle to feed remnant or unneeded power back onto the grid when required by the grid operator. In the future, therefore, advancements in power electronics will allow PHEVs to serve as mobile energy-storage devices for smart grid systems. Thus, consumers will be able to view their vehicles as revenue generators.

### 1.2.4 ELECTRIC VEHICLES

EVs are powered entirely by their electric propulsion motors and battery packs. Their batteries are charged from the electrical grid system at homes at night, at workplaces during working hours, and at public or commercially owned drive-in charging stations that use grid-connected chargers.

The key benefit associated with EVs is their extreme efficiency as compared to ICEs or even hybridized vehicles. Electric motors can be made to operate with efficiency levels above 90% com- pared to less than 30% for ICEs. A second benefit associated with EVs is their low lifetime mainte- nance costs, which is due in part to lower electricity costs for fuel.

Most importantly, for policy makers, climate scientists, and environmentally minded consum- ers is the fact that EVs produce zero emissions when they operate using electricity drawn from renewable electricity systems, including solar, wind, hydro, and tidal power. Judged from a source- to-wheel (STW) perspective, the only GHGs produced in the entire life cycle of an EV are those associated with the manufacture, assembly, and transportation of the vehicle parts to the assembler and dealer. Thus, EVs offer hope for a form of motive power that will radically reduce current levels of GHG emissions in transportation.

BEVs are limited in their range when compared to HEVs and PHEVs. However, their range is expected to increase significantly due to ongoing advanced research and development in the areas of lightweight materials, high-energy-density storage devices, and electric propulsion motor drives.

Examples of EVs include the Tesla Model S, Ford Focus Electric, Nissan Leaf, Mitsubishi i-MiEV, Fiat 500E, Chevy Spark EV, and the Smart ED, along with a series of recently manufac- tured all EVs made in China, including the Chery QQ3 EV.

## 1.3 ENABLING TECHNOLOGIES FOR TRANSPORTATION ELECTRIFICATION

Electrified vehicles include integrated electromechanical power trains, electric machines, power electronics, embedded software and controllers, and batteries and energy-storage devices. Innovations in these areas, along with lightweight materials technologies, hold the key to disruptive shifts in the transportation paradigm. In fact, mass commercialization of electrified vehicles will require the development of power-train components and controls that are low-cost, rugged, reliable, light in weight, low in volume, and scalable.

AC motors constitute the most common form of electric machines in EVs today. They con- vert electromagnetism into mechanical motion. Common types of electric motors used in electri- fied power trains include permanent magnet motors (which rely on rare-earth metals), induction machines, and switched reluctance motors (SRMs). Each of these machines has its benefits and disadvantages.

Power electronics is used to convert and control electric power. Power electronic converters include DC/DC converters, DC/AC inverters, and AC/DC rectifiers. The future of powerful electrified power trains resides in the development of powerful power electronics and control systems that are high in energy and power density, while also low in cost.

Batteries intended for electrified vehicles are judged by the metrics of power density, energy density, weight, volume, life cycle, temperature range, and of course, cost. Three types of battery devices have been typical of the electric automotive industry over the past 20 years. They include lead–acid batteries (used typically for starting, lighting, and ignition [SLI] applications in vehicles), nickel–metal hydride (NiMH), and lithium-ion (Li-ion) batteries used for traction applications. Some manufacturers of hybrid and fully EVs have recently begun to use Li-ion batteries, because of their higher specific energy as compared to NiMH.

Both NiMH and Li-ion batteries face other obstacles requiring technological fixes in the future. Engineers today are working to improve and optimize the energy-storage capacity of batteries for use in more powerful EVs to ensure longer ranges than those manifest in today's models. In addition, batteries in both PHEVs and EVs must perform well in the long term despite frequent deep discharges. Finally, any new battery design must be scalable to ensure mass production and low costs for manufacturers and consumers alike.

Allied to the emerging battery technologies for use in hybrids and all EVs are ultracapacitors (sometimes called supercapacitors). Ultracapacitors are devices that store and discharge energy rapidly. The development of advanced hybrid energy-storage systems (HESS), which hybridize batteries with ultracapacitors, has allowed engineers to improve the performance of electrically driven vehicles overall by combining batteries as the main source of energy with ultracapacitors as the main source of power.

Ultracapacitors are used in a range of differing motive applications as a means of complementing and supplementing ICEs, batteries, and fuel cells—all of which work best as sources of continuous power, but none of which performs well as a source of sudden power. Ultracapacitors are also used to gather power from regenerative braking systems efficiently. In addition, they can serve as storage devices by recharging rapidly during peak or unexpected periods of surging power generation. In fact, because of their rapid discharge and recharge characteristics, ultracapacitors are versatile in their transportation applications. They can release power to help hybrid vehicles accelerate quickly. They can help vehicles start in very cold weather. They can also provide starting power for start–stop systems in hybrids.

## 1.4   ELECTRICAL GRID SYSTEMS

Plug-in hybrids and fully EVs will have a significant impact on grid systems, both as loads and as energy-storage devices. The Transportation 2.0 paradigm engenders a fundamental shift in grid-side innovations—one that will embrace microgrids, smart grids, advanced energy-storage devices, and advanced energy management systems (EMSs). These innovative technologies will define many of the capabilities and possibilities associated with future electric drives by making grid systems more robust, reliable, and intelligent.

To explain why, let us consider the history of the grid system as it has developed over the past century. The first DC supply system installed for public consumption was Thomas Edison's 1882 system at Pearl Street in New York City. Edison's DC network was not without its opponents. Nikola Tesla and George Westinghouse promoted their AC system as superior to Edison's DC network. Proponents of both systems argued over efficiencies, long-distance transmission capabilities, load requirements, safety levels, and price points.

These internecine debates soon became public policy "wars," resulting in a hodgepodge of electrical supply systems between 1882 and 1914, as state, regional, and municipal governments struggled to determine which system to support, subsidize, and install. The "wars" also brought lighting, telegraphy, and electrical automotive manufacturers to the debating table—each of which advocated for the local system that would best serve the needs of their local industry.

By the turn of the twentieth century, however, installations of AC systems had surpassed those of DC systems in their ubiquity and preference by local and state authorities. This was largely due to the utilization of transformers, which allowed AC to be stepped up or stepped down in voltage depending on need.

Transformers allowed the voltage of ACs to be stepped up for the purposes of long-distance transmission (i.e., from remote generators to regional load centers) and they allowed the voltage to be stepped down into low voltage for distribution to residential, commercial, and industrial loads, such as heating appliances, fans, pumps, and—most importantly—lighting. They served as a crucial enabling technology allowing Tesla and Westinghouse to win a series of public relations battles, and creating ripe conditions for Westinghouse's installation of private AC systems across growing North American cities.

Over the past century, both public and private utilities have incorporated transformers, along with inverters and converters, to supply AC across vast networks that supply power for both AC and DC loads in residential and industrial contexts. Over the same time period, AC networks have become more robust. They evolved from original three-phase systems transmitting 250 kW of power around 1900 to mega-transmission systems that can carry 2000 MW of power in the 2000s. Rather than transporting electricity a few kilometers at a time, as was the norm in the nineteenth century, contemporary AC transmission systems now carry electricity over hundreds or thousands of kilometers.

However, the development of advanced power electronics from the 1970s onward has reopened the wounds associated with the "war of currents," leading engineers to question the long-term viability of AC-dominant systems. First, the rise of electronics and digital communications industries over the past 30 years has meant that both residential and industrial loads are increasingly DC in nature. Everything from computers, printers, and mobile phones to EVs constitute DC loads on the electrical grid system. Second, to accommodate these loads, AC systems have been equipped with converters to transform AC into DC for particular applications. Converters are now common throughout AC systems, but their use as band-aid solutions raises questions about the system's overall efficiency, especially given that the number of DC applications is still rising (and has already surpassed the number and type of AC loads on most electrical grid systems). Consider, for example, highly efficient light-emitting diode (LED) lighting, which is a DC load and which is slowly replacing AC fluorescent lighting. Adjustable speed drives are another important DC load to consider in this transformative process.

### 1.4.1  Microgrids and Renewable Energy Supplies

Direct current systems have always been touted for their simplicity, modularity, and safety. Today, high-voltage DC transmission systems enabled by advanced power electronics are also praised for their higher power ratings and their superior and highly efficient control over power flow, especially in emergency conditions where disruptive natural or sociopolitical events (such as hurricanes, tornadoes, snow storms, or terrorist acts of sabotage) can lead to widespread and dangerous blackouts. Most importantly, major industrialized nations are pushing for ever-greater levels of renewable electricity supplies, and the electricity created by photovoltaic (PV) panels, wind turbine generators (WTGs), fuel cell generators, ESS, and EVs is all DC. Thus, the shift toward renewable energy supplies and zero-emissions mobility requires a fundamental reconsideration of the role played by DC networks in the grid system.

Today, microgrids composed of PV panels or WTGs require inverters to connect to local AC distribution networks. Every inverter requires control circuitry to synchronize loads with 60 or 50 Hz AC systems. These inverters help to provide high-quality AC current without causing mains interference, also known as harmonics. If the final load requires DC, the AC must be converted back again into DC. Thus, the incorporation of renewable distributed generation currently requires multiple stages of conversion, engendering multiple stages of inefficiency and loss.

In addition, both PHEVs and EVs connected to the grid system require AC/DC charging controllers to charge the vehicle's battery, while the use of PHEVs and EVs as energy-storage devices to feed electricity back into the grid system (i.e., to smooth peak curves) also hinges on the use of controllers to manage charging and discharging cycles. These V2G interfaces create multiple stages of conversion.

Perhaps not surprisingly, DC systems have come back into vogue, as the rising prominence of DC loads, electric mobility, energy-storage devices, and renewable distributed generators has put pressure on AC systems built in the last century. One solution is to upgrade aged AC systems by incorporating advanced conversion devices. Alternatively, AC systems can be hybridized with DC systems to support differing load types more efficiently. Hybridized AC and DC systems offer multiple advantages. They eliminate unnecessary multiconversion pathways from generation to final load, thereby reducing the total loss experienced throughout the system. Hybrid systems can also simplify equipment requirements for local systems, as the segregation of DC loads with DC supplies improves the provision of high-quality AC in the grid by reducing the harmonics associated with synchronization.

The growing prominence of PHEVs and EVs, along with the growing interest in sustainable energy supplies, energy-storage devices, and microgrids, will continue to motivate further investigations into improved power electronic converters and EMSs for hybridized grid systems.

### 1.4.2 Smart Grid

A final variable in the electrical infrastructure equation is the integration of smart grid systems. Smart grids are particularly important for the future of electrically driven vehicles, as they ensure that new loads associated with PHEVs and EVs will not overburden electrical grid systems. They also ensure the lowest possible price for fuel for PHEV and EV owners and drivers.

The concept of a smart grid means different things to different users. Here, we use the characterization offered by the Institution of Engineering and Technology (IET), which argues a smart grid system is required for industrialized societies that hope to transition to low-carbon electrical networks without compromising security, stability, or low cost. In brief, smart grid systems are meant to integrate demand management systems with distributed generation to achieve the most efficient utilization of existing infrastructure, while also operating seamlessly in conjunction with new large-scale power generators.

While most twentieth-century power grids were designed to ensure that generation sources responded to user needs on demand, smart grid technologies enable more effective supply of loads with electrical production from conventional sources of power, as well as intermittent sources of power such as solar and wind. This is done by using intelligent control systems to monitor both loads and sources and to feed electricity back onto the grid from energy-storage devices when needed. To do this, grid operators will need to efficiently incorporate intermittent and variable renewable energy supplies, such as solar and wind, with energy-storage devices. In addition, smart V2G systems allow PHEVs and EVs to communicate with the electrical grid dynamically.

## 1.5 TRANSPORTATION ELECTRIFICATION IS A PARADIGM SHIFT

The combination of newly electrified mobility with emerging renewable energy systems entails a future unrecognizable to many contemporary motorists, travelers, public transit users, and policy makers. However, fundamental transformations such as the ones we are proposing with Transportation 2.0 are not without precedence.

Consider the revolution in telecommunications technology that has occurred over the past century. The Communications 1.0 paradigm started with Alexander Graham Bell, who helped to develop direct home-to-home communication lines and later home-to-network and business-to-network telephony systems. While the shift to electrically powered telegraphy and telephony liberated people from the slowness of letter-writing and horse-powered postal telegraphs, early telegraphs and telephones were highly limited in their scope and reach. The shift away from human operators of phone lines to digitized communication signals has ushered in an era of virtual networking, digitized communications, and global interconnectivity. This telecommunications revolution has had far-reaching effects on human behavior. It has created deep-seated cultural norms unbeknownst to previous generations. For example, the ability to call someone overseas from a hand-held mobile device and to view the real-time video of that person (who might themselves be on a mobile device in, say, a remote region of the world) has become second nature to many of us. We barely think of the novelty of the situation. Similarly, the ability to post a message to a public-messaging site such as Twitter.com where hundreds of thousands and even millions of people can view it immediately and engage in a conversation or action based on the posting is incredible; yet, it seems a standard practice these days.

Instantaneous, digitized, and dynamic communication systems would have been inconceivable for Graham Bell and his contemporaries just a century ago. But, digitized telephony, wireless and mobile telephony, smart phones, and integrated Internet and communication services are a natural part of twenty-first century life. Maintenance of these systems is also seemingly unproblematic, even though they require new behavioral patterns. Today, most mobile phone users do not think twice about plugging their mobile phones into sockets at night to charge; so, they can be used anywhere anytime the following day. Yet, just 30 years ago, citizens of eastern bloc states and developing economies still had to walk to local village shops to use single-community telephone lines to communicate with the outside world.

Digitized communications have changed the world fundamentally, even though it all seems so natural now. The same could be said of the Transportation 2.0 paradigm. Whether it is the practice of plugging in an EV for overnight charging, communicating with local road infrastructure to optimize transportation routes for commuters, or measuring dynamic demands on a grid system to manage energy needs and automate discharge cycles for cars as storage devices, the Transportation 2.0 world will involve a radical shift away from current practices. Yet, it will also constitute a natural pathway for social existence in a sustainable world.

Transportation 2.0 will change the way we transit, it will change the way we transport goods, and it will alter our personal behavior. The end result will be an interconnected world of highly efficient, clean, and electrified transportation. By the time it happens, it will seem as though the polluting, inefficient, insecure, disconnected, and cumbersome world of ICE-powered transportation was a relic of the past best left to the last century.

It is important to note that although transportation electrification is a clear paradigm shift and, therefore, a revolutionary force in itself, it is going to happen over time due to the size, complexity, and multidimensional nature of the transportation and energy systems. This paradigm shift has already begun and is gaining momentum. It includes different electrification technologies from more electric and lightly hybridized vehicles with low electrification levels, to full hybrids with medium electrification levels, to plug-in hybrid and all EVs with high electrification levels.

## 1.6 CONCLUSION

As the world shifts to a Transportation 2.0 future, the nature of our transportation networks and connective infrastructure will shift as well. This book will teach readers the skills they require to excel in today's rapidly shifting automotive industry—an industry fit for the twenty-first century, which embraces and embodies the concepts of electrification and sustainability. In sum, readers will learn how to envision and concretize a Transportation 2.0 reality.

## REFERENCES

Emadi, A. 2011. Transportation 2.0. *IEEE Power and Energy Magazine* 9 (4):18–29. doi: 10.1109/MPR.2011.
    941320.
Electric Vehicles. 2008. In *The Gale Encyclopedia of Science*, 4th Edition, edited by K. L. Lerner and B. W.
    Lerner. Detroit: Gale Cengage Learning, Vol. 2, pp. 1474–1477.
Prince Albert of Monaco unveils for the first time the Venturi VBB-3 the world's most powerful electric car.
    2013. *Canada News Wire*, September 25.

# 2 Fundamentals of Conventional Vehicles and Powertrains

*William Long and Berker Bilgin*

## CONTENTS

## 2.1 LONGITUDINAL VEHICLE MODEL

In practical terms, a vehicle not only travels on a level road but also up and down the slope of a roadway as well as around corners. In order to model this motion, the description of the roadway can be simplified by considering a straight roadway with two-dimensional movement. This two-dimensional model will focus on vehicle performance, including acceleration, speed, and gradeability, as well as braking performance.

Figure 2.1 shows the forces acting on a vehicle as it travels at a given speed along a roadway with a specific grade. Fundamental principles of mechanical systems can be used to express the relationship between the vehicle acceleration and the forces acting on the vehicle body as:

$$ma = F_t - F_w - F_g - F_r \tag{2.1}$$

where $m$ is the vehicle mass, $a$ is the acceleration of the vehicle. $F_t$ is the total tractive force acting upon the vehicle body, $F_w$ is the aerodynamic drag force, $F_g$ is the grading resistance force, and $F_r$ is the rolling resistance force.

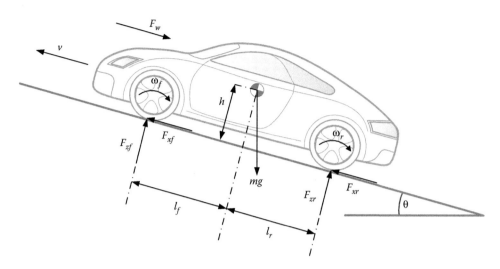

**FIGURE 2.1**   Forces acting on a vehicle.

## 2.2   LONGITUDINAL RESISTANCE

### 2.2.1   Aerodynamic Drag

As air travels over the body of the vehicle, it generates normal pressure and shear stress on the vehicle's body. The external aerodynamic resistance is comprised of two components, shape drag and skin friction. The shape drag arises from high-pressure areas in front of the vehicle and low-pressure areas behind the vehicle that are created as the vehicle propels itself through the air. These high-and low-pressure zones act against the motion of the vehicle, while the skin friction is due to the shear stress in the boundary layer on the surface of the body of the vehicle. In comparison, shape drag is much larger in magnitude than skin friction and constitutes more than 90% of the total external aerodynamic drag of a vehicle. Aerodynamic drag is a function of effective vehicle frontal area, $A$, and the aerodynamic drag coefficient, $C_d$, which are highly dependent on the design of the vehicle body:

$$F_w = \frac{1}{2}\rho A C_d (V + V_w)^2 \tag{2.2}$$

where $\rho$ is the air density, $V$ is the vehicle longitudinal speed, and $V_w$ is the wind speed.

### 2.2.2   Grading Resistance

As a vehicle travels up or down an incline, gravity acting on the vehicle produces a force which is always directed downward, as shown in Figure 2.1. This force opposes the forward motion during grade climbing and aids in the forward motion during grade descending. In typical vehicle performance models, only uphill operation is considered as it resists the total tractive force. The equation for this force is a function of the road angle $\theta$, vehicle mass $m$, and the gravitational acceleration $g$:

$$F_g = mg\sin(\theta) \tag{2.3}$$

For a relatively small angle of $\theta$, $\tan\theta = \sin\theta$. Using this approximation, the grade resistance can be approximated by $mg\tan\theta$, or $mgG$, where $G$ is the slope of the grade.

### 2.2.3 ROLLING RESISTANCE

Rolling resistance force is a result of the hysteresis of the tire at the contact patch as it rolls along the roadway. In a stationary tire, the normal force due to the road balances the force due to the weight of the vehicle through the contact patch which is in line with the center of the tire. When the tire rolls, as a result of tire distortion or hysteresis, the normal pressure in the leading half of the contact patch is higher than that in the trailing half. The normal force due to the road is shifted from the center of the tire in the direction of motion. This shift produces a moment that exerts a retarding torque on the wheel. The rolling resistance force is the force due to the moment, which opposes the motion of the wheel, and always assists in braking or retarding the motion of the vehicle. The equation for this force is a function of the normal load $F_z$ and the rolling resistance coefficient $f_r$, which is derived by dividing the distance the normal force due to the road is shifted by the effective radius of the tire $r_d$.

$$F_r = F_z f_r \cos(\theta) \tag{2.4}$$

## 2.3 TOTAL TRACTIVE FORCE

Equation 2.1 shows the factors affecting vehicle performance with a particular interest in the overall tractive force of the vehicle.

$$ma = F_t - F_w - F_g - F_r \Rightarrow ma = (F_{tf} + F_{tr}) - (F_w + F_g + F_{rf} + F_{rr}) \tag{2.5}$$

By rearranging Equation 2.1 we arrive at an equation that expresses longitudinal vehicle motion as a combination of total tractive effort minus the resistance. In order to determine the total tractive effort, the normal forces, $F_{zf}$ and $F_{zr}$, need to be determined. The front and rear tire contact points should satisfy the equilibrium equations for moments:

$$\sum M_r = 0, \quad \sum M_f = 0 \tag{2.6}$$

Therefore,

$$F_{zf}(l_f + l_r) + F_w h_w + (mg \sin(\theta)h) + (mah) - (mg \cos(\theta)l_r) = 0 \tag{2.7}$$

and

$$F_{zr}(l_f + l_r) - F_w h_w - (mg \sin(\theta)h) - (mah) - (mg \cos(\theta)l_f) = 0 \tag{2.8}$$

where $F_{zf}$ and $F_{zr}$ are the normal forces on the front and rear tires, $l_f$ and $l_r$ are the distances between the front and rear axles and vehicle center of gravity, respectively. $h_w$ is the height for effective aerodynamic drag force and $h$ is the height of vehicle center of gravity. For simplicity, usually $h_w$ is assumed to be equal to $h$. Equations 2.7 and 2.8 can be rearranged to solve for the normal forces on the front and rear tires:

$$F_{zf} = \frac{-F_w h - mg \sin(\theta)h - mah + mg \cos(\theta)l_r}{l_f + l_r} \tag{2.9}$$

$$F_{zr} = \frac{F_w h + mg \sin(\theta)h + mah + mg \cos(\theta)l_f}{l_f + l_r} \tag{2.10}$$

The total tractive force can be expressed as the tractive forces acting on each tire:

$$F_t = F_{xf} + F_{xr} \tag{2.11}$$

where $F_{xf}$ and $F_{xr}$ are the longitudinal forces on the front and rear tires, respectively. The friction generated between the tire–road contact patch creates the longitudinal force. Therefore, the longitudinal force generated on each tire can be represented as a function of the tire friction coefficient and the normal force:

$$F_{xf} = \mu_f F_{zf}, \quad F_{xr} = \mu_r F_{zr} \tag{2.12}$$

where $F_{xf}$ and $F_{zr}$ are the normal forces on the front and rear tires given by Equations 2.9 and 2.10 and $\mu_f$ and $\mu_r$ are the friction coefficients on the front and rear tires, respectively.

## 2.4   MAXIMUM TRACTIVE EFFORT AND POWERTRAIN TRACTIVE EFFORT

The maximum tractive effort of the vehicle is proportional to the slip ratio of the tire, which represents the difference between the angular tire speed and the vehicle speed. During acceleration, the slip ratio of the front and rear tires can be expressed as:

$$\sigma_r = \frac{r_{wr}\omega_r - V}{r_{wr}\omega_r}, \quad \sigma_f = \frac{r_{wf}\omega_f - V}{r_{wf}\omega_f} \tag{2.13}$$

where $r_{wf}$ and $r_{wr}$ are the radii of the front and rear tires, and $\omega_f$ and $\omega_r$ represent their angular speed. Slip-friction coefficient characteristics of a tire have a nonlinear relationship and depend on the road surface conditions, as shown in Figure 2.2. The Pacejka Tire Model is widely used to define these characteristics:

$$\mu_{f/r} = D\sin(C a\tan(B\sigma_{f/r} - E(B\sigma_{f/r} - a\tan(B\sigma_{f/r})))) \tag{2.14}$$

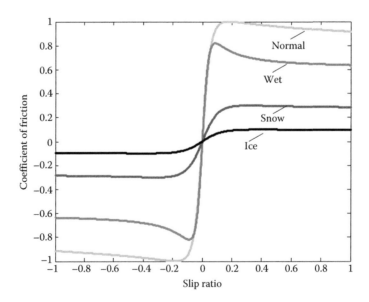

**FIGURE 2.2**   Typical tire slip ratio–friction coefficient characteristics.

where $\mu_{f/r}$ and $\sigma_{f/r}$ are the friction coefficient and slip ratio of the front or rear tire. $B$, $C$, $D$, and $E$ are tire coefficients and their values depend on the road surface conditions.

The sum of the torques on each wheel determines the rate of change of wheel speeds:

$$J_f \frac{d\omega_f}{dt} = T_{ef} - T_{rf}, \quad J_r \frac{d\omega_r}{dt} = T_{er} - T_{rr} \tag{2.15}$$

where $J_f$ and $J_r$ are the inertias, $T_{ef}$ and $T_{er}$ are the traction torques delivered from the drivetrain, $T_{rf}$ and $T_{rr}$ are the reaction torques due to the tractive force of the front and rear tires, respectively. The reaction and friction torques on the front and rear tires can be defined as:

$$T_{rf} = r_{wf} F_{xf}, \quad T_{rr} = r_{wr} F_{xr} \tag{2.16}$$

where $\omega_f$ and $\omega_r$ are the angular speed, and $C_f$ and $C_r$ are the friction coefficients of the front and rear tires, respectively.

The traction torques on the tires, $T_{ef}$ and $T_{er}$ are provided from the powertrain. If the vehicle is rear wheel or front wheel driven, the nondriven wheel provides no traction torque. Therefore, for nondriven wheels, Equation 2.15 can be expressed as:

$$J_r \frac{d\omega_r}{dt} = -r_{wr} F_{xr} \tag{2.17}$$

In conventional vehicles, the source of the traction torque is the internal combustion engine and the output power of the engine is supplied to the tires through the clutch, the transmission, and the differential. Therefore, the traction torque applied for either a front or a rear wheel drive vehicle can be expressed as:

$$T_p = T_{en} i_t i_0 \eta_p \tag{2.18}$$

where $T_{en}$ is the torque from the engine, $i_t$ is the gear ratio of the transmission, $i_0$ is the gear ratio of the differential, and $\eta_p$ is the total efficiency of the powertrain.

## 2.5 VEHICLE PERFORMANCE

Performance characteristics of a road vehicle refer to its capability to both accelerate and decelerate, and negotiate grades in a straight-line motion. These characteristics are different depending on the vehicle's type and size. Mass of the vehicle is of great importance to vehicle performance. By researching not only vehicle electrification, but lightweight materials as well, all aspects of vehicle performance would be improved, including fuel economy. The tractive and braking effort developed by the tires and the resisting forces acting on the vehicle determine the performance potential of the vehicle. Typically, overall vehicle performance is also concerned with cornering ability, but as this is mainly a function of suspension geometry and vehicle design it is outside the scope of this chapter.

### 2.5.1 MAXIMUM SPEED OF A VEHICLE

The maximum speed of a vehicle is the highest constant cruising speed that the vehicle can achieve at full power on a level road. The maximum speed of a vehicle is calculated with full torque from the traction source on a flat road when the tractive force and the resistive force are at equilibrium.

Since the vehicle acceleration and road gradient are zero at this point, the equilibrium can be represented as:

$$F_t = F_w + F_r \tag{2.19}$$

Considering that the wheel speed is also constant, the tractive force can be expressed in terms of the torque applied to the wheels:

$$F_t = \frac{T_p}{r_d} \tag{2.20}$$

where $T_p$ can be expressed as in Equation 2.18.

The aerodynamic drag force $F_w$ and the rolling resistance force $F_r$ have been derived in Equations 2.2 and 2.4, respectively.

Combining Equations 2.19, 2.20, 2.2, 2.4, and 2.18 yields:

$$\frac{T_{en}i_t i_0 \eta_p}{r_d} = mgf_r + \frac{1}{2}\rho A C_d V^2 \tag{2.21}$$

$$V = \sqrt{\frac{2(((T_{en}i_t i_0 \eta_p)/r_d) - mgf_r)}{\rho A C_d}} \tag{2.22}$$

### 2.5.2 GRADEABILITY

The gradeability of a vehicle is the maximum gradient on which the vehicle can start climbing from stand-still with all the wheels of the vehicle on the gradient at the time of start.

As a vehicle drives on a road with a small grade and constant speed, the tractive effort and resistance equilibrium can be written as an extension of Equations 2.19 and 2.21 to include gradeability for small angles, as calculated in Equation 2.3.

$$F_t = F_w + F_r + F_g \tag{2.23}$$

$$\frac{T_{en}i_t i_0 \eta_p}{r_r} = mgf_r + \frac{1}{2}\rho A C_d V^2 + mgG \tag{2.24}$$

$$G = \frac{((T_{en}i_t i_0 \eta_p)/r_r) - mgf_r - (1/2)\rho A C_d V^2}{mg} \tag{2.25}$$

### 2.5.3 ACCELERATION PERFORMANCE

When high-performance vehicles are compared to one another, one of the first statistics to be reviewed is the acceleration performance. It is most often referred to as a vehicle's 0–60 mph time. Referring to Equation 2.1, the acceleration of the vehicle on level ground can be written as:

$$a = \frac{F_t - F_r - F_w}{m\delta} \tag{2.26}$$

where $\delta$ is the mass factor which takes into account the mass moments of inertia of the rotating components involved during a change of acceleration.

Combining Equations 2.26, 2.20, 2.2, 2.4, and 2.18 yields:

$$a = \frac{((T_{en}i_ti_0\eta_p)/r_d) - mgf_r - (1/2)\rho AC_dAV^2}{m\delta} \tag{2.27}$$

By integrating Equations 2.28 and 2.29 from zero to potentially 60 mph, the predicted acceleration time and distance for a vehicle can be calculated:

$$t = m\delta\int_{V1}^{V_2} \frac{V}{((T_{en}i_ti_0\eta_p)/r_d) - mgf_r - (1/2)\rho AC_dAV^2} \, dV \tag{2.28}$$

$$S = m\delta\int_{V1}^{V_2} \frac{1}{((T_{en}i_ti_0\eta_p)/r_d) - mgf_r - (1/2)\rho AC_dAV^2} \, dV \tag{2.29}$$

The torque of the engine during acceleration is not constant which makes these equations very difficult to solve analytically, thus numerical methods are typically used. These methods are outside the scope of this chapter.

## 2.6 BRAKING PERFORMANCE AND DISTRIBUTION

Conventional brakes, disc or drum, are the single most important safety device on any vehicle. By transferring kinetic energy into thermal energy through friction between a rotating surface and a stationary brake pad, the vehicle speed is decreased, but that thermal energy is typically wasted in conventional vehicles. With the introduction of vehicle electrification, through regenerative braking with electric motors, the kinetic energy can be recovered into stored electrical energy and it can be reused rather than being wasted. This can greatly increase the overall efficiency of the vehicle. Careful attention should be paid to the brake balance of the vehicle when designing a powertrain with regenerative braking capability. In high-performance vehicles, up to 80% of the braking force may be on the front axle, which creates a much larger potential to recapture energy at the front wheels when compared with a rear wheel setup.

Proper brake balance during the braking cycle is extremely important as the vehicle will not achieve the maximum braking deceleration unless all four tires are brought to the peak friction level simultaneously. Improper brake balance will cause either the front or rear wheels to lock up prematurely and these stationary wheels will lose cornering traction. Proper brake balance is a function of the loads on the wheels, which is in turn a function of the deceleration.

### 2.6.1 BRAKING FORCE

The braking force $F_b$, due to the brake system, which is developed on the interface between the road and the tire, is the primary braking force. When the braking force is below the tire–road adhesion limit, the braking force is given by:

$$F_b = \frac{T_b - \sum I\alpha_{an}}{r} \tag{2.30}$$

where $T_b$ is the applied brake torque, $I$ is the rotating inertia connected with the wheel being decelerated, $\alpha_{an}$ is the corresponding angular deceleration, and $r$ is the rolling radius of the tire. Once

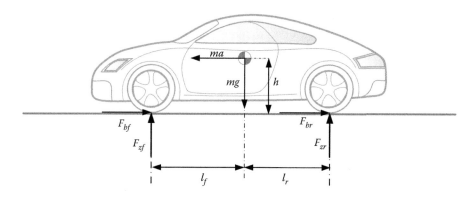

FIGURE 2.3   Vehicle braking forces on a level road.

the braking force reaches the limit of tire–road adhesion, it cannot increase any further. In addition to the braking force, the rolling resistance of tires, aerodynamic resistance, grade resistance, and powertrain resistance, as discussed previously, also affect vehicle motion during braking. For the purpose of this vehicle model, it is assumed that these are minor losses and can be neglected.

### 2.6.2   BRAKING CHARACTERISTICS OF A TWO-AXLE VEHICLE

When a vehicle is under braking or deceleration, an inertial reaction force is developed that is similar to the centrifugal force. Figure 2.3 shows the braking forces acting on a vehicle on a level road.

The braking force is directly proportional to the normal load acting on the tire, which is proportional to the tire–road adhesion. Similar to when total tractive force was calculated, the front and rear tire contact points should once again satisfy the equilibrium equations for moments:

$$F_{zf} = \frac{mg}{(l_f + l_r)}\left(l_r + \frac{ha}{g}\right), \quad F_{zr} = \frac{mg}{(l_f + l_r)}\left(l_f - \frac{ha}{g}\right) \tag{2.31}$$

The maximum braking force that the tire–road adhesion will support can be determined by multiplying the normal force at the front and rear wheels by the coefficient of road adhesion, μ, as expressed in Equation 2.12.

## 2.7   VEHICLE POWER PLANT AND TRANSMISSION CHARACTERISTICS

The two factors that limit conventional vehicle performance are the maximum tractive effort that the tire–ground adhesion can support, and the tractive effort that the overall powertrain can provide. The performance limits of the vehicle are dictated by the lesser of the tractive efforts. With the transmission in low gear and the engine throttle at maximum, the tractive effort may be limited by the nature of tire–road adhesion and loss of traction. As the transmission is shifted into higher gears, the tractive effort is often determined by the engine and transmission characteristics, which have to be taken into consideration when predicting the overall performance of a road vehicle.

### 2.7.1   POWER PLANT CHARACTERISTICS

For conventional vehicles, the ideal performance characteristic of a power plant is an unchanging power output across the entire operating range. At low speeds, motor torque is forced to maintain a constant value so as not to exceed the adhesion limit between the tire–ground contact area. After

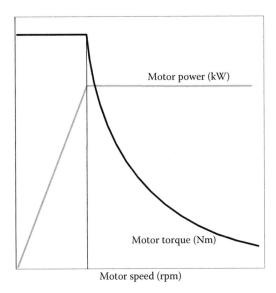

Motor power (kW)

Motor torque (Nm)

Motor speed (rpm)

FIGURE 2.4    Ideal performance characteristics for vehicular power plants.

the constant torque low-speed range, the torque varies with speed steeply, as shown in Figure 2.4. Constant power characteristics provide conventional vehicles with a high tractive effort at low speeds, which improves acceleration, grade climbing, and towing capacity. The internal combustion engine has been the standard power plant in conventional vehicles for over a century because of its relatively high power-to-weight ratio and low cost, but it is not without its shortcomings. Typically, an internal combustion engine has torque–speed characteristics that do not directly correlate with the ideal performance characteristics required by traction.

When compared with an ideal torque–speed profile, as shown in Figure 2.4, typical internal combustion engines have a relatively flat profile, which is why a multiple gear transmission is usually employed to modify it. As an internal combustion engine moves through its typical torque and power curves, representative characteristics are developed in each speed range that can be used to predict vehicle performance. The engine begins operating smoothly at idle speed, and as the engine speed approaches an intermediate range, good combustion quality and maximum engine torque are reached. As the speed increases further, the increasing losses in the air-induction manifold and the steady decline in engine torque cause the mean effective pressure to decrease. Power output continues to increase to its maximum at a specific high engine speed. Beyond this peak power point the engine torque decreases more rapidly as the engine speed is increased, which results in the decline of engine power output.

### 2.7.2    Transmission Characteristics

The term "transmission" refers to all of the systems or subsystems used for transmitting the engine power to the driven wheels or sprockets. The principal requirements for the transmission are to attain the desired maximum vehicle speed with an appropriate engine, to be able to move the vehicle on a steep slope as well as maintain speed on a gentle slope in high gear, and to properly match the engine characteristics to achieve the desired operating fuel economy and acceleration rate.

The two most common types of transmissions for conventional vehicles are the manual gear transmission and the automatic transmission with a torque converter. Other types of transmissions, such as the continuous variable transmission, are also in use and are beginning to gain popularity due to their relatively high overall efficiency.

### 2.7.2.1  Manual Transmission

Manual transmissions were the first gearbox designs used in conventional vehicles and were used for decades before automatic transmissions were introduced. These transmissions are still popular due to their simplicity, low cost, and high efficiency. The term "manual" implies that the driver must perform the shifting from gear to gear manually. A manual gear transmission consists of a clutch, gearbox, final drive, and driveshaft. The gearbox provides a number of gear reduction ratios, between three and five for passenger cars, and more for heavy commercial vehicles.

In determining maximum and minimum gear ratios, we can think back to the two factors that limit conventional vehicle performance. The maximum speed requirement determines the ratio of the highest gear. On the other hand, the gear ratio of the lowest gear is determined by the requirement of the maximum tractive effort or the gradeability, which is often assumed to be 33%. Choosing the ratios between high and low gear should be spaced in such a way that they will provide the tractive effort–speed characteristics as close to the ideal as possible, but unfortunately this is not always achievable. Traditionally, in conventional vehicles, the gear ratios are often chosen to minimize the time required to reach a specific speed or the maximum speed of the vehicle. With fuel economy moving to the forefront of vehicle design, a geometric progression approach, where consecutive gear ratios are very close to one another, could be used. The basis for this is to have the engine operating within the same speed range in each gear. This would ensure that in each gear, the operating fuel economy is similar.

### 2.7.2.2  Automatic Transmission

With an automatic transmission a driver no longer needs to actively change gears during driving, thus making the vehicle easier to drive. Automatic transmissions use fluid to transmit power in the form of torque and speed and are widely used in conventional passenger vehicles. In a conventional automatic transmission, the clutch is replaced with a fluid coupling or torque converter to eliminate engaging and disengaging action during gear changes. The torque converter's three major components, shown in Figure 2.5 are the pump impeller, the turbine runner, and the stator. When the impeller is driven by the engine, the fluid in the impeller rotates with it, and as the speed increases, centrifugal force causes the fluid to flow into the turbine. The hydraulic fluid in the converter transfers torque through the kinetic energy of the transmission fluid as it is forced from the pump impeller to the turbine. The higher the engine speed the greater the torque applied to the turbine. The stator is located between the pump impeller and the turbine. The vanes of the stator catch the fluid as it leaves the turbine runner and redirects it so that it strikes the back of the vanes of the pump impeller giving it added torque. The major advantages of the automatic transmission are that it

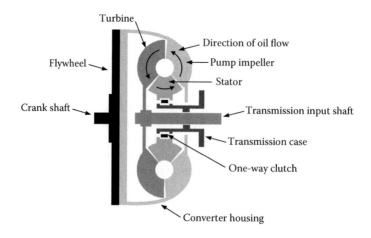

FIGURE 2.5   Torque converter cutaway schematic.

prohibits the engine from stalling, and it provides torque–speed characteristics that approach the ideal characteristics discussed previously. The major disadvantages of the automatic transmission are its low efficiency in a stop and go drive cycle and its very complicated construction.

### 2.7.2.3 Continuously Variable Transmission

As the interest in improving the fuel economy of automotive vehicles continues to grow, continuously variable transmissions (CVT) have attracted a great deal of attention. This type of transmission provides a continuously variable reduction ratio within a certain range, thus providing an infinite number of gear ratios. This allows the engine to operate under the most economical conditions over a wide range of vehicle speeds. It is therefore possible to achieve an ideal torque–speed profile, because any engine power output to the transmission can be applied to the wheels at any speed.

A belt CVT system is similar to standard belt–pulley drive, with one pulley connected to the engine shaft and the other connected to the output shaft. The exception, however, is that the half pulleys are not fixed and are able to move apart. Both pulleys have fixed axes of rotation at a distance from each other. The sides of each pulley are controlled so that they may move apart or together laterally, varying the effective diameter on which the belt grips. The overall lateral displacement of each pulley is the opposite of the other.

### CHAPTER REVIEW PROBLEMS

#### Problem 1

On a 20°C sunny day with no wind speed, and an air density of 1.2 kg/m³, a 1500 kg vehicle travels along an asphalt roadway with a 6° grade at 100 km/h and a rolling resistance coefficient of 0.013. The vehicle center of gravity is located 0.6 m from the ground in the center of the 2.5 m wheelbase with 0.66 m wheels and has a frontal area of 2.05 m² and a drag coefficient of 0.32.

Calculate the aerodynamic, grading, and rolling resistance.

#### Problem 2

After some engine and transmission performance upgrades, the same vehicle is taken to a test track to find out the new vehicle top speed. The upgrades have increased engine torque to 450 Nm and engine horsepower to 300 kW and an overall powertrain efficiency of 88%. After the upgrades, the minimum gear ratio of the transmission is 0.9, and the differential gear ratio is 3.21.

Calculate the maximum speed of the vehicle.

#### Problem 3

The same vehicle travels along a level asphalt roadway at 100 km/h, with a coefficient of road adhesion of 0.72, and encounters a large obstacle 50 m ahead and needs to stop to avoid striking the obstacle.

Calculate the braking force at the front and rear wheels that is required to stop just before the obstacle.

# 3 Internal Combustion Engines

*Fengjun Yan*

## CONTENTS

## 3.1 INTRODUCTION

Internal combustion (IC) engines are a type of devices which convert the chemical fuel energy into mechanical work. There are many types of IC engine classifications ([1], pp. 7–9). Most typically, IC engines can be classified into two types: spark-ignition (SI) engines and compression–ignition

(CI) engines. The energy source of SI engines is usually gasoline. Correspondingly, diesel fuel is usually the energy source of CI engines. These two types of IC engines are the mainstreams nowadays and therefore are the objectives we mainly focus on in this chapter. However, it does not mean that in real practice, the fuel types are confined in above two kinds. Instead, other fuels, such as natural gas, ethanol, and hydrogen, can also be selected as the energy sources in IC engines.

IC engines convert chemical energy into mechanical work through combustion and power strokes, that is, at each time of fuel combustion, it will provide a power stroke which provides the mechanical work. In the operation, both SI engines and CI engines operate on *cycles*. Specifically, *four-stroke* cycle and *two-stroke* cycle are used.

*Four-stroke* cycle is typically used for engines in most of the automobiles due to its high efficiency, while *two-stroke* cycle has a good feature, that is, it can provide higher power output for a given engine size. Therefore, it is typically used in the situation where the engine size is of significance, such as motorcycles. Since the *two-stroke* cycle has much lower fuel efficiency than that of the *four-stroke* one, it is seldom adopted in passenger cars and other fuel efficiency-oriented vehicles. In this chapter, we mainly focus on the *four-stroke* cycle, which is commonly used for both SI and CI engines.

As can be seen in Figure 3.1, crank shaft revolutions are indicated in the bottom. When the cylinder volume achieves maximum (minimum), the crank position is called bottom-dead center (top-dead center). In the four-stroke cycle engine, each power stroke will take two crank shaft revolutions. That means each cycle will take about half-crank shaft revolution. The four strokes, in Figure 3.2, are:

1. Intake stroke: The intake stroke is from Top Dead Centre (TDC) to Bottom Dead Centre (BDC). In this stroke, fresh charge is drawn into the cylinder through intake valves. Typically, the intake valve opens slightly before TDC and closes after BDC to increase the inducted charge.
2. Compression stroke: The compression stroke is right after the intake stroke. In this stroke, both intake and exhaust valves are closed and charge in the cylinder is compressed. The cylinder pressure increases during the compression. When it is close to the end of the compression stroke (near TDC), combustion is initiated through SI engines or fuel injection and compression (CI engines).

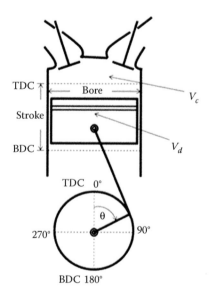

FIGURE 3.1    Basic geometry of the IC engines.

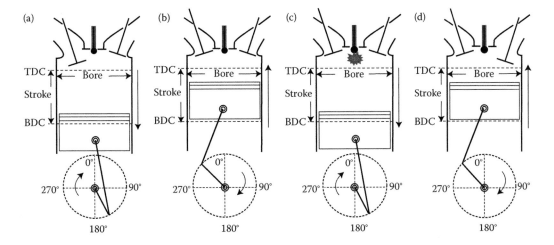

FIGURE 3.2    Four-stroke cycle. (a) Intake; (b) compression; (c) power; (d) exhaust.

3. Power stroke: After compression stroke, the cylinder pressure increases rapidly due to the combustion. The mixture gas with high pressure and temperature pushes the piston downward and enhances the crank rotation. In this stroke, the power is generated through combustion and transferred into the mechanical work in the form of crank rotation.
4. Exhaust stroke: At the end of the power stroke, the piston is around BDC. Then the exhaust valve starts to open and the exhaust gas is pushed out when the piston moves from BDC to TDC.

### 3.1.1    OPERATING CYCLES

In this section, we investigate in detail the relationships between engine-operating cycles and their performance. To facilitate the cycle analysis, pressure–volume diagrams are commonly used, as in Figure 3.3.

### 3.1.1.1    Ideal Cycles

As we discussed, the engine operation in one cycle can be divided into four sequences, that is, intake, compression, power, and exhaust. The power process includes combustion and expansion. On the basis of some reasonable approximations, these sequences can be ideally described and

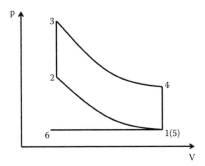

FIGURE 3.3    Otto cycle [1].

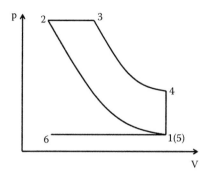

FIGURE 3.4   Diesel cycle.

analyzed through pressure–volume (p–V) diagrams. Commonly used approximations include constant-volume cycle (Otto cycle) in Figure 3.3 and constant-pressure cycle (Diesel cycle) in Figure 3.4. We call the cycles with these approximations ideal cycles.

In the ideal cycles, the intake and exhaust process are assumed as adiabatic and valves are assumed to open instantly. The compression and expansion process is assumed as isentropic.

The difference between the Otto cycle and Diesel cycle is in the combustion process. In Otto cycle, the combustion is assumed at constant volume. In Diesel cycle, the combustion is assumed at constant pressure.

The cycle efficiency plays important roles in the overall engine combustion efficiency. Without loss of generality, let us analyze the efficiency for a limited-pressure cycle (as is shown in Figure 3.5).

The cycle fuel conversion efficiency is defined as

$$\eta = \frac{W_c}{m_f Q_{LHV}} \tag{3.1}$$

where $W_c$ is the indicated work per cycle, $m_f$ is the fuel mass amount injected per cycle, and $Q_{LHV}$ is the lower heat value of the fuel.

In the following, the subscripts 1–5 represent the corresponding value in Figure 3.5.

The compression work produced in the limited-pressure cycle is

$$W_{com} = mc_v(T_1 - T_2) \tag{3.2}$$

where $c_v$ is the constant-volume heat capacity.

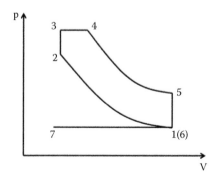

FIGURE 3.5   Limited-pressure cycle.

The expansion work is

$$W_{\text{exp}} = m[c_v(T_4 - T_1) + p_3(V_4 - V_3)] \tag{3.3}$$

For the combustion process,

$$\begin{aligned} m_{f,2-3}Q_{LHV} &= mc_v(T_3 - T_2) \\ m_{f,3-4}Q_{LHV} &= mc_p(T_4 - T_3) \end{aligned} \tag{3.4}$$

where $c_p$ is the constant pressure heat capacity.

With the relation that

$$W_c = W_{com} + W_{\text{exp}} \tag{3.5}$$

and

$$m_f = m_{f,2-3} + m_{f,3-4} \tag{3.6}$$

the efficiency in the limited-pressure cycle is

$$\eta = 1 - \frac{T_5 - T_1}{(T_3 - T_2) + \gamma(T_4 - T_3)} \tag{3.7}$$

where $\gamma$ is the heat capacity ratio.

With the assumptions that process 1–2 and process 4–5 in Figure 3.5 are isentropic, we have

$$\eta = 1 - \frac{\kappa_p \kappa_v^\gamma - 1}{R_c^{\gamma-1}[\kappa_p\gamma(\kappa_v - 1) + \kappa_p - 1]} \tag{3.8}$$

where $R_c$ is the compression ratio, $\kappa_p$ is the pressure ratio, defined as

$$\kappa_p = \frac{p_3}{p_2} \tag{3.9}$$

and $\kappa_v$ is the volume ratio, defined as

$$\kappa_v = \frac{V_4}{V_3} \tag{3.10}$$

As can be seen, when $\kappa_p = 1$, it is Diesel cycle; when $\kappa_v = 1$, it is Otto cycle.

### 3.1.1.2 Real Cycles

In real operation, the engine-operating cycle is different from the ideal cycles due to the following considerations:

1. Combustion time: In real engine operation, the combustion does not occur instantly as in the ideal cycle.
2. Heat transfer: After the combustion, the in-cylinder gas temperature increases significantly. The temperature difference between gas temperature and cylinder wall temperature will lead to a heat loss.

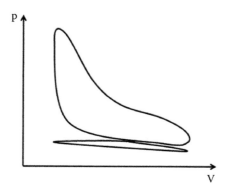

FIGURE 3.6   Real cycle.

3. Exhaust blowdown: When the exhaust valve opens before the exhaust stroke, the in-cylinder pressure will reduce below the isentropic line in the ideal cycle. The in-cylinder pressure reduction will decrease the expansion work.
4. Incomplete combustion and leakage: Incomplete combustion and leakage in the real practice are the other differences between ideal and real cycles.

Owing to these factors, the real cycle (Figure 3.6) is different from the ideal cycle.

## 3.2   CONCEPTS

### 3.2.1   BASIC GEOMETRICAL PROPERTIES

#### 3.2.1.1   Cylinder Volume

The volume that the piston swept by in a single movement from TDC to BDC in a single cylinder is defined as the *displaced volume*, referred to as $V_d$. The volume swept by the pistons in all the cylinders in the movement from TDC to BDC is called the *engine displacement*, $V_{ed}$. As can be easily noted, the *engine displacement*

$$V_{ed} = V_d \cdot n_{cyl} \tag{3.11}$$

where $n_{cyl}$ is the number of the cylinders.

The cylinder volume when the piston is at TDC is called clearance volume, $V_c$. Note that $V_c$ is the minimum cylinder. According to Figure 3.7, the connecting rod length is $l$, the cylinder bore is $B$, the stroke length is $L$, and the crank radius is $a$. Therefore, the distance between the piston pin axis and the crank axis, referred to as $s$, can be calculated at any crank angle $\theta$ (the initial position of $\theta$ is defined as in Figure 3.7) by

$$s(\theta) = a\cos\theta + (l^2 - a^2 \sin^2\theta)^{1/2} \tag{3.12}$$

In each cycle, the cylinder volume changes as the piston movement. By these geometry parameters, we can determine the cylinder volume at $\theta$ by

$$V(\theta) = V_c + \frac{\pi B^2}{4}(l + a - s(\theta)) \tag{3.13}$$

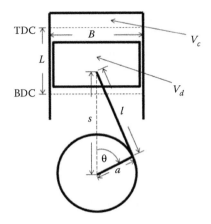

FIGURE 3.7 Geometry of the cylinder.

Note that, the maximum cylinder volume is at the crank angle $\theta = \pi$, where $s = -a + l$ and $V = V_c + ((\pi B^2 a)/2) = V_c + V_d$. The minimum cylinder volume is at the crank angle $\theta = 0$, where $s = a + l$ and $V = V_c$.

### 3.2.1.2 Compression Ratio

The compression ratio, $r_c$, is defined as the ratio between the maximum cylinder volume $V_c + V_d$ and the minimum cylinder volume $V_c$, that is,

$$r_c = \frac{V_c + V_d}{V_c} \tag{3.14}$$

The compression ratio is an important parameter that represents how much the charge can be compressed for combustion. Typically, the compression ratio is smaller in the SI engine than that of the CI engine.

### 3.2.1.3 Other Parameters

With the information given above, other basic parameters can be derived straightforwardly.

The surface area of the combustion chamber, $A$, is calculated by

$$A = A_h + A_p + \pi B(l + a - s) \tag{3.15}$$

where $A_h$ and $A_p$ are the cylinder head surface area and the piston crown surface area, respectively.

The ratio of connecting rod length to the crank radius is defined as

$$R = \frac{l}{a} \tag{3.16}$$

The ratio of cylinder bore to stroke length is defined as

$$R_{bs} = \frac{B}{L} \tag{3.17}$$

The *instantaneous piston speed $S_p$* is derived by differentiating Equation 3.2

$$S_p = \frac{ds}{dt} = LN\pi\sin\theta\left[1 + \frac{\cos\theta}{(R^2 - \sin^2\theta)^{1/2}}\right]$$

(3.18)

where $N$ is the crank shaft rotation speed (rev/s).

The *mean piston speed* is defined as

$$\bar{S}_p = 2LN$$

(3.19)

### 3.2.2 Performance Index

Besides the basic parameters of IC engines, some important concepts related to the performance are introduced here.

#### 3.2.2.1 Torque and Power

Before the introduction of the torque and power of IC engines, two notations are provided, that is, "brake" and "indicated." "*Brake*" normally refers to the usable portion of a specific term. "*Indicated*" refers to something that can be generated directly through a device or indicator and usually contains the "*brake*" and the friction portion.

The brake torque, $T_b$, refers to the useful torque output of the engine, which is normally measured by a dynamometer. Correspondingly, the brake power, $P_b$, is defined as

$$P_b = 2\pi N T_b$$

(3.20)

Here, $N$ is the crank shaft rotation speed (rev/s).

The *indicated work per cycle* can be obtained through cylinder pressure versus cylinder volume integration as

$$W_{c,i} = \oint p\,dV$$

(3.21)

If the integration only covers the compression and power strokes, it is called *gross indicated work per cycle*, $W_{c,ig}$. If the integration covers the entire four strokes, it is called *net indicated work per cycle*, $W_{c,in}$. The difference between these two is the work cost for the gas intake and exhaust process. This portion of work is defined as *pumping work*.

With the *indicated work per cycle*, we can calculate the *indicated power* as

$$p_i = \frac{W_{c,i}N}{2}$$

(3.22)

Note that 2 means that there are 2 revolutions per cycle in the four-stroke engine. Similar to considering the *pumping work* in the *indicated work per cycle*, the *indicated power* can be distinguished as *net indicated power* or *gross indicated power* by considering or not considering the pumping loss.

#### 3.2.2.2 Mean Effective Pressure

The *mean effective pressure (MEP)* is defined by the *work per cycle* dividing the cylinder *displaced volume*

$$MEP = \frac{W_c}{V_d}$$

(3.23)

where $W_c$ refers to *work per cycle*.

The *MEP* is an essential measurement especially for engine design and evaluation, since the *work per cycle* or torque is only for the engine ability measurement regardless of the difference in the various cylinder *displaced volumes*.

Again, the MEP can be classified as the *indicated mean effective pressure* (*IMEP*) and the *brake mean effective pressure* (*BMEP*) according to whether the *work per cycle* used in Equation 3.13 is *indicated* or *brake*.

### 3.2.2.3 Specific Fuel Consumption

Another important term for measuring the efficiency of the engine is called *specific fuel consumption* (*SFC*). It describes the fuel flow rate per unit of output power as

$$SFC = \frac{\dot{m}_f}{P} \tag{3.24}$$

where $\dot{m}_f$ is the fuel flow rate. Obviously, a lower value of *SFC* indicates better fuel efficiency of the engine.

### 3.2.2.4 Air/Fuel or Fuel/Air Ratios

Before combustion occurs in the cylinder, air and fuel are mixed. The ratio of these two parts will largely determine the combustion process after mixing. The air/fuel ratio (A/F) and fuel/air ratio (A/F) are defined as

$$\frac{A}{F} = \frac{\dot{m}_a}{\dot{m}_f} \tag{3.25}$$

$$\frac{F}{A} = \frac{\dot{m}_f}{\dot{m}_a} \tag{3.26}$$

where $\dot{m}_a$ is the air flow rate.

In conventional SI engines, the air and fuel are normally mixed homogeneously. The *A/F* in the SI engine is relatively small and close to the value that allows the oxygen in the air to be completely consumed after combustion. We call the combustion with high *A/F* "rich" burn, indicating that the portion of fuel is large in the mixture. In the CI engines with diesel fuel, the air and fuel are not homogeneously mixed and the *A/F* is relatively large. We call this mixture "lean" burn, indicating that the portion of fuel is small in the mixture.

### 3.2.2.5 Volumetric Efficiency

Another important term in the IC engines is the *volumetric efficiency*. In the intake stroke, charge is inducted into the cylinder through intake manifold. The *volumetric efficiency* is a measurement of the efficiency of the intake system, that is, indicating the inducted charge amount per cycle for a particular cylinder volume.

The definition of the *volumetric efficiency* is

$$\eta_v = \frac{m_a}{\rho_i V_d} \tag{3.27}$$

where $m_a$ is the mass of charge inducted into the cylinder per cycle and $\rho_i$ is the intake charge density. As can be seen, when the charge is inducted into the cylinder without the density reduction, that is, the inducted charge density in the cylinder is maintained as high as the one in the intake manifold, the *volumetric efficiency* is 1. When considering the intake charge mass flow rate, Equation 3.27 can be written as

$$\eta_v = \frac{2\dot{m}_a}{\rho_i V_d N} \tag{3.28}$$

## 3.3   AIR-PATH LOOP

In this section, the engine-breathing issues along the routines of the charge and the exhaust gas in IC engines are discussed. The charge here refers to the gases before entering the cylinder, including (1) the fresh air; (2) the mixture of fuel and air; (3) the mixture of the recirculated exhaust gas and fresh air; and (4) the mixture of the recirculated exhaust gas, fresh air, and fuel.

The air-path loop includes intake and exhaust systems. The charge is introduced into the engine cylinder through intake systems and the exhaust gas goes out to ambience through exhaust systems after combustion. Properly designed air-path loop systems provide air/mixed gas at a desired gas initial condition in the cylinder and gas-exchange process. Among the factors affecting the air-path loop system, the structure and the mass flow rates control throughout each air-path loop section are of importance. The structure/geometry design typically considers the engine space limitation, the fluid dynamics, and the special purpose on intake gases. In conventional engine intake systems, the intake gas/mixture goes to intake manifold through a throttle. In more complex engine intake systems, the exhaust gas recirculation (EGR), turbocharger, and/or supercharger are frequently used based on the intake gas condition requirements. From the control application viewpoint, the throttles and valves along the air-path loop enable the intake gas mass flow rate control.

### 3.3.1   THROTTLES AND VALVES

The mass flow rate through the throttle body or valve can be calculated by orifice equations [2,3]. When the flow through the throttle body or valve is unchocked, that is, the pressure ratio across the throttle/valve, $p_d/p_u$, satisfies $p_d/p_u > (2/(\gamma + 1))^{\gamma/\gamma-1}$, then

$$\dot{m}_v = \frac{C_d A_v p_u}{\sqrt{RT_u}} \left(\frac{p_d}{p_u}\right)^{1/r} \left\{\frac{2\gamma}{\gamma - 1}\left[1 - \left(\frac{p_d}{p_u}\right)^{(\gamma-1)/\gamma}\right]\right\}^{1/2} \tag{3.29}$$

When the flow through the throttle body or valve is chocked, that is, the pressure ratio satisfies $p_d/p_u < (2/(\gamma + 1))^{\gamma/\gamma-1}$, then

$$\dot{m}_v = \frac{C_d A_v p_u}{\sqrt{RT_u}} \gamma^{1/2} \left(\frac{2}{\gamma + 1}\right)^{(\gamma+1)/2(\gamma-1)} \tag{3.30}$$

where $p_d$ is the pressure upstream of the valve and $p_u$ is the pressure downstream of the valve; $T_u$ is the temperature upstream of the valve; $\gamma$ is the ratio of specific heats; $R$ is the ideal gas constant; $C_d$ is the discharge coefficient, which can be derived through experimental calibration; and $A_v$ is the valve effective open area. The critical pressure ratio is $(2/\gamma + 1)^{\gamma/\gamma-1} \approx 0.528$.

### 3.3.2   MANIFOLDS

The charge will enter the intake manifolds before going into the cylinder. In the intake manifold, the charge is evenly distributed into each cylinder. Even distribution helps the optimal design of the engine in terms of efficiency and performance.

After combustion, the exhaust gas from all the cylinders will be collected to the exhaust manifold before exiting through the exhaust pipe.

### 3.3.3 POWER BOOSTING

For a given size of engine, the maximal injected fuel amount is related to the intake charge amount into the cylinder. Therefore, the maximal power for a given size of engine can increase, if the amount of intake charge increases. For this purpose, power-boosting techniques are commonly used to increase the power density of IC engines. Through power boosting, the intake charge will be compressed to a high density (high pressure) before entering into the cylinder.

In general, there are two types of power-boosting techniques, that is, turbocharging and supercharging. A turbocharging system [4] consists of a turbine and a compressor. These two components are connected through a common shaft. The energy from the exhaust gas will be used to drive the turbine and is transferred to the compressor. The compressor will therefore boost the intake charge to a high density. The good attribute of the turbocharging system is that it uses the exhaust gas energy to boost the engine and increases the energy efficiency. However, the available energy from the exhaust gas largely depends on the engine-operating conditions. In some cases, in the low engine speed and low load conditions, for instance, the turbocharging system cannot provide sufficient boosting energy as required.

To overcome the shortcomings of the turbocharging system, supercharging system provides a viable solution. In the supercharging system, the boosting power is from the engine's crank shaft through mechanical connections. The mechanical power is typically transferred into the compressor through a belt, gear, shaft, or chain from the engine crank shaft. Since the engine power was used in the intake charge boosting, the efficiency of the engine with the supercharging system is sacrificed to some extent.

## 3.4 FUEL-PATH LOOP

In IC engines, fuel is delivered into the combustion chamber through different ways.

### 3.4.1 SI ENGINE FUEL INJECTION SYSTEMS

In conventional port fuel injection (PFI) SI engines [3], the fuel is injected into the intake port of each cylinder upstream of the intake valve. In this way, the amount of the fuel can be controlled by measuring/estimating the amount of the inlet air, and thus forms a generally homogeneous and stoichiometric mixture. The stoichiometric mixture is the one with which the fuel and the oxygen in the mixture can be just completely consumed. The stoichiometric mixture benefits the emission control in the after-treatment system (a three-way catalyst system) in SI engines.

In direct injection SI engines, the fuel is injected into the cylinder directly. The advantages of direct injection engines are: (1) the mixture is directly formed in the cylinder and therefore, a more accurate air/fuel ratio can be achieved; (2) the injection pressure is higher than that of the PFI engine and therefore, a more complete mixing can be achieved; and (3) the fuel injector design can be combined with the cylinder and piston shape to achieve specific mixing forms that enable extremely high-efficiency combustion.

### 3.4.2 CI ENGINE FUEL INJECTION SYSTEMS

In most of the CI engines, fuel is directly injected into the chamber through the fuel injector. The mass flow rate of the fuel, $\dot{m}_f$, can be calculated by

$$\dot{m}_f = C_D A_n \sqrt{2\rho_f \Delta p} \tag{3.31}$$

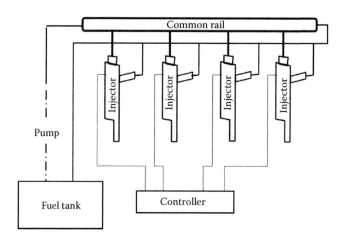

FIGURE 3.8   Common rail fuel injection systems.

where $C_D$ is the discharge coefficient, $A_n$ is the minimum area of the nozzle, and $\Delta p$ is the pressure difference across the nozzle.

As can be seen from Equation 3.21, when the other parameters are fixed, the main source to increase the fuel flow rate is, $\Delta p$, the pressure difference across the nozzle. In conventional fuel injection system, the fuel pressure before the injector is built through the fuel pump. To significantly increase the fuel pressure before injection and therefore to increase $\Delta p$ across the nozzle of the injector, common rail fuel injection systems, as in Figure 3.8, are used. In common rail fuel injection systems, all the injectors share a high-pressure common rail, where the fuel pressure is built up through a fuel pump and regulated by a pressure-control valve. The pressure accumulated in the high-pressure common rail can be up to around 2000 bars.

## 3.5   COMBUSTION

In this section, the combustion processes in SI engines and CI engines are described respectively.

### 3.5.1   COMBUSTION IN SI ENGINES

In conventional SI engines, the fuel and air are mixed before the combustion chamber. When the mixed gas is induced into the engine cylinder, it is mixed with the residual gas and is then compressed by the piston during the compression cycle. The combustion starts when SI occurs at the spark plug about the end of the compression cycle. This is the first step of the combustion, called inflammation. After it is initiated by the sparking (inflammation), the mixed gas releases its energy through rapid flame development and propagation. The flame development and propagation differ from cycle to cycle. This is why the in-cylinder conditions, including pressure, temperature and species concentration, and space distributions, are significantly different. The speed of the flame propagation largely depends on the unburned mixture concentration, that is, the higher the unburned mixture density, the faster the flame propagates. However, the flame front is approximately in a circular shape in an SI engine, since it is with relatively homogeneous unburned gas distributions.

The combustion typically happens around the TDC at the end of the compression cycle. Proper combustion timing is of importance for the combustion efficiency. When the combustion happens ahead of TDC, the piston pushes the mixed gas and therefore, the kinetic energy from the piston reduces. On the other hand, when the combustion occurs after TDC, the peak in-cylinder pressure delays and therefore, the work transfer from the gas to the piston reduces. Thus, the combustion timing cannot be too early or too late and there exists optimal combustion timing. The torque produced by the optimal combustion timing is called maximum brake torque (MBT).

### 3.5.2 COMBUSTION IN CI ENGINES

In CI engines, combustion occurs with several crank angles delay after the fuel injection, which typically happens near the end of the compression cycle. The fuel is injected and atomized into the combustion chamber at high pressure. After fuel injection, the fuel is vaporized and mixed with the air in the chamber. With high pressure and temperature at the time of fuel injection, the diesel fuel reaches its ignition point and starts to ignite spontaneously. After the start of ignition, the unburned fuel/air mixture will keep on combusting for the rest of the expansion cycle. Since the combustion in CI engines happens spontaneously and with multiple points, its process is much more complex than that of SI engines and is related to the fuel property, mixture homogeneity, and the shape of the engine chambers.

Since the combustion in CI engines is with autoignition, the duration between combustion initiator (fuel injection timing) and the timing of ignition is crucial. Such duration is called ignition delay in CI engines. Empirically, the ignition delay can be approximated as a function of in-cylinder pressure and temperature as follows:

$$\tau_d = Ap^{-n}\exp\left(\frac{E_a}{RT}\right) \tag{3.32}$$

where $\tau_d$ is the ignition delay; $E_a$ is an apparent activation energy for the fuel autoignition process; $R$ is the gas constant; $A$ and $n$ are constant parameters related to the fuel and other species concentration.

## 3.6 EMISSIONS

### 3.6.1 EMISSION FORMATION

In SI engines (gasoline engines), the exhaust gases contain nitrogen oxides ($NO_x$), including nitrogen oxide (NO) and nitrogen dioxide ($NO_2$), carbon monoxide (CO), and unburned (or partially burned) hydrocarbons (HC) [5]. In CI engines (diesel engines), nitrogen oxides emissions are at the same level of that in SI engines. Hydrocarbons and particular emissions in diesel engines are more significant than that of SI engines. The level of carbon monoxide in diesel engines is lower than that in SI engines [6].

#### 3.6.1.1 Nitrogen Oxides

The majority of nitrogen oxides are produced by the chemical reactions between the atmospheric nitrogen and oxygen. There are several factors that contribute the formation of nitrogen oxides. The most critical factor is the peak combustion temperature [7]. In general, the higher the peak combustion temperature, the more likely nitrogen oxides will produce. Another important factor is the oxygen concentrations in the air/fuel mixture. Specifically, the fuel/air equivalence ratio, burned gas fraction, the exhaust gas recirculation rate, and the spark timing/fuel injection timing will largely determine the above two factors and therefore influence the nitrogen oxides productions.

#### 3.6.1.2 Carbon Monoxide

Carbon monoxide is produced due to the rich combustion. When excessive fuel exists locally, the carbon in the fuel tends to burn insufficiently. Therefore, the fuel/air equivalent ratio is the most significant factor influencing the carbon monoxide productions. The fuel in SI engines is typically richer than that in CI engines. Thus, the carbon monoxide emission is more serious in SI engines than that in CI engines.

### 3.6.1.3 Unburned Hydrocarbon

Similar to carbon monoxide, the unburned hydrocarbon is produced as a result of incomplete combustion of IC engines on one aspect. On the other aspect, the fuel composition is another factor related to the unburned hydrocarbon production.

### 3.6.1.4 Particulate Matter

Because of the high percentage, particular matter emission is more significant in CI engines than that in SI engines.

### 3.6.2 Emission Control Strategy

There are two classes of emission control strategies: through combustion control and after-treatment systems. Combustion control is through the control of the initial in-cylinder conditions at combustion and/or the combustion itself.

An effective way to control $NO_x$ emission is exhaust gas recirculation (EGR) [8]. In general, $NO_x$ production is closely related to the peak combustion temperature. The higher the peak combustion temperature, the more $NO_x$ is produced. By introducing exhaust gas into the cylinder, the peak combustion temperature can be reduced.

### PROBLEMS

3.1 Explain the difference between IMEP and BMEP. Under what operating conditions, the maximal BMEP for a given IC engine can be achieved?

3.2 Discuss what are the design factors influencing the fuel efficiency for an IC engine.

3.3 Discuss what are the design factors influencing the maximal power for an IC engine.

3.4 For a four-cylinder SI engine, the required MBT is 160 Nm at the rotation speed of 2500 rpm. Assume that:
   1. The BMEP is 950 kPa at this operating point;
   2. The lengths of the bore and stroke are equal.
   What are the desired engine displacement and the length of the bore? What is the maximal brake power?

3.5 For a four-cylinder diesel engine, the displaced volume is 6 L, bore is 110 mm, stroke is 120 mm, and the compression ratio is 15.2. If the volumetric efficiency is 0.9 and the mean piston speed is 7 m/s, what is the air flow rate?

3.6 Calculate the stoichiometric air/fuel ratios on a mass basis for gasoline and diesel. Assume that the main component in gasoline is octane ($C_8H_{18}$) and the average chemical formula for diesel fuel is $C_{12}H_{23}$.

3.7 Calculate the average gas mass flow rate, when the air is going through a 30-mm-diameter valve with 30° opening at atmospheric condition. The upstream pressure is 1.5 bar and the downstream pressure is 1.2 bar.

3.8 For a conventional engine, discuss the nitrogen oxides emission changes during the process where the EGR ratio increases from zero to maximum.

### REFERENCES

1. Heywood, John. *Internal Combustion Engine Fundamentals*. McGraw-Hill, New York, 1988.
2. Bicen, A. F., Vafidis, C., and Whitelaw, J. H. Steady and unsteady airflow through the intake valve of a reciprocating engine. *ASME Transactions, Journal of Fluids Engine* 107, 1985: 413–420.

3. Cook, J. A., Sun, J., Buckland, J. H., Kolmanovsky, I. V., Peng, H., and Grizzle, J. W. Automotive powertrain control—A survey. *Asian Journal of Control* 8, no. 3, 2006: 237–260.

4. Watson, N. and Janota, M. S. *Turbocharging the Internal Combustion Engine*. John Wiley, New York, 1982.

5. Abdel-Rahman, A. A. On the emissions from internal-combustion engines: A review. *International Journal of Energy Research* 22, no. 6, 1998: 483–513.

6. Sher, E. *Handbook of Air Pollution from Internal Combustion Engines Pollutant Formation and Control*. Academic Press, San Diego, USA, 1998.

7. Lavoie, G. A., Heywood, J. B., and Keck, J. C. Experimental and theoretical study of nitric oxide formation in internal combustion engines. *Combustion Science and Technology* 1, no. 4, 1970: 313–326.

8. Abd-Alla, G. H. Using exhaust gas recirculation in internal combustion engines: A review. *Energy Conversion and Management* 43, no. 8, 2002: 1027–1042.

# 4 Fundamentals of Power Electronics

*Pierre Magne, Xiaodong Shi, and Mahesh Krishnamurthy*

## CONTENTS

Power electronics can be described as the technology that combines contributions from electronic, magnetic, and electrochemical components to control and convert electric power. While analog and digital electronics are used to transmit and transform data and information, power electronics correspond to the conversion and transport of energy and power. The first power electronics system was developed in 1902 by Peter Cooper Hewitt. It consisted of a mercury arc rectifier to convert alternating current (ac) into direct current (dc) [1]. Modern power electronics started at the end of the 1940s with the invention of the first transistors. From then onward, performances of designed converters are regularly enhanced, thanks to the improvement of components, such as switching devices, magnetic material, or cold plates to name a few.

Power electronics defines the technology that enables the management of electrical energy. In an electrified vehicle, electric traction motors provide mechanical power to the wheels. Energy that powers the motors is either available from an electrochemical energy source (battery, ultracapacitor) or generated by the engine. Therefore, to operate a motor, it is necessary to transfer energy from onboard source(s) to the motor. This energy needs to be transformed in a controlled manner, to create a rotating magnetic field in the motor to match the load requirements. All these phases (transfer, transformation, and control) are enabled by the power electronic systems. Power converters (dc–dc, ac–dc, and dc–ac) are used to convert electrical energy and manage the power flow through the vehicle. This power flow can be from the battery to the wheels (in both directions), from the grid to the battery (in both directions), and also from the engine to the battery or wheels.

This chapter first presents the fundamentals of switch-mode dc–dc converter. Electrical circuit and steady-state operation of the most popular converters are introduced and semiconductor devices are briefly introduced. The second section deals with switch-mode ac–dc converters. Single-phase and three-phase inverters are presented as well as the pulse-width-modulation (PWM) method. The third section discusses the fundamental concepts of ac–dc rectifiers. Both uncontrolled and controlled rectifiers are introduced in single-phase and three-phase configurations. The last section discusses design recommendations for power converter design. Loss evaluation and selection process of the power module is presented and illustrated with examples. Gate driver designs, as well as snubbers and busbar designs, are also discussed.

This chapter is aimed at helping the reader to understand and analyze most of the power electronics circuits. The design and selection of electrical parameters of the most popular dc–dc and ac–dc converters are introduced, as well as their operation principles. In some cases, assumptions or simplifications are assumed to perform the analysis, which are explicitly mentioned. In these cases, references are proposed to invite the interested reader to go further of the assumptions considered.

## 4.1 SWITCH-MODE DC–DC CONVERTER

### 4.1.1 DC–DC CONVERTER USE IN MODERN ELECTRIFIED VEHICLE

Dc–dc converter enables the generation of a controlled output dc voltage from any input dc voltage. In other words, using a dc–dc converter gives the possibility to transform a dc input voltage into a different dc output voltage. According to the used converter (Boost, Buck, and Buck–Boost), the input voltage can be either stepped up or stepped down. In applications such as power generation (e.g., wind turbine) or domestic application, dc–dc converter transforms unregulated rectified voltage into a controlled dc-voltage. The value of the output voltage is chosen according to the application (540 or 28 V in aerospace systems, 225–650 V in hybrid vehicle, etc.). In an electrified vehicle, a high-voltage battery is conventionally used as the electrical energy source of the vehicle. This battery can be sized differently according to the type of vehicle. The more the battery capacity, the more the electrical energy that can be stored within the vehicle and used to power it. In some cars (Toyota Hybrid System I, Fiat 500e), the battery is directly connected to the traction inverter. In this configuration, battery output is imposed to the drive system, which can be, in some cases, a constraint as it limits the performances of the electric motor (especially in terms of maximum speed). To avoid this, other configurations (Toyota Hybrid System II: Camry 2007, Prius 2010) have a dc–dc converter between the battery and the traction inverter. The dc–dc converter steps up the battery voltage to obtain the required dc-bus voltage. For instance, in the Prius 2010, three different dc-link voltages are used, thanks to the step-up dc–dc converter [2]. By adjusting this value, performances of the drive system can be modified in terms of performances as well as efficiency. Also, the use of a dc–dc converter gives flexibility in the system design; different battery packs with different voltage ratings can be used for the same motor.

Another application of the dc–dc converter in an electrified vehicle is the realization of an auxiliary power unit to power the low-voltage 12-V (or 48 V) dc grid. This dc grid supplies power to all the electronic components of the vehicle (air-conditioning system, microprocessor, lights, etc.).

This section of the chapter focuses on the presentation of three different dc–dc converter topologies: Buck, Boost, and Buck–Boost converters. Their steady-state operation will be presented first. Then, an overview of switching devices used in power electronics is proposed. It contains the basic information and details required to explain the operation of switching dc–dc, ac–dc, and dc–ac converters. To obtain more details about switching devices and their operation, interested readers are invited to refer to Reference 3. Then, mathematical models required to perform dynamic analysis and parameters selection of the converters are presented.

### 4.1.2 STEADY-STATE OPERATION OF SWITCH-MODE DC–DC CONVERTER

In switch-mode dc–dc converter, the term "switch" comes from the fact that the converter comprises at least one switching device. This device is a semiconductor component enabling the circulation (on state) or not (off state) of a current through it. The state (on and off) of the semiconductor is either defined by the electric circuit (e.g., diode) or controlled by a periodic gate signal $u$, which is characterized by its duty cycle $d$ over a period $T$ (e.g., MOSFET and IGBT). The value of $d$ represents the average value of $u$ over the period $T$, as represented in Figure 4.1. The steady-state operation of a dc–dc converter corresponds to the expression of the input/output currents and voltages according to $d$. It specifies the root main square (RMS) output voltage of the converter according to the RMS input voltage and the duty cycle. In the section presented here, it is assumed that all the converters operate in continuous conduction mode (CCM). This means that the current within the inductor is always higher than 0.

#### 4.1.2.1 Buck Converter

An electrical circuit of the Buck converter with ideal switches is represented in Figure 4.2. The Buck converter enables to step down the input voltage. When $S_1$ is on, $V_L = V_{in} - V_{out}$, and when $S_1$ is off, $V_L = -V_{out}$. Over a period of time, the average value of $V_L$ is null in the steady state. If not, it means that the steady state is not reached yet. Hence, relation (4.1) can be expressed to obtain the transformation ratio between $V_{in}$ and $V_{out}$. Considering Equation 4.1, the transformation ratio between $V_{in}$ and $V_{out}$ in the steady state is given by relation (4.2), where $d$ is the duty cycle of $S_1$.

$$V_L = \frac{1}{T}\left(\int_0^{dT} V_{in} - V_{out}\,dt + \int_{dT}^{T} -V_{out}\,dt\right) = 0 \tag{4.1}$$

$$\Rightarrow d.V_{in} = V_{out} \tag{4.2}$$

As the duty cycle is comprised between 0 ($S_1$ always open) and 1 ($S_1$ always closed), $V_{out}$ has a value ranging between 0 and $V_{in}$.

If we consider that all the components are perfect, there is no loss within the converter, and the input power (4.3) is equal to the output power (4.4) in the steady state. Then, by considering Equations 4.2, 4.4, and 4.5, the relation between the input current and the output current can be expressed as Equation 4.6.

$$P_{in} = V_{in} \cdot i_{in}, \tag{4.3}$$

$$P_{out} = V_{out} \cdot i_{out} \tag{4.4}$$

$$P_{in} = P_{out} \tag{4.5}$$

$$\Rightarrow d \cdot i_{out} = i_{in} \tag{4.6}$$

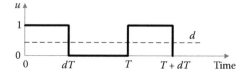

FIGURE 4.1   Gate signal $u$.

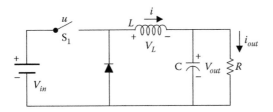

FIGURE 4.2 Buck converter.

### 4.1.2.2 Boost Converter

Boost converter steps up the input voltage. Considering ideal switches, its electrical circuit is represented in Figure 4.3. When $S_1$ is on, the diode is reverse biased and the inductor $L$ is charged by the source ($V_{in}$) and $V_L = V_{in}$. Then, when $S_1$ is off, energy stored within the inductance is transmitted to the dc-link capacitor and $V_L = V_{in} - V_{out}$. This energy transfer yields the generation of an output voltage higher than the input voltage. In an electrified vehicle, such a converter can be used between the battery pack and the traction inverter. This is the case in the Toyota Prius 2010.

Over an entire period, the average value of $V_L$ is equal to 0 in a steady-state operation. Hence, relation (4.7) can be expressed to obtain the transformation ratio between $V_{in}$ and $V_{out}$. The steady-state relation between $V_{in}$ and $V_{out}$ is given by relation (4.8), where $d$ is the duty cycle of $S_1$.

$$V_L = \frac{1}{T}\left( \int_0^{dT} V_{in}dt + \int_{dT}^{T} V_{in} - V_{out}dt \right) = 0 \tag{4.7}$$

$$\Rightarrow V_{in} = (1 - d) \cdot V_{out} \tag{4.8}$$

As was done for the Buck converter, relations (4.3) through (4.5) and (4.8) can be used to express Equation 4.9 that expresses the current levels of Boost converter in the steady state.

$$i_{out} = (1 - d) \cdot i_{in} \tag{4.9}$$

In the ideal case, Equation 4.8 shows that $V_{out}$ can be theoretically boosted up to infinity (case $d = 1$). However, an infinite output voltage is obviously not possible in practice. Indeed, parasitic resistive elements of the circuit limit the maximum output voltage reachable by the converter. For instance, if the series resistance $r_L$ of the inductor windings is considered, the steady-state relation becomes Equation 4.10.

$$V_{in} = \left( 1 - d + \frac{r_L}{R \cdot (1 - d)} \right) \cdot V_{out} \tag{4.10}$$

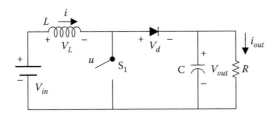

FIGURE 4.3 Boost converter.

### 4.1.2.3 Buck–Boost Converter

Buck–Boost converter enables to either step up or step down the output voltage. Considering an ideal switch, its electrical topology is given in Figure 4.4. The switch $S_1$ permits to store energy within the inductor. When $S_1$ is "on-state," the voltage across the inductor is $V_L = V_{in}$, and the current is circulating from the input voltage source to the inductor. Then, when $S_1$ is "off-state," the energy stored within the inductor is transmitted to the dc-link capacitor through the diode. During this phase, $V_L = -V_{out}$. In steady-state operation, over one period, as explained above, the average value of $V_L$ is zero, and hence relation (4.11) can be expressed to obtain the transformation ratio between $V_{in}$ and $V_{out}$. From Equation 4.12, it can be seen that if $d < 0.5$, then $V_{out} < V_{in}$ and the converter operates in the Buck mode. Inversely, if $d > 0.5$, then $V_{out} > V_{in}$ and the converter operates in the Boost mode.

$$V_L = \frac{1}{T}\left( \int_0^{dT} V_{in}dt + \int_{dT}^{T} -V_{out}dt \right) = 0 \tag{4.11}$$

$$\Rightarrow V_{in} \cdot d = (1 - d) \cdot V_{out} \tag{4.12}$$

Similarly, for the Buck and Boost converters, if the converter is considered as ideal, and hence without any loss, Equations 4.3 through 4.5 and 4.12 can be considered to express Equation 4.13.

$$i_{out} \cdot d = (1 - d) \cdot i_{in} \tag{4.13}$$

From Equation 4.12, it can be seen that the ideal Buck–Boost converter can theoretically generate an infinite output voltage (case $d = 1$). As it has been mentioned for the Boost converter, this is not achievable in practice due to the parasitic resistivity of the components. Indeed, considering the series resistance $r_L$ of the inductor, Equation 4.12 becomes Equation 4.14.

$$d \cdot V_{in} = \left( 1 - d + \frac{r_L}{R \cdot (1 - d)} \right) \cdot V_{out} \tag{4.14}$$

### 4.1.2.4 Summary of Steady-State Characteristics of Presented dc–dc Converters

In Figure 4.5, the steady-state characteristics of each converter are plotted, which are given by Equation 4.1 for the Buck, Equations 4.8 and 4.10 for the Boost, and Equations 4.12 and 4.14 for the Buck–Boost. When selecting a topology, the designer should keep in mind that the more the duty cycle, the less the efficiency. This point is more detailed in Section 4.3.2.3. Thus, even if a Boost or a Buck–Boost offers theoretically very high step-up capabilities, they are limited in practice because of a too low efficiency in these operating areas.

### 4.1.3 OVERVIEW OF SWITCHING DEVICES

In the previous section, steady-state operations of three different dc–dc converters were presented. As it has been mentioned, these converters operate by way of semiconductors. Owing to their

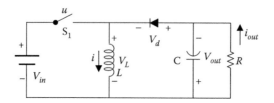

FIGURE 4.4   Buck–Boost converter.

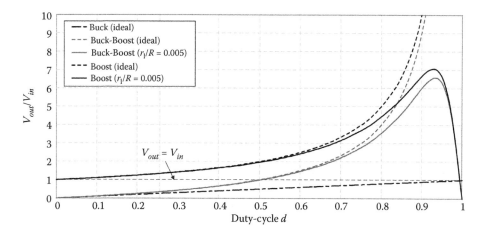

FIGURE 4.5 Steady-state characteristics of the Buck, Buck–Boost, and Boost converters.

complexity, semiconductors are very often considered as ideal in power electronics circuit analysis. This approach simplifies a lot the analysis of the circuit and lets the designer focus on the electrical operation of the converter instead of the semiconductor itself. Nevertheless, it is important to have an understanding of how to select a semiconductor for an application as the use of an inappropriate device reduces the good operation of the converter and often leads to its failure. First, this section proposes an overview of the electrical characteristics of a switch, and then presents ideal characteristics of most used semiconductors in power electronics design. More detailed information about nonideal characteristics and properties of semiconductor devices can be found in References 3,4.

### 4.1.3.1 Electrical Characteristics of a Switch

In the power converter, switching devices have to open and close an electrical circuit. Hence, they need to behave as an electrical conductor to close the circuit, as well as an electrical insulator to open it. This double characteristic defines what a semiconductor is: a device able to conduct current in an efficient way, as well as block it. Semiconductors are rated in terms of the maximum voltage they can handle and still behave as an insulator, and the maximum current that can circulate through them without damaging the device. Maximum allowed current does not only depend on the module rating but also on the thermal properties of the semiconductor. Thus, according to the power module packaging, as well as the used heat sink, maximum allowed current can vary for the same device.

#### 4.1.3.1.1 Current–Voltage Characteristic

Figure 4.6 shows the current–voltage characteristic of the four possible quadrants ($Q_1$, $Q_2$, $Q_3$, and $Q_4$) of an ideal switch shown in Figure 4.7. For every switch, at least one of these quadrants

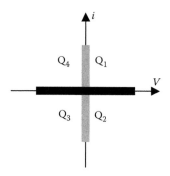

FIGURE 4.6 Current–voltage characteristic.

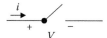

FIGURE 4.7    Ideal switch.

corresponds to the electrical characteristic of the device. This characteristic is an essential property of the switch as it defines in which application the switch can be used. For instance, in a bidirectional application, the switch should handle both positive and negative currents. Hence, at least two quadrants (e.g., $Q_1$, and $Q_2$) are mandatory to make it a possible candidate for the application. Such a configuration can be obtained by using two different switches in parallel.

The next section presents ideal electrical characteristics of some of the most used devices in automotive application: diode, metal oxide semiconductor field effect transistor (MOSFET), and insulated gate bipolar transistor (IGBT). The thyristor is also presented.

### 4.1.3.1.2 Switching Characteristics

Three families of semiconductors can be identified according to how their states (on and off) are controlled: uncontrolled, turn-on state is controlled, and turn-on and turn-off states are controlled. These properties are very important in the selection of a semiconductor as they define how the semiconductor can be used and controlled, and hence how it can behave in an electrical circuit.

### 4.1.3.2 Most Popular Switches Used in Switch-Mode Power Converter

In this section, the electrical characteristics of four widely used switches, diode, thyristor, MOSFET, and IGBT, are presented. Other components, such as the transistor, gate turn-off thyristors (GTO), bipolar junction transistor (BJT), or junction field-effect transistors (JFET), to name a few, also exist and are not mentioned here. More details related to all these semiconductors can be found in Reference 3.

Figure 4.8 shows the power rating and switching frequency range of popular power devices. MOSFET is often used in applications with power rating less than the kilowatt level. Usually, low-power MOSFET is cheap, low in power loss, and high in switching frequency (up to 1000 kHz). As a result, MOSFET dominates the low-power switch market. With the advance of semiconductor technology, more and more high-power-rating MOSFET devices are also available in the market. However, the cost is still much higher than IGBT with the same power rating.

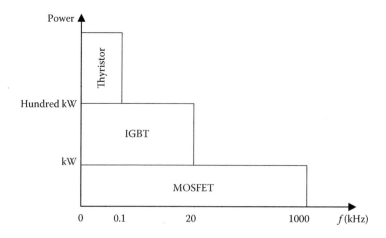

FIGURE 4.8    Power rating and switching frequency range of switching devices.

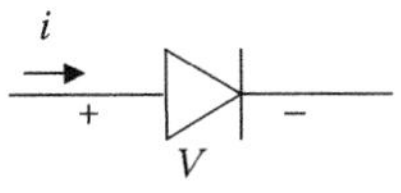

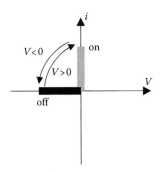

FIGURE 4.9   Diode.

From kilowatt level up to a 100-kW level, IGBT is the best candidate. The popular IGBT devices can handle the collector–emitter voltage up to kilovolt level and collector–emitter current up to kiloampere level. Beyond that, the cost of the IGBT is very high compared to a thyristor. The maximum switching frequency of IGBT is usually around 20 kHz, which is relatively low compared to MOSFET but can meet the requirement of most vehicle power applications such as dc–dc converters and dc–ac inverters. One of the main disadvantages of IGBT is the turn-off current tail that leads to relatively high switching losses compared to MOSFET.

Thyristor devices are often used in megawatt-level applications. Unlike MOSFET and IGBT (voltage-controlled switch), thyristor is a current-controlled device that can be turned on by a controlled current but is hard to be turned off. The switching frequency of a thyristor is below 100 Hz. The thyristor is heavily used in megawatt-level ac–dc rectifier applications.

An example is given in Section 4.4.3.2 of this chapter to show how to determine the type of switching device based on voltage, current, and switching frequency requirements of the application.

### 4.1.3.2.1   Diode

Diode is a unidirectional uncontrolled component. The electrical symbol of a diode is represented in Figure 4.9. When the voltage across the diode is lower than 0, the diode is reverse biased and there is no current circulating through it. Hence, the diode is in its "off-state," and is equivalent to an open switch. When the voltage across it is higher than 0, the diode is in its "on-state," and current can circulate through the device. In practice, a voltage drop, called forward voltage, exists across the diode in the "on-state." Current–voltage characteristic of an ideal diode is given in Figure 4.10.

### 4.1.3.2.2   Thyristor

Thyristor is an "on-state" controlled component. The electrical symbol of a thyristor is represented in Figure 4.11. Its "off-state" behaves in a similar way to a diode: when $V_{AK}$ is lower than 0, the component is reverse biased and there is no current circulating through it. However, when $V_{AK}$ becomes higher than 0, the device does not switch to its "on-state" as a diode would do. To turn on a thyristor, it has to be triggered by the gate signal. As a result, to switch from the "off-state" to the "on-state," two conditions have to be respected: (1) $V_{AK} > 0$ and (2) a pulsed current has to be injected into the device through the gate port. In Figure 4.12, the electrical characteristic of an ideal thyristor is represented. In practice, a voltage drop, called forward voltage, exists across the thyristor in the "on-state" as is also for the diode. The nonideal behavior of a thyristor can be found in Reference 3.

FIGURE 4.10   Current–voltage characteristic of ideal diode.

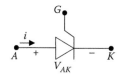

FIGURE 4.11   Thyristor.

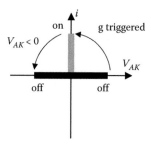

FIGURE 4.12   Current–voltage characteristic of ideal thyristor.

### 4.1.3.2.3   MOSFET

MOSFET is a fully controllable switch. This means that both the "on-state" and the "off-state" are controlled by a gate signal, and thus by the user. When the gate signal is active, MOSFET is in the "on-state" and a current can circulate through it. When the gate signal is inactive (0 V), the device is in the "off-state" and no current can flow across it. In practice, the amplitude of the gate signal is impacting the current rating of the MOSFET: the higher the gate voltage, the higher the saturation current. Also, an "on-state" resistance exists within the device. This yields a voltage drop across the MOSFET. More details regarding MOSFET operation and its nonideal characteristics can be found in Reference 3.

The electrical symbol of a MOSFET is represented in Figure 4.13. An ideal characteristic of a MOSFET is given in Figure 4.14. Voltage rating of power MOSFET can be as large as 900 V. They are usually preferred over IGBT for low-voltage application (<50 V) as they propose lower

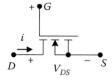

FIGURE 4.13   MOSFET (*n*-channel type).

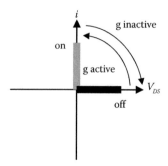

FIGURE 4.14   Current–voltage characteristic of ideal MOSFET.

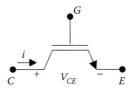

FIGURE 4.15   IGBT.

conduction loss and can be operated for higher-switching frequency. This is especially profitable in low-voltage power supply design, where using a high switching frequency yields the reduction of the magnetic component size (inductor, transformer).

### 4.1.3.2.4   IGBT

Similar to the MOSFET, IGBT is a fully controllable switch. The electrical symbol of an IGBT is represented in Figure 4.15 and its corresponding electrical characteristic IGBT is given in Figure 4.16. Voltage rating of IGBT can be as large as 1600 V. They are usually preferred to MOSFET for high-power application (grid–tie inverter, drive system, etc.) because of their better current conduction capability at high voltage (400–1200 V). In automotive electrified traction systems, they are the conventionally used component for the traction system (Boost, inverter) due to their voltage and current ratings.

The electrical characteristic of a nonideal IGBT is different from the ideal one presented in Figure 4.16. Especially, current is also dependent on the gate voltage. Also, breakdown voltage and reverse blocking capability are not mentioned by the ideal model. More information regarding nonideal behavior and characteristic of IGBT can be found in Reference 3.

### 4.1.3.2.5   Example: Switching Selection for a Unidirectional Buck Converter

In this example, the operation of the Buck converter is considered to identify switching characteristics required for its two switches. From these characteristics, semiconductors that enable the proper operation of the converter are identified.

Let us consider the Buck converter given in Figure 4.17 comprising two nonidentified ideal switches. Electrical requirements imposed by the circuit operation to $S_1$ and $S_2$ are given in Table 4.1. From the defined overall characteristics of $S_1$, it can be seen from Figures 4.14 and 4.16 that MOSFET and IGBT enable the operation of the circuit. The thyristor cannot be used for $S_1$ as it does not enable the commutation from the "on-state" to the "off-state." Also, it is obvious that characteristics defined for $S_2$ are exactly those of a diode. Hence, the use of these semiconductors enables the proper operation of a unidirectional Buck converter.

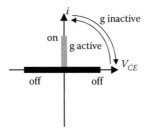

FIGURE 4.16   Current–voltage characteristic of ideal IGBT.

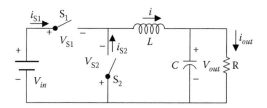

FIGURE 4.17    Buck converter with two unidentified ideal switches.

TABLE 4.1

**Electrical Requirements of Switches in a Unidirectional Buck Converter**

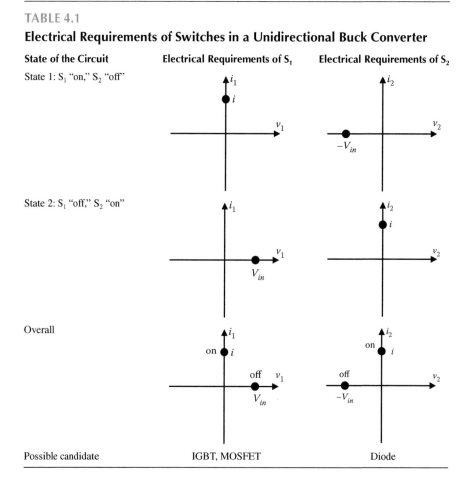

| State of the Circuit | Electrical Requirements of $S_1$ | Electrical Requirements of $S_2$ |
|---|---|---|
| State 1: $S_1$ "on," $S_2$ "off" | | |
| State 2: $S_1$ "off," $S_2$ "on" | | |
| Overall | | |
| Possible candidate | IGBT, MOSFET | Diode |

### 4.1.4  STATE-SPACE REPRESENTATION OF SWITCH-MODE DC–DC CONVERTER

In this section, the state-space models of the Buck (Figure 4.2), Boost (Figure 4.3), and Buck–Boost (Figure 4.4) converters are presented. These representations are essential for the designer as they enable the dynamic consideration of the converters. From these models, it is possible to determine the dynamic properties of the converter required for parameter selection, such as inductance, capacitance, and switching frequency. Moreover, they are also required to design any control loop for the converter.

In the following section, switching devices are considered as ideal and no parasitic resistance is taken into account. The method to obtain the state-space representation of the Buck converter is presented in detail. For the other converter (Boost and Buck–Boost), corresponding state-space representations are directly given. They can be obtained following a similar method as the one presented for the Buck converter.

### 4.1.4.1 State-Space Representation of the Buck Converter

The electrical circuit of the Buck converter is given in Figure 4.2. To obtain its corresponding state-space representation, it is necessary to express the state variables (current across the inductor and voltage across the capacitor) as a function of the converter parameters and the gate signal $u$. To do so, the electrical circuit of Figure 4.2 is considered separately for the two values of $u$: $u = 1$ and $u = 0$. Then, from the analysis of these two configurations, a unique state-space model is identified for the converter.

#### 4.1.4.1.1 State-Space Representation of Buck Converter: Case u = 1

When $u = 1$, switch $S_1$ is in the "on-state" and voltage across the diode is $V_d = -V_{i \cdot n}$. Hence, the diode is reverse biased and the electrical circuit of the Buck corresponding to $u = 1$ is the one in Figure 4.18. From the electrical configuration, state-space Equations 4.15 and 4.16 can be expressed.

$$\left\{ L\frac{di}{dt} = V_{in} - V_{out} \right. \tag{4.15}$$

$$\left\{ C\frac{dV_{out}}{dt} = i - \frac{V_{out}}{R} \right. \tag{4.16}$$

As it has been mentioned in Section 4.1.2.1 by Equation 4.2, $V_{in} > V_{out}$ for a Buck converter. Thus, it can be seen from Equation 4.15 that the sign of $Ldi/dt$ is positive, and thus, the voltage across $L$ is increasing when $u = 1$. Also, the output current is higher than the input current, as explained by Equation 4.6. Input current is the one circulating through the inductor in this case. Hence, it can be seen from Equation 4.16 that $V_{out}$ is decreasing when $u = 1$. From these two observations, we can say that when $u = 1$, the inductor $L$ is charged by the voltage source whereas the capacitor $C$ is discharged by the load.

#### 4.1.4.1.2 State-Space Representation of Buck Converter: Case u = 0

When $u = 0$, the switch $S_1$ is in the "off-state." Energy stored within the inductance $L$ (during the state $u = 1$) is transferred to the capacitor $C$ and the load ($R$ here) by the current $i$ circulating through the diode. Figure 4.19 gives the electrical configuration of this case. The state-space model of the Buck when $u = 0$ is given by Equations 4.17 and 4.18. During this state, current $i$ coming from $L$ is higher than the load current. Hence, $C$ is charged and $V_{out}$ is increasing.

$$\left\{ L\frac{di}{dt} = -V_{out} \right. \tag{4.17}$$

$$\left\{ C\frac{dV_{out}}{dt} = i - \frac{V_{out}}{R} \right. \tag{4.18}$$

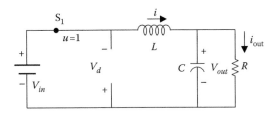

**FIGURE 4.18**  Buck converter configuration when $u = 1$.

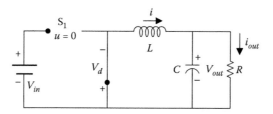

**FIGURE 4.19** Buck converter configuration when $u = 0$.

*4.1.4.1.3 State-Space Representation of Buck Converter*

To have a unique state-space representation of the circuit, the two state-space models corresponding to the cases $u = 0$ and $u = 1$ have to be combined together. For this, the value of $u$ is integrated to Equations 4.15 through 4.18, and the obtained representation is given by Equations 4.19 and 4.20. By replacing $u$ by 1 or 0, it is easy to see that Equation 4.19 is equivalent to Equation 4.15 when $u = 1$, and Equation 4.17 when $u = 0$. Also, by considering the duty cycle $d$ of $u$, the steady state of Equation 4.19 yields relation (4.2).

$$\left\{ L\frac{di}{dt} = u \cdot V_{in} - V_{out} \right. \tag{4.19}$$

$$\left\{ C\frac{dV_{out}}{dt} = i - \frac{V_{out}}{R} \right. \tag{4.20}$$

In Figure 4.20, waveforms of $V_{out}$ and $i$ are plotted considering Equations 4.19 and 4.20, and the parameters given in Table 4.2.

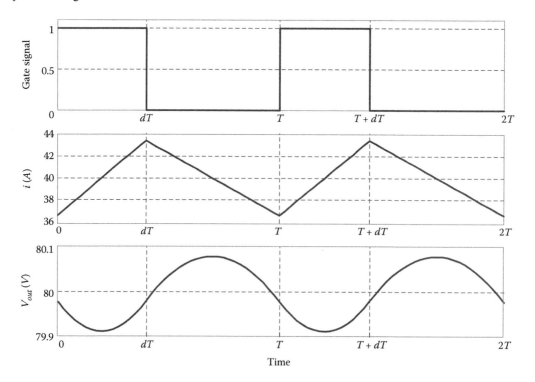

**FIGURE 4.20** Gate signal, current, and voltage waveforms of Buck converter.

**TABLE 4.2**

**Parameters of the Buck Converter**

| $V_{in}$ (V) | $d$ | $L$ (µH) | $C$ (µF) | $f_{sw}$ (kHz) | $R$ (Ω) |
|---|---|---|---|---|---|
| 200 | 0.4 | 700 | 500 | 10 | 2 |

#### 4.1.4.2 State-Space Representation of the Boost Converter

Following a similar method as for the Buck converter, state-space model of the Boost converter represented in Figure 4.3 can be obtained. Considering the two possible configurations of the Boost converter ($u = 1$ and $u = 0$), the corresponding state-space model is expressed in Equation 4.21.

$$\begin{cases} L\dfrac{di}{dt} = V_{in} - (1-u)V_{out} \\ C\dfrac{dV_{out}}{dt} = (1-u)i - \dfrac{V_{out}}{R} \end{cases}$$

(4.21)

In Figure 4.21, $i$ and $V_{out}$ given by Equation 4.21 are plotted over two periods for the set of parameters given in Table 4.3.

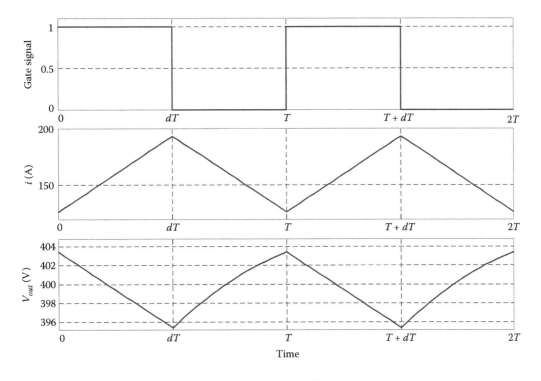

**FIGURE 4.21** Gate signal, current, and voltage waveforms of Boost converter.

**TABLE 4.3**

**Parameters of the Boost Converter**

| $V_{in}$ (V) | $d$ | $L$ (µH) | $C$ (µF) | $f_{sw}$ (kHz) | $R$ (Ω) |
|---|---|---|---|---|---|
| 200 | 0.5 | 150 | 500 | 10 | 5 |

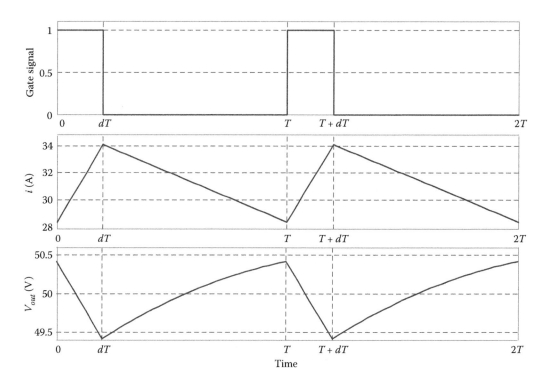

**FIGURE 4.22**   Gate signal, current, and voltage waveforms of Buck–Boost converter.

### 4.1.4.3   State-Space Representation of the Buck–Boost Converter

Similar to that for the Buck converter, state-space model of the Buck–Boost converter represented in Figure 4.4 can be obtained. Considering the two possible configurations of the converter ($u = 1$ and $u = 0$), the corresponding state-space model is expressed in Equation 4.22. In Figure 4.22, waveforms of the converter obtained with the parameters given in Table 4.4 are plotted.

$$\begin{cases} L\dfrac{di}{dt} = u \cdot V_{in} - (1-u)V_{out} \\ C\dfrac{dV_{out}}{dt} = (1-u)i - \dfrac{V_{out}}{R} \end{cases} \tag{4.22}$$

### 4.1.5   Inductance and Capacitance Selection

In the switching-mode dc–dc converter, the selection of proper inductance and capacitance values is very important for several reasons: the inductance is mandatory to enable power transfer through the converter as well as voltage transformation, as the inductor is the intermediary energy tank of the converter. Also, the inductance value has a direct impact on the input current ripple level, whereas the capacitance value mostly affects the output voltage ripple. These points are detailed above. For both the input current and the output voltage, specific levels often have to be respected

**TABLE 4.4**

**Parameters of the Buck–Boost Converter**

| $V_{in}$ (V) | $d$ | $L$ (μH) | $C$ (μF) | $f_{sw}$ (kHz) | $R$ (Ω) |
|---|---|---|---|---|---|
| 200 | 0.2 | 700 | 500 | 10 | 2 |

for several reasons: power quality (standards, i.e., MIL-STD-704 for aerospace application) and/or preserving the health of electrochemical energy source (i.e., output current ripple of battery or fuel cell). To select proper values for both the inductance and the capacitance, it is then required to express the current and voltage ripple expressions. As it has been done for the state-space expression of converters, the method to obtain ripple expressions will be detailed for the Buck only.

### 4.1.5.1 Current and Voltage Ripples of Buck Converter

#### 4.1.5.1.1 Current Ripple Evaluation

As it is represented in Figure 4.20 and explained in Section 4.1.4.1, inductor current $i$ behavior is different according to the state of $S_1$ in a Buck converter. For each state, $i$ is ruled by a different set of equations: (4.15), (4.16), (4.17), and (4.18). Current ripple $\Delta i$ can be determined from any of these equations. Meanwhile, consideration of Equation 4.17 is easier as it does not include $u$ and $V_{in}$, and $\Delta i$ can be evaluated by the integration of the current. To simplify the integration, it is assumed that the capacitance is big enough to make variations of $V_{out}$ negligible in front of the current variations. As a result, $V_{out}$ is considered to be equal to its operating point: $V_{out0} = d \cdot V_{in}$ (see Equation 4.2) during this phase. Considering this, expression of the current when $u = 0$ can be expressed as Equation 4.23, and $\Delta i$ is expressed as Equation 4.24.

$$L \frac{di}{dt} = -V_{out0} \tag{4.23}$$

$$\Delta i = \frac{1}{L} \int_{d.T}^{T} V_{out0} dt \Rightarrow \Delta i = \frac{V_{out0}(1 - d)T}{L} \tag{4.24}$$

#### 4.1.5.1.2 Inductance Selection

To select the inductance value for a specific application, the required maximal current ripple $\Delta i_{max}$ has to be defined. For $V_{in}$ = constant, the maximum value of this latter is obtained for $d = 0.5$, and is expressed by relation (4.25). Consideration of Equation 4.25 gives the minimum value of the inductance to ensure $\Delta i_{max}$ for a given output voltage and time period (or switching frequency). It is worth noting that reducing $T$, and hence increasing the switching frequency enables the reduction of the required inductance. However, in practice, the switching frequency cannot be increased too much because of the switching loss generated at each turn-on and turn-off phases of the semiconductors.

$$\Delta i_{max} = \frac{V_{in} \cdot T}{4 \cdot L_{min}} \tag{4.25}$$

#### 4.1.5.1.3 Voltage Ripple Evaluation

Voltage ripple level $\Delta v$ across a capacitor $C$ is expressed by Equation 4.26, where $\Delta Q$, defined by Equation 4.27, is the charge variation within the dc-link capacitor.

$$\Delta v = \frac{\Delta Q}{C} \tag{4.26}$$

$$\Delta Q = \int i dt \tag{4.27}$$

The capacitor is charged by the current $i$ when the latter is higher than the load current $i_{out}$. However, in Buck converter with resistive load only, the load current is equal to the average value of the inductor current. Then, $\Delta Q$ corresponds to the area represented in Figure 4.23, which is expressed by Equation 4.28. Therefore, from Equations 4.26 and 4.28, expression of $\Delta v$ for the Buck converter in the considered case is given by Equation 4.29.

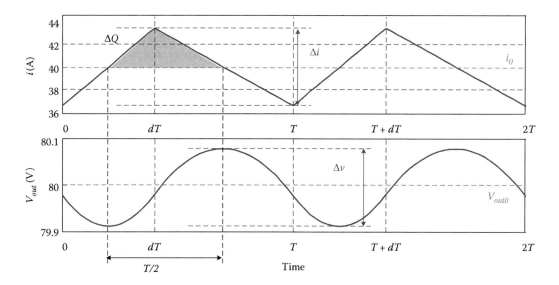

FIGURE 4.23   Current and voltage ripple of Buck converter.

$$\Delta Q = \frac{1}{2} \times \frac{T}{2} \times \frac{\Delta i}{2} \tag{4.28}$$

$$\Delta v = \frac{T \Delta i}{8C} \tag{4.29}$$

### 4.1.5.1.4   Capacitance Selection

The selection of the capacitance is related to the maximum voltage ripple $\Delta v_{max}$, as well as $\Delta i_{max}$ defined by the specification of the system. Relation (4.30) expresses the relation used to define the minimum capacitance value for a specific application. Similar to that for the inductance selection, the required capacitance can be reduced by increasing the switching frequency (reducing $T$).

$$\Delta v_{max} = \frac{T \Delta i_{max}}{8 C_{min}} \tag{4.30}$$

## 4.1.5.2   Current and Voltage Ripples of Boost Converter

### 4.1.5.2.1   Inductance Selection

As was done for the Buck converter, $\Delta i$ can be expressed as Equation 4.31. Considering a constant value of the output voltage, the worst case corresponds to $d = 0.5$, and the inductance selection can be achieved considering relation (4.32).

$$\Delta i = \frac{(1 - d) \cdot d \cdot V_{out0} \cdot T}{L} \tag{4.31}$$

$$\Delta i_{max} = \frac{V_{out0} \cdot T}{4 \cdot L_{min}} \tag{4.32}$$

### 4.1.5.2.2   Capacitance Selection

Assuming that the capacitor is charged by current $i$, and discharged by the load current $i_{out}$ only (this means that $i_0 - \Delta i/2 > i_{out}$), then following a similar method as for the Buck converter, $\Delta Q$

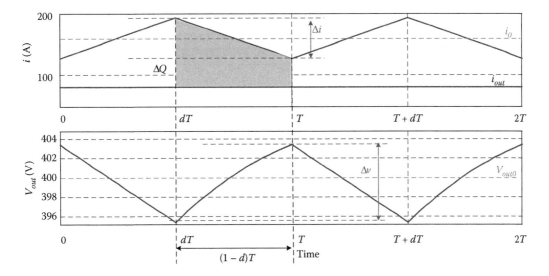

FIGURE 4.24   Current and voltage ripple of Boost converter.

corresponding to the gray area in Figure 4.24 is expressed by Equation 4.33. Then, from Equations 4.33, 4.9, and 4.26, the expression of $\Delta v$ for the Boost converter in the considered case is given by Equation 4.34.

$$\Delta Q = \frac{1}{2}(1 - d)T\Delta i + (1 - d)T\left( i_0 - \frac{\Delta i}{2} - i_{out} \right) \tag{4.33}$$

$$\Delta v = \frac{d \cdot T \cdot V_{out0}}{RC} \tag{4.34}$$

Considering the worst case of Equation 4.34 given for $d = 1$, and the requested value $\Delta v_{max}$, expression (4.35) gives the minimum capacitance value ensuring the requirements. It should be noted that if a different load is used (other than a pure resistor $R$), then the load current is not constant and Equations 4.33 through 4.35 have to be reevaluated.

$$\Delta v_{max} = \frac{T \cdot V_{out0}}{RC_{min}} \tag{4.35}$$

### 4.1.5.3   Current and Voltage Ripples of Buck–Boost Converter

#### 4.1.5.3.1   Inductance Selection

Similar to that for the Buck and the Boost converters, $\Delta i$ expressed by Equation 4.36 is used to select the inductance. Considering the worst case defined by $d = 0$, the inductance selection can be achieved using relation (4.37).

$$\Delta i = \frac{V_{out0} \cdot (1 - d) \cdot T}{L} \tag{4.36}$$

$$\Delta i_{max} = \frac{V_{out0} \cdot T}{L_{min}} \tag{4.37}$$

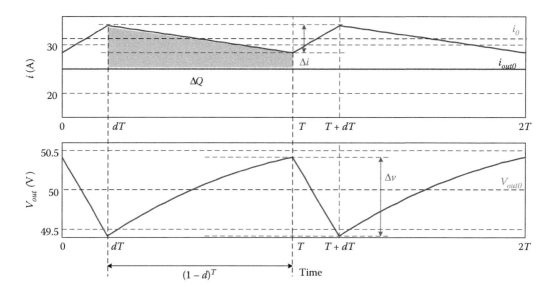

**FIGURE 4.25**   Current and voltage ripple of Buck–Boost converter.

### 4.1.5.3.2   Capacitance Selection

Assuming that the capacitor is charged by the current $i$, and discharged by the load current $i_{out}$ only (this means that $i_0 - \Delta i/2 > i_{out}$), then by following a similar method as for the Buck converter, $\Delta Q$ corresponding to the area represented in Figure 4.25 is expressed by Equation 4.38. Then, from Equations 4.38, 4.36, and 4.26, the expression of $\Delta v$ for the Buck–Boost converter in the considered case is given by Equation 4.39.

$$\Delta Q = \frac{1}{2}(1-d)T\Delta i + (1-d)T\left(i_0 - \frac{\Delta i}{2} - i_{out}\right) \tag{4.38}$$

$$\Delta v = \frac{d \cdot T \cdot V_{out0}}{RC} \tag{4.39}$$

Similar to that for the Boost converter, expression (4.40) gives the minimum capacitance value ensuring the requirement. Also, if a different load is used (other than a pure resistor $R$), then $\Delta Q$ has to be reexpressed and the analysis reconsidered.

$$\Delta v_{max} = \frac{T \cdot V_{out0}}{RC_{min}} \tag{4.40}$$

### 4.1.6   Limit Continuous Conduction Mode–Discontinuous Conduction Mode

Current and voltage waveforms presented above for Buck, Boost, and Buck–Boost converters assume that the current within the inductor is always higher than zero. However, as it is shown by Equations 4.24, 4.31, and 4.36, the current ripple is independent of the average value $i_{in0}$ of the current circulating through the inductor. Hence, if $i_{in0}$ is too low, then the current can become null during the phase corresponding to the discharge of the inductor. This operation of the converter is named: discontinuous conduction mode (DCM). An example of corresponding current waveform of the Buck converter operating in DCM is given in Figure 4.26.

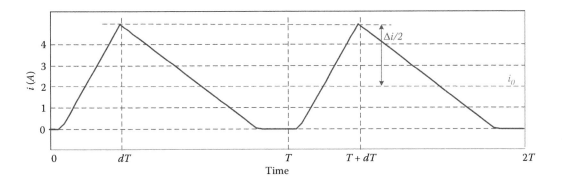

FIGURE 4.26   Current of Buck converter in DCM mode.

For each presented converter, the limit between CCM and DCM is defined by relation (4.41).

$$\frac{\Delta i}{2} = i_0 \tag{4.41}$$

For instance, considering the Buck converter case, corresponding $\Delta i$ given by Equation 4.24 yields relation (4.42) for the average value of the inductor current $i_{0B}$ at the boundary. Boundary current corresponding to the Boost and Buck–Boost converter can be expressed considering Equations 4.31 and 4.36, respectively. When one of the converters operates in the DCM mode, its step-up/step-down characteristics are different and the analysis presented in this chapter has to be reconsidered accordingly.

$$i_{0B} = \frac{V_{out0}(1-d)dT}{2L} \tag{4.42}$$

## 4.2   SWITCH-MODE DC–AC INVERTERS

### 4.2.1   SINGLE-PHASE INVERTER

Inverter converts a continuous (dc) current into an alternative (ac) one. Single-phase inverter generates, as expressed by their name, a single-phase ac output from a dc source. They are widely used in microgrid application as they enable the connection of solar panels, battery, or any other dc source to the ac grid. They are also used to drive single-phase ac motor.

#### 4.2.1.1   Electrical Circuit

The electrical circuit of a single-phase inverter is given in Figure 4.27. The required electrical characteristic of each switch is given in Figure 4.28. This can be achieved using either a MOSFET or an IGBT with an antiparallel diode, as shown in Figure 4.29. Owing to its electrical configuration, the two switches of the same leg cannot be closed simultaneously because this case short-circuits the voltage source.

For leg 1 (controlled by $u_1$), when $u_1 = 1$ (and $\bar{u}_1 = 0$), the voltage $V_{AN}$ between $A$ and $N$ is equal to $V_{DC}$. When $u_1 = 0$, the switch at the bottom is closed and $V_{AN} = 0$. General relations between $V_{DC}$, $V_{AN}$, $V_{BN}$, and $V_{AB}$ are given by Equation 4.43.

$$V_{AN} = u_1 V_{DC}$$
$$V_{BN} = u_2 V_{DC} \tag{4.43}$$
$$V_{AB} = V_{AN} - V_{BN} = (u_1 - u_2)V_{DC}$$

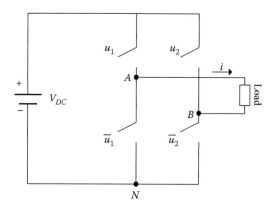

FIGURE 4.27   Single-phase inverter.

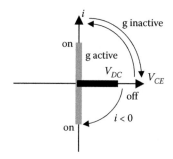

FIGURE 4.28   Current–voltage characteristic of ideal bidirectional switch for drive inverter.

In most applications, such as tie–grid inverter, a sinusoidal output is required. The single-phase inverter can generate such an output if a proper control scheme is applied. The most used method to generate sinusoidal output is the PWM.

### 4.2.1.2   Bipolar PWM

$$u_2 = 1 - u_1 \tag{4.44}$$

$$\begin{cases} u_1 = 1 & \text{if } V_{ref} > V_{carrier} \\ u_1 = 0 & \text{if } V_{ref} \leq V_{carrier} \end{cases} \tag{4.45}$$

The principle of PWM is to compare a given reference signal ($V_{ref}$) to a carrier ($V_{carrier}$) in order to generate the gate signal of the inverter switches, as represented in Figure 4.30. In bipolar PWM, only one reference signal is used to control the two legs. Relation between the control signal $u_2$ of leg

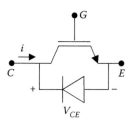

FIGURE 4.29   Bidirectional switch (IGBT + diode).

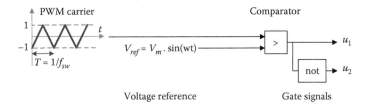

FIGURE 4.30   Bipolar PWM scheme for single-phase inverter.

2 and $u_1$ is given in Equation 4.44. Equation 4.45 gives the relations followed by the bipolar PWM method. Figure 4.31 represents reference and carrier signals, as well as the corresponding generated gate signal. The amplitude of the carrier signal is designed as $V_c$.

To generate a sinusoidal output with an inverter, the reference signal is chosen sinusoidal as specified in Equation 4.47. Also, it is assumed in this chapter, that the frequency of $V_{carrier}$ is much larger than the frequency of $V_{ref}$, and that the amplitude of $V_{ref}$ is lower than the amplitude of $V_c$. If these two conditions are not respected, then the analysis conducted here varies. Details regarding these different cases can be found in Reference 5. Figure 4.32 shows the obtained output voltage using a sinusoidal reference signal $V_{ref}$ given in Equation 4.47. In accordance with Equations 4.43 and 4.44, $V_{AB}$ is expressed by Equation 4.46 and can only be equal to $+V_{dc}$ and $-V_{dc}$.

$$
\begin{aligned}
V_{AN} &= u_1 V_{DC} \\
V_{BN} &= (1 - u_1)V_{DC} \\
V_{AB} &= V_{AN} - V_{BN} = (2u_1 - 1)V_{DC}
\end{aligned}
\qquad (4.46)
$$

By observing the shape of $V_{AB}$, it is not obvious to identify the sinusoidal pattern generated by the PWM. To observe it, the Fourier transform of $V_{AB}$ is plotted in Figure 4.33. It can be seen that the fundamental of $V_{AB}$ is generated at the same frequency as $V_{ref}$, and has its amplitude equal to $V_m$ multiplied by the dc-bus voltage. $V_m$ is commonly named modulation index. Also, a significant harmonic is generated at the switching frequency $f_{sw}$. This harmonic decreases the quality of the generated output. In some applications, such as grid–tie inverter, standards ask for specific power quality requirements in terms of generated harmonics. If generated high-frequency harmonics are too important regarding power

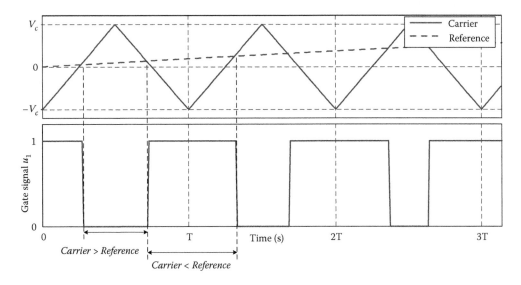

FIGURE 4.31   PWM signals.

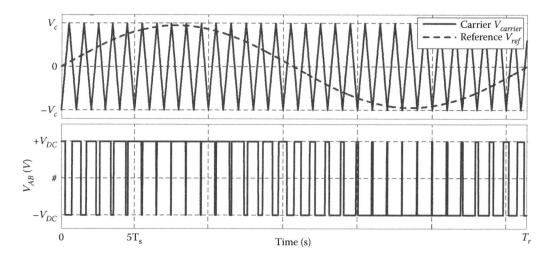

**FIGURE 4.32** Output voltage generated by PWM.

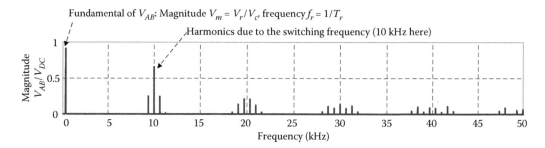

**FIGURE 4.33** Fourier transform of $V_{AB}$.

quality standards, additional low-pass filter can be added at the output of the inverter. The size of this filter decreases with its cutoff frequency. Thus, when the switching frequency is the highest, the output filter is the smallest. As a result, increasing the switching frequency yields the reduction of the filter. However, this also increases the switching loss of the inverter. Another solution to increase the frequency of the harmonics without increasing the switching frequency is to use a unipolar PWM scheme.

$$V_{ref} = V_r \sin(2\pi f_r t) \tag{4.47}$$

### 4.2.1.3 Unipolar PWM

In unipolar PWM scheme, the two legs of the inverter are controlled separately as represented in Figure 4.34. The two gate signals $u_1$ and $u_2$ follow relations specified in Equation 4.48, where $V_{ref}$ is

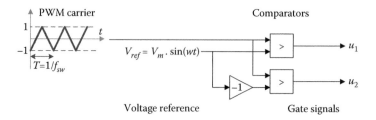

**FIGURE 4.34** Unipolar PWM scheme for single-phase inverter.

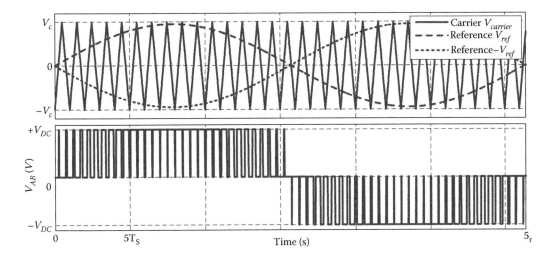

**FIGURE 4.35**   Output voltage generated by PWM.

defined as in Equation 4.47. Figure 4.35 shows the generated phase voltage $V_{AB}$ according to $V_{carrier}$ and $V_{ref}$. Using this control scheme enables the generation of three values for $V_{AB}$ ($+V_{dc}$, 0, and $-V_{dc}$), whereas the previous method generates only two ($+V_{dc}$ and $-V_{dc}$). Thanks to this, the power quality of $V_{AB}$ is improved. In Figure 4.36, the Fourier transform of $V_{AB}$ is given. It can be seen that the first harmonic is generated at a frequency of 20 kHz, which is twice the switching frequency used here. The use of this PWM scheme enables the generation of a sinusoidal output with a better power quality without increasing the switching frequency of the switches.

$$\begin{cases} u_1 = 1 & \text{if} \quad V_{ref} > V_{carrier} \\ u_1 = 0 & \text{if} \quad V_{ref} \leq V_{carrier} \\ u_2 = 1 & \text{if} \quad -V_{ref} > V_{carrier} \\ u_2 = 0 & \text{if} \quad -V_{ref} \leq V_{carrier} \end{cases} \tag{4.48}$$

#### 4.2.1.4   Output Power of Single-Phase Inverter

To analyze the output power of the single-phase inverter, we assume that perfect sinusoidal output voltage and current, given by Equation 4.49, are generated across the load.

$$V_{AB} = V_{rms}\sqrt{2}\,\cos(wt)\, i = I_{rms}\sqrt{2}\,\cos(wt + \varphi) \tag{4.49}$$

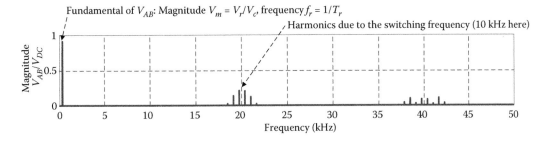

**FIGURE 4.36**   Fourier transform of $V_{AB}$.

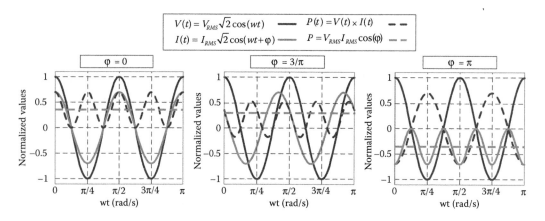

**FIGURE 4.37**   Load voltage, current, and power generated by single-phase inverter.

Then, the "true" power in watts can be found from Equation 4.50, where $\cos(\varphi)$ is defined as the power factor of the inverter. As shown in Figure 4.37, power factor represents the phase angle between the voltage and the current of the load. The maximum transmitted power is reached when the power factor is equal to 1. In this case, current and voltage are exactly in phase. In a similar way, if the power factor is equal to $-1$, then the output power is negative. This means that energy is transmitted from the load to the dc-side of the inverter. In Figure 4.38, the output power versus power factor is plotted.

$$P = V_{rms} I_{rms} \cos(\varphi) \tag{4.50}$$

If no feedback control is applied to the inverter, the power factor is determined by the load. If the load is inductive (as for a motor), or capacitive, then phase currents lag behind phase voltages and the power factor is lower than one. If the load is purely resistive, then current and voltage are in phase and the power factor is equal to 1. For a load with an impedance $Z$ given by Equation 4.51, phase angle $\varphi$ defining the power factor is expressed by Equation 4.52.

$$Z = R + jX \tag{4.51}$$

$$\varphi = \tan^{-1}\left(\frac{X}{R}\right) \tag{4.52}$$

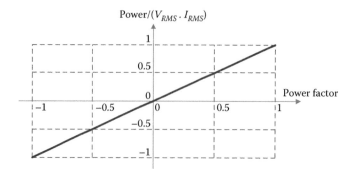

**FIGURE 4.38**   Power generated versus power factor.

## 4.2.2 THREE-PHASE INVERTER

Three-phase inverters are used to transform a dc-voltage source into a three-phase ac output. The two main applications of this type of inverter are: high-power grid–tie inverter for electrical energy transportation and three-phase ac drive system. Three-phase ac drive system is very important in any transportation application as almost every motor used for traction (automotive, railway, ship) are three-phase ac.

### 4.2.2.1 Electrical Circuit

Electrical circuit of a three-phase inverter is presented in Figure 4.39. It is composed of three legs in parallel to the dc voltage source. The middle point of each leg corresponds to one of the three outputs of the converter. In motor drive application, the three phases of the motor are connected in Y configuration, and are connected to the middle point of each leg, as in Figure 4.39. Three-phase inverter is composed of six bidirectional switches, such as MOSFET or IGBT with an antiparallel diode. Electrical characteristics of the switches are the same as for the single-phase configuration. Also, similar to that for the single-phase inverter, the two switches of one leg cannot be closed simultaneously. If this happens, the dc-voltage source is short-circuited.

### 4.2.2.2 Line-to-Line and Phase Voltages

Considering electrical configuration of Figure 4.39, the line-to-line and phase voltages of the inverter's load can be expressed according to the gate signal of the switches and the dc-bus voltage. In Table 4.5, the different line-to-line voltage possibilities are given.

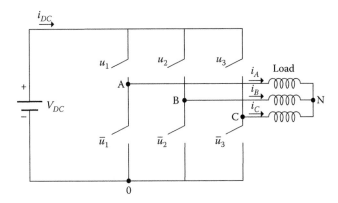

FIGURE 4.39 Electrical circuit of three-phase inverter.

TABLE 4.5
**Line-to-Line Voltage**

| $u_1$ | $u_2$ | $u_3$ | $V_{AB}$ | $V_{BC}$ | $V_{CA}$ |
|---|---|---|---|---|---|
| 1 | 1 | 0 | 0 | $V_{dc}$ | $-V_{dc}$ |
| 1 | 0 | 1 | $V_{dc}$ | $-V_{dc}$ | 0 |
| 0 | 1 | 1 | $-V_{dc}$ | 0 | $V_{dc}$ |
| 0 | 0 | 1 | 0 | $-V_{dc}$ | $V_{dc}$ |
| 0 | 1 | 0 | $-V_{dc}$ | $V_{dc}$ | 0 |
| 1 | 0 | 0 | $V_{dc}$ | 0 | $-V_{dc}$ |
| 1 | 1 | 1 | 0 | 0 | 0 |
| 0 | 0 | 0 | 0 | 0 | 0 |

TABLE 4.6

**Line-to-Neutral Voltage**

| $u_1$ | $u_2$ | $u_3$ | $V_{A0}$ | $V_{B0}$ | $V_{C0}$ |
|---|---|---|---|---|---|
| 1 | 1 | 0 | $V_{dc}$ | $V_{dc}$ | 0 |
| 1 | 0 | 1 | $V_{dc}$ | 0 | $V_{dc}$ |
| 0 | 1 | 1 | 0 | $V_{dc}$ | $V_{dc}$ |
| 0 | 0 | 1 | 0 | 0 | $V_{dc}$ |
| 0 | 1 | 0 | 0 | $V_{dc}$ | 0 |
| 1 | 0 | 0 | $V_{dc}$ | 0 | 0 |
| 1 | 1 | 1 | 0 | 0 | 0 |
| 0 | 0 | 0 | $V_{dc}$ | $V_{dc}$ | $V_{dc}$ |

In Table 4.6, the line-to-neutral voltages are given. These voltages can be easily obtained from the electrical circuit of the inverter. In an opposite way, phase voltages of the load are not obvious to determine from the electrical circuit. Nevertheless, it can be demonstrated that they are linked to the line-to-neutral voltage by relation (4.53). Obtained phase voltages are expressed in Table 4.7. It can be noticed that the maximum voltage that can be applied to a phase of the load is equal to $|2/3 \times V_{dc}|$.

$$\begin{bmatrix} V_{AN} \\ V_{BN} \\ V_{CN} \end{bmatrix} = \frac{1}{3} \begin{bmatrix} 2 & -1 & -1 \\ -1 & 2 & -1 \\ -1 & -1 & 2 \end{bmatrix} \begin{bmatrix} V_{A0} \\ V_{B0} \\ V_{C0} \end{bmatrix} \tag{4.53}$$

### 4.2.2.3   dc-Side Current

The dc-side current of a three-*phase* inverter is expressed by Equation 4.54. Considering relation (4.55), which corresponds to the Y configuration of the load, and relation (4.53), dc-side currents corresponding to each state configuration of the inverter are expressed in Table 4.8.

$$i_{DC} = u_1 \times i_A + u_2 \times i_B + u_3 \times i_C \tag{4.54}$$

$$0 = i_A + i_B + i_C \tag{4.55}$$

### 4.2.2.4   PWM in Three-Phase Inverter

The principal function of a three-phase inverter is to generate a three-phase ac output from a dc input. In motor drive application, the frequency of the output, as well as its amplitude, is used to control the speed and torque of the motor. To generate both frequency-and amplitude-variable outputs, a

TABLE 4.7

**Phase Voltage**

| $u_1$ | $u_2$ | $u_3$ | $V_{AN}$ | $V_{BN}$ | $V_{CN}$ |
|---|---|---|---|---|---|
| 1 | 1 | 0 | $1/3 \times V_{dc}$ | $1/3 \times V_{dc}$ | $-2/3 \times V_{dc}$ |
| 1 | 0 | 1 | $1/3 \times V_{dc}$ | $-2/3 \times V_{dc}$ | $1/3 \times V_{dc}$ |
| 0 | 1 | 1 | $-2/3 \times V_{dc}$ | $1/3 \times V_{dc}$ | $1/3 \times V_{dc}$ |
| 0 | 0 | 1 | $-1/3 \times V_{dc}$ | $-1/3 \times V_{dc}$ | $2/3 \times V_{dc}$ |
| 0 | 1 | 0 | $-1/3 \times V_{dc}$ | $2/3 \times V_{dc}$ | $-1/3 \times V_{dc}$ |
| 1 | 0 | 0 | $2/3 \times V_{dc}$ | $-1/3 \times V_{dc}$ | $-1/3 \times V_{dc}$ |
| 1 | 1 | 1 | 0 | 0 | 0 |
| 0 | 0 | 0 | 0 | 0 | 0 |

**TABLE 4.8**

**dc-Side Current**

| $u_1$ | $u_2$ | $u_3$ | $i_{DC}$ |
|-------|-------|-------|----------|
| 1 | 1 | 0 | $-i_C$ |
| 1 | 0 | 1 | $-i_B$ |
| 0 | 1 | 1 | $-i_A$ |
| 0 | 0 | 1 | $i_C$ |
| 0 | 1 | 0 | $i_B$ |
| 1 | 0 | 0 | $i_A$ |

three-phase PWM scheme can be used, as represented in Figure 4.40. As was detailed in the previous section, when using a sinusoidal reference, the generated output has its fundamental at the frequency of the reference signal, and its amplitude is proportional to the modulation index (defined as the ratio between the reference signal's amplitude and the carrier's amplitude). Considering signals defined in Figure 4.40, RMS values of the load's phase voltage is given by Equation 4.56.

$$V_{phase\ rms} = \frac{V_m}{\sqrt{2}} \times \frac{V_{DC}}{2} \tag{4.56}$$

In Figure 4.41, line-to-line and phase voltages obtained using the PWM scheme of Figure 4.40 are plotted. It can be seen that values of the line-to-line and phase voltage are in accordance with the ones given in Tables 4.5 and 4.7, respectively.

Corresponding dc-side and phase current waveforms are shown in Figure 4.42. The value of the dc-side current is always equal to one of the phase current or its opposite. Relations given in Table 4.8 specify the value of $i_{DC}$ according to the gate signals. The RMS value of the current is given by Equation 4.57, where $P$ is the input power of the inverter.

$$i_{DCrms} = \frac{P}{V_{DC}} \tag{4.57}$$

Current flowing through one of the switches is the one circulating through its corresponding phase when its gate signal is 1 (switch in "on-state"). This state of period and pattern is defined by the duty cycle, and thus by the controller. The duty cycle is not constant; it varies continuously as it follows the PWM.

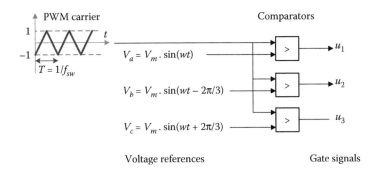

**FIGURE 4.40** PWM scheme for three-phase inverter.

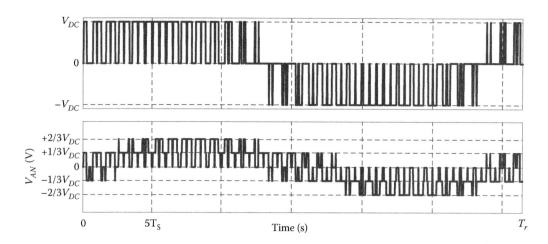

**FIGURE 4.41**  Line-to-line and phase voltages generated by three-phase inverter and PWM scheme.

### 4.2.2.5  Output Power of Three-Phase Inverter

If we assume that the phase voltages and current across the load are the one defined in Equations 4.58 and 4.59, respectively, then the true output power (expressed in watts) transmitted to the load by the inverter is given by Equation 4.60, and can be simplified into Equation 4.61. As for the single-inverter case, $\cos(\varphi)$ is defined as the power factor of the load, which can be obtained from the load by Equation 4.52. However, in motor drive application, current feedback is used to control the motor and, according to the control algorithm, the power factor of the motor can be changed to improve the efficiency/controllability of the drive system. Conventionally used control method such as space-vector-control enables to keep the angle between the magnetic fields of the rotor and the stator in a way (typically around 90°) to keep the power factor at a high value, and thus to ensure high efficiency of the motor. This is further discussed in Chapter 6.

$$V_{AN} = V_{RMS}\sqrt{2}\cos(wt)$$

$$V_{BN} = V_{RMS}\sqrt{2}\cos\left(wt - \frac{2\pi}{3}\right)$$  (4.58)

$$V_{CN} = V_{RMS}\sqrt{2}\cos\left(wt + \frac{2\pi}{3}\right)$$

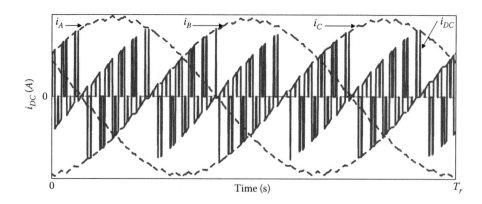

**FIGURE 4.42**  Phase and dc-side currents generated by three-phase inverter and PWM scheme.

$$i_A = I_{RMS}\sqrt{2}\cos(wt + \varphi), \quad i_B = I_{RMS}\sqrt{2}\cos\left(wt - \frac{2\pi}{3} + \varphi\right), \quad i_C = I_{RMS}\sqrt{2}\cos\left(wt + \frac{2\pi}{3} + \varphi\right)$$

(4.59)

$$P = V_{AN}i_A + V_{BN}i_B + V_{CN}i_C$$

(4.60)

$$P = 3V_{RMS}I_{RMS}\cos(\varphi)$$

(4.61)

## 4.3 SWITCH-MODE AC–DC CONVERTERS

Single-phase ac/dc rectifiers can be classified into two main categories—half-wave rectifiers and full-wave rectifiers. In this chapter, diodes and switches have been considered to be ideal for the sake of simplicity. In other words, they are assumed to be short-circuited (zero voltage drop) when they are forward-biased with instantaneous reverse recovery.

### 4.3.1 SINGLE-PHASE HALF-WAVE RECTIFIER

A half-wave rectifier is one of the simplest circuits. It essentially allows a portion of the positive input half-cycle to pass while blocking the negative half-cycle. Typically, such a circuit uses a single switch or diode for rectification and produces an output with pulsating characteristic and single polarity. This heavily pulsating signal requires significant filtering to eliminate harmonics and provide a constant dc output. Needless to say, the efficiency of the half-wave rectifier is limited since only half of the input sinusoidal waveform is converted to dc.

#### 4.3.1.1 Uncontrolled Half-Wave Rectifier

An uncontrolled single-phase rectifier consists of a single diode connected in series with the ac source, as shown in Figure 4.43. Applying Kirchoff's voltage law to the loop, voltages of the circuit can be expressed by Equation 4.62.

$$\begin{cases} V_s = V_d + V_l \\ V_l = Zi \end{cases}$$

(4.62)

If $V_s > V_l$, then the diode is forward-biased causing the current to flow through the load. In this case, $V_d = 0$ (assuming the diode is ideal), and we can see that the voltage across the load is exactly equal to the source. If $V_s < V_l$, the anode of the diode is at a lower potential than the cathode. This reverse biases the diode and the load is disconnected from the source.

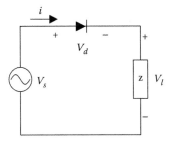

FIGURE 4.43 Uncontrolled half-wave rectifier.

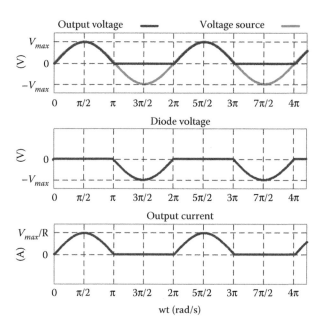

**FIGURE 4.44** Waveforms corresponding to uncontrolled half-wave rectifier; case $Z = R$.

### 4.3.1.2 Uncontrolled Half-Wave Rectifier with Resistive Load

In this case, the voltage source of the circuit is assumed to be a perfect sinus, as defined in Equation 4.63, and the load is purely resistive: $Z = R$.

$$V_s(\omega t) = V_{max}\sin(\omega t)\ 0 \leq \omega t \leq 2\pi \tag{4.63}$$

Associated waveforms of voltage and current for this case are shown in Figure 4.44. It can be seen that the rectified voltage and current always have a positive polarity. Since the diode cannot conduct when it is reverse biased, it stays OFF during the negative half-cycle. No current is circulating through the load and $V_l = 0$. During the positive half-cycle, the diode is ON and the load is directly connected in parallel to the source. Expressions of $V_l$ and $i$ are given in Equations 4.64 and 4.65, respectively. Average and RMS values of $V_l$ can be determined as shown in Equations 4.66 and 4.67, respectively. A similar calculation can be done for the current. Average and RMS values of $i$ are given in Equations 4.68 and 4.69, respectively.

$$\begin{cases} V_l(\omega t) = V_{max}\sin(\omega t) & 0 \leq \omega t \leq \pi \\ V_l(\omega t) = 0 & \pi \leq \omega t \leq 2\pi \end{cases} \tag{4.64}$$

$$\begin{cases} i(t) = \dfrac{V_l(t)}{Z} = \dfrac{V_{max}}{R}\sin(\omega t) & 0 \leq \omega t \leq \pi \\ i(t) = 0 & \pi \leq \omega t \leq 2\pi \end{cases} \tag{4.65}$$

$$V_{l,avg} = \frac{\int_0^T v_l(t)dt}{T} = \frac{\int_0^\pi v_l(\omega t)d\omega t}{2\pi} = \frac{V_{max}}{\pi} \tag{4.66}$$

$$V_{l,rms} = \sqrt{\frac{\int_0^T v_l^2(t)dt}{T}} = \sqrt{\frac{\int_0^\pi v_l^2(\omega t)d\omega t}{2\pi}} = \frac{V_{max}}{2} \tag{4.67}$$

$$I_{avg} = \frac{V_{l,avg}}{R} = \frac{V_{max}}{\pi R} \tag{4.68}$$

$$I_{rms} = \frac{V_{l,rms}}{R} = \frac{V_{max}}{2R} \tag{4.69}$$

### 4.3.1.3 Uncontrolled Half-Wave Rectifier with RL Load

An uncontrolled half-wave rectifier with the series resistive-inductive (RL) load defined in Equation 4.70 is now considered. The main difference between a purely resistive load and an inductive load is that energy is stored within the inductance when a current is circulating through it (as a reminder, this energy is equal to $E = 0.5Li^2$). When $V_s$ goes negative, the current within the inductive load is not null yet. As a result, the diode in Figure 4.45 continues to conduct until no current flows through the series circuit. This will be further detailed in the following.

$$Z = R + j\omega L = |Z|\, e^{-j\theta}, \quad \text{where } |Z| = \sqrt{R^2 + (\omega L)^2} \text{ and } \theta = \tan^{-1}\left(\frac{\omega L}{R}\right) \tag{4.70}$$

Considering the voltage source defined in Equation 4.62, and assuming that there is no current flowing through the diode at $t = 0$, relation (4.71) can be expressed. Similar to that in the case with a purely resistive load, the diode is conducting during the interval $[0;\beta]$, where $\beta$ needs to be determined.

$$\begin{cases} v_s(t) = V_{max}\sin(\omega t) = Ri_l(t) + L\dfrac{di_l(t)}{dt}, & 0 \le wt < \beta \\ v_s(t) = 0, & \beta \le wt < 2\pi \end{cases} \tag{4.71}$$

By solving the first-order differential Equation 4.71, and considering the aforementioned initial condition for the current, this latter can be expressed as in Equation 4.72 during the interval $[0;\beta]$.

$$\begin{cases} i_l(t) = \dfrac{V_{max}}{Z}[\sin(\omega t - \theta) + \sin(\theta)e^{\frac{(-R\omega t)}{\omega L}}], & 0 \le wt < \beta \\ i_l(t) = 0, & \beta \le wt < 2\pi \end{cases} \tag{4.72}$$

Because of the energy stored in the inductor, the diode is forced to conduct until an angle $\beta$, which is called the extinction angle. To determine $\beta$, we need to identify when the diode stops conducting. This is equivalent to determining when the current $i$ reaches 0. This can be calculated by solving Equation 4.73. Owing to the nonlinear properties of Equation 4.73, numerical methods are recommended to evaluate $\beta$.

$$i_l(\beta) = \frac{V_{max}}{Z}\left[\sin(\beta - \theta) + \sin(\theta)e^{\frac{-R\beta}{\omega L}}\right] = 0 \tag{4.73}$$

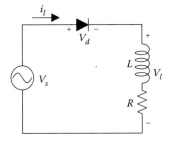

FIGURE 4.45   Uncontrolled half-wave rectifier with RL load.

Finally, from the output voltage 4.71 and the extinction angle defined by Equation 4.73, average and RMS voltages of the circuit are expressed as in Equations 4.74 and 4.75, respectively.

$$V_{l,avg} = \frac{\int_0^T v_l(t)dt}{T} = \frac{\int_0^\beta v_l(\omega t)d\omega t}{2\pi} = \frac{V_{max}}{2\pi}(1 - \cos\beta) \tag{4.74}$$

$$V_{l,rms} = \sqrt{\frac{\int_0^T v_l^2(t)dt}{T}} = \sqrt{\frac{\int_0^\beta v_l^2(\omega t)d\omega t}{2\pi}} = \frac{V_{max}}{2\sqrt{\pi}}\sqrt{\beta - \frac{\sin 2\beta}{2}} \tag{4.75}$$

Associated waveforms of the circuit are plotted in Figure 4.46. The influence of the inductor can clearly be seen in Figure 4.46, where the output voltage goes negative for a period of time. Moreover, owing to the inductive properties of the load, the current flowing through the series-connected circuit is "out of phase" with the voltage and has a distorted sinusoidal shape.

### 4.3.1.4 Half-Wave Controlled Rectifier

The single-phase controlled rectifier has a topology that is very similar to the uncontrolled rectifier defined in the previous section. The only difference is that in this circuit, a controlled switch is used instead of a diode. The electrical configuration of the converter is shown in Figure 4.47.

Replacing the diodes with controllable switches can control the output of a rectifier. Now the switch does not conduct unless a voltage is applied to the gate terminal. In the case where the switch is a thyristor, only a triggering pulse needs to be applied to the terminal to turn it on. Therefore, by controlling the firing angle of the switch, the output voltage of the circuit can also be controlled.

### 4.3.1.5 Half-Wave Controlled Rectifier with R Load

In this case, the voltage source of the circuit is assumed to be a perfect sinus as defined in Equation 4.62 and the load is purely resistive: $Z = R$. Gate signal $d$ of the switch is given in Equation 4.76. As

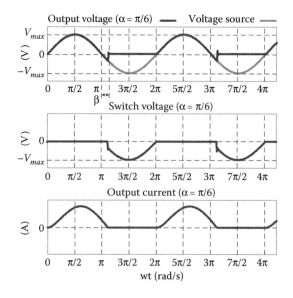

**FIGURE 4.46**   Waveforms corresponding to uncontrolled half-wave rectifier; case $Z = RL$.

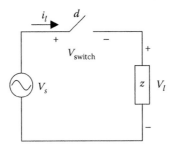

FIGURE 4.47    Controlled half-bridge rectifier.

mentioned previously and detailed in Section 4.1.3.2.2, if a thyristor is used, only a pulsed signal is required to turn on the switch. Similar to that for a diode, the turn-off phase is controlled by the electrical circuit. Hence, relation (4.76) does not correspond to the use of a thyristor.

Corresponding load voltage and current are expressed in Equations 4.77 and 4.78, respectively. Waveforms of the controlled rectifier are shown in Figure 4.48.

$$\begin{cases} d(wt) = 0, & 0 \leq \omega t \leq \alpha \\ d(wt) = 1, & \alpha \leq \omega t \leq \pi \\ d(wt) = 0, & \pi \leq \omega t \leq 2\pi \end{cases} \tag{4.76}$$

$$\begin{cases} v_l(\omega t) = 0, & 0 \leq \omega t \leq \alpha \\ v_l(\omega t) = V_{max} \sin(\omega t), & \alpha \leq \omega t \leq \pi \\ v_l(\omega t) = 0, & \pi \leq \omega t \leq 2\pi \end{cases} \tag{4.77}$$

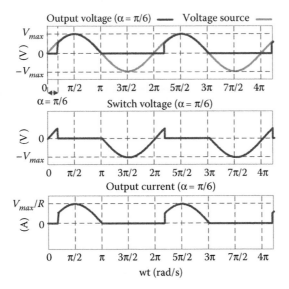

FIGURE 4.48    Waveforms corresponding to controlled half-bridge rectifier; case $Z = R$.

$$\begin{cases} i_l(\omega t) = \dfrac{v_l(t)}{Z_l} = 0, & 0 \le \omega t \le \alpha \\[2mm] i_l(\omega t) = \dfrac{v_l(t)}{Z_l} = \dfrac{V_{max}}{R}\sin(\omega t), & \alpha \le \omega t \le \pi \\[2mm] i_l(\omega t) = \dfrac{v_l(t)}{Z_l} = 0, & \pi \le \omega t \le 2\pi \end{cases} \quad (4.78)$$

Similar to the half-wave uncontrolled rectifier in Equations 4.66 and 4.67, average and RMS values of the output voltage are expressed in Equations 4.79 and 4.80. It can be seen that these values are dependent on the firing angle $\alpha$. The maximum value of the average and RMS output voltage are reached for $\alpha = 0$, and they are zero for $\alpha = \pi$. As a result, the output voltage of the rectifier can be controlled by adapting the firing angle $\alpha$ accordingly.

$$V_{l,avg} = \frac{\displaystyle\int_\alpha^\pi v_l(\omega t)d\omega t}{2\pi} = \frac{V_{max}}{2\pi}(1 + \cos\alpha) \quad (4.79)$$

$$V_{l,rms} = \sqrt{\frac{\displaystyle\int_\alpha^\pi v_l^2(\omega t)dt}{2\pi}} = \frac{V_{max}}{2}\sqrt{1 - \frac{\alpha}{\pi} + \frac{\sin(2\alpha)}{2\pi}} \quad (4.80)$$

Average and RMS values of the current can be easily obtained from Equations 4.79 and 4.80. They are expressed in Equations 4.81 and 4.82, respectively.

$$I_{l,avg} = \frac{V_{l,avg}}{R} = \frac{V_{max}}{\pi R}(1 + \cos\alpha) \quad (4.81)$$

$$I_{l,rms} = \frac{V_{l,rms}}{R} = \frac{V_{max}}{2R}\sqrt{1 - \frac{\alpha}{\pi} + \frac{\sin(2\alpha)}{2\pi}} \quad (4.82)$$

### 4.3.2  SINGLE-PHASE FULL-WAVE RECTIFIER

A full-wave rectifier is capable of converting the alternating input voltage to an output with unipolar characteristic. In essence, this circuit allows the positive half-cycle of the input waveform to pass once the requirements for turning the switch ON are met. During the negative half-cycle, this configuration is capable of providing an alternate path to rectify the input waveform, that is, invert it to maintain the same polarity at the output. This circuit is inherently more efficient than its half-wave counterpart as it uses both the positive and negative portions of the input voltage to generate the output dc voltage instead of every other half-cycle. Operation of the converter will be presented for the purely resistive case only. However, the case corresponding to the RL load can be determined following a similar analysis method like the one presented for the single-phase half-wave inverter.

#### 4.3.2.1  Single-Phase Full-Wave Uncontrolled Bridge Rectifier

The single-phase bridge rectifier configuration uses four diodes connected in a bridge configuration to produce dc output voltage. The ac input voltage is applied across the diagonal ends of the bridge while the load is connected between other two ends of the bridge. In Figure 4.49, four diodes $D_1$–$D_4$ are arranged such that only two diodes conduct during each half-cycle. During the positive half,

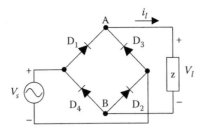

**FIGURE 4.49**   Uncontrolled full-bridge rectifier.

diodes $D_1$ and $D_2$ conduct, while diodes $D_3$ and $D_4$ are reverse biased. Current flows through the load from point A to point B, as shown in Figure 4.50. During the negative half, diodes $D_3$ and $D_4$ conduct, while $D_1$ and $D_2$ are reverse biased. In this case also, current flows from point A to point B of the circuit. This results in a rectified output voltage with a ripple frequency that is twice that of the half-wave circuit. The following sections show the circuit operation and associated equations for a purely resistive load profile.

### 4.3.2.2   Single-Phase Uncontrolled Bridge Rectifier with R Load

In this case, the voltage source of the circuit is assumed to be a perfect sinus as defined in Equation 4.62, and the load is purely resistive: $Z = R$. Associated output voltage and current are expressed in Equations 4.83 and 4.84, respectively. Waveforms of the converter are shown in Figure 4.51.

$$v_l(\omega t) = |v_s(\omega t)| = V_{max} \, |\sin(\omega t)|, \quad 0 \le \omega t \le 2\pi \tag{4.83}$$

$$i_l(\omega t) = \frac{v_l(t)}{Z} = \frac{V_{max}}{R} \, |\sin(\omega t)|, \quad 0 \le \omega t \le 2\pi \tag{4.84}$$

From relation (4.83), the average and RMS values of the output voltage can be determined. These are given in Equations 4.85 and 4.86, respectively.

$$V_{l,avg} = \frac{\int_0^{2\pi} v_l(\omega t) d\omega t}{2\pi} = \frac{\int_0^{\pi} v_l(\omega t) d\omega t}{\pi} = \frac{2V_{max}}{\pi} \tag{4.85}$$

$$V_{l,rms} = \sqrt{\frac{\int_0^{2\pi} v_l^2(\omega t) dt}{2\pi}} = \sqrt{\frac{\int_0^{\pi} v_l^2(\omega t) dt}{\pi}} = \frac{V_{max}}{\sqrt{2}} \tag{4.86}$$

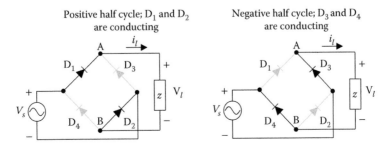

**FIGURE 4.50**   Conducting half-cycles of uncontrolled full-bridge rectifier.

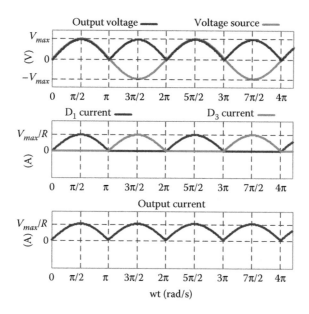

FIGURE 4.51   Waveforms corresponding to uncontrolled full-bridge rectifier; case $Z = R$.

In a similar way, characteristics (4.87) and (4.88) of the current can be determined from Equation 4.84.

$$I_{l,avg} = \frac{V_{l,avg}}{R} = \frac{2V_{max}}{\pi R} = \frac{2I_{max}}{\pi} \tag{4.87}$$

$$I_{l,rms} = \frac{V_{l,rms}}{R} = \frac{V_{max}}{\sqrt{2}R} = \frac{I_{max}}{\sqrt{2}} \tag{4.88}$$

### 4.3.2.3   Single-Phase Full-Wave Uncontrolled Rectifier with RC Load

As mentioned above, a rectifier is used to convert an ac input into a dc output. In some applications, there is a specific requirement for the ripple of the output voltage. However, as can be seen in all the waveforms given previously in this section, output voltages generated by rectifiers have significant harmonics when debiting on a resistive or inductive load. Hence, to improve the power quality, a capacitor is added at the output of the rectifier, as shown in Figure 4.52. This capacitor acts as an energy buffer and smoothes the output voltage. Associated voltage waveforms of the converter in Figure 4.52 are shown in Figure 4.53. Similar behavior can be observed for other topologies (center-tapped, half-bridge, etc.).

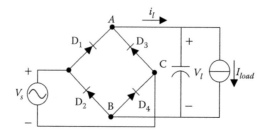

FIGURE 4.52   Full-bridge rectifier with dc-link capacitance.

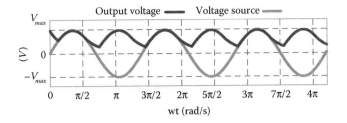

FIGURE 4.53 Waveforms corresponding to uncontrolled full-bridge rectifier; case $Z = RC$.

#### 4.3.2.4 Single-Phase Full-Wave Controlled Rectifier

As in the previous case, diodes can be replaced in the circuit by a switch to control the output voltage. Configuration of the controlled bridge rectifier is shown in Figure 4.54.

#### 4.3.2.5 Single-Phase Full-Wave Controlled Rectifier with R Load

Similar to the previous circuits, the voltage source of the circuit is assumed to be a perfect sinusoidal as defined in Equation 4.62 with a purely resistive load: $Z = R$. Gate signal $d_1$ controlling the positive diagonal is given in Equation 4.89. Similarly, gate signal $d_2$ controlling the negative diagonal is given in Equation 4.90. Also, if switches are thyristor, only the turn-on has to be controlled by a pulsed signal. Corresponding load voltage and current are expressed in Equations 4.91 and 4.92, respectively. Waveforms of the controlled rectifier are shown in Figure 4.55.

$$
\begin{cases}
d_1(wt) = 0, & 0 \le \omega t \le \alpha \\
d_1(wt) = 1, & \alpha \le \omega t \le \pi \\
d_1(wt) = 0, & \pi \le \omega t \le 2\pi
\end{cases}
\tag{4.89}
$$

$$
\begin{cases}
d_2(wt) = 0, & 0 \le \omega t \le \pi + \alpha \\
d_2(wt) = 1, & \pi + \alpha \le \omega t \le 2\pi
\end{cases}
\tag{4.90}
$$

$$
\begin{cases}
v_l(\omega t) = 0, & 0 \le \omega t \le \alpha \\
v_l(\omega t) = V_{max} \,|\sin(\omega t)|, & \alpha \le \omega t \le \pi \\
v_l(\omega t) = 0, & \pi \le \omega t \le \pi + \alpha \\
v_l(\omega t) = V_{max} \,|\sin(\omega t)|, & \pi + \alpha \le \omega t \le 2\pi
\end{cases}
\tag{4.91}
$$

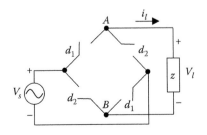

FIGURE 4.54 Controlled full-bridge rectifier.

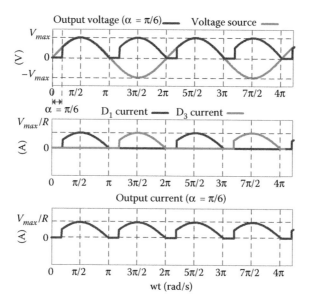

**FIGURE 4.55** Waveforms corresponding to controlled full-bridge rectifier; case $Z = R$.

$$
\begin{cases}
i_l(\omega t) = \dfrac{v_l(t)}{Z_l} = 0, & 0 \le \omega t \le \alpha \\[2mm]
i_l(\omega t) = \dfrac{v_l(t)}{Z_l} = \dfrac{V_{max}}{R} |\sin(\omega t)|, & \alpha \le \omega t \le \pi \\[2mm]
i_l(\omega t) = \dfrac{v_l(t)}{Z_l} = 0, & \pi \le \omega t \le \pi + \alpha \\[2mm]
i_l(\omega t) = \dfrac{v_l(t)}{Z_l} = \dfrac{V_{max}}{R} |\sin(\omega t)|, & \pi + \alpha \le \omega t \le 2\pi
\end{cases}
\tag{4.92}
$$

Expressions of the average and RMS output voltage are given in Equations 4.93 and 4.94, respectively, where $V_l$ is the one defined in Equation 4.91. As it was observed for the half-wave-controlled rectifier, the output characteristics of the converter depend on the firing angle $\alpha$. By adjusting $\alpha$, the RMS output voltage of the rectifier can be controlled.

$$
V_{l,avg} = \frac{\displaystyle\int_0^{2\pi} v_l(\omega t)\, d\omega t}{2\pi} = \frac{\displaystyle\int_\alpha^\pi v_l(\omega t)\, d\omega t}{\pi} = \frac{V_{max}}{\pi}(1 + \cos\alpha)
\tag{4.93}
$$

$$
V_{l,rms} = \sqrt{\frac{\displaystyle\int_0^{2\pi} v_l^2(\omega t)\, dt}{2\pi}} = \sqrt{\frac{\displaystyle\int_\alpha^\pi v_l^2(\omega t)\, dt}{\pi}} = \frac{V_{max}}{\sqrt{2}}\sqrt{1 - \frac{\alpha}{\pi} + \frac{\sin 2\alpha}{2\pi}}
\tag{4.94}
$$

Expressions of the average and RMS output current can be deduced from Equations 4.92, 4.93, and 4.94. They are given in Equations 4.95 and 4.96.

$$
I_{l,avg} = \frac{1}{Z}\frac{\displaystyle\int_0^{2\pi} v_l(\omega t)\, d\omega t}{2\pi} = \frac{1}{ZR}\frac{\displaystyle\int_\alpha^\pi v_l(\omega t)\, d\omega t}{\pi} = \frac{V_{max}}{R\pi}(1 + \cos\alpha)
\tag{4.95}
$$

$$I_{l,rms} = \frac{1}{Z}\sqrt{\frac{\int_0^{2\pi} v_l^2(\omega t)dt}{2\pi}} = \frac{1}{R}\sqrt{\frac{\int_\alpha^\pi v_l^2(\omega t)dt}{\pi}} = \frac{V_{max}}{\sqrt{2}R}\sqrt{1 - \frac{\alpha}{\pi} + \frac{\sin 2\alpha}{2\pi}} \qquad (4.96)$$

#### 4.3.2.6 Bridge versus Center-Tapped Transformer Configurations

An alternate configuration that could be used for full-wave rectification uses a center-tapped transformer, as shown in Figure 4.56. When point A of the transformer is positive with respect to the center tap C, diode $D_1$ is forward-biased and it conducts. On the other hand, during the negative half-cycle, point B of the transformer is positive with respect to C. At this point, diode $D_2$ is forward-biased. In each case, current flows through the load Z in the same direction. This produces a unipolar output during both half-cycles. The following subsection briefly discusses some of the advantages and disadvantages of the two rectifier topologies.

✓ Advantages

- No center tap is required in the secondary winding of the transformer. Therefore, a transformer is only needed if the voltage level needs to be stepped up or down or to provide isolation. This typically leads to a more compact design.
- The peak inverse voltage is one half that of a center-tap rectifier. Hence, the bridge rectifier is highly suited for high-voltage applications.
- In case of a bridge rectifier, transformer utilization factor is higher than that of a center-tap rectifier.
- For a given power output, power transformer of smaller size can be used in case of the bridge rectifier because current in both (primary and secondary) windings of the supply transformer flow for the entire ac cycle.

✗ Disadvantages

- The main drawback is that it needs four diodes, two of which conduct in alternate half-cycles. Because of this, the total voltage drop in diodes becomes double of that in case of a center-tap rectifier.
- Another drawback of a bridge rectifier is that the load resistor $R_L$ and the supply source have no common point, which may be grounded.

#### 4.3.3 THREE-PHASE RECTIFIER

Three-phase rectifiers are commonly used in industry to produce a dc voltage and current for large loads. In hybrid electric vehicle, they are usually used to power the electrified traction system from the internal combustion engine. The engine is mechanically connected to a three-phase ac generator, which generates electrical power for the system. This ac power is then rectified by a three-phase converter.

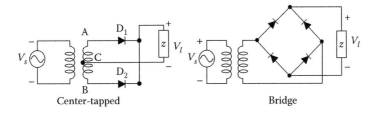

Center-tapped                                             Bridge

FIGURE 4.56  Full-wave rectifier with center-tapped transformer and bridge rectifier with transformer.

### 4.3.3.1   Uncontrolled Rectifier

Figure 4.57 shows the circuit configuration for a three-phase uncontrolled bridge rectifier. In this circuit, $D_1$ and $D_4$, $D_3$ and $D_6$, or $D_5$ and $D_2$ cannot conduct simultaneously. The diodes that are ON are determined by which line-to-line voltage is the highest at that instant. The fundamental frequency of the output voltage is six times the line frequency.

### 4.3.3.2   Three-Phase Uncontrolled Rectifier with R Load

In this case, the load is assumed to be purely resistive. The three-phase ac voltage source defined by Equation 4.98 is applied to the system. From the voltage source expression, line-to-line voltages can be deduced, as given in Equation 4.99.

The relation between the amplitude of the line-to-line voltage and the amplitude of the phase voltage is $V_{max,L-L} = \sqrt{3}V_{max}$. It can be noticed that the amplitude of the line-to-line voltage is higher than that of the phase voltage.

$$\begin{cases} V_{AN} = V_{max}\sin(\omega t) \\ V_{BN} = V_{max}\sin(\omega t - 2\pi/3) \\ V_{CN} = V_{max}\sin(\omega t + 2\pi/3) \end{cases} \tag{4.97}$$

$$\begin{cases} V_{AB} = V_{AN} - V_{BN} = V_{max,L-L}\sin(\omega t + \pi/6) \\ V_{BC} = V_{BN} - V_{CN} = V_{max,L-L}\sin(\omega t - 2\pi/3 + \pi/6) \\ V_{CA} = V_{CN} - V_{AN} = V_{max,L-L}\sin(\omega t + 2\pi/3 + \pi/6) \end{cases} \tag{4.98}$$

With the ac voltage source and the rectifier configuration of Figure 4.57, the output voltage can be expressed by Equation 4.99. A continuous representation of $V_l$ is given in Equation 4.100 where $V_{l,avg}$ is expressed in Equation 4.101 and $V_n$ in Equation 4.102.

$$V_l(\omega t) = \begin{cases} -V_{bc} & 0 < \omega t < \dfrac{\pi}{3} \\ V_{ab} & \dfrac{\pi}{3} < \omega t < \dfrac{2\pi}{3} \\ -V_{ca} & \dfrac{2\pi}{3} < \omega t < \pi \\ V_{bc} & \pi < \omega t < \dfrac{4\pi}{3} \\ -V_{ab} & \dfrac{4\pi}{3} < \omega t < \dfrac{5\pi}{3} \\ V_{ca} & \dfrac{5\pi}{3} < \omega t < 2\pi \end{cases} \tag{4.99}$$

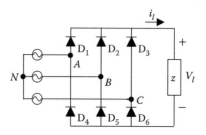

**FIGURE 4.57**   Uncontrolled three-phase rectifier.

$$V_l(t) = V_{l,avg} + \sum_{n=6,12,18,\ldots}^{\infty} V_n \cos(n\omega_0 t + \pi) \qquad (4.100)$$

$$V_{l,avg} = \frac{3V_{max,L-L}}{\pi} \qquad (4.101)$$

$$V_n = \frac{6V_{max,L-L}}{\pi(n^2 - 1)} \qquad (4.102)$$

As was done for single-rectifier circuits, RMS expression of the output voltage can be determined and is given in Equation 4.103.

$$V_{l,rms} = \sqrt{\sum_{n=0,6,12,18,\ldots}^{N} V_{n,rms}^2} = \sqrt{V_{l,avg}^2 + \sum_{n=6,12,18,\ldots}^{\infty} \frac{V_n^2}{2}} = V_{max}\sqrt{\frac{3}{2} + \frac{9\sqrt{3}}{4\pi}} \qquad (4.103)$$

Expressions of the current can be deduced from the output voltage and the load. They are expressed in Equations 4.104 through 4.106. Associated waveforms are shown in Figure 4.58.

$$i_l(t) = \frac{v_l(t)}{R} = \frac{V_{l,avg}}{R} + \sum_{n=2,4,\ldots}^{\infty} \frac{V_n}{R} \cos(n\omega_0 t + \pi) \qquad (4.104)$$

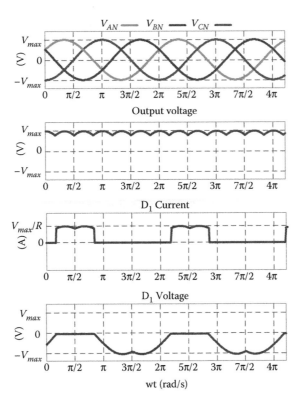

FIGURE 4.58 Waveforms corresponding to three-phase uncontrolled full-bridge rectifier; case $Z = R$.

$$I_{l,avg} = \frac{3V_{max,L-L}}{\pi R} \tag{4.105}$$

$$I_{l,rms} = \sqrt{\sum_{n=0,6,12,18,\ldots}^{N} I_{n,rms}^2} = \sqrt{\left(\frac{V_{l,avg}}{R}\right)^2 + \sum_{n=6,12,18,\ldots}^{\infty} \left(\frac{V_{l,avg}}{\sqrt{2}R}\right)^2} \tag{4.106}$$

### 4.3.3.3   Three-Phase Controlled Rectifier

With controlled switches, conduction does not begin until a gate signal is applied and the device is forward-biased. Delay angle for a switch is referred to as the time from where it begins to conduct in a manner similar to a diode. Delay angle is defined as the interval between when the switch becomes forward-biased and the gate signal is applied.

### 4.3.3.4   Three-Phase Controlled Rectifier with R Load

The three-phase ac voltage source defined in Equation 4.97 is applied to the system. To ensure CCM operation of the converter, the delay angle should be lower than the angle defining the interval between when the switch is forward-biased. In the considered case, this yields $\alpha \leq \pi/3$. Considering that the delay angle of the switches respects this condition and the line-to-line voltages expressed in Equation 4.98, the output voltage is given by Equation 4.107. Average and RMS values of the output voltage are given in Equations 4.108 and 4.109, respectively. As observed in single-phase rectifiers, characteristics of the output voltage depend on $\alpha$. Hence, by adjusting $\alpha$, it is possible to control the output voltage. Average and RMS of the current, deduced from the voltages expressions, are given in Equations 4.110 and 4.111, respectively. Associated waveforms are shown in Figure 4.59.

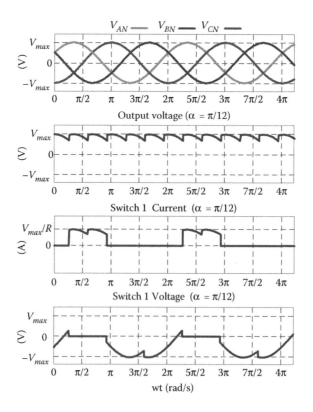

FIGURE 4.59   Waveforms corresponding to three-phase controlled full-bridge rectifier; case $Z = R$.

$$
v_l(\omega t) = \begin{cases} -v_{bc}, & \alpha < \omega t < \dfrac{\pi}{3} + \alpha \\[2mm] v_{ab}, & \dfrac{\pi}{3} + \alpha < \omega t < \dfrac{2\pi}{3} + \alpha \\[2mm] -v_{ca}, & \dfrac{2\pi}{3} + \alpha < \omega t < \pi + \alpha \\[2mm] v_{bc}, & \pi + \alpha < \omega t < \dfrac{4\pi}{3} + \alpha \\[2mm] -v_{ab}, & \dfrac{4\pi}{3} + \alpha < \omega t < \dfrac{5\pi}{3} + \alpha \\[2mm] v_{ca}, & \dfrac{5\pi}{3} + \alpha < \omega t < 2\pi + \alpha \end{cases}
\tag{4.107}
$$

$$
V_{l,avg} = \frac{3V_{max,L-L}}{\pi}\cos\alpha
\tag{4.108}
$$

$$
V_{l,rms} = V_{max,L-L}\sqrt{\frac{3}{2} + \frac{9\sqrt{3}}{8\pi}(1 + \cos 2\alpha) - \frac{1}{2\pi}(\alpha + \sin 2\alpha)}
\tag{4.109}
$$

$$
I_{l,avg} = \frac{3V_{max,L-L}}{\pi R}\cos\alpha
\tag{4.110}
$$

$$
V_{l,rms} = \frac{V_{max,L-L}}{R}\sqrt{\frac{3}{2} + \frac{9\sqrt{3}}{8\pi}(1 + \cos 2\alpha) - \frac{1}{2\pi}(\alpha + \sin 2\alpha)}
\tag{4.111}
$$

## 4.4 PRACTICAL ASPECTS OF POWER CONVERTER DESIGN

### 4.4.1 INTRODUCTION

In this section, different practical aspects of power electronics design to help the designer to implement its theoretical design are presented. The points considered in this section are evaluation of semiconductor loss, selection of a power module for a specific application, design and implementation of a snubber circuit on the power module to reduce emitted switching loss, and gate driver design.

### 4.4.2 EVALUATION OF LOSSES IN SEMICONDUCTOR

In power converter, most of the losses are produced by semiconductors. The other main loss source is magnetic component such as inductor and transformer. Information regarding loss in magnetic components can be found in Reference 6. Loss in semiconductors can be divided into two different types: conduction loss and switching loss. Loss distribution during the operation of the switch is shown in Figure 4.60. Total loss within the semiconductor ($P_{loss}$) corresponds to the sum of the conducted ($P_{cond}$) and switching ($P_{sw}$) losses as expressed in Equation 4.112.

$$
P_{loss} = P_{cond} + P_{sw}
\tag{4.112}
$$

In this section, a method to estimate losses generated by a semiconductor switch only using the characteristics provided by the datasheet is described. The method is based on the linear

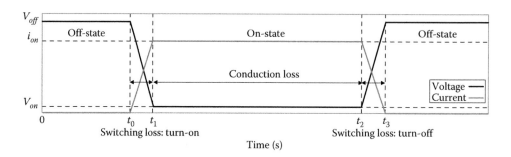

FIGURE 4.60   Electrical waveforms and generated loss.

approximation of the semiconductor's characteristics and does not require any specific software for its application. Some semiconductor manufacturers provide evaluation tools to perform this type of analysis (IPOSIM by Infineon, SemiSel by Semikron). Unfortunately, electrical topologies proposed by these tools are limited and the provided outputs are not always extractable for postprocessing analysis. Thus, in some cases, it is necessary to evaluate loss generated by a switch using a different way to perform semiconductor selection and the efficiency evaluation of a converter.

The method proposed in the following is based on data conventionally available in the datasheet provided by manufacturers. However, these data correspond to specific operating conditions defined in the datasheet (driver board, circuitry, etc.). As a result, for more accurate evaluation of losses, the best way is to perform experimental measurements with the hardware to be evaluated. Like this, data corresponding to the evaluated configuration can be obtained and used in the proposed method. Meanwhile, the method proposed is usually sufficient enough to perform power module selection and efficiency investigation.

### 4.4.2.1  Evaluation of Conduction Loss

Conduction losses are produced when the semiconductor is on-state. In this case, current $i_{on}$ circulates within it while there is still a voltage ($V_{on}$) across the switch. This is due to the collector–emitter voltage saturation within the semiconductor. Energy dissipated within the semiconductor during the "on-state" period ($T_{on} = t_2 - t_1$) is expressed by relation (4.113).

$$E_{cond} = \int_{T_{on}} V_{on} i_{on} dt \tag{4.113}$$

While the on-state current $i_{on}$ is defined by the electric circuit in which the switch is used, the voltage $V_{on}$ depends on the switch characteristics. This latter depends on many factors such as the gate voltage, junction temperature, or the current $i_{on}$ to name a few. To simplify the loss evaluation, junction temperature is considered as the maximum one for which the characteristic $V_{on} = f(i_{on})$ is available. Moreover, only the influence of the on-state current is considered; the impact of other factors, such as the gate voltage, is neglected. Then, the relation $V_{on} = f(i_{on})$ can be approximated, for instance, by using linear approximation, and is used to estimate $V_{on}$ first, and then the conduction loss. Considering the on-state current constant during $T_{on}$, the conduction loss defined in Equation 4.113 can be expressed in terms of power (in watts) by Equation 4.114, where $f_{sw}$ is the switching frequency of the switch.

$$P_{cond} = V_{on}(i_{on}) i_{on} T_{on} f_{sw} \tag{4.114}$$

### 4.4.2.2  Evaluation of Switching Loss

As it can be seen in Figure 4.60, switching losses are generated during the turn-on and turn-off phases of the semiconductor. As a result, $P_{sw}$ can be expressed as Equation 4.115.

$$P_{sw} = P_{sw\_on} + P_{sw\_off} \tag{4.115}$$

They correspond to the product of the collector current and the collector–emitter voltage. As these current and voltage waveforms during the switching state may be difficult to estimate or obtain, the energy dissipated during the switching phases for one or several given operating conditions are provided in semiconductor datasheet. Conventionally provided relations are $E_{on} = f(i_{on})$ and $E_{off} = f(i_{on})$ corresponding to a constant $V_{off0}$ value, as well as constant junction temperature and gate voltage. From these relations, switching loss can be estimated by Equations 4.116 and 4.117, where $f_{sw}$ is the switching frequency of the switch. If relations $E_{on} = f(i_{on})$ and $E_{off} = f(i_{on})$ are provided for different junction temperature $T_j$ values, it is better to consider the one corresponding to the maximum value of $T_j$, especially if the analysis is done to validate the selection of a power module (see Section 4.4.3). Also, it is always possible to use an experimental setup and characterize the switching losses using experiments.

$$P_{sw\_on} = E_{on}(i_{on}) \frac{V_{off}}{V_{off0}} f_{sw} \qquad (4.116)$$

$$P_{sw\_off} = E_{off}(i_{on}) \frac{V_{off}}{V_{off0}} f_{sw} \qquad (4.117)$$

Also, in the case of a diode, only turn-off loss is considered. Turn-off loss is called recovery loss for a diode. Similar to switching loss of IGBT, the recovery loss characteristic of diode is conventionally provided by the datasheet.

### 4.4.2.3 Example: Loss Evaluation of a Boost Converter

We want to evaluate losses generated by the IGBT of the Boost converter of Figure 4.61. Specifications of the Boost are listed in Table 4.9. In Figure 4.62 are given $V_{on} = f(i_{on})$, $E_{on} = f(i_{on})$, and $E_{off} = f(i_{on})$ characteristics of the IGBT. From the given information, determine the loss generated by the IGBT for $V_{out0} = 150$ V, 250 V, and 500 V can be determined. Diodes are assumed to be ideal.

*4.4.2.3.1 Solution*

First, electrical waveforms of the IGBT have to be defined. According to Equation 4.21 of Section 4.1, these waveforms correspond to the ones plotted in Figure 4.63 where $i_0$ and $v_{out0}$ are expressed by Equations 4.118 and 4.119.

$$V_{out0} = \frac{V_{in}}{1 - d} \qquad (4.118)$$

$$i_0 = \frac{V_{out0}}{R} \qquad (4.119)$$

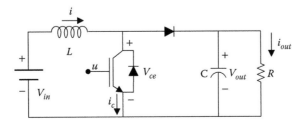

FIGURE 4.61   Boost converter.

**TABLE 4.9**

**Boost Converter Specifications**

| $V_{in}$ (V) | Capacitance (μF) | Inductance (mH) | $R$ (Ω) | Switching Frequency (kHz) |
|---|---|---|---|---|
| 100 | 1000 | 10 | 15 | 10 |

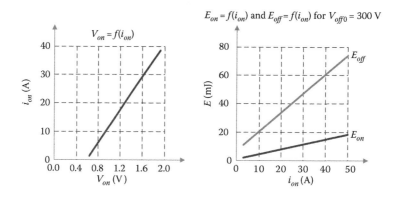

**FIGURE 4.62**   Conduction and switching characteristics.

From Equations 4.118 and 4.119, and the three considered output voltages, $V_{out0}$, values of $i_0$ and conduction time $T_{on}$ can be deduced. As can be seen in Figure 4.63, $V_{ce}$ is equal to $V_{out}$ when the IGBT is off. As the voltage ripple is usually lower than 10% in this kind of application, it can be assumed that the voltage $V_{ce}$ across the IGBT is constant and equal to $V_{out0}$ when the switch is off. If it is not the case or if a more accurate evaluation is required, the ripple of the voltage has to be considered to evaluate the voltage across the IGBT at turn-on and turn-off. Finally, using the loss characteristics in Figure 4.62, corresponding $V_{on}$, $E_{off}$, and $E_{on}$ are identified. Then, from these values and relations (4.114), (4.116), and (4.117), both conduction and switching loss can be estimated. All mentioned values are listed in Table 4.10. From the obtained results, it can be noted that the overall

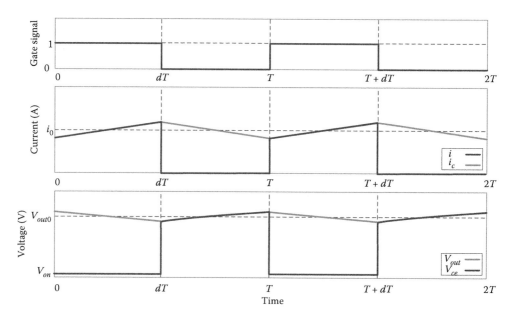

**FIGURE 4.63**   Boost and IGBT electrical waveforms.

TABLE 4.10

**Identified Values for Loss Evaluation**

| $V_{out0}$ | 150 V | 250 V | 500 V |
|---|---|---|---|
| Duty cycle | 0.33 | 0.6 | 0.8 |
| $i_0$ | 10 A | 16.66 A | 33.33 A |
| $V_{on}$ | 0.9 V | 1.1 V | 1.7 V |
| $E_{off}$ ($V_{off0}$ = 300 V) | 20 mJ | 30 mJ | 50 mJ |
| $E_{on}$ ($V_{off0}$ = 300 V) | 5 mJ | 7.5 mJ | 12 mJ |
| Conduction time $T_{on} = dT_{sw}$ | 0.033 ms | 0.06 ms | 0.08 ms |
| $P_{cond}$ | 2.97 W | 11 W | 45.33 W |
| $P_{sw}$ | 125 W | 312.5 W | 1033 W |
| $P_{loss}$ | 127.97 W | 323.5 W | 1078.33 W |

loss is increasing with the duty cycle. This observation is generally true in any dc–dc converter and should be taken into account when designing a system.

### 4.4.3 POWER MODULE SELECTION

There are many manufacturers such as Powerex, Infineon, Semikron, and International Rectifier to name a few, which propose different power modules. When you look at their catalog, you can see that the same type of component is proposed for different voltage and current ratings, as well as for different packagings. All proposed power modules present different characteristics in terms of forward voltage, switching loss, and thermal impedance of their packaging. The purpose of this section is to provide inputs to select the good power module for a specific application.

To select a proper power module, the first thing to do is to perform a similar analysis as the one presented in Section 4.1.3.2.5 to identify the electrical characteristics required for the switch to be selected. The first information to obtain from this (other than the type of semiconductor required) is the voltage rating of the component. Owing to voltage overshoot that can happen during turn-off phase of the switch caused by the stray inductance of circuitry, it is recommended to select a power module rated for twice the voltage specified by the operation of the circuit. Then, current rating of the power module also has to be respected according to the use of the switch. However, only considering the current rating is not enough. Indeed, every power module is also rated for a maximal junction temperature. If the junction temperature becomes higher than the maximal one specified by the manufacturer, the device is damaged and will most probably fail. The junction temperature of a power module depends on the current circulating through the device, as well as the thermal properties of the system. This means that for a different packaging or used heat sink, the maximum current a switch can handle is different. A cross-sectional view of a power module is shown in Figure 4.64.

To check the proper use of a module, it should be verified that the maximum power loss ($P_{loss}$) dissipated by the semiconductor in the considered application, and the thermal properties of the system (packaging, heat sink), do not yield a junction temperature higher than the maximal allowed one for the module. To do so, junction temperature can be estimated using Equation 4.120, where

- $P_{diss\,max}$ (W): maximum dissipated power by the device (conduction loss and switching loss)
- $T_j$ (°C): junction temperature
- $T_a$ (°C): ambient temperature
- $R_{th\_jc}$ (K/W): thermal resistance junction to case
- $R_{th\_ch}$ (K/W): thermal resistance case to heat sink
- $R_{th\_ha}$ (K/W): thermal resistance heat sink to ambient

$$T_j = P_{diss\,max} \cdot (R_{th\_jc} + R_{th\_ch} + R_{th\_ha}) + T_a \qquad (4.120)$$

Junction temperature $T_j$                    Dissipated power by the switch

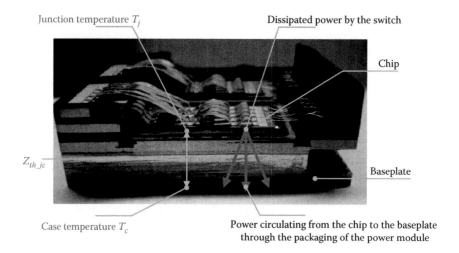

Chip

$Z_{th\_jc}$

Baseplate

Case temperature $T_c$                     Power circulating from the chip to the baseplate
                                           through the packaging of the power module

**FIGURE 4.64**   Cross-sectional view of an IGBT power module. (From M. Olszewski, Evaluation of the 2010 Toyota Prius hybrid synergy drive system, in Oak Ridge National Laboratory Report, March 2011.)

This method is sufficient enough to estimate the maximum junction temperature and validate or not the power module selection. However, to improve and optimize the design of the system (especially the heat sink selection and/or design), a dynamic consideration of the problem is required. This latter can be performed considering equivalent thermal network of the components (power module, heat sink), or computational fluid dynamics (CFD) analysis using appropriate simulation software.

Most of the manufacturers (Infineon, Semikron, International Rectifier, and Powerex) propose on their website simulation software to select power module [7–9]. For instance, IPOSIM from Infineon [7] estimates power loss, as well as the junction temperature, when a power module is selected, its application is specified (inverter, dc–dc, power rating, etc.), and performances of the heat sink is defined. This permits to easily and quickly identify a potential candidate for the design of a converter.

### 4.4.3.1   Example: Power Module Selection: Case of a Traction Inverter

The inverter for an 80 kW ac motor used in a hybrid car has six switches to convert dc power to ac power. The maximum dc-link voltage is 300 V and the maximum continuous peak phase current of the ac motor is 500 A. The maximum switching frequency required for the inverter is 12 kHz. Determine the type of switching device and its maximum voltage and current required for this application.

#### *4.4.3.1.1   Solution*

The type of switching device is determined by its power rating and switching frequency. Based on the requirement of the inverter, the power rating of the switching device is in the range of 100 kW. At the same time, the switching frequency must be at least 12 kHz. As a result, as detailed in Subsections 4.2.2.1 and 4.1.3, IGBT is the type of switch that can meet the requirements of this application. The voltage rating of the IGBT is determined by the maximum voltage that could be applied to it. Also, it is recommended to use an IGBT that has twice the voltage rating of the application in case there is a turn-off voltage overshoot. Since the maximum dc-link voltage is 300 V, 600 V is the desired voltage rating for IGBT.

Moreover, current rating is determined by the maximum current of the application. Since the maximum current 500 A is a continuous current, it is possible that the phase current might sometimes go beyond 500 A. As a result, the rated current of the target IGBT should be continuous 600 A without causing a high junction temperature (this point is rather detailed in the next example). Based on the above analysis, the IGBT with 600 V voltage rating and 600 A current rating is the desired IGBT.

Besides voltage, current and switching frequency, compactness of power device, ease of installation (wiring and heat sink), ease of maintenance (replace parts if damaged), thermal characteristics, and cost are also important for switching device selection. The goal is to find a switching device with compact design, easy heat dissipation, easy maintenance, and low cost. However, in reality, these conditions actually contradict with each other. A compact switching device can be heated up more quickly than a normal-size device. As a result, a compact switching device usually has lower power rating. If the power rating is too high, the use of a compact design with integrated devices might not be possible due to heat dissipation constraint. A compact power module often has multiple power switches integrated together. If one of the switches is damaged, the whole module needs to be replaced. The compact module with multiple switches is cheaper than the cost of multiple single switches added up together. However, the potential maintenance fee of the compact module actually is higher than single switches. Figure 4.65 shows three different popular package types for IGBT modules and their corresponding circuit symbols. Table 4.11 shows the detailed comparison using different IGBT packages to form a same power-rating inverter. It can be noted that each packaging has its own advantages and disadvantages and can meet requirements for different applications.

### 4.4.3.2  Example: Power Module Selection: Case of a dc–dc Boost Converter

In this example, we need to select an IGBT power module for a 50 kW Boost converter. Specifications of the converter are given in Table 4.12. Power module candidates are listed in Table 4.13, in which are also specified the loss dissipated by each component for the operating condition given by the specifications.

#### 4.4.3.2.1  Solution

Voltage rating of the power module is chosen equal to at least $3 \times V_{out}/2 = 600$ V to handle the maximum value of the induced voltage across the inductor. Then, according to the specifications,

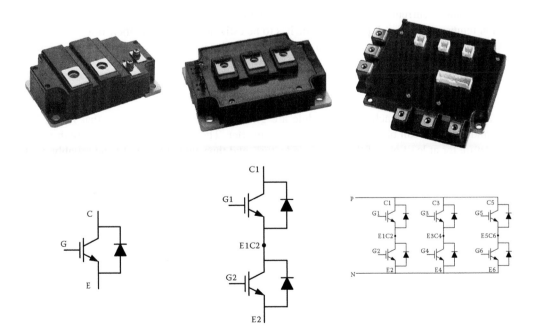

FIGURE 4.65  Three different package types of IGBT modules and their corresponding symbols. (Photos courtesy of Powerex, Inc.)

TABLE 4.11

**Comparison of dc–ac Inverter Using Different IGBT Packages**

| dc–ac Inverter | Single-IGBT Module | Dual-IGBT Module | Six-IGBT Module |
|---|---|---|---|
| Cost of single IGBT | High | Medium | Low |
| Number of modules | 6 | 3 | 1 |
| Total cost | High | Medium | Low |
| Compactness | Bad | Medium | Good |
| Heat dissipation | Good | Medium | Bad |
| Ease of installation | Bad | Medium | Good |
| Ease of replacement | Good | Medium | Bad |
| Maintenance cost | Low | Medium | High |

TABLE 4.12

**Boost Converter Specifications**

| $V_{in}$ (V) | $V_{out}$ (V) | Inductance (mH) | Power (kW) | Switching Frequency (kHz) | Temperature of the Heat Sink (°C) |
|---|---|---|---|---|---|
| 200 | 400 | 10 | 50 | 10 | 50 |

the maximum current is $I_{max} = P_{max}/V_{out} = 125$ A. From the data given in Table 4.13, all the power modules present a higher current rating than the one required here. Nevertheless, it is very important to note that the current rating is valid for a case temperature of 25°C. Unfortunately, as specified in the specifications, the considered application has to operate with a heat sink temperature of 50°C. Thus, continuous current rating specified by the datasheet cannot be considered anymore and the junction temperature of each power module has to be estimated to validate their use.

As explained previously, the junction temperature can be estimated using the following relation:

$$T_j = P_{diss} \cdot (Z_{th\_jc} + R_{th\_ch}) + T_c \tag{4.121}$$

Results obtained for each power module are given in Table 4.14. It appears that PM#2 is the only candidate that can be used for this Boost converter. The other two present too high junction temperature for at least one of their semiconductors, and thus cannot be safely and reliably used in the converter.

TABLE 4.13

**Power Module Candidates**

| Power Module | $V_{ce\,max}$ (V) | Continuous $I_c$ (A) (@$T_c = 25$°C, $T_{vj} = 175$°C) | Loss IGBT (W) | Loss Diode (W) | $R_{th\_jc}$ IGBT (K/W) | $R_{th\_ch}$ IGBT (K/W) | $R_{th\_jc}$ Diode (K/W) | $R_{th\_ch}$ Diode (K/W) | $T_{jmax}$ (for Diode and IGBT) |
|---|---|---|---|---|---|---|---|---|---|
| PM#1 | 600 | 260 | 430 | 290 | 0.22 | 0.03 | 0.42 | 0.06 | 150°C |
| PM#2 | 600 | 550 | 320 | 230 | 0.120 | 0.03 | 0.22 | 0.06 | 150°C |
| PM#3 | 600 | 400 | 360 | 270 | 0.16 | 0.03 | 0.32 | 0.06 | 150°C |

TABLE 4.14

**Estimated Junction Temperatures**

| Power Module | $T_j$ IGBT (°C) | $T_j$ Diode (°C) | IGBT OK? | Diode OK? | Power Module OK? |
|---|---|---|---|---|---|
| PM#1 | 157 | 189 | No | No | No |
| PM#2 | 98 | 115 | Yes | Yes | Yes |
| PM#3 | 119 | 152 | Yes | No | No |

#### 4.4.4 DRIVE CIRCUIT FOR SWITCHING DEVICES

#### 4.4.4.1 MOSFET Gate Drive Circuits

MOSFET is a voltage-controlled switching device and its "on-state" and "off-state" can be controlled by a gate-to-source voltage signal. If the gate-to-source voltage exceeds a threshold voltage, the MOSFET is fully turned on. If the gate-to-source voltage drops to zero, the MOSFET is completely turned off. The gate-to-source voltage of MOSFETs varies depending on their power rating and customer's application need. Typically, 10–20 V is the most common voltage level to completely turn on a power MOSFET. Detailed gate-to-source voltage information regarding a specific MOSFET can always be found in its datasheet.

A MOSFET gate drive circuit is capable of amplifying PWM signal from a microcontroller to match the turn-on and turn-off voltage level required by MOSFET. Depending on the location of MOSFET in the application circuit, generally, there are two kinds of gate drive circuits for MOSFET: low-side gate driver and high-side gate driver. Low-side gate driver is required by applications such as Boost converter since it uses a low-side MOSFET with its source terminal connected to the circuit ground. High-side gate driver is required by applications such as Buck converter because it uses a high-side MOSFET with its source terminal connected to a floating voltage point. For some applications, both low- and high-side gate drivers are required such as in dc–ac inverter.

Figure 4.66 shows a typical low-side gate drive circuit. $S_1$ and $S_2$ are logical MOSFET. The COM port of the gate driver is connected to the same circuit ground as the source terminal of the power MOSFET, the voltage source of the gate driver has the same voltage level required to turn on the MOSFET, and the LO output port is connected to the gate terminal of the MOSFET. When the

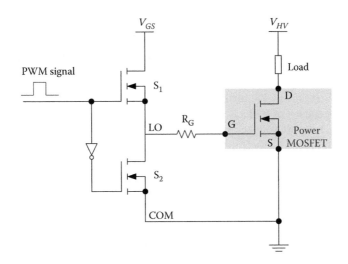

FIGURE 4.66   Low-side gate drive circuit for MOSFET.

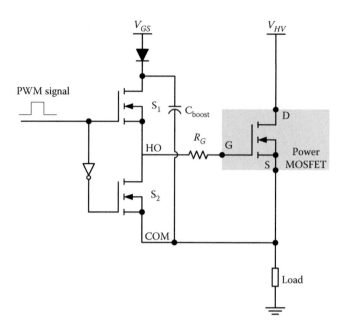

FIGURE 4.67   High-side gate drive circuit for MOSFET.

PWM signal is logical high, $S_1$ is on and $S_2$ is off; as a result, the gate-to-source voltage is $V_{GS}$ and the MOSFET is turned on. When PWM signal is logical low, $S_1$ is off and $S_2$ is on; as a result, the gate-to-source voltage is zero and the MOSFET is turned off.

Figure 4.67 shows a typical high-side gate drive circuit. The COM port of the gate driver and the source terminal of the MOSFET are both connected to the floating voltage point. In order to turn on the MOSFET, the gate voltage must be higher than the source terminal with at least a voltage level of $V_{GS}$. Therefore, a Boost capacitor is often used to charge the voltage up. When the PWM signal is logical low, $S_1$ is off and $S_2$ is on. As a result, the gate-to-source voltage is zero and the MOSFET is turned off. At the same time, the Boost capacitor is charged to voltage level $V_{GS}$. When PWM signal is logical high, $S_1$ is on and $S_2$ is off. As a result, the gate-to-source voltage is $V_{GS}$ and the MOSFET is turned on. The S terminal of the MOSFET immediately becomes $V_{HV}$ because MOSFET is on. At the same time, the gate terminal voltage of the MOSFET rises to $V_{HV} + V_{GS}$ because the Boost capacitor holds $V_{GS}$ across it. As a result, the MOSFET is still turned on.

### 4.4.4.2   IGBT Gate Drive Circuits

IGBT is also a fully voltage-controlled switching device. The overall operation principle is very similar to MOSFET. Typically, the turn-on voltage level of IGBT is 15 V. The major difference is that the switching frequency of IGBT is lower than MOSFET due to the turn-off tailing collect current. As a result, the turn-off voltage usually is −8 V instead of 0 V to force the collect-to-emitter voltage drop faster to reduce turn-off losses and safely turn off the IGBT.

Figure 4.68 shows a typical gate driver circuit for the IGBT module. This circuit works for both high-side and low-side IGBT applications. Two voltage sources (15 and 8 V) are used with their common point connected to the emitter of the IGBT. NPN and PNP transistor modules are used, and their output is connected to the gate of the IGBT through the gate resistor $R_G$. When the PWM signal is logical high, the $S_1$ transistor is turned on and the $S_2$ is turned off. As a result, the gate is pulled up to 15 V and the IGBT is turned on. When PWM signal is logical low, the transistor $S_1$ is turned off and $S_2$ is turned on. As a result, the gate is pulled down to −8 V and the IGBT is fully turned off.

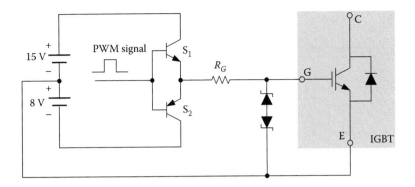

FIGURE 4.68   Gate drive circuit for IGBT.

### 4.4.4.3   Gate Drive IC

When designing a gate driver circuit, a convenient and time-effective option is to use gate driver ICs (integrated circuit) from the market. A lot of semiconductor companies sell gate driver modules for their switching devices. Some companies integrate the gate driver and the switches in the same module. In this section, two examples are given to show some typical gate driver ICs for MOSFET and IGBT.

IR2110 is one of the commonly used cost-effective gate driver IC for MOSFET and IGBT. IR2110 is often used in relatively low-power-rating applications and its maximum voltage supported is around 500 V. One IR2110 chip can drive two MOSFET or IGBT switches. Figure 4.69 shows a typical IR2110 circuit layout to drive one phase leg with two MOSFETs. The phase leg is connected to the high-voltage source, which is the only power source at the high-voltage side. The IR2110 typically needs two voltage levels, which can be converted from the high-voltage source $V_{HV}$ using dc–dc converter. 15 V is used as the gate drive voltage and 5 V is used as input logical signal processing. HIN and LIN are input ports, which receive PWM signals and SD is the port to accept shutdown request signal. These three signals are originally coming from the microcontroller. In order to isolate the microcontroller board from the high-voltage system, optocouplers are used in between to ensure no circuit connection between control board and the high-voltage side. For detailed values regarding peripheral components, refer to the IR2110 datasheet.

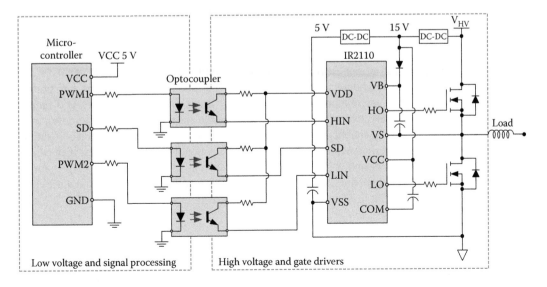

FIGURE 4.69   IR2110 gate drive circuit for MOSFET.

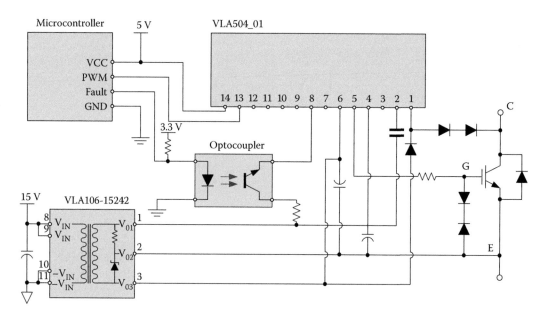

FIGURE 4.70   VLA504_01 gate drive circuit for IGBT.

Powerex gate driver VLA504_01 is one of the high-end gate drivers for IGBT. VLA106-15242 is the isolated dc–dc power supply often used in pair with VLA504_01 to provide 15 V turn-on and −8 V turn-off voltage. As shown in Figure 4.70, Pin 13 and 14 of VLA504_01 are built-in optocoupler inputs that are connected to 5 V and PWM signal of the microcontroller board. The fault signal of the control board is connected with the gate driver through an outside optocoupler. As a result, the control board is totally isolated from the gate driver circuit. The VLA106-15242 provides power for VLA504_01 from 15 V. Since VLA106-15242 is an isolated dc–dc power converter, the 15 V power source is isolated from its output $V_{01}$ (15 V) and $V_{02}$ (−8 V) with the reference $V_{02}$ connected to the emitter of the IGBT. As a result, the potential floating voltage at the emitter of IGBT will not affect the 15 V power source. The isolation of the control board and the gate driver power supply makes this gate driver circuit very stable. For detailed values regarding peripheral components, refer to the VLA504_01 and VLA106_15242 datasheet.

### 4.4.5   SNUBBER CIRCUIT

Snubbers are widely used in power electronics circuits to control circuit resonance. Snubbers can effectively reduce voltage overshoot and current overshoot at switch turn-off and turn-on to protect switches from switching stress. Snubbers can also greatly improve switching losses and reduce EMI. With snubbers, power electronics circuits are more reliable and more efficient. Moreover, even if the additional cost by adding snubbers is low, obtained benefits versus implementation cost should be considered prior to final implementation. Based on the functionality of the snubber, they can be categorized as current snubber and voltage snubber. Basic current RL snubber and voltage parallel resistive-capacitive (RC) snubber are shown in Figure 4.71. Since voltage snubber is used more commonly in a switching circuit, it is discussed in detail in this section.

#### 4.4.5.1   RC Snubber Design

Voltage snubber is used to reduce voltage overshoot and damp the high-frequency voltage ringing across the switch at turn-off. Figure 4.72a represents a Boost converter connected to a resistive load. The analysis performed here can be adapted to almost all the other commonly existing leg configuration (inverter, Buck, etc.). Figure 4.72b shows a typical switch $S_1$ voltage waveform during

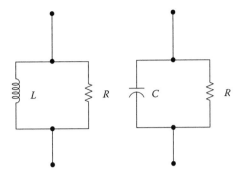

FIGURE 4.71    RL snubber and RC snubber.

switching without adding a snubber to the circuit. The voltage overshoot and ringing effect across $S_1$ when switch $D_2$ turns on can be explained by Figure 4.73.

The assumption is that the current is already built up and flowing in the inductor $L$ and $S_1$. When $S_1$ is turned off, the current is directed to flow through $D_2$ and $L$. The voltage $V_{S1}$ jumps from 0 (ground) to $V_C$. However, owing to the capacitance of the switch $S_1$ and the stray inductance existing in the circuit, a resonant circuit is formed and the voltage across $S_1$ rings. Corresponding equivalent circuit can be represented as in Figure 4.74. When $S_1$ is off, the energy stored in the inductor $L_{stray}$ will cause a resonant effect in this parallel RL circuit. The frequency of this voltage ringing is calculated by Equation 4.120. By adding a proper voltage snubber, the overshoot can be significantly reduced and the voltage ringing can be damped out. Similar effect exists when current is flowing in $L$ and $D_2$, and then $S_1$ is turned on and switch $D_2$ is turned off. As a result, voltage snubber can be used for both switches and switches $S_1$ and $D_2$.

$$f_{ring} = \frac{1}{2\pi\sqrt{L_{stray}C_{switch}}} \tag{4.122}$$

RC snubber is one of the simplest and most widely used voltage snubber circuits. RC snubber can be connected in parallel with the switch. As a result, at switch turn off, the excessive current that is used to cause the voltage overshoot and ringing is directed into the RC snubber circuit to charge the capacitor, and the rising time of the turn-off voltage is determined by the capacitance of the capacitor. By increasing the capacitance, the rising time of the turn-off voltage tends to be longer and the switching loss is smaller. However, if the capacitance is too big, the power dissipated from the resistor of the snubber tends to increase, which should be avoided.

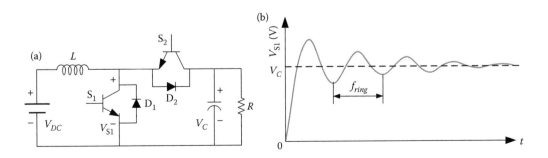

FIGURE 4.72    (a) Boost converter and (b) turn-off voltage at phase node A.

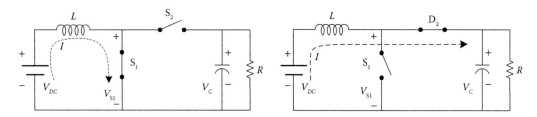

FIGURE 4.73   Two switching states of Boost converter.

The selection of a capacitor and a resistor of the RC snubber is determined by the resonant characteristics of the power circuit. Taking the Boost converter displayed in Figure 4.72 as an example, in order to properly design an RC snubber for the switch $S_2$, the capacitance of the switch $S_1$ $C_{switch}$ and the stray inductance of the circuit need to be calculated. From Equation 4.122, the frequency of the voltage ringing $f_{ring}$ can be measured and this frequency can be called as $f_{ring1}$. The next step is to add a known capacitor $C_{known}$ across the switch $S_1$, as shown in Figure 4.75, and measure the ringing frequency $f_{ring2}$. This frequency $f_{ring2}$ is different from $f_{ring1}$ because the resonant characteristic is changed by $C_{known}$, and $f_{ring2}$ can be expressed by Equation 4.116. Based on Equations 4.123 and 4.124, the two unknown parameters can be calculated.

$$f_{ring1} = \frac{1}{2\pi\sqrt{L_{stray}C_{switch}}} \tag{4.123}$$

$$f_{ring2} = \frac{1}{2\pi\sqrt{L_{stray}(C_{switch} + C_{known})}} \tag{4.124}$$

The resistance $R$ of the snubber can be determined from the characteristic impedance $Z$ of the LC ring, which is expressed by Equation 4.125. It is recommended to select $R$ no larger than Z. A good range is between $0.5 \times Z$ and Z. The resistance can be calculated using Equation 4.126 with k chosen between 0.5 and 1.

When the switch turns on, we want its voltage to be stabilized before the switch turns off again. Thus, the time constant RC of the snubber has to be smaller than the shortest off-state time $t_{off\_min}$ of the converter. For a dc–dc converter or an inverter, this time can reasonably be assumed to be 0.5% of the switching period. As a result, $C$ can be calculated using Equation 4.127, in which $k_1$ is chosen between 5 and 10. Also, it is recommended to select the value of $C$ larger than $C_{switch}$. Thus, $C$ also has to respect Equation 4.128. As large snubber capacitance increases loss, the value of $C$ should be selected at the low end of the range defined by Equations 4.119 and 4.120. Once the initial values of the RC snubber is defined, the actual values can always be adjusted based on experimental measurements to obtain the optimal value.

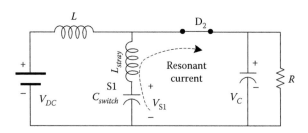

FIGURE 4.74   Corresponding resonant circuit when $S_2$ is off.

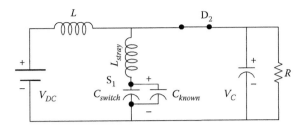

FIGURE 4.75    Circuit with the addition of $C_{known}$ for parameter identification.

$$Z = \sqrt{\frac{L_{stray}}{C_{switch}}} \tag{4.125}$$

$$R = k\sqrt{\frac{L_{stray}}{C_{switch}}} \tag{4.126}$$

$$C < \frac{t_{off\_min}}{k_1 R} \tag{4.127}$$

$$C > C_{switch} \tag{4.128}$$

Other snubber circuits also exist and can be used to reduce electrical stress across switches. These are not detailed here, but interested readers are invited to refer to Reference 10 for more detailed information.

### 4.4.5.2    Example: Design of an RC Snubber

We want to design an RC snubber circuit for the converter presented in Figure 4.72a. The switching frequency of the converter is 50 kHz. The original behavior of the voltage across the switch $S_1$ is given in Figure 4.76 for a duty cycle of 50%. It can be seen that $V_{S1}$ is strongly ringing. To design the snubber, two experimental measurements have been performed and are presented in Figure 4.77. Using results provided in these figures, first, identify the parameters of the ring, and then propose a design for the RC snubber.

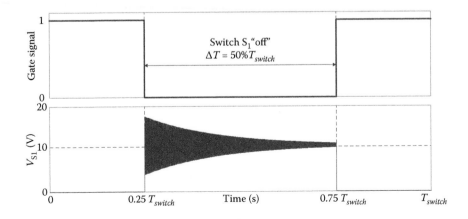

FIGURE 4.76    Original voltage waveform across the switch $S_1$.

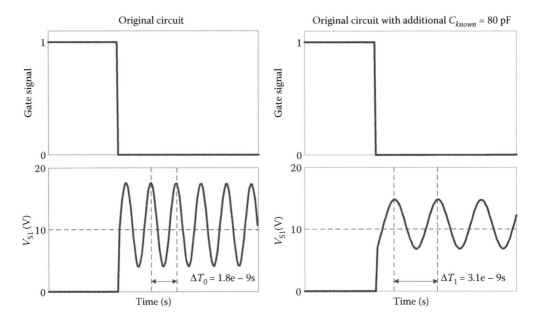

**FIGURE 4.77**   Voltage waveforms across the switch $S_1$ with and without $C_{known}$.

### 4.4.5.2.1   Solution

Parameter identification: From the two oscillation periods given in Figure 4.77 and relations (4.123) and (4.124), the capacitor of the switch and the stray inductance can be calculated using relations (4.129) and (4.130). This yields $C_{switch} = 40$ pF and $L_{stray} = 2$ nH.

$$C_{switch} = \frac{\Delta T_0^2 C_{known}}{(\Delta T_1^2 - \Delta T_0^2)} \tag{4.129}$$

$$L_{stray} = \frac{\Delta T_0^2}{4\pi^2 C_{switch}} \tag{4.130}$$

Selection of $R$: From identified $C_{switch}$ and $L_{stray}$ and Equation 4.118, the value of $R$ can be identified; choosing $k = 0.75$ yields R = 5.3 $\Omega$.

Selection of $C$: Equations 4.127 and 4.128 define an interval for $C$. We choose to define $t_{off\_min}$ as 0.5% of the switching period and $k_1$ is set to 10 to assure that the voltage is very well damped even for very short conduction period of the switch. As a result, $C$ has to respect Equations 4.129 and 4.130. As recommended, the final value of $C$ is chosen at the low end of the range defined by Equations 4.131 and 4.132, and $C = 60$ pF is selected.

$$C < \frac{0.05 T_{switch}}{10R} = 188\,\mu\text{F} \tag{4.131}$$

$$C > C_{switch} = 40\,\text{pF} \tag{4.132}$$

The obtained voltage across the switch with the designed snubber is shown in Figure 4.78. It can be seen that the overshoot, as well as the oscillations, is significantly reduced by the snubber.

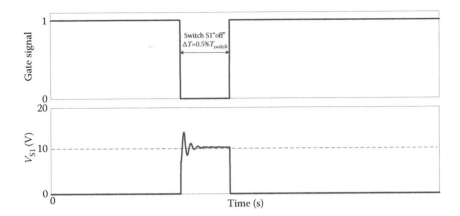

FIGURE 4.78    Voltage waveform across the switch $S_1$ with the designed snubber.

## 4.4.6    BUSBAR DESIGN

It has been described in this chapter that power converters consist of the assembly of several components, such as semiconductor switches, capacitor, and eventually inductor. All these components have to be electrically connected together. This is achieved by a busbar that corresponds to the assembly of conducting bars. Copper is the preferred material for busbar because of its low resistivity. In recent power converter design, laminated busbars are increasingly used instead of conventional ones. Laminated busbar is made of at least two superposed conduction plates separated by an insulation layer.

The most critical use of busbar is the connection between the power module and the dc-link capacitor. Indeed, as mentioned in the previous section, series stray inductance of semiconductors generates resonant loop when the switch turns off. Thus, stray capacitance of the busbar has to be as low as possible.

For each power converter design, the shape of the busbar has to be adapted to the used component. For instance, in Figure 4.79, a dc-link capacitor and three IGBT power modules are shown. These two components have to be connected together to form an inverter. A possible busbar design is shown in Figure 4.80.

To design a busbar, an efficient way to proceed is to build 3D models of the component to be connected together first, and then, based on these models, the busbar can be built. The dimension

6-IGBT Power Module                                                       dc-link capacitor

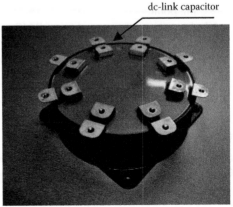

FIGURE 4.79    IGBT power module and dc-link capacitor to be connected together.

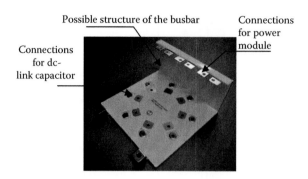

**FIGURE 4.80** Possible structure for the busbar to connect components of Figure 4.79.

of the busbar can be adjusted to reduce stray capacitance, as well as to enable a compact design of the converter to increase its power density. To evaluate stray parameters of the busbar, analytical method or specific simulation software can be used.

## QUESTIONS

4.1  In the 2010 Toyota Prius, a Boost converter is used to step up the output voltage of the battery (assumed to be constant and equal to 200 V). Three dc-bus voltage levels are used: $V_{DC} = 250$ V, $V_{DC} = 500$ V, and $V_{DC} = 650$ V. Assuming the Boost converter to be ideal, what are the three duty cycles required to meet the steady-state requirements?

4.2  Assuming that the power of the Boost converter in 4.1 is 20 kW, define the input and output currents corresponding to each case. And the same question with 40 kW.

4.3  We want to size the inductor to ensure maximum input current ripple of 10%. What are the allowed ripple levels (peak-to-peak) and the requirements for the inductance for the Boost in 4.1 if the switching frequency is 10 kHz (case 20 kW)?

4.4  If the load of the Boost in 4.1 is a pure resistor, what should be its value to absorb 20 kW if $V_{DC} = 650$ V?

4.5  Assuming that the inductance found in 4.3 and the load resistor found in 4.4 are selected, what is the required capacitance to ensure less than 10% voltage ripple?

4.6  We propose to evaluate the loss generated by the switches and the inductor of the Boost converter of Figure 4.3. Only conduction loss of the inductor is considered here. Also, the saturation voltage of both the diode and the switch are supposed to be constant (they do not depend on the current). For the parameters given in Table 4.15, estimate the conduction and switching loss as well as the efficiency of the converter for the considered operating point.

4.7  An ac motor has a maximum RMS phase voltage equal to $V_{RMS\,max} = 200$ V. Considering that this motor is driven by an inverter controlled by a PWM scheme similar to the one in Figure 4.40, what is the minimum value of the dc-bus voltage required to provide $V_{RMS\,max}$ to the motor?

**TABLE 4.15**

**Converter's Parameters**

| Input Power | Input Voltage | Output Voltage | Resistivity of the Inductor | Switching Loss | Saturation Voltage | Switching Frequency |
|---|---|---|---|---|---|---|
| $P = 20$ kW | $V_{in} = 200$ V | $V_{out} = 300$ V | 0.01 Ω | $E_{on} = 9$ mJ | $V_{ce} = 1.3$ V | $F_{sw} = 10$ kHz |
| | | | | $E_{off} = 14$ mJ | $V_f = 1.4$ V (diode) | |
| | | | | $E_{rec} = 4$ mJ (diode) | | |

4.8 Considering a 20 kW Boost converter operating at the conditions listed in Table 4.16 and the power module parameters listed in Table 4.17, estimate the junction temperature of both the diode and the IGBT. Conclude on the selection of this power module for this application.

TABLE 4.16

**Boost Converter Specifications**

| $V_{in}$ (V) | $V_{out}$ (V) | Inductance (mH) | Power (kW) | Switching Frequency (kHz) | Temperature of the Heat Sink (°C) |
|---|---|---|---|---|---|
| 200 | 400 | 10 | 20 | 10 | 50 |

TABLE 4.17

**Power Module Candidates**

| Power Module | Loss IGBT (W) | Loss Diode (W) | $R_{th\_jc}$ IGBT (K/W) | $R_{th\_ch}$ IGBT (K/W) | $R_{th\_jc}$ Diode (K/W) | $R_{th\_ch}$ Diode (K/W) | $T_{jmax}$ (for Diode and IGBT) |
|---|---|---|---|---|---|---|---|
| PM#1 | 272.9 | 133.8 | 0.22 | 0.03 | 0.42 | 0.06 | 150°C |

## REFERENCES

1. *Wikipedia: The Free Encyclopedia*. Wikimedia Foundation Inc. Updated July 22, 2004, 10:55 UTC. Encyclopedia on-line. Available from http://en.wikipedia.org/wiki/Power_electronics. Internet. Retrieved October 08, 2013.
2. M. Olszewski, Evaluation of the 2010 Toyota Prius hybrid synergy drive system, in Oak Ridge National Laboratory Report, March 2011.
3. Z.J. Shen. Automotive Power Semiconductor Devices, in *Handbook of Automotive Power Electronics and Motor Drives*, A. Emadi, Ed. Boca Raton, FL: CRC Press, 2005, pp. 117–158.
4. N. Mohan, T.M. Undeland, and W.P. Robbins. *Power Electronics: Converters, Applications, and Design*. 3rd ed., New York: Wiley, 2003.
5. B.K. Bose. *Power Electronics and Motor Drives: Advances and Trends*. Burlington, MA: Academic Press, 2006.
6. W.G. Hurley and W.H. Wolfle, *Transformers and Inductors for Power Electronics Theory, Design and Applications*, West Sussex, United Kingdom: John Wiley & Sons, 2013.
7. *Infineon IPOSIM* [Online]. Available from www.infineon.com/cms/iposimonlinetool [August 08, 2013].
8. *Semikron SEMISEL* [Online]. Available from http://semisel.semikron.com/License.asp [August 08, 2013].
9. *International Rectifier IGBT Selection Tool* [Online]. Available from http://igbttool.irf.com [August 08, 2013].
10. M.H. Rashid, *Power Electronics: Circuits, Devices, and Applications*. 3rd ed., Upper Saddle River, NJ: Pearson Education, 2004.

# 5 Fundamentals of Electric Machines

*Berker Bilgin and Anand Sathyan*

## CONTENTS

## 5.1   INTRODUCTION

Electric motors convert electrical energy into mechanical energy by means of electromechanical energy conversion. They play a fundamental role in our industry, for power generation and also for the electric drive vehicles. Millions of electric machines are manufactured everyday from a few fractional horsepower to megawatt range and they are used in our everyday applications such as fans, pumps, household supplies, power tools, computers, vehicles, and so on. In the industry, more than 65% of the energy is consumed by electric machines. Almost all of the energy generation is performed by electric generators. Including the traction motors for electrified vehicles, the global electric machine market is projected to be US$96.5 billion by 2018. By 2023, it is projected to have a need for 147.7 million electric motors just for traction applications, including cars, bikes, and also military vehicles [1].

The wide range of applications, drive cycles, operating environment, and the cost constraints bring different challenges in selecting the right electric traction machine. Various parameters including torque-speed characteristics, peak power condition, temperature, volume, and efficiency constraints affect the electric machine design process, from defining the number of phases, number of poles, winding configuration to the selection of lamination material, shape of the coil, use of permanent magnets (PMs) and PM material, and so on.

Successful integration of electric machines in electrified vehicles requires achieving many targets in terms of volume, weight, high-temperature operation, performance, reliability, and, especially, cost. Electric traction motors are desired to be designed with higher power per unit weight (specific power, kW/kg), higher power per unit volume (power density, kW/L), lower cost ($/kW), and higher efficiency. In addition, reliable operation under harsh environmental and temperature conditions is a must, along with structural integrity. However, in terms of machine design, some of these requirements contradict with each other. For example, an electric motor with higher power and smaller volume results in higher magnetic and electric loading and this might reduce its efficiency. An electric machine designer is asked to be knowledgeable in the various aspects of electric machines and optimize the design considering the operational requirements and performance targets. It is significantly important that the designer is aware of how the machine performance will be affected when a certain parameter in the machine is modified. In addition, the designer should select the correct materials and consider the mechanical design along with the powertrain integration to achieve reliability, structural integrity, and lifetime targets.

This chapter is organized to provide guidance for electric traction motor operation and design. It includes the fundamental information on electric machines and also covers the design considerations for traction applications and some practical aspects. The information provided for in-vehicle operation will help the reader to understand the multidisciplinary nature of an electric traction motor and the drive system, where electromagnetic, thermal, structural, and material conditions have significant influence on each other and, all together, they define the performance of the machine.

## 5.2 FUNDAMENTALS OF ELECTROMAGNETICS

In electric machines, motion is created by the electromagnetic force. For this reason, the fundamentals of operation need to be understood to analyze electric machines. The electromagnetic theory has been packaged in Maxwell's equations and it basically states that electricity and magnetism cannot be regarded as separate entities, but rather, they should be considered as two interdependent phenomena [5].

### 5.2.1 DIVERGENCE AND CURL OF MAGNETIC FIELD

It has been quantified by the French physicist André-Marie Ampere in 1826 that a magnetic field is generated around a current-carrying conductor. The direction of the magnetic field can be identified by the right-hand rule, where the thumb shows the direction of the current and the remaining four fingers show the direction of the magnetic field. As shown in Figure 5.1, when the current flows into the page, it leads to a magnetic field that curls in clockwise direction. The magnetic field is represented by two vector quantities: magnetic flux density $\vec{B}$ and magnetic field intensity $\vec{H}$.

The curl of magnetic field is explained by Ampere's law. It states that the line integral around the surface that the total current passes through is a constant:

$$\nabla \times \vec{H} = \vec{J} \tag{5.1}$$

where $\vec{J}$ is the current density and $\nabla \times \vec{H}$ is the curl operator for magnetic field intensity vector. Equation 5.1 expresses Ampere's law in differential form. It can also be defined in integral form using Stoke's theorem:

$$\int (\nabla \times \vec{H}) d\vec{S} = \int_{\ell} \vec{H} d\vec{\ell} = \int_{S} \vec{J} d\vec{S} = I_{enc} \tag{5.2}$$

where $d\vec{\ell}$ is the integration component along the circumferential of the amperian loop and $I_{enc}$ is the total current that passes through the amperian loop. The magnitude of the magnetic field decreases as the distance from the current source increases. In its form given in Equation 5.2, Ampere's law defines the magnetic field only in case of steady currents, where a continuous flow of electrons

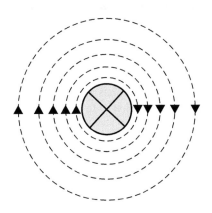

FIGURE 5.1   Magnetic field generated by a current-carrying conductor.

through a conductor is considered without picking up new electrons into the motion. James Clerk Maxwell introduced the displacement current in Ampere's law, which states that a time-changing electric field also induces a magnetic field [6]:

$$\int_\ell \vec{H} \cdot d\vec{\ell} = \int_S \vec{J} \cdot d\vec{S} + \frac{d}{dt} \int_S \vec{D} \cdot d\vec{S} \tag{5.3}$$

where $\vec{D}$ is the electric displacement. The right-most term in Equation 5.3 is the rate of change of electric flux and it is usually neglected in electric machine applications, since the operating frequency is not high enough to encounter its effect.

Magnetic flux density and magnetic field intensity have a close relationship, which is represented by the characteristics of the medium:

$$\vec{B} = \mu\vec{H} \tag{5.4}$$

where $\mu$ is the permeability of the magnetic material and is expressed as

$$\mu = \mu_r\mu_0 \tag{5.5}$$

$\mu_r$ is the relative permeability and equals to 1 for air. Depending on the properties of the magnetic materials, $\mu_r$ can be thousandfold higher, which shows that the magnetic field inside the material is much stronger than that in air. $\mu_0$ is a constant and defined as the permeability of free space:

$$\mu_0 = 4\pi \times 10^{-7}\,\text{H/m} \tag{5.6}$$

As shown in Figure 5.1, for a point inside the magnetic field, magnetic field lines return back to the point where they started. This shows that there is no source for magnetic flux and it is expressed in Gauss's law in integral form as

$$\int_S \vec{B} \cdot d\vec{S} = 0 \tag{5.7}$$

In differential form, this can be expressed as the divergence of the magnetic field, which is the measure of how much a vector spreads out (diverges) from the point it started. Since the magnetic field has no source, the magnetic flux lines end up where they started; so, they have zero divergence:

$$\nabla \cdot \vec{B} = 0 \tag{5.8}$$

Unlike the magnetic field, the electric field has zero curl and nonzero divergence. This shows that, the electric field lines move from one charge to another while magnetic fields do not have a start or end point:

$$\nabla \cdot \vec{E} = \frac{1}{\varepsilon_0}\rho \tag{5.9}$$

$$\nabla \times \vec{E} = 0 \tag{5.10}$$

where $\vec{E}$ is the electric field vector, $\varepsilon_0$ is the permittivity of free space, and $\rho$ is the charge density. The electric field in a dielectric is represented by the electric displacement:

$$\vec{D} = \varepsilon_0 \vec{E} \tag{5.11}$$

Therefore, Gauss's law for electric fields in integral form can be represented as

$$\int_S \vec{D} \cdot d\vec{S} = \int_V \rho_v \, dV = Q \tag{5.12}$$

where $Q$ is the instantaneous net charge inside the closed surface $\vec{S}$.

### 5.2.2 Lorentz's Force Law

In 1892, the Dutch physicist Hendrik Antoon Lorentz quantified the physical phenomena that the magnetic fields created by moving charges generate a force:

$$\vec{F}_{mag} = q[\vec{E} + (\vec{v} \times \vec{B})] \tag{5.13}$$

where $q$ represents the charge moving with a velocity of $\vec{v}$. The direction of magnetic force can be explained by the magnetic fields generated by two current-carrying conductors as shown in Figure 5.2. When the currents are in opposite direction, their magnetic fields also curl in the opposite direction. Therefore, the same magnetic poles are created between the conductors and naturally, they repel each other. When the currents are in the same direction, their magnetic fields curl in the same direction and this results in opposite magnetic poles attracting each other.

Equation 5.13 shows the force exerted on a moving charge and it is composed of electric and magnetic components. Since the velocity of the moving charge and the magnetic field density are three-dimensional (3D) vectors, their cross-product ends up with zero, which states that magnetic fields do no work, they cannot change the kinetic energy of a charge, but they can change its

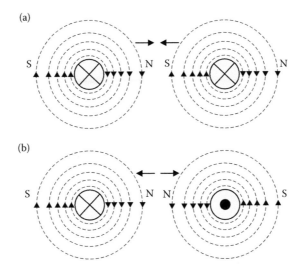

FIGURE 5.2  Representation of magnetic forces. (a) Currents in the same direction attract and (b) currents in the opposite direction repel.

direction. However, as it will be shown in the next section, a time-changing magnetic field induces an electric field, which causes the acceleration of charges.

### 5.2.3 ELECTROMAGNETIC INDUCTION AND FARADAY'S LAW

In 1831, the British physicist Michael Faraday discovered that time-changing magnetic field generates current in a loop of wire that the magnetic field lines go through. As shown in Figure 5.3, the time-changing flux flowing through the open surface creates an electric field around the closed loop of that surface:

$$\int_\ell \vec{E} \cdot d\vec{\ell} = -\frac{d}{dt}\left(\int_S \vec{B} \cdot d\vec{S}\right) \tag{5.14}$$

When the time-changing magnetic flux flows through the closed surface, an electromotive force (EMF) is generated across the loop. When the wire loop is short circuited or connected to a load, a current flows around the wire loop in a certain direction where flux generated by the induced current opposes the magnetic flux that creates it. Therefore, the induced EMF tends to maintain the existing flux. This is known as Lenz's law and it is shown with the negative sign in Equation 5.14. Using Stoke's theorem, Faraday's law can also be written in differential form:

$$\nabla \times \vec{E} = -\frac{\partial \vec{B}}{\partial t} \tag{5.15}$$

As stated by Gauss's law for electric fields, charge-based electrostatic fields originate from positive charge and terminate on negative charge; therefore they have non-zero divergence. Equation (5.15) shows that electric fields induced by changing magnetic fields have non-zero curl; that is the electric field lines form closed loops and, hence, they have zero divergence. In electrical engineering applications [7], Equation 5.14 is usually expressed as

$$\varepsilon = -\frac{d\phi}{dt} \tag{5.16}$$

where $\varepsilon$ is the potential difference or induced voltage and $\phi$ is the magnetic flux flowing through the open surface.

Potential difference is quantified as the line integral of the electric field:

$$\vec{E} = \frac{\varepsilon}{\ell} \tag{5.17}$$

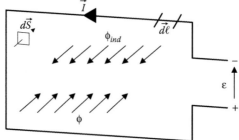

FIGURE 5.3   Physical representation of Faraday's law.

where $\ell$ is the length of the wire loop. As quantified by the French physicist Charles Augustin de Coulomb in 1783, the electric field applies a force; so, the free electrons in the conductor move:

$$\overrightarrow{F_e} = q\overrightarrow{E} \tag{5.18}$$

where $q$ is the charge of an electron. The electrons move through the conductor to make the electric field zero; however, this would not be achievable if the time-changing magnetic flux keeps flowing through the surface of the wire loop. The electrostatic force defined in Equation 5.18 accelerates the free electrons:

$$\overrightarrow{a_e} = \frac{\overrightarrow{F_e}}{m_e} \tag{5.19}$$

This is Newton's second law of motion and $m_e$ is the mass of an electron. Therefore, current is defined as the total number of free electrons flowing through the conductor surface area $S_e$:

$$\overrightarrow{I} = \overrightarrow{v_d} q n_e S_e \tag{5.20}$$

where $v_d$ is the drift velocity inside the conductor and is a function of the acceleration and the time collusion of free electrons ($\tau$):

$$\overrightarrow{v_d} = \overrightarrow{a_e}\tau \tag{5.21}$$

Combining Equations 5.17 through 5.21, the relation between the induced voltage and current can be derived as follows:

$$\frac{\varepsilon}{I} = \frac{m_e}{q^2 n_e \tau}\frac{\ell}{S_e} \tag{5.22}$$

where the first term is a constant and related to the properties of the current-conducting medium. It is defined as resistivity ($\rho$). This linear relationship is called Ohm's law and was quantified by the German physicist Georg Ohm in 1827:

$$V = IR \tag{5.23}$$

where

$$R = \rho\frac{\ell}{S_e} \tag{5.24}$$

## 5.2.4 INDUCTANCE AND MAGNETIC FIELD ENERGY

Faraday's law states that when a time-varying magnetic flux flows through the surface enclosed by a conductor, an EMF is generated across the conductor. Lenz's law states that when the current that is created by this EMF flows through the conductor, it generates an opposing flux. Therefore, the induced flux resists the change of the magnetic field. This signifies that the magnetic field has inertia and the induced current cannot change instantaneously.

When the conductor is composed of $N$ turns sharing a similar surface area, the flux flowing through the conductor surfaces links with each turn. The flux linkage is defined as

$$\lambda = N\phi \tag{5.25}$$

In case of a linear magnetic medium without saturation, the flux linkage is proportional to the current and the proportionality constant is called the inductance:

$$L = \frac{\lambda}{i} \tag{5.26}$$

The unit of inductance is in henries and 1 H of inductance states that for a rate of change of current of 1 A/s, the induced voltage is 1 V.

In the presence of an inductance, a work must be done against inertia of the system to generate current. Therefore, energy will be stored in the magnetic circuit, which is calculated as

$$p_{mag} = ie = i\frac{d\lambda}{dt} \Rightarrow W_{mag} = \int_{t_1}^{t_2} p_{mag}\, dt = \int_{\lambda_1}^{\lambda_2} i\, d\lambda = \int_0^i L i\, di = \frac{1}{2}L i^2 \tag{5.27}$$

## 5.3 LOSSES IN ELECTRIC MACHINES

### 5.3.1 INTRODUCTION

Electric machines convert the input electrical power into mechanical power. During this conversion process, some of the input power is lost and dissipated inside the machine in the form of heat. Figure 5.4 shows the distribution of losses in an electric machine. When the total losses are subtracted from the input power, the mechanical output power and, hence, the efficiency of the system can be calculated as

$$\eta = \frac{P_{in} - P_{loss}}{P_{in}} \times 100\%$$

$$\eta = \frac{P_{mech}}{P_{in}} \times 100\% \tag{5.28}$$

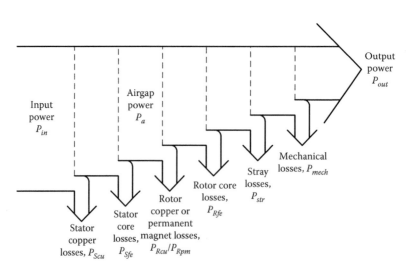

FIGURE 5.4 General loss distribution diagram in electric machines for motoring mode of operation.

where $P_{in}$ is the electrical input power, $P_{loss}$ is the total power loss, and $P_{mech}$ is the mechanical output power. In general, losses in electric machines are dominated by the copper losses, core losses, mechanical losses, and stray losses.

## 5.3.2  COPPER LOSSES

In a current-carrying conductor, a power loss occurs due to current distribution inside the conductor. This power loss is called as the copper or joule loss and dissipated as heat inside the electric machines, usually by means of convection. As shown in Equation 5.24, there is a relationship between the voltage and current, which is called the resistance. Therefore, the copper losses can be calculated as

$$P_{cu} = VI = (IR)I = I^2 R \tag{5.29}$$

Resistance is inversely proportional to the surface area, $S$ and it is directly proportional to the resistivity of the conductor, $\rho$ and its length, $\ell$. Current density is defined as the current per unit cross-section area of the conductor in units of A/mm$^2$; therefore, for a round conductor with a radius of $r_0$, the copper loss can be derived as

$$P_{cu} = I^2 R = I^2 \rho \frac{\ell}{S} = (JS)^2 \rho \frac{\ell}{S} = 2\pi \ell \rho \int_0^{r_0} J^2(r) r \, dr \tag{5.30}$$

The resistivity $\rho$ is defined in units of $\Omega$m and its reciprocal is defined as conductivity, $\sigma$ in units of S/m. The term $J(r)$ represents the distribution of current density in the conductor cross-section area. For a uniform current density distribution, Equation 5.30 can be simplified to the form in Equation 5.29:

$$P_{cu} = \frac{2\pi \ell}{\sigma} \int_0^{r_0} J^2 r \, dr = \frac{2\pi \ell}{\sigma} J^2 \frac{r^2}{2} \Big|_{r=0}^{r=r_0} = \frac{\ell}{\sigma} J^2 S = \frac{\ell}{\sigma S}(JS)^2 = I^2 R_{DC} \tag{5.31}$$

Resistivity of a conductor changes with temperature and it is a function of the resistivity at room temperature (20°C), $\rho_{20}$ and the temperature coefficient, $\alpha_{20}$:

$$\rho = \rho_{20}[1 + \alpha_{20}(T - 20°C)] \tag{5.32}$$

For copper conductors, $\alpha = 0.004041°C^{-1}$ at 20°C. Therefore, for a given value of length and cross-section area, the resistance of the conductor and, hence, the copper loss increases with temperature.

Equation 5.30 shows that the copper loss is a function of the current density distribution. As derived in Equation 5.31, for a homogeneous current density distribution, it is proportional to the DC resistance. In case of alternating currents (AC), the current density inside the conductor might not be homogeneous. The distribution of the current density becomes a function of the frequency, magnetic flux density, and the slot and conductor geometry. In high-frequency operation, skin and proximity effects cause nonuniform current density distribution, leading to additional AC copper losses.

The skin effect is the tendency of the current to flow on the surface of the conductor. This results in a nonuniform distribution of the current density, which will be higher at the surface of the

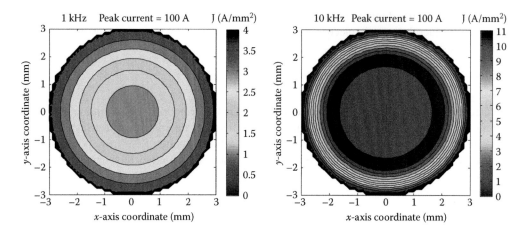

**FIGURE 5.5** Current distribution in a circular conductor for different excitation frequencies.

conductors and decreases toward the center as shown in Figure 5.5. It can be observed that as the frequency of the current increases, the depth of the highest current density from the outer surface of the conductor decreases. This is called the skin depth and is a function of the electrical and magnetic properties of the conducting medium and the excitation frequency:

$$\delta = \frac{1}{\sqrt{\pi \sigma \mu f}} \tag{5.33}$$

where $f$ is the excitation frequency, and $\sigma$ and $\mu$ are the conductivity and permeability of the conducting medium. For copper, conductivity is $5.8140 \times 10^7$ S/m. Since it is not a magnetic material ($\mu_r = 1$), permeability equals $\mu = \mu_r \mu_0 = 4\pi 10^{-7}$. Therefore, the skin depth at 1 kHz will be 2.0873 mm and at 10 kHz, it will be 0.66 mm. It is clear that as the frequency increases, skin depth decreases and the effective cross-section area of the conductor reduces. As shown in Figure 5.6, at higher frequencies, the current density is smaller at the center of the conductor and higher on its surface. This leads to an increase in the effective resistance of the conductor and, hence, higher copper losses [2].

Skin effect occurs due to eddy currents, which is a direct result of Faraday's law. As it will be discussed later in this section, eddy currents are also the reason of magnet losses and a significant part

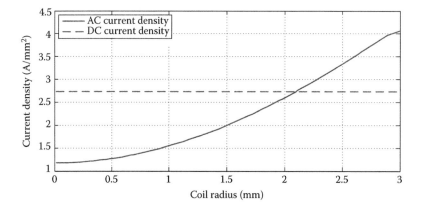

**FIGURE 5.6** Current density distribution inside the coil ($f = 1$ kHz).

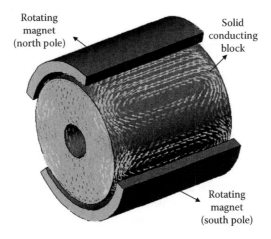

**FIGURE 5.7**   Eddy currents on a conductive solid block.

of the core losses. According to Faraday's law, AC current in the conductor creates a time-changing magnetic field, which creates a time-changing electric field opposing the magnetic field created by the AC current. This opposing electric field induces a voltage that is stronger at the center of the conductor and, hence, the electrons are pushed to flow on the surface of the conductor. Figure 5.7 shows the finite-element analysis (FEA) results for eddy currents in a solid, stationary block made of a highly conductive material. JMAG software has been used for FEA. The time-changing magnetic field is created by the rotating magnets outside of the rotor. To visualize the eddy currents, a small part of the upper magnet is removed from the figure.

Skin effect is the result of eddy currents that are generated by the time-changing magnetic field created by the conductor itself. Proximity effect, on the other hand, is due to the eddy currents caused by an external magnetic field, which are usually the other conductors in the neighborhood. Therefore, proximity effect occurs at high frequencies and usually with multilayer windings. Figure 5.8 shows the proximity effect on a winding, which is composed of four-stranded circular coils. The same AC is applied to the coils with a frequency of 1 kHz and amplitude of 100 A. To mitigate the high-frequency effect on conductors and, hence, to reduce the effective copper losses, Litz wires are widely used in electric machine applications, which can suppress the eddy currents due to their twisted geometry [8].

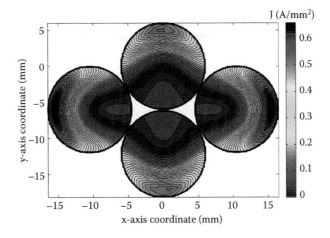

**FIGURE 5.8**   Proximity effect on a stranded conductor.

### 5.3.3 CORE LOSSES

The magnetic core in electric machines is usually made up of ferromagnetic materials. When alternating magnetic field is applied to the core material, hysteresis and eddy current losses occur.

As it has been described in the previous section, magnetic fields always come in dipoles. The same principle applies in the atomic scale as well. Electrons spinning around their axis create small currents, which create a magnetic dipole moment. Atoms or molecules with magnetic dipole moments can be regarded as small magnets with north and south poles. If these materials are exposed to an external magnetic field, the magnetic field will exert a torque on the magnetic dipoles and these dipoles in atomic scales will be aligned in the direction of the external magnetic field.

The electrical steel used in traction motors is a ferromagnetic material. The magnetization characteristics of a ferromagnetic material are described in terms of B–H magnetization curve as shown in Figure 5.9, where B is the magnetic flux density in tesla and H is the magnetic field intensity in A/m. Ferromagnetic materials are made of domains whereby the magnetic dipoles are 100% aligned. But, the net magnetic field is zero due to the random orientation of these magnetic domains as shown in Figure 5.10a. When an external magnetic field is applied, these magnetic domains try to align themselves in parallel to the axis of the applied magnetic field as shown in Figure 5.10b and c. The operating points are referred to the ones given in Figure 5.9. This creates a much stronger magnetic field inside the material. The stronger the external magnetic field, the more the domains align. It should be noted that the increase in the amount of external field does not create an infinite increase in the magnetic flux density. Saturation occurs when practically all the domains are lined up; hence, any further increase in the applied field cannot cause further alignment of the domains. Saturation limits the maximum magnetic field achievable in the ferromagnetic core around at 2 T. At this point, the permeability of the ferromagnetic material (slope of the BH curve) becomes same as that of air. As shown in Figure 5.10d and e, when the core saturates, an increase in the external magnetic field has little or no effect in the magnetic flux of the steel, since most or all the domains are already aligned with the external magnetic field. This puts a limit on the minimum size of the traction motor cores and it is the reason why high-power motors and generators are physically larger, as they must have large magnetic cores. Operating the electrical steel in the saturation region causes more losses.

In case of an alternating external field, the entire BH loop given in Figure 5.9 is covered and the magnetic material is exposed to a periodic magnetization. In this process, the area under the BH loop represents the energy, which is required to change the orientation of domains and expanded in the form of heat. This energy is called as the hysteresis loss and it is dependent on the strength of the magnetic flux density and the excitation frequency.

Another type of core loss, eddy current losses occur in a similar way as the skin effect. Keeping in mind that their conductivity is not as high as that of copper, magnetic materials are also electrically conductive. Therefore, when the core material is exposed to an alternating magnetic field,

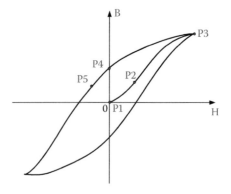

FIGURE 5.9 Typical hysteresis loop for a ferromagnetic material.

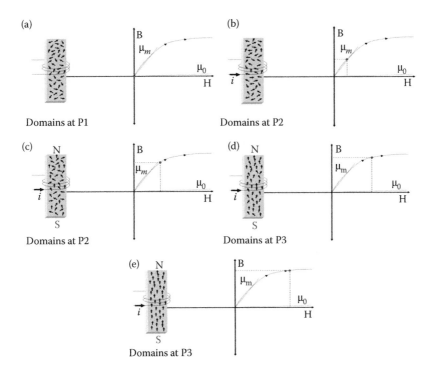

**FIGURE 5.10**  Orientation of magnetic domains according to the strength of the external magnetic field: (a) non-magnetized condition (b) under external magnetic field (c) under stronger magnetic field (d) at the knee point of the magnetization curve (e) saturation of the magnetic material.

eddy currents are induced in the core, which create a magnetic field opposing the external field. In practice, the core materials are laminated to reduce the eddy current loss. As shown in Figure 5.11, this reduces the conductive path of eddy currents and increases the resistivity of the core. Therefore, for a given induced EMF due to the time-changing magnetic field, the eddy currents will be lower in a laminated core, because of the higher equivalent core resistance. Similar to the case in skin effect, the lamination thickness is related to the skin depth of the core and the eddy current losses are a function of the frequency and the magnetic field strength.

**FIGURE 5.11**  Eddy currents and laminations.

### 5.3.4 Losses in PMs

PMs are widely used in synchronous machines to generate the excitation field on the rotor. Similar to the case in lamination materials and conductors, losses in PMs are mainly due to eddy currents.

In the ideal case, the field in a synchronous machine rotates with the same speed as the rotor. For a sinusoidal field distribution, the flux density that the rotor circuit sees does not change in time and, therefore, no eddy currents would be induced. In practice, however, nonideal conditions create harmonics in the armature field, which induces eddy currents on the rotor circuit. These include the space harmonics due to the nonsinusoidal ampere–conductor distribution, time harmonics due to the pulse width modulation (PWM) frequency effect on the stator current [3], and slot harmonics due to the changing airgap permeance when the rotor is aligned with the stator tooth and the slot opening.

If a magnet is exposed to a time-changing magnetic field, the induced eddy currents create losses and the amount of the losses is related to the conductivity of the PM material. Magnet losses are usually quite small as compared to the stator and rotor core losses. When ferrite magnets are used, the losses would be much lower, since ferrite has high resistivity. However, in the case of high-energy density rare-earth magnets, such as neodymium–iron–boron (NdFeB), the material has high conductivity and the eddy currents might end up in relatively higher values. This will cause an increase in the magnet temperature, weaken its magnetization, and lead to a reduction in the performance of the motor.

Segmentation of magnets in the axial direction is an effective way of reducing the magnet eddy currents and, hence, the losses in PMs. This way, the path for eddy currents is reduced and the effective resistance is increased as shown in Figure 5.12. It can also be observed that the eddy current concentrates at the edge of the magnet. This is also a result of skin effect.

### 5.3.5 Mechanical Losses

Mechanical losses in an electric machine mainly consist of friction losses on the bearing and windage losses. Windage losses are also due to friction between air and the rotating parts of the machine. Similar to the core and copper losses, windage losses also cause an increase in the temperature

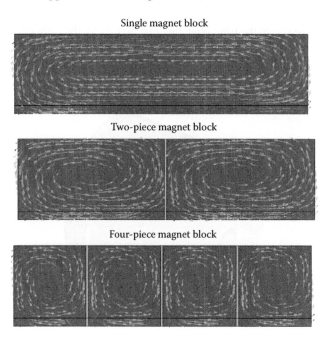

FIGURE 5.12  Magnet eddy currents and the current density distribution in axially segmented magnets.

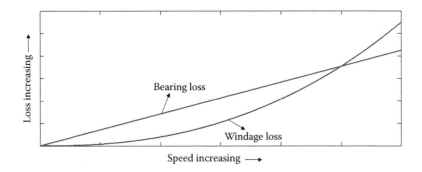

FIGURE 5.13   Change of windage and bearing losses with speed.

of the rotor. As shown in Figure 5.13, windage losses increase nonlinearly as the rotational speed increases and apply a drag torque on the rotor. Bearing losses are dependent on factors such as dimension of the bearings, the vertical load, friction coefficient, and also the bearing type. Bearing losses increase linearly with speed as depicted in Figure 5.13.

## 5.4   WINDINGS IN ELECTRIC MACHINES

The operation of electric machines is based on the interaction of the rotational magnetic field, which is usually created using windings, and the magnetic circuit. Therefore, windings play a fundamental role in electric machines by creating the magnetomotive force (mmf) distribution and they contribute to both torque generation and losses.

There are many different types of windings used in electric machines. Here, the concentration will be given on the most frequently used machines in electrified drivetrain applications. These include the distributed and concentrated windings, which are widely used in AC machines; and also the salient pole windings, which are mainly used in switched reluctance machines (SRMs).

### 5.4.1   AC Machine Windings

Distributed windings are mainly used in the stator of AC machines, including induction and synchronous machines. The main purpose of distributed windings is to create a sinusoidal rotating mmf distribution and, hence, magnetic field in the airgap. This is achieved by applying sinusoidal currents to the coils that are distributed around the airgap by means of slots opened in the stator core. This process can be explained on a simple case where the stator has six slots, three phases, and two poles as shown in Figure 5.14.

The configuration of the winding is shown in Figure 5.14b. For a given number of slots, $Q$, the mechanical angle between two arbitrary slots is defined as

$$\theta_m = \frac{360}{Q}[\text{deg}] \tag{5.34}$$

It can be observed that each phase occupies two slots to create a coil. The change in the direction of the current in each coil side creates different poles. Each pole occupies 180° electrical in the winding configuration and, hence, each pole pair covers 360° electrical. Since the poles are distributed around the circumference of the airgap, for a given number of poles, $p$, the mechanical angle of 360° corresponds to an electrical degree of $(p/2)360°$. Therefore, the electrical angle can be calculated as

$$\theta_e = \left(\frac{p}{2}\right)\theta_m \tag{5.35}$$

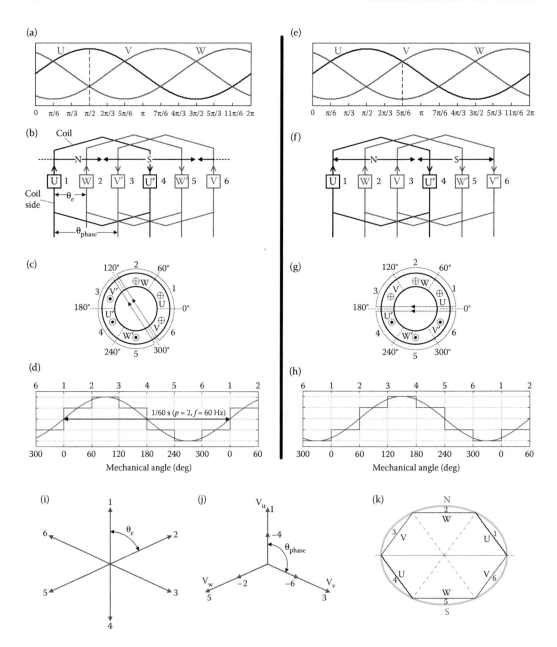

**FIGURE 5.14** Analysis for distributed winding for six slots, three phases, and two poles ($q = 1$). (a) Stator currents instantaneous value at $\theta = \pi/2$; (b) winding configuration at $\theta = \pi/2$; (c) magnetic poles and direction of coil currents $\theta = \pi/2$; (d) current linkage distribution $\theta = \pi/2$; (e) stator currents instantaneous value at $\theta = 5\pi/6$; (f) winding configuration at $\theta = 5\pi/6$; (g) magnetic poles and direction of coil currents $\theta = 5\pi/6$; (h) current linkage distribution $\theta = 5\pi/6$; (i) voltage phasor diagram; (j) vector sum of the voltages in each coil; (k) distribution of coils for each phase in each slot under single pole pair.

The number of slots that each phase occupies under each slot, $q$, is an important parameter in AC-distributed windings. It is calculated as shown in Equation 5.36, where $m$ represents the number of phases. For integer values of $q$, the winding configuration is called integral slot winding. For fractional values of $q$, distributed winding design is still possible and it will be investigated later as fractional slot windings.

$$q = \frac{Q}{mp} \tag{5.36}$$

Under each pole, each phase occupies an electrical angle of $q\theta_e$. In an integral slot, three-phase winding, the electrical angle that each phase covers under a single pole equals to 60°:

$$\theta_b = q\theta_e = q\frac{p}{2}\frac{360°}{Q} = \frac{Q}{mp}\frac{p}{2}\frac{360°}{Q} \Rightarrow \text{for } m = 3 \Rightarrow \theta_b = 60° \tag{5.37}$$

This shows that for $m = 3$, the coils of each phase are distributed 120° electrical apart from each other under one pole pair. Once the winding is constructed this way and it is fed by three-phase current, since at each time instant, the currents satisfy the condition $i_u + i_v + i_w = 0$ as shown in Figure 5.14a, the direction of the currents in $qm$ number of coil sides becomes the same. As shown in Figure 5.14c, for the slots having the same current direction, a current sheet is created around the circumference of the airgap. This current sheet creates a magnetic field whose direction can be found using the right-hand corkscrew rule. The distance that a magnetic pole covers in the airgap is called the pole pitch and it can be defined as a function of the bore diameter, $D$ or the number of slots:

$$\tau_p = \frac{\pi D}{p} \quad \tau_p = \frac{Q}{p} \tag{5.38}$$

Once the coil sides are distributed as shown in Figure 5.14b and three-phase symmetric currents are applied as shown in Figure 5.14a, a staircase-like current linkage distribution appears around the airgap as shown in Figure 5.14d. It can be observed that the current linkage distribution contains some harmonics, but its fundamental looks similar to a sine waveform. These harmonics are due to the nonsinusoidal distribution of the coils around the stator and are called space harmonics.

Figure 5.14f–h shows the direction of currents in each slot, the distribution of the poles around the airgap, and the current linkage waveform for the same winding, respectively, but at a different time instant where the instantaneous values of the current are different as shown in Figure 5.14e. It can be observed that the magnetic poles have moved around the airgap and the peak of the current linkage distribution is now at a different mechanical angle. When this analysis is applied for the entire current waveform, it would be observed that the current linkage waveform travels around the airgap, which creates the rotating magnetic field. The rotational speed of the current linkage waveform is dependent on the frequency of the current waveform and is called the synchronous speed:

$$n = \frac{f}{p/2}[\text{rps}] = \frac{120f}{p}[\text{rpm}] \tag{5.39}$$

Figure 5.14i shows the voltage phasor diagram for the given winding configuration. If each coil side has the same number of conductors, the magnitude of the voltage induced in these coil sides would be the same with a difference in phase due to the electrical angle between them. Since the coil sides are connected in series, the voltage of the phase is represented as the vector sum of the voltages in each coil side as shown in Figure 5.14j.

Figure 5.14k shows how the coils for each phase in each slot are distributed under a single pole pair. Since each phase under each pole is created using a single slot ($q = 1$), the vector representing the phase voltage is the same as the vector sum of the voltages in each slot. In the next example, a winding configuration with $q = 2$ is analyzed and the effect of winding distribution factor can be observed by comparing the corresponding figures in these two examples.

Figure 5.15 shows the design of the winding for a 24-slot, three-phase, and four-pole configuration. Figure 5.15a–g shows the schematics of the winding, the current waveform, distribution of the poles around the airgap, the current linkage waveform, voltage phasor diagram, the calculation of phase voltages, and the derivation of the distribution factor, respectively. The analysis for this winding is very similar to the one given in Figure 5.14, so that it would not be repeated.

As it was shown in Figure 5.14h, the current linkage distribution for the configuration with $q = 1$ deviates from the sinusoidal waveform and contains a significant amount of harmonics in it. By increasing the number of slots per-phase-per-pole, the number of coils is increased, which leads to a current linkage waveform closer to a sine wave as shown in Figure 5.15d. However, this increases the number of slots and, hence, for the given stator bore diameter, the slot width reduces. A very high value of $q$ means a more sinusoidal current linkage, but it also leads to a higher number of coils and higher manufacturing cost. In practice, two to four coils per pole per phase are usually desirable in medium-size machines.

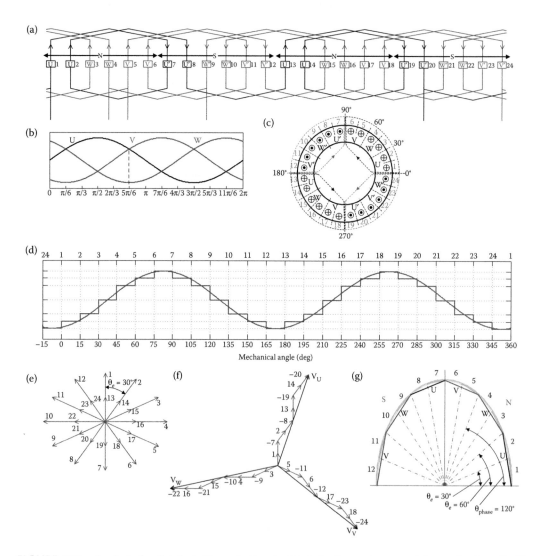

FIGURE 5.15  Analysis for distributed lap winding for 24 slots, three phases, and four poles ($q = 2$). (a) Winding schematics; (b) stator currents; (c) magnetic poles and direction of coil currents; (d) current linkage distribution; (e) voltage phasor diagram; (f) vector sum of the voltages in each coil; (g) distribution of coils for each phase in each slot under a single pole pair.

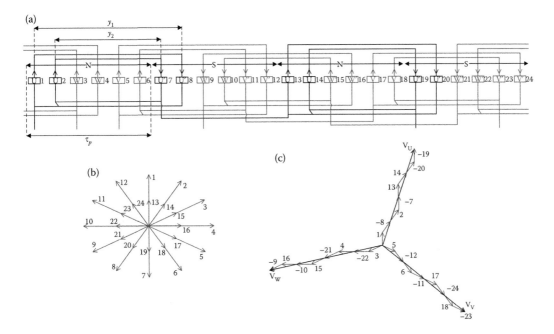

FIGURE 5.16 Analysis for concentric winding for 24 slots, three phases, and four poles. (a) Winding schematics; (b) voltage phasor diagram; (c) vector sum of the voltages in each coil.

For $q = 2$, the phase voltage under each pole is created by two coil sides whose voltage vectors are apart from each other by $\theta_e$ as shown in Figure 5.15e. Therefore, the phase voltage, which equals to the vector sum of the voltage in each slot, is now smaller from its arithmetic sum as shown in Figure 5.15f. This was not the case for $q = 1$, where the vector and arithmetic sum of the voltages in each slot were the same as shown in Figure 5.14j. This is due to the phase shift between the voltage vectors and the distribution of the phase winding in more than one slot. The reduction in phase voltage is expressed by the distribution factor and for the fundamental, it can be calculated using Figure 5.15g:

$$k_d = \frac{\sin(q(\theta_e/2))}{q\sin(\theta_e/2)} \tag{5.40}$$

For the 24-slot, four-pole winding in Figure 5.15, using Equation 5.38, the pole pitch can be calculated as six slots. In the configuration given in Figure 5.15a, the distance between each coil side is also six slots and it is called as full-pitch coil. The same phase voltage can be achieved with a different coil configuration as shown in Figure 5.16. In this case, the same pole pitch is maintained by applying different slot pitches ($y_1 = 7$, $y_2 = 5$). This configuration is called as concentric winding and it can be observed from Figure 5.16a that the end turns for each coil in a phase do not cross over each other. This makes coil-insertion process and end-turn forming simpler. However, in concentric windings, since the slot pitches are either lower or higher than the pole pitch, this reduces the induced voltage. In concentric windings, when the number of coils per phase increases, this effect gets more significant on the inner and outermost coils.

Short pitching, where the slot pitch is smaller than pole pitch is widely applied in distributed lap windings, especially to reduce the harmonic content in the airgap flux density waveform. Figure 5.17 illustrates a ⅚th short-pitched winding for 24 slots, four poles, and three-phase configuration. For the full-pitch configuration, the slot pitch corresponds to six as shown in Figure 5.15a. For a ⅚th short pitch, the slot pitch corresponds to five instead. This configuration reduces to fifth and seventh harmonics in the airgap flux density waveform; however, it also reduces the utilization of the stator.

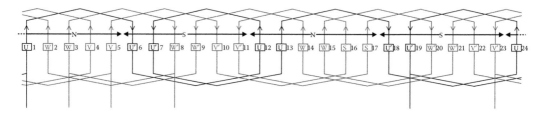

FIGURE 5.17   Winding configuration for 24 slots, three phases, and four poles with ⅚th short pitch.

Therefore, for the same kilowatt output, more copper and steel might be required in short-pitch windings as compared to the full-pitch ones depending on the design.

To remove certain harmonics in the airgap flux density waveform, fractional slot windings can also be applied, where the number of slots per-phase-per pole, $q$ is not an integer anymore. Fractional slot windings can be applied in distributed windings, but in electrified vehicular applications, they are widely applied in concentrated windings, either with single or double layers [4]. Figure 5.18 shows the configuration of a PM machine with concentrated single-layer fractional-slot windings with 12 slots, 10 poles, and three phases, which correspond to $q = 0.4$.

Concentrated windings offer a higher fill factor and simpler manufacturing and also better thermal and electrical isolation between phases. Especially in permanent magnet synchronous machines (PMSMs), depending on the rotor design, concentrated windings can offer better performance, but the reluctance torque component might end up being lower due to lower saliency [9].

### 5.4.2   Salient Pole Stator Windings

Concentrated windings are widely used in salient pole machines, such as SRMs. In these configurations, the series or parallel connection of the coils wound around each pole creates the phase winding and, since each phase is electrically and magnetically isolated from each other, the rotational magnetic field is generated by electronically controlled commutation between the phases as shown in Figure 5.19. In salient pole or, for this case, SRMs, different flux paths can be maintained depending on the direction of coils. Figure 5.19a shows one of these configurations on a three-phase SRM with 12 stator and eight rotor poles. Each phase is made up of four stator poles and, therefore, four coils concentrated around them. By applying the right-hand rule, it can be observed that the flux generated by the coil pairs U1–U2, U1–U4, U3–U2, and U3–U4 is facing the same direction; whereas the flux generated by coil pairs U1–U3 and U2–U4 is facing the opposite direction. For a symmetric machine, it can be assumed that the reluctances along the given flux paths are the same. This divides the flux path into two as shown in Figure 5.19a. The winding diagram for the configuration given in Figure 5.19a is shown in Figure 5.20.

By changing the coil polarities, the flux path can be modified as shown in Figure 5.19b, where the polarity of the coils around U2 and U3 is reversed. When the right-hand rule is applied, it can be observed that now, the flux generated by coils U1–U2 and U3–U4 is facing the opposite direction. Theoretically, the flux paths given in Figure 5.19a and b create the same output torque. However, for multiple-phase conduction, these two configurations might end up in different flux paths during

FIGURE 5.18   Concentrated fractional slot winding configuration with $q = 0.4$.

(a)                           (b)

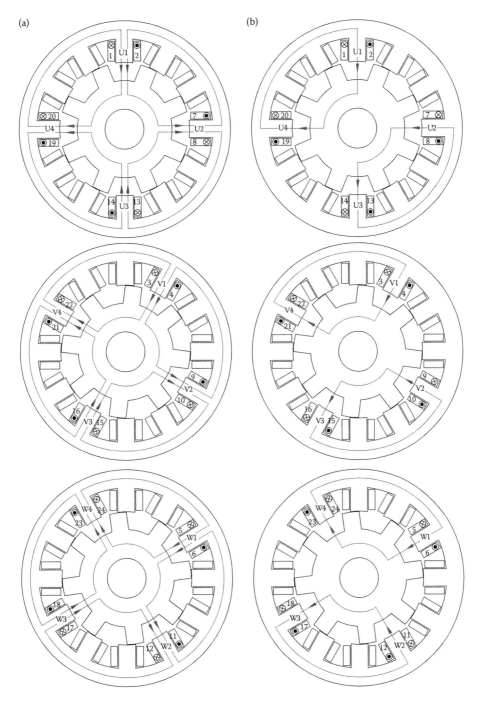

**FIGURE 5.19** Salient pole-winding configuration with different flux paths shown on a 12/8 SRM (a) coils with same polarity in each phase (b) coils with opposite polarity on each side.

phase commutation. Depending on the length of the flux path during the commutation, this might affect the stator core losses [10].

Different flux paths can be maintained by modifying the polarity of the coils in a single phase. In Figure 5.19b, it can be observed that all three phases have the same flux polarity. In SRM, the torque is independent of the direction of the phase current; therefore, it is possible to reverse the coil

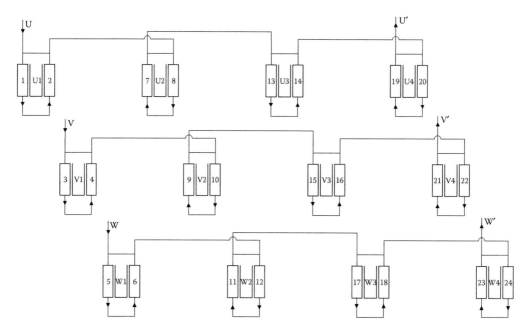

FIGURE 5.20   Winding configuration for 12/8 SRM.

polarities of one of the phases, as shown in Figure 5.21a for phase W. This configuration also has an effect on flux paths during phase commutation in multiphase excitation [11].

In Figure 5.19, it can be observed that the flux from each coil is linking with the same phase. This means that the flux from the coils of phase U is not flowing through the stator poles belonging to other phases. In this case, the mutual inductance is very low as compared to self-inductance and, therefore, the phases of SRM are regarded as magnetically isolated from each other. This improves the fault-tolerant operation capability of the machine. If the coil directions are modified, as shown in Figure 5.21b, the flux of all the coils faces the opposite directions. This can be validated again using the right-hand rule. In this case, since the flux from U1 and U2 opposes each other, they

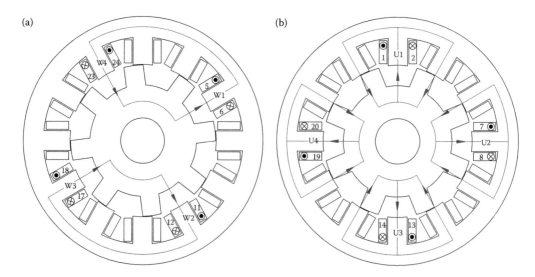

FIGURE 5.21   (a) Coil configuration for an opposite flux pattern in phase W and (b) opposite polarities for all the coils, resulting in higher mutual inductance.

complete their path using the stator poles of adjacent phases, which results in a higher mutual inductance between phases and it cannot be neglected anymore. SRM can still generate torque under this coil configuration, but it requires a more complicated control algorithm, which takes the mutual inductance into account [12].

### 5.4.3 COIL DESIGN

The design of the windings has a significant effect on the performance and efficiency of the drivetrain and the construction of the coils is highly dependent on the operational requirements of the electrical machine (i.e., speed, torque, heat dissipation capability, etc.). The distributed winding shown in Figure 5.14 is composed of three phases and each phase is made of $q$ number of coils. Depending on the output torque and speed of the machine, the coils can be connected in parallel or series or with a combination of these two to determine the induced voltage waveform. Each coil is made of many turns and the current rating, thermal requirements, loss characteristics, and, also, the manufacturing capabilities define the way the coils are constructed.

The distributed windings can be designed either with stranded or bar-wound construction. Bar wound offers higher slot fill factor than the stranded design. Together with the shorter end turns, bar-wound construction has lower DC resistance and better packaging [13]. Owing to the geometry of the coils, construction of the slot, and larger surface area in the end turns, bar-wound design has better heat dissipation. This enables better performance in transient conditions, which is highly important in electrified power train applications [14].

On the other hand, in the bar-wound design, skin and proximity effects are considerably higher as compared to the stranded design, especially at higher speeds. This increases the AC resistance and, hence, the stator copper losses as described in Section 5.2. However, depending on the speed–torque requirements of the machine and the duty cycle of high-speed operation for the given driving profile, a traction motor design could be possible, where the advantages of bar-wound design are utilized. As shown in Figure 5.22, in low-to-medium speed range, the AC resistance of bar-wound design is lower than the stranded design. However, at higher speeds, the AC resistance of bar-wound construction is higher since the eddy current losses due to the skin and proximity effect start dominating. In practice, in most of the drive cycles for today's road and traffic conditions, the electric

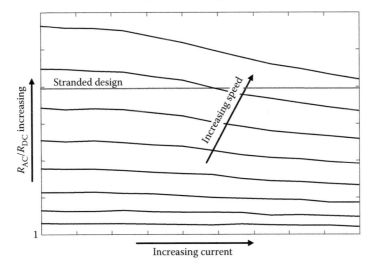

**FIGURE 5.22** Variation of AC resistance in bar-wound constructed coils for different speed and current. (Adapted from K. Rahman et al., Design and performance of electrical propulsion system of extended range electric vehicle (EREV) Chevrolet Voltec, in *Proceedings of the Energy Conversion Congress and Exposition*, Raleigh, NC, September 2012, pp. 4152–4159.)

traction motors operate with much less torque at lower speeds than their rated values most of the time. For this reason, bar-wound design can also be considered for electric traction motors, provided that the electrical, mechanical, and also the thermal requirements are satisfied [15].

## 5.5 MATERIALS IN ELECTRIC MACHINES

### 5.5.1 CORE MATERIALS

The core materials in traction motors play a very critical role in meeting torque-speed and efficiency targets. Figure 5.23 illustrates the speed–torque characteristics required for hybrid electric vehicle (HEV) traction motor, which are in turn reflected in the required properties of the electrical steel used for the motor core. High torque is required for quick starting of the vehicle and hill climbing, high efficiency in the medium-speed region for city driving, and high-speed operation for highway driving. In addition, the traction motor needs to be especially efficient at the speed range in which the vehicle is most frequently driven. Also, in a HEV, due to space constraints, the traction motor must be compact in size, light in weight, and economical.

Figure 5.24 summarizes the requirements of electrical steel used in traction motors. To meet the high torque requirements of traction motors, the electrical steel is required to have high flux density. Traction motors are required to be compact and deliver high power. This makes the motor to be operated in high speeds. When traction motors are prone to high-speed operation, the rotor is prone to large centrifugal force. The electrical steel sheet in the rotor must be capable of handling this large force; at the same time, have minimum iron loss to get high efficiency for highway driving.

#### 5.5.1.1 Electrical Steel Requirements for HEV Traction Motor

A nonoriented (NO) electrical steel is usually used for traction motors. Figure 5.25 shows the important characteristics of an NO steel that influence its magnetizing properties, and in turn are reflected in the efficiency of the traction motor.

The magnetizing properties required for a traction motor are realized through measures such as purification of steel, controlling of alloying elements, grain orientation, and grain size. Since efficiency is an important factor for most traction motors, it is necessary to reduce the resistivity of the steel lamination, so that the eddy current loss in the steel is reduced. This results in reduction of iron loss and hence better efficiency. Silicon (Si) is used as an alloying element in electrical steel

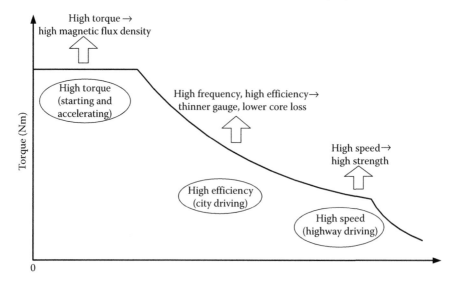

FIGURE 5.23 Traction motor requirements.

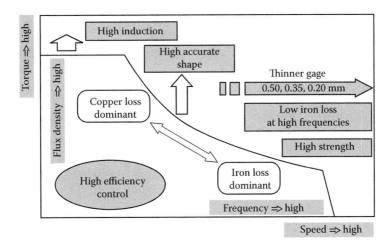

**FIGURE 5.24** Characteristics of electrical steel used in traction motors.

to reduce the resistivity. As it can be seen from Figure 5.25, adding Si reduces the resistivity, but at the same time reduces the saturation magnetic flux density. Hence, when defining the Si content, it is very important to control the iron loss and saturation magnetic flux density in an optimal way.

The other important factor to be considered in the design of traction motors is space. The packaging constraints limit the motor size but still require the motor to achieve a high torque and power density. To have a high-torque-density motor, the electrical steel sheet is required to achieve a high flux density with lower current. On the other hand, the current generation traction motors operate at very high speeds to get a high output power. The electrical steel sheet used should be capable of providing the required strength so as to be used in this high-speed region. High-speed operation increases the excitation frequency of the electrical steel and, hence, the core loss increases.

As it has been described in the previous section, to reduce the eddy current losses, magnetic cores are made up of laminated steel, which are insulated from each other. Owing to the lower resistivity, the eddy currents in the laminated core are much less than the solid core. Typical lamination thickness used in traction motors ranges from 0.2 to 0.35 mm.

Although lamination helps to reduce the eddy current loss, it reduces the stacking factor. Stacking factor depends on the lamination thickness and also on the manufacturing process such as punching and stamping of the lamination steels. Stacking factor of <100% reduces the flux-carrying capacity

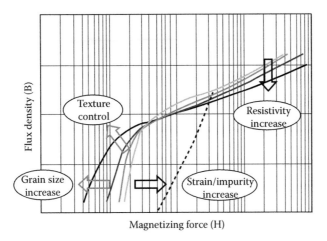

**FIGURE 5.25** Magnetizing characteristics of electrical steel.

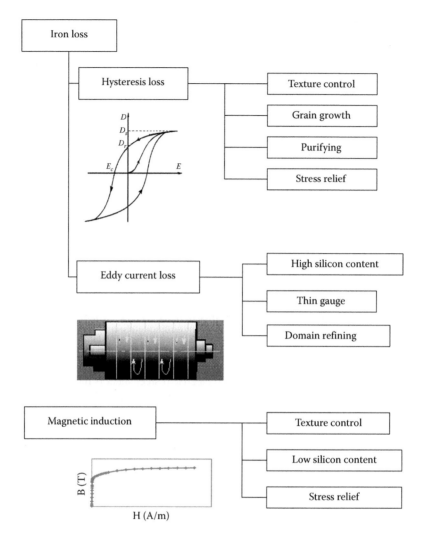

FIGURE 5.26 Methods to reduce the eddy current loss, hysteresis loss, and improve magnetic induction during electrical steel manufacturing.

of the electrical steel, which in turn requires a higher current to get the same torque when compared to a solid core motor. This can increase the copper loss. The typical stacking factor for a lamination thickness of 0.1 mm is 89.6%, while for 0.2 mm, it is 92.8% and for 0.5 mm, it is 95.8%.

Figure 5.26 summarizes the ways to reduce the hysteresis loss, eddy current loss, and improve magnetic induction during electrical steel manufacturing. Figure 5.27 shows the variation of copper loss and core loss as a function of motor RPM. In the constant torque region, the copper loss is constant, since the current required to get the peak torque is the same in this region. The core loss is a function of speed and current, and hence, the core loss increases in this region. In the flux-weakening region, due to the voltage limitation, the current has to be reduced. Hence, the copper loss reduces and the core loss increases. The dominant loss in the high-speed region is core loss.

### 5.5.1.2 Manufacturer Core Loss Data

Electrical steel manufacturers provide the B–H curve of the electrical steel, core loss at various frequencies, and the mechanical properties of the steel. The B–H curve of typical electrical steel for traction motor application is shown in Figure 5.28. The figure shows that as the value of magnetic

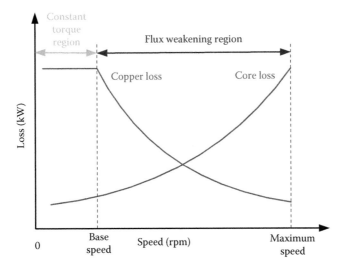

FIGURE 5.27   Variation of copper and core loss in traction motor.

field intensity, H or in other terms if the current is increased, the magnetic flux density (B) increases linearly up to a certain value of H. Beyond this value, the rate of increase in B gets smaller for the same rate of increase in current or H and the core starts saturating. Figure 5.29 shows the comparison of two different electrical steels. It is easy to infer from the figure that to achieve the same electrical flux density, Steel 1 requires less current than Steel 2. Choosing Steel 1 will result in lower copper loss and higher efficiency.

The second important data that manufacturers provide are the core loss data. Figure 5.30 shows that the core loss increases with frequency. The core loss data are shown from 50 to 1000 Hz. Figure 5.31 shows that the core loss increases with increasing current or flux density. For the same frequency, the core loss is much higher for 1.5 T when compared with 0.3 T.

When selecting between different materials and different material grades, electrical properties usually have higher preference, and the mechanical behavior is tuned to the application. The

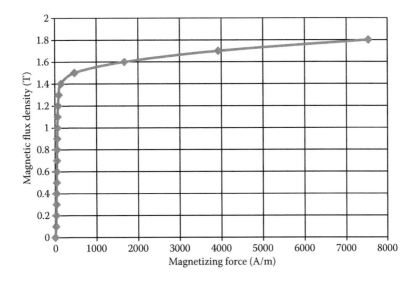

FIGURE 5.28   B–H curve of electrical steel lamination.

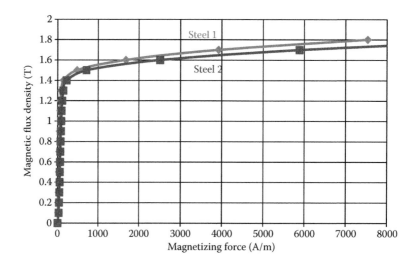

FIGURE 5.29   Comparing B–H curve of two different electrical steels.

mechanical properties of electrical steel can be inferred from the stress–strain curve shown in Figure 5.32.

There are outstanding points and regions in such a curve corresponding to various stress and strain stages. The first strain stage is the elastic deformation. Here, the strain is not permanent, and the material returns to its original shape upon unloading. As stress increases, the plastic deformation begins where the original shape becomes partially unrecoverable. For most materials, this elastic limit is marked on the stress–strain curve by a deviation from linearity.

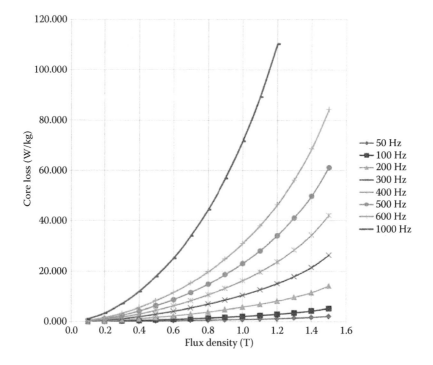

FIGURE 5.30   Core loss versus frequency for electrical steels.

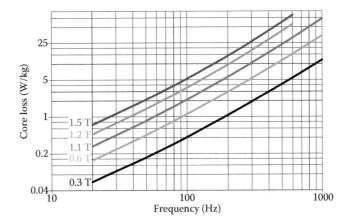

FIGURE 5.31 Variation of core loss with flux density and frequency.

Most electrical steels gradually depart from the linear-elastic region and noticeable yielding might not occur until a bit past the earliest plastic portion of the curve; so, the yield strength is set slightly above the elastic limit. The ultimate (tensile) strength, which is the highest point on the curve, is usually well into the plastic region. Stress in electrical steel should be well within the elastic limit.

The typical value of yield strength and tensile strength of the electrical steel used in traction motors is around 450 and 560 Mpa, respectively. These values can change depending on the electrical steel material grade. Figure 5.33 summarizes the yield strength of different grades of electrical steels. The figure shows that the lower-grade steel has higher strength (low iron loss too). This is due to the high Si content.

From the design standpoint for interior PM machines, high-strength electrical steel can be helpful to improve the motor efficiency by reducing the rib thickness. As shown in Figure 5.34, the rotor rib is the small portion of the steel core between the magnet and the airgap. In terms of electromagnetic operation, the rib is expected to be small enough, so that it saturates quickly and behaves like an airgap. From the mechanical perspective, the strength of the rib should be high enough to prevent failure. During high-speed operation, the rib area is prone to high stresses. If a low-strength steel is used, the rib has to be much thicker. This will move the magnet away from the air gap, and cause reduction in efficiency. Last but not the least, good electrical steel is required to have good punchability. Punchability is the combination of different properties of electrical steel that maintain long tool life and help to minimize the cost.

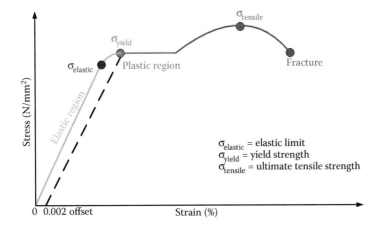

FIGURE 5.32 Stress–strain curve for typical electrical steel.

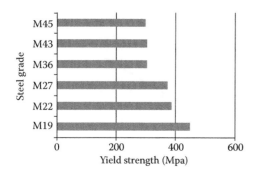

| Mechanical properties of electrical steel | | |
|---|---|---|
| Grade | Yield strength (Mpa) | Tensile strength (Mpa) |
| M19 | 450 | 565 |
| M22 | 385 | 500 |
| M27 | 374 | 494 |
| M36 | 305 | 450 |
| M43 | 305 | 445 |
| M45 | 300 | 434 |

FIGURE 5.33   Mechanical properties of typical electrical steels.

### 5.5.1.3   Effect of Manufacturing Process on Core Loss

To get the best motor efficiency, a careful selection of electrical steel is not just sufficient. The motor-manufacturing process plays a very vital role too. Figure 5.35 shows the various steps in the stator-manufacturing process. The stator lamination is first stamped. Interlocking is done to secure all the stator cores, after which, it goes through the process of stress relief annealing. Copper wires are then inserted into the stator slots, after which, squeezing and heat shrinking is done on the stator. Undesirable stresses are created adjacent to the cut edge by stamping, shearing, or slitting operations. These result from the distortion of the crystal structure that is caused by the cutting operation. Stress relief annealing is a heat-treatment process that relieves some of these stresses and helps to reduce the core loss.

Figure 5.36 shows the changes in the B–H curve of an electrical steel when it was sheared in parallel with the rolling direction into two or four sections. The magnetic permeability decreases and the magnetizing force required to secure a certain magnetic flux density is increased.

Figure 5.37 shows the effect of compression and tension on the core loss of electrical steel lamination. High-grade steel contains a higher percentage of Si and the grain size is larger when compared to a low-grade steel. Stress-related increase in core loss is much higher for low-grade steel than a higher-grade steel.

Figure 5.38 shows the influence of elastic stress over the iron loss of electrical steel. Iron loss increases significantly under a compressive stress. Also, the rate of iron loss increase becomes larger as the magnetic flux density is lowered. This is one of the main reasons why the core loss factor that has to be taken into account during simulation is much higher in low-flux-density region

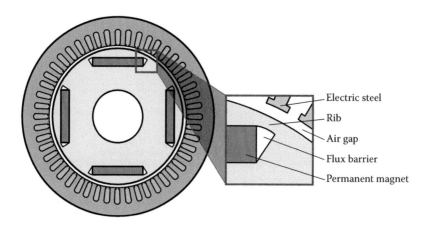

FIGURE 5.34   Effect of rib thickness on motor efficiency.

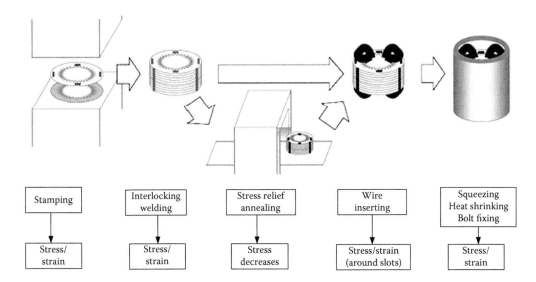

| Stamping | Interlocking welding | Stress relief annealing | Wire inserting | Squeezing Heat shrinking Bolt fixing |
|---|---|---|---|---|
| Stress/ strain | Stress/ strain | Stress decreases | Stress/strain (around slots) | Stress/ strain |

FIGURE 5.35    Stator manufacturing process.

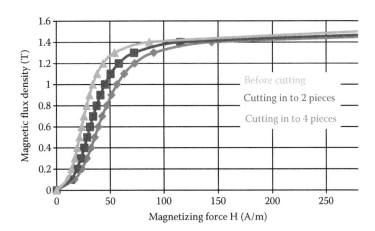

FIGURE 5.36    Effect on the B–H curve due to shearing.

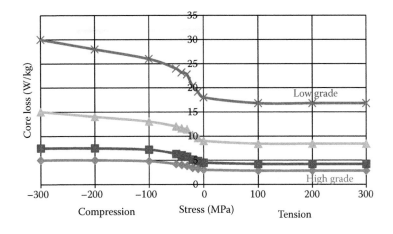

FIGURE 5.37    Effect on core loss due to compression and tension.

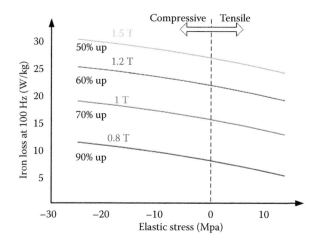

FIGURE 5.38   Effect on core loss due to elastic stress.

than high-flux-density regions. The increase in iron loss will be very significant due to the residual compressive stress that is the result of press fitting or heat shrink fitting.

During the manufacturing process, interlocking/welding is done to secure the stator/rotor laminations together. Interlocking/welding introduces local strain in the area where it is performed and this can increase the iron loss. As the number of interlock/weld points increases, the core loss increases as shown in Figure 5.39.

### 5.5.2   PERMANENT MAGNETS

Owing to their high efficiency and high-power density, PM machines are very attractive in electrified power train applications. In AC synchronous and brushless DC machines, PMs are generally utilized to provide the excitation field on the rotor. As compared to excitation with field windings,

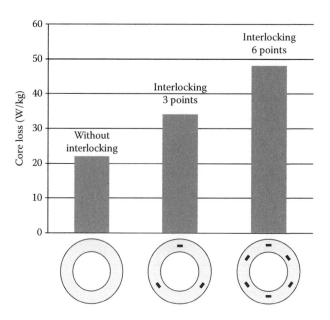

FIGURE 5.39   Effect of interlocking on core loss.

PMs provide a lower loss excitation and, hence, improve the efficiency. High-energy-density magnets have higher energy product per unit volume. Along with high coercivity, this enables higher power density. These are usually iron-and cobalt-based rare-earth magnets, such as NdFeB and samarium–cobalt (SmCo) magnets.

The field inside the PM is the result of the magnetization of the material and the field applied externally. The relationship between the magnetic flux density and the applied magnetic field intensity of a PM is represented as

$$B_m = \mu_0 H_m + J \tag{5.41}$$

The magnetization of an unmagnetized magnet is accomplished by applying a strong external magnetic field to align the magnetic dipoles. This result in the alignment of the dipoles and the material becomes magnetically polarized or "magnetized." Once the material is saturated (all the dipoles are aligned) and the external field is removed, the measured magnetic field density is called the remanence, $B_r$. This value can be measured by applying the magnet in a closed magnetic circuit (see Figure 5.42). In this case, the magnet is said to be "keepered" and its poles are short circuited. Assuming that the magnetic field intensity around the magnetic circuit could be neglected due to the high permeability of the magnetic material shortening the magnet, this results in $H_m = 0$ and, hence, $B_m = J$, which represents the intrinsic properties of the magnet. Therefore, the relationship between $H_m$ and $J$ shown in Figure 5.40 (intrinsic curve) represents how the external field affects the intrinsic magnetization of the material. To completely demagnetize the magnet, the negative external field of $H_{ci}$ should be applied, which is called as the intrinsic coercive force and it shows the resistance of the magnet to demagnetization. This is the reason why the second-quadrant characteristics are generally used in analyzing the behavior of PMs.

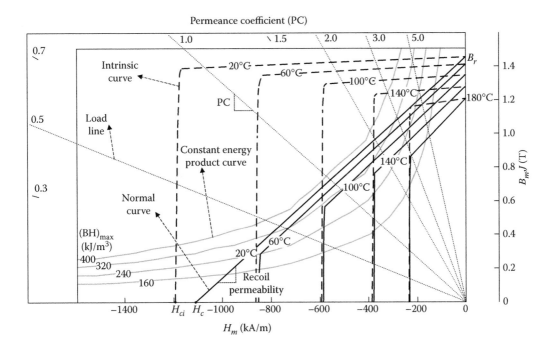

FIGURE 5.40 Demagnetization and characteristic curves of a PM (TDI Neorec53B iron-based rare-earth magnet). (Adapted from TDK Corporation, NEOREC series neodymium iron boron magnet datasheet, May, 2011. Online. Available: http://tdk.co.jp/)

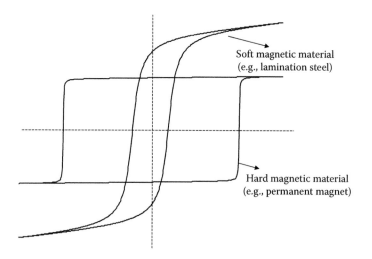

FIGURE 5.41   Comparison of the hysteresis loop for a PM and magnetic steel.

If the applied field on a PM is varied in four quadrants, the hysteresis loop of the material is obtained. In Figure 5.41, the magnetization curves of a PM and magnetic steel are shown together. It can be observed that the area of the hysteresis loop is much larger than that of magnetic steel. This shows that the PMs require higher magnetic force to be demagnetized and the area under the loop shows the stored magnetic energy. For the electric steel, a small amount of magnetic force can saturate the material. The area under the loop is much smaller and it represents the hysteresis loss. For this reason, magnets are called as hard magnetic materials and electrical steels are called as soft magnetic materials.

The normal curve in Figure 5.40 designates the relationship between the coercive force ($H_m$) and the field inside the PM ($B_m$) and it represents the operating point of the magnet. Figure 5.42 shows the case when an airgap is introduced in the magnetic circuit. Considering that the leakage flux in the circuit and mmf drop on the iron are negligible, the magnetizing force inside the PM and the iron is uniform; Ampere's law can be applied to the circuit. Since there are no external ampere turns, the line integral of magnetic field intensity around the circuit will be zero:

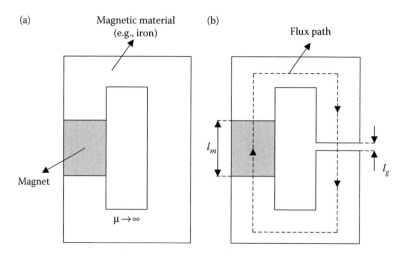

FIGURE 5.42   Magnetic circuit with a PM (a) closed circuit (keepered) and (b) with airgap.

$$H_m l_m + H_g l_g = 0 \Rightarrow H_m = -\frac{l_g}{l_m} H_g \tag{5.42}$$

where $H_m$, $l_m$ and $H_g$, $l_g$ are the magnetic field intensity and the length of the magnet and airgap, respectively. Since the flux in the circuit is the same, the magnetic field density of the magnet can be calculated:

$$B_m = B_g \frac{A_g}{A_m} \quad B_g = \mu_0 H_g \Rightarrow B_m = \mu_0 \frac{A_g}{A_m} H_g \tag{5.43}$$

where $A_m$ and $A_g$ are the cross section of the magnet and the airgap, respectively. Combining Equation 5.42 with Equation 5.43 results in:

$$\frac{B_m}{H_m} = -\mu_0 \underbrace{\frac{A_g}{A_m} \frac{l_m}{l_g}}_{\text{PC}} \tag{5.44}$$

When there is no external mmf source in the circuit, Equation 5.44 represents a straight line starting from the origin whose slope is named as the permeance coefficient (PC). It is called as the load line and its intersection with the normal curve represents the operating point of the circuit. The magnetic field intensity in the airgap can be expressed as a function of the energy product:

$$H_m = -\frac{l_g}{l_m} H_g \quad B_m = \mu_0 \frac{A_g}{A_m} H_g \quad \lambda_g = \frac{\mu_0 A_g}{l_g} \Rightarrow H_g = \frac{1}{l_g} \sqrt{\frac{\left| -H_m B_m \right| V_m}{\lambda_g}} \tag{5.45}$$

where $V_m$ is the volume of the magnet and $\lambda_g$ is the permeance of the airgap.

PC and energy product provide important information about the design of the magnetic circuit:

- Higher values of PC result in higher magnet flux density, $B_m$. In this case, the risk of demagnetization reduces.
- Higher PC means larger airgap area and smaller magnet pole area. For the given $B_m$ at the operating point of the normal curve and load line, this results in lower airgap flux density and, hence, reduces the force in the airgap and also increases the leakage flux.
- As the magnet thickness is increased, it increases the PC and reduces the risk of demagnetization.
- The energy that is required to magnetize the volume of the airgap is inversely proportional to the volume of the magnet. However, a design utilizing the highest energy product does not always signify the best performance.
- For a certain $B_m$ and $H_m$, the magnetic energy has its highest value per unit volume as shown in Figure 5.40. For a design with higher magnetic flux density than the one at the maximum energy point, magnet volume should be increased.

When an external field is applied to a PM, its operational segment is defined by the recoil permeability ($\mu_{rec}$). It is an important parameter that defines the normal and intrinsic demagnetization characteristics of a PM material. For high-coercivity magnets, recoil permeability can be defined as the slope of the normal curve as shown in Figure 5.40. Neglecting the effect of permeability of free space for simplification ($\mu_0 = 4\pi \times 10^{-7}$), Equation 5.41 can be rewritten as

$$B_m = H_m + J \tag{5.46}$$

This holds at every point on the demagnetization curve and it shows that the intrinsic values can be calculated automatically if the normal values are available. For the operation in the second quadrant, Equation 5.46 can be reorganized as

$$B_m = J - H_m \tag{5.47}$$

Since $\mu_{rec}$ is the slope of the normal curve, neglecting $\mu_0$ for simplification, the equation for the normal curve can be derived as

$$B_m = \mu_{rec} H_m + B_r \tag{5.48}$$

To relate Equations 5.47 and 5.48, the following modification can be applied and the equation for the intrinsic curve can be calculated as follows:

$$B_m = (-H_m + H_m) + \mu_{rec} H_m + B_r \Rightarrow B_m = H_m + \underbrace{(\mu_{rec} - 1)H_m + B_r}_{J} \tag{5.49}$$

Equations 5.48 and 5.49 highlight a few important points about the recoil permeability:

- The slope of the intrinsic curve is related to the recoil permeability of the normal curve. If the recoil permeability equals to 1, $J$ would be constant over the entire range of $H_m$.
- For iron-based rare-earth magnets, the recoil permeability is usually in the range of 1.05–1. This is why there is a small decrease in the intrinsic curve as the coercive force increases. This can also be observed in Figure 5.40. For higher recoil permeability, the drop in the intrinsic curve will also be higher.
- For iron-based rare-earth magnets, since the recoil permeability is close to 1, the total permeability of the magnet $\mu_{rec}\mu_0$ is almost same as air. For this reason, PMs would have an effect on the reluctance of the magnetic path [17].

The relationship between the normal and intrinsic curves is also reflected in the PC, which was calculated in Equation 5.44. Since PC represents the slope of the load line, which starts from the origin, it can be defined as the ratio between the magnetic flux density and the coercive force. Combining with Equation 5.47, the relationship between the PC of the normal curve (PC) and the intrinsic curve ($PC_i$) can be derived as

$$PC = \frac{B_m}{H_m} = \frac{J - H_m}{H_m} = \underbrace{\frac{J}{H_m}}_{PC_i} - 1 \tag{5.50}$$

In PM machine design, the demagnetization curves should be analyzed in detail, considering the PC of the magnetic circuit, operating temperature, and also the capability of tolerating the negative field from the stator windings. Figure 5.43 shows an example where different operating conditions are presented. If the magnetic circuit is designed for a PC of PC1, at 20°C, the operation point will be at point A, which corresponds to flux density of $B_{mA}$. In this case, if the temperature increases to 140°C and, due to the demagnetization characteristics of the magnet, the flux density will reduce to $B_{mB}$. At PC1, since the normal curves for both temperatures have the same slope, the change is approximately linear and, hence, when the temperature goes back to 20°C, the magnetic flux density will increase back to $B_{mA}$. This is called a reversible loss and, in general, it is represented by two temperature coefficients, $\alpha$ and $\beta$:

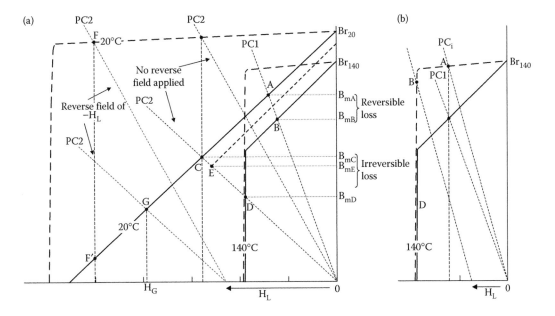

**FIGURE 5.43** Characteristics demagnetization curves and operating points (a) normal and intrinsic curves at different conditions (b) reverse field condition at high temperature.

$$\alpha = \left( \frac{1}{B_r} \frac{\Delta B_r}{\Delta T} \right) \times 100 \, [\%]$$

$$\beta = \left( \frac{1}{H_{ci}} \frac{\Delta H_{ci}}{\Delta T} \right) \times 100 \, [\%]$$

(5.51)

where $B_r$ is the remanence, $H_{ci}$ is the intrinsic coercive force, and $T$ is the temperature. It can observed from Equation 5.51 that $\alpha$ represents the reversible temperature coefficient for the remanence, whereas $\beta$ is the reversible temperature coefficient for the intrinsic coercivity [18].

PC1 represents a relatively high value of PC. Depending on the design, it might require a higher magnet volume to achieve PC1 and this may cause an inefficient use of the magnetic energy. If the magnetic circuit is designed for a lower PC (PC2), the operating point will be at point C at 20°C. When the temperature increases to 140°C, the magnetic circuit operates at point D. At this point, if the coercive force decreases, the operating point will recoil, originating from point D with the slope of $\mu_{rec}$. If the temperature goes back to 20°C, the operating point will not go back to point C, but instead, it will be at point E, which has a lower magnetic flux density and remanence. In this case, this means that the magnet has lost some of its magnetization. This is an irreversible loss and the magnet should be remagnetized to get back to its original operating point.

Besides the operating temperature, irreversible loss depends on how long the magnet has been exposed to elevated temperature, along with the PC and the applied reverse magnetic field. As described in Reference 18, irreversible losses due to temperature are caused by the reversal of domains that are aligned in the direction of magnetization. The longer the material is exposed to elevated temperature, the chance for the reversal of the domains increases.

Normal curves define the operating point of the magnet. However, when an external reverse field is applied, for example, PM machine operating in flux-weakening mode, intrinsic curves are used to define the coercive force and, hence, the operating point on the normal curve. As shown in Figure 5.43a, when there is no reverse field applied, the load lines of PC2 and PC$_i$2 intersect the intrinsic and normal curves at the same value of the coercive force. However, the effect of external magnetic forces is related to the intrinsic characteristics of the material. This is shown in Figure 5.43a where a reverse field of $-H_L$ is applied and the coercive force is calculated at point F where the $PC_i$ load line intersects with the intrinsic curve. Therefore, the operating point on the normal curve will be at point F'.

When designing a PM machine, analyzing the reverse field capability of the magnetic circuit is very important to define the performance of the machine for the given operating temperature. Figure 5.43b shows an example when a certain reverse field is applied at a higher temperature. For the given negative coercive force, it can be noticed that the $PC_i$ intersects the intrinsic curve below the knee point. As mentioned previously, this will cause irreversible demagnetization of the PM material, which will result in a performance reduction in the machine. Similar to the case in temperature, the longer the material is exposed to the reverse field, the higher the change of the reversal of the domains and this leads to higher irreversible losses.

In electric machines, irreversible loss causes a loss in magnet flux that leads to a reduction in torque. During the design process, the operating temperature should be carefully investigated by taking the operating conditions such as load line and the reverse field into account. In addition, the irreversible loss data provided by the magnet manufacturers should be evaluated. Figure 5.44 shows the temperature dependence of irreversible magnetization for different PCs [19]. It can be observed that for a lower value of PC, the irreversible demagnetization factor is higher for the same temperature. In addition, the chance in the flux loss is not linear with the increase in the temperature. For example, in rare-earth magnets, the change in magnetic flux density and, hence, the irreversible loss is higher from 80°C to 110°C, and then it is from 20°C to 80°C.

In electric machine applications, maximum operating temperature should be defined to maintain stable operation of the magnet. As discussed previously, PMs might become unstable due to the nonlinear nature of the demagnetization curve and also the irreversible losses. The designer can consider a design with high PC, which might require higher magnet volume. This might end up with inefficient use of the material. A higher coercivity magnet, which has linear demagnetization characteristics around the desired operating point might also be considered. However, this might increase the rare-earth content and, hence, the cost of the machine.

To verify the lifetime of the magnets in harsh operating conditions in a traction application, corrosion resistance needs to be considered. NdFeB magnets are prone to corrosion, which reduces the performance and lifetime. With the recent improvements in the chemistry of the materials and the surface treatment, the corrosion resistance of iron-based rare-earth magnets has been increased significantly. In addition, different surface treatment techniques improve the durability of the adhesive material and the quality of the insulation, which are significantly important in electric traction motors [20]. Table 5.1 summarizes the specifications of the surface treatment

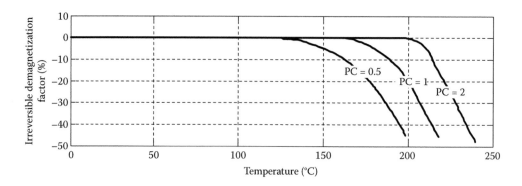

**FIGURE 5.44** Variation of irreversible demagnetization factor with temperature (shown for TDI Neorec series NdFeB magnets).

**TABLE 5.1**

**Standard Specifications for Surface Treatment Techniques for PMs**

| Surface Treatment | Standard Coating Thickness (μm) | Moisture Resistance | Salt Water Resistance | Bonding Durability | Insulation Proof | Surface Cleanness | Dimensional Accuracy |
|---|---|---|---|---|---|---|---|
| Aluminum coating | 5–20 | ✓ | | ✓ | | | ✓ |
| Nickel coating | 10–20 | ✓ | ✓ | | | ✓ | ✓ |
| Titanium nitride coating | 5–7 | | | | | | ✓ |
| Simple rust prevention | <2 | | | | | ✓ | ✓ |
| Electro-deposition coating | 10–30 | ✓ | ✓ | ✓ | ✓ | | |

*Source:* Adapted from Hitachi-Metals, Neodymium–iron–boron magnets Neomax, Retrieved on August, 2013. Online. Available: http://www.hitachi-metals.co.jp/

techniques, the thickness of the coating layer, and the effect of the treatment on the performance of the magnet.

Depending on the performance requirements, torque, speed, and operating temperature, different magnet materials can be utilized in electric machines. Table 5.2 compares some of the important characteristics of the most frequently used magnet materials in electric machines. The properties of different magnet materials can be summarized as follows:

- NdFeB is widely used in electric machines for traction applications due to their high-energy product and high coercivity. High-energy density enables a higher power density, whereas high coercivity brings better performance when the external field is applied.
- The coercivity of NdFeB drops rapidly as the temperature increases.
- SmCo is a cobalt-based rare-earth magnet and it has better heat resistance than NdFeB. It can operate in elevated temperatures up to 250°C.
- The corrosion resistance of SmCo is better than NdFeB and it is less prone to brittle.
- SmCo is higher in price as compared to NdFeB.
- Ceramic magnets are usually ferrite and they have the lowest energy product.

**TABLE 5.2**

**Comparison of Typical Properties of Typical Magnets**

| | Unit | AlNiCo | Ceramic | SmCo | NdFeB |
|---|---|---|---|---|---|
| Remanence | T | 1.35 | 0.41 | 1.06 | 1.2 |
| Coercivity | kA/m | 60 | 325 | 850 | 1000 |
| Energy product $(BH)_{max}$ | kJ/m$^3$ | 60 | 30 | 210 | 250 |
| Recoil permeability $\mu_{rec}$ | | 1.9 | 1.1 | 1.03 | 1.1 |
| Specific gravity | | 7.3 | 4.8 | 8.2 | 7.4 |
| Resistivity | μΩ cm | 47 | >10$^4$ | 86 | 150 |

- Similar to other magnet materials, the remanence of ceramic magnet decreases with temperature, but their resistance to demagnetization increases with temperature.
- Ceramic or ferrite magnets are the cheapest among all other magnets and they are the most widely used magnets in electric machine applications in general.
- Ceramic magnets have positive temperature coefficient ($\beta$) unlike other types of magnets. Therefore, the intrinsic coercivity of ceramic magnets reduces with lower temperature.
- AlNiCo magnets are aluminum–nickel–cobalt alloys. They can operate at high temperatures, but they have low coercivity.
- AlNiCo magnets can be demagnetized if they are taken out of the rotor. They have to be keepered.

### 5.5.3 INSULATION IN ELECTRIC MACHINES

Insulation, especially in powertrain applications, is significantly important in terms of dimensioning, reliable operation, and also the lifetime of the electric machine. The main purpose of insulation is to provide a nonconducting medium between different components with different potentials. In addition, a proper insulation improves the heat dissipation in the windings and, hence, helps to increase the peak power capability of the machine.

Figure 5.45 shows sample stator slot insulation represented on an SRM. It can be observed that there are many layers of insulation and, depending on the operational characteristics of the machine, different insulating materials can be used. As discussed previously, stator windings are made up of coils that are composed of many turns. Therefore, the potential difference between each turn in a slot varies and this makes it necessary to use insulation between the conductors. Magnet wires are widely used in electric machines and they are manufactured with a varnish-like thermoplastic insulation around them. The thickness of the insulation is dependent on the conductor size, voltage, and thermal requirements and it comes as a single, double, triple, or quadruple built.

Slot insulator or slot liner provides galvanic isolation between the winding and the core. It is usually an insulating foil with high tensile strength. Owing to their thermal and mechanical properties, polyester film and aramid paper are widely used as slot insulators in electric machines. In high-voltage applications (over 1 kV) Mika is the main material for slot insulation. Mika has high thermal endurance, and good chemical resistance. High dielectric strength of Mika makes it very suitable in applications where partial discharges occur due to the high voltage or high rate of change of phase voltage, which can be the case in inverter-fed motor drives.

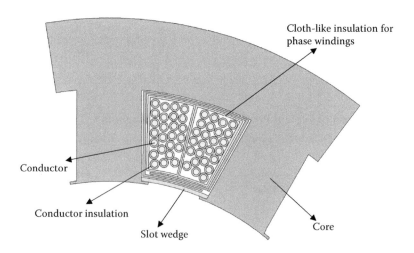

FIGURE 5.45   Insulation of stator slot in an SRM.

Once the wires are placed into the slot, both sides of the slot insulator are folded up on each other as shown in Figure 5.45. Then, the slot is closed using the slot wedge, which is usually made up of epoxy resin. In electric machines, different phases are usually on different potentials in terms of the root mean square (RMS) voltage. Therefore, windings of different phases located in the same slot are covered with a cloth-type flexible insulation to maintain galvanic isolation. In AC machines with single-layer windings, this is applied in the end turns.

It can be observed from Figure 5.45 that the insulation materials take significant amount of space in the slot and limit the fill factor for the conductor. For this reason, depending on the operational requirements of the electric machine to be designed, the space the insulation occupies should be taken into account during the dimensioning of the magnetic circuit and the design of the windings.

Insulation system plays a vital role in the reliable operation and lifetime of the electric machine. One of the main causes of failure in electric machines is due to the breakdown of the insulation system. Once the stator structure is finalized, there are inherent voids between the coils, and also between the insulation and the core, which reduces the dielectric strength and heat dissipation capability. In this case, partial electric discharges, which often occur in high-voltage machines and also the inverter-driven machines (due to the high rate of change of voltage), cause progressive deterioration of the insulation system, leading to degradation in its dielectric strength in time and a reduction in the breakdown voltage of the insulation system. This reduces the lifetime of the machine. In addition to proper dimensioning of the insulation material, it is also very important to fill the voids in the stator structure. This is usually accomplished by vacuum pressure impregnation (VPI) where the stator core together with the windings is filled with resin. VPI has many processes such as preheating, resin filling, wet vacuuming, and so on, and it reduces the physical voids in the stator and creates a solid structure, guaranteeing a better heat dissipation and proof against leakage. It also helps to protect the windings from external influences and increases the lifetime of the machine.

## 5.6  OPERATIONAL PRINCIPLES OF ELECTRIC MACHINES

### 5.6.1  PERMANENT MAGNET SYNCHRONOUS MACHINES

The PMSM originates from the synchronous motor with PMs replacing the field circuit. This modification eliminates the rotor copper loss as well as the need for the maintenance of the field-excitation circuit. Thus, a PMSM has high efficiency and enables an easier design for the cooling system. Moreover, the use of rare-earth magnet materials increases the flux density in the air gap and accordingly increases the motor power density and torque-to-inertia ratio. In demanding motion control applications, the PMSM can provide fast response, compact motor structure, and high efficiency. There are several advantages of PM motors compared to its counterparts:

- Operates at a higher power factor compared to the induction motor due to the absence of magnetizing current.
- Does not require regular brush maintenance such as conventional wound-rotor synchronous machines.
- The rotor does not require any supply and rotor losses are very low.
- Low noise and vibration than switched reluctance and induction machines.
- Lower rotor inertia and hence fast response.
- Larger energy density and compact structure.

The main disadvantage of PMSM is the high cost of the PMs, and its sensitivity to temperature and load conditions. Figure 5.46 shows that a PM machine is composed of two main parts, a stator and a rotor. The stator is the part that is mechanically fixed and connected to the external circuitry. In most traction applications, usually, the stator has a three-phase star-connected winding. The stator can be broken down into an iron part, which is "magnetically conductive," and winding slots,

(a)                                                        (b)

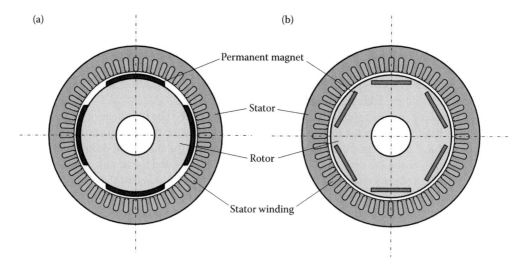

FIGURE 5.46 Cross section view of permanent magnet machines (a) surface permanent magnet machine (SPM) (b) interior permanent magnet machine (IPM).

which contain electrical (copper) windings that generate the stator magnetic flux. The rotor, on the other hand, is the part that is mechanically free to rotate and is attached to the rotor shaft with bearings. The rotor is also made of two parts: iron that conducts the magnetic flux, and PMs that produce the rotor magnetic flux. The rotor magnets are placed as alternate north and south poles. These magnets generate flux in the radial direction to flow through the air gap. Mmf generated due to the stator currents crosses the air gap and links the PM flux.

The interaction of the PM flux and stator mmf causes the rotor to rotate. As the rotor moves, the flux linkage varies and, hence, induces back EMF in the stator windings. Finally, the interaction between the stator phase currents and the corresponding back EMFs produces the electromagnetic torque. The rotor has no windings or electrical connections to the stator. To operate the machine properly, the rotor position has to be known, either from the feedback given by a position sensor, or from a position estimation algorithm.

### 5.6.1.1 PMSM Operation

Figure 5.47 illustrates the four-quadrant-operating region of a PMSM machine. Similar characteristics also apply for other electric machines. Each quadrant has a constant torque region from 0 to ±nominal speed, $\omega_b$ and a region where the torque decreases inversely with the speed from $\omega_b$ to $\omega_{max}$. The region where the torque drops inversely with speed is called the constant power region that is obtained by decreasing the rotor magnetic flux.

### 5.6.1.2 Classification of PM Machines

On the basis of excitation current and back EMF shape, PMSM machines can be classified as trapezoidal-and sinusoidal-type machines. Sinusoidal-type PMSM machines can be further classified as surface mount (SPM), surface inset, or interior permanent magnet (IPM) types as shown in Figures 5.48 and 5.49.

In an SPM machine, the magnets are glued on the rotor surface. The rotor cannot operate at high speeds due to the in adequate mechanical strength of the magnets; however, manufacturing is simple for surface magnets. SPM machine can be either sinusoidal or trapezoidal type, but the inset and IPM type are usually sinusoidal-type machines. The $d$-and $q$-axis inductances are almost the same in an SPM machine. This is due to the fact that the length of air gap is equal to that of the magnet, which has almost the same permeability as air. As it can be seen from Figure 5.49b and c, in the inset type, the magnets are put right inside the rotor surface and in the IPM type, the magnets

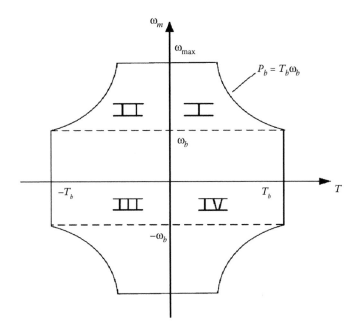

FIGURE 5.47 Four-quadrant operation of a PMSM.

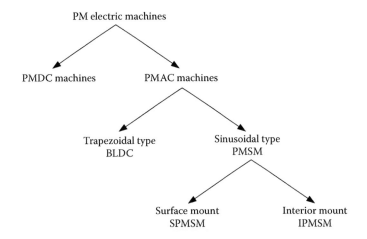

FIGURE 5.48 PM machine classification.

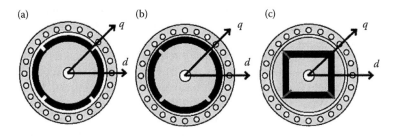

FIGURE 5.49 Different types of PMSM. (a) Surface PM motor; (b) inset PM motor; (c) IPM motor.

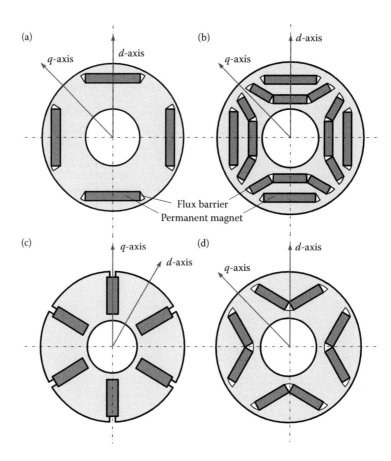

**FIGURE 5.50**  Different types of IPM rotor configuration: (a) single layer; (b) double layer; (c) spoke-type (tangential magnetization); and (d) V-shape.

are buried well inside the rotor. The $q$-axis inductance is much larger than the $d$-axis inductance in these types of machines. The space occupied by the magnet in the $d$-axis is occupied by iron in the q-axis. This means that in addition to the electromagnetic torque (mutual torque), a reluctance torque also exists in IPM-and inset-type PM machines.

Figure 5.50 shows different magnet configurations for IPM machines. The configuration given in Figure 5.50c is called spoke-type machine, where the magnetization is in tangential direction. For the other machines, the magnetization is usually in radial direction.

### 5.6.1.3  Saliency in PM Machines

IPM machines use both PM and mechanical rotor saliency for electromagnetic energy conversion. Rotor construction in IPM machines provides an inherent saliency (difference in the $d$-and $q$-axis inductances). Saliency produces additional reluctance torque that is used for improvement of air gap torque production. Saliency also facilitates field weakening. To understand the effect of saliency, a good understanding of $d$-and $q$-axis inductances is necessary. Figure 5.51 shows the PM machine in stator and rotor reference frames. The letters $d$ and $q$ refer to the $d$-and $q$-axis, whereas the letters s and r represent stator and rotor reference frames. By definition, d-axis or direct axis is the axis that is aligned with the PM field. In designs without magnets, d-axis is defined to be aligned with the high-inductance axis. $q$-Axis or quadrature axis is $90°$ electrically advanced in the counterclockwise (positive) direction from the positive d-axis.

Figure 5.52a and b shows the $d$-and $q$-axis inductance path for an SPM machine. Since PMs have a low permeability, which can be approximated to that of air, the effective airgap in the flux

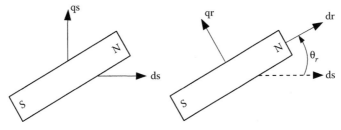

IPM machine in stator reference frame          IPM machine in rotor reference frame

FIGURE 5.51   PM machine in different reference frames.

path of $L_d$ and $L_q$ is the same in SPM motors. Therefore, an SPM motor has negligible saliency. In the stator three-phase reference frame, the inductance measured at the motor terminal is constant regardless of the motor position. In the rotor reference frame, the d-axis flux path can be represented as magnet→steel→magnet, whereas $q$-axis path is air→steel→air. Since the permeability of air and magnet is almost the same, $L_d = L_q$ for SPM.

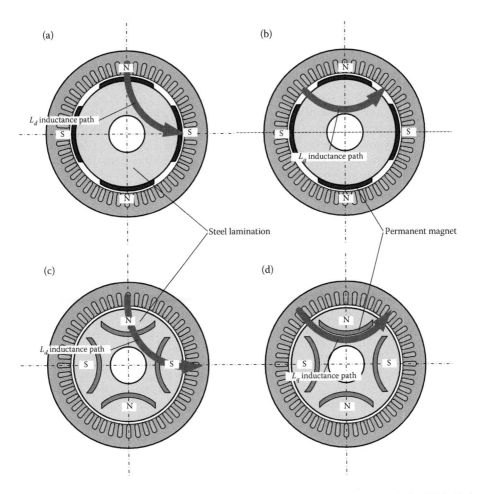

FIGURE 5.52   $d$- and $q$-axis inductances for SPM and IPM: (a) $d$-axis for SPM; (b) $q$-axis for SPM; (c) $d$-axis for IPM; and (d) $q$-axis for IPM.

Figure 5.52c and d shows the *d*-and *q*-axis inductance paths in an IPM machine. The effective airgap in the flux path of $L_d$ and $L_q$ varies according to the rotor position. This magnetic saliency results in the variation of inductance at the motor terminal according to rotor position. Similar to the case in SPM, in the stator three-phase reference frame, the inductance measured at the motor terminal is constant regardless of the motor position. In rotor reference frame, the *d*-axis flux path is steel→magnet→steel →magnet→steel, but q-axis flux path is all steel as shown in Figure 5.52d. Hence, $L_q > L_d$ in this type of machine. There are IPM designs where the opposite is also true, but they are not covered in this chapter.

### 5.6.1.4  Inductance Variation due to Magnetic Saturation

The d-and q-axis inductances vary due to magnetic saturation in the iron. The electrical resistance is not affected by saturation, but the torque constant will be affected. When current passes through an inductor wound over an iron core, initially, the flux increases linearly but later, the magnetic flux increases much slowly (see Figure 5.53), due to magnetic saturation. The inductance is the slope of this curve at any given point. In the linear portion of the curve, the inductance remains constant, after which, the inductance starts dropping due to saturation of the iron core. Also, the inductance changes in response to the change in current in any electrical motor much quicker than temperature-related parameter variations that have a slower time constant.

In IPM machines, the q-axis inductance is affected by saturation much more than the d-axis inductance. The important differences in $L_d$ and $L_q$ due to magnetic saturation are:

The d-axis path includes iron as well as magnet. The PM is the main source of saturation even when the machine is not excited (similar to L2 in Figure 5.53). The currents associated with this axis ($i_d$) are usually oriented to oppose the magnetic flux, either to exploit the reluctance torque or to achieve flux weakening. Depending on the intensity of these currents, a very small change can occur to the d-axis inductance. The increased air gap thickness, due to the presence of magnets along this axis, reduces the sensitivity of $L_d$ toward $i_d$.

The q-axis inductance path is all steel and does not include any PMs. This axis is not excited when the machine is at rest. For small q-axis currents (low torque), the q-axis magnetic circuit operates in the linear region (L1 in Figure 5.53) of the electrical steel. However, for large q-axis currents (high torque), the magnetic path on this axis will become saturated and the q-axis inductance will drop drastically. This is summarized in Figure 5.54.

Saliency ratio is defined as the ratio of the q-axis to d-axis inductance:

$$\xi = \frac{L_q}{L_d} \tag{5.52}$$

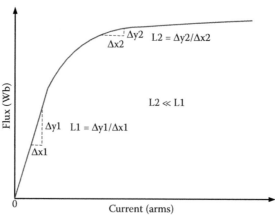

FIGURE 5.53  Saturation in electrical steel and its effect on inductance.

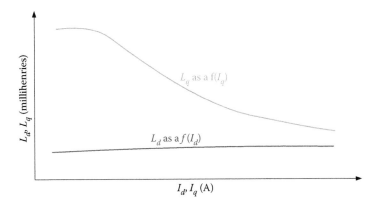

FIGURE 5.54  $L_d$ and $L_q$ as a function of $I_d$ and $I_q$, respectively.

A high saliency ratio makes it possible to achieve a wide field-weakening range without suffering from excessive back EMF in high-speed operation:

$$T_e = \frac{3}{2}p(\psi_m i_{sq}) + \frac{3}{2}p(L_d - L_q)i_{sd}i_{sq}$$

$$i_{sd} = \frac{-\psi_m + \sqrt{\psi_m^2 + 8(L_d - L_q)^2 I_{s\,max}^2}}{4(L_d - L_q)} \qquad (5.53)$$

$$i_{sq} = \sqrt{I_{s\,max}^2 - i_{sd}^2}$$

where $\psi_m$ is the PM flux linkage, $p$ is the number of poles, $i_{sd}$ and $i_{sq}$ are the $d$-and $q$-axis stator currents, $L_d$ and $L_q$ are the $d$-and $q$-axis inductances, and $I_{s\,max}$ is the maximum stator current. Taking the IPM motor torque equation from Equation 5.53 and plotting it on the $d$–$q$ plane results in constant torque loci as shown in Figure 5.55. Torque is zero along d-axis ($i_q = 0$) and along the $q$-axis, the following expression holds:

$$i_{sd} = \frac{\psi_m}{L_q - L_d} \qquad (5.54)$$

### 5.6.1.5  Maximum Torque per Ampere Operation

The maximum torque per ampere (MTPA) control strategy is applied in the constant torque region as shown in Figure 5.56. MTPA assures that for a required torque, the minimum stator current is applied. By doing so, the copper loss is minimized and the motor efficiency is increased. As shown in Figure 5.57, there are multiple stator current vectors (is1, is2, is, etc.) that can produce the desired torque, but, considering the efficiency, the smallest current should be applied that can guarantee this torque. All the points given by the intersection of the minimum current vectors and the corresponding torque levels provide the MTPA curve as depicted in Figure 5.58.

The set of equations that give the MTPA curve of an interior permanent magnet synchronous machine (IPMSM) and the relationship between the reference torque and the corresponding stator currents is obtained by solving $dT/di_{sd} = 0$ in Equation 5.53. In Figure 5.58, the motor parameters $\psi_m$, $L_d$, and $L_q$ that determine the shape of the MTPA curve have been considered constant. Since $L_d$ and $L_q$ vary with current or due to saturation as shown in Figure 5.54, this can influence the shape of the MTPA curve in real applications. Figures 5.59 and 5.60 show the influence of the variation of $L_q$ and $L_d$ on the MTPA curve, respectively.

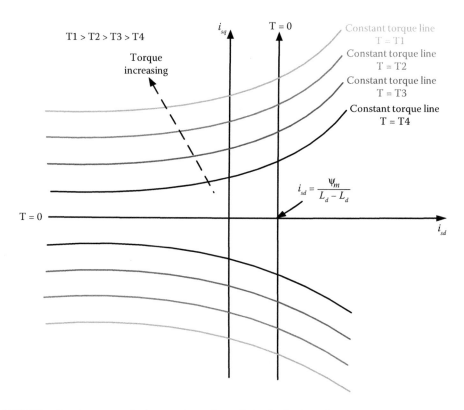

**FIGURE 5.55** Constant torque loci appearing as hyperbolas in the d–q plane.

From Figure 5.59, it can be inferred that if $L_q$ is decreasing, the slope of the MTPA curve is increasing. This means that to get the same torque, $i_q$ has to be increased and $i_d$ has to be decreased. When $L_q$ is increasing, the slope of the MTPA curve is decreasing; therefore, to get the same torque, $i_q$ should be reduced and $i_d$ should be increased. Variation of the MTPA curve is larger with decreasing $L_q$. From Figure 5.60, it can be observed that if $L_d$ is increased, the slope of the MTPA curve increases. Hence, to get the same torque, $i_q$ has to be increased and $i_d$ has to

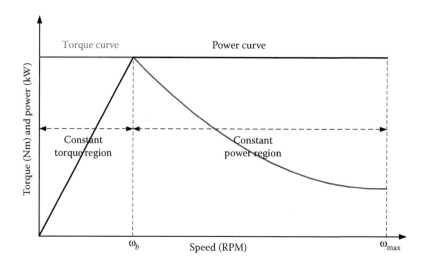

**FIGURE 5.56** Torque-speed characteristics for a PMSM traction motor.

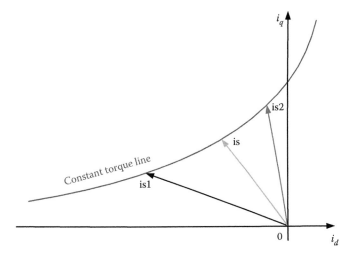

**FIGURE 5.57** Minimum stator current for a required torque level.

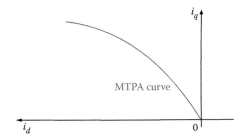

**FIGURE 5.58** MTPA curve in the $d$–$q$ plane.

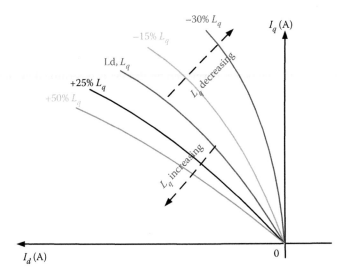

**FIGURE 5.59** Sensitivity of MTPA curve for variations in $L_q$.

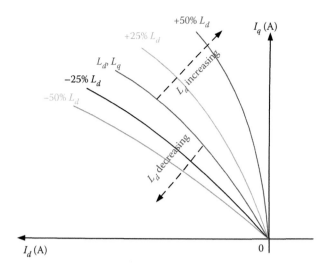

**FIGURE 5.60**   Sensitivity of MTPA curve for variations in $L_d$.

be reduced. On the other hand, if $L_d$ is reduced, the slope of the MTPA curve also reduces. In this case, to get the same torque, $i_q$ is reduced and $i_d$ is increased. MTPA curve is more sensitive to an increase in $L_d$.

### 5.6.1.6   Operational Characteristics of PM Machines

In $d$–$q$ coordinates, the voltage equation of the IPM machine can be expressed as

$$v_{sd} = R_s i_{sd} + L_d \frac{di_{sd}}{dt} - \omega L_q i_{sq}$$
$$v_{sq} = R_s i_{sq} + L_q \frac{di_{sq}}{dt} + \omega(L_d i_{sd} + \psi_m) \tag{5.55}$$

Under steady state, Equation 5.55 simplifies into

$$v_{sd} = R_s i_{sd} - \omega L_q i_{sq}$$
$$v_{sq} = R_s i_{sq} + \omega(L_d i_{sd} + \psi_m) \tag{5.56}$$

It can be observed from Equation 5.56 that to increase the speed, a higher voltage is required. The maximum output torque and power developed by PMSM is determined by the maximum current and voltage that the inverter can supply to the motor. The maximum stator voltage is dependent on the DC bus voltage and also on the PWM scheme applied, and it can be expressed as

$$v_{sd}^2 + v_{sq}^2 \leq V_{s\,max}^2 \tag{5.57}$$

Combining Equations 5.56 and 5.57 and neglecting the voltage drop on the stator resistance:

$$(-\omega L_q I_{sq})^2 + (\omega L_d I_{sd} + \omega\psi_m)^2 \leq V_{s\,max}^2$$
$$\Rightarrow \left(\frac{I_{sq}}{L_d/L_q}\right)^2 + \left(I_{sd} + \frac{\psi_m}{L_d}\right)^2 \leq \left(\frac{V_{s\,max}^2}{\omega L_d}\right) \tag{5.58}$$
$$\Rightarrow \left(\frac{I_{sq}}{a}\right)^2 + \left(\frac{I_{sd} + \psi_m/L_d}{b}\right)^2 \leq 1$$

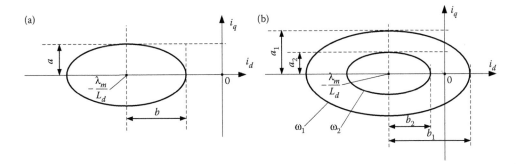

FIGURE 5.61 Voltage-limiting ellipses for IPM machines. (a) Voltage limiting ellipses and (b) voltage-limiting ellipses for different speeds.

where

$$a = \frac{V_{s\,max}}{\omega L_q} \quad b = \frac{V_{s\,max}}{\omega L_d} \tag{5.59}$$

For the given $V_{s\,max}$ and $\omega$, Equation 5.58 represents an ellipse as shown in Figure 5.61a. The center of the ellipse is located at $(-(\psi_m/L_d), 0)$, which is the characteristic current of the machine. The eccentricity of the ellipse can be represented as

$$e = \frac{\sqrt{b^2 - a^2}}{b}$$

$$e = \frac{\sqrt{(V_{s\,max}/\omega L_d)^2 - (V_{s\,max}/\omega L_q)^2}}{V_{s\,max}/\omega L_d} = \sqrt{1 - \left(\frac{L_d}{L_q}\right)^2} \tag{5.60}$$

As shown in Figure 5.61b, the ellipse shrinks inversely with the rotor speed. The shape of the ellipse depends on the saliency ratio. In addition, the d-and q-axis currents must satisfy:

$$i_{sd}^2 + i_{sq}^2 \le I_{s\,max}^2 \tag{5.61}$$

Equation 5.61 represents a current circle centered at the origin with a radius of $I_{s\,max}$ as shown in Figure 5.62a. Unlike the voltage-limiting ellipses, current-limiting circles remain constant for any speed. Since both Equations 5.57 and 5.61 should be satisfied during the operation, for a given rotor

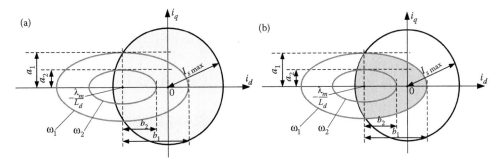

FIGURE 5.62 (a) Voltage-limiting ellipses and current-limiting circles for IPM machines and (b) overlap area between voltage limiting ellipses and current-limiting circles.

speed, the current vector can be located anywhere inside or on the boundary of the overlap area between the voltage-limiting ellipse and current-limiting circle as shown in Figure 5.62b. The overlap area becomes smaller when the rotor speed keeps increasing indicating progressively smaller changes for current vector in the flux-weakening region.

Below base speed, where the phase voltage is less than $V_{s\,max}$ for the rated current, the operation of the IPM machine is based on the control of $q$-axis current as shown in Figure 5.63a. In the flux-weakening region, negative $d$-axis current is applied as shown in Figure 5.63b. This creates a negative voltage vector on the $d$-axis, which opposes the one induced by the PM flux linkage, so that the motor can speed up. In this case, the stator current vector has both $d$-and $q$-axis components, and the magnitude of the $q$-axis current vector reduces resulting in lower torque.

The maximum torque and output power developed by PM machines is ultimately dependent on the allowable inverter current rating and maximum output voltage that the inverter can supply to the machine. In a PM operating at a given speed and torque, optimal efficiency can be obtained by applying optimal voltage that minimizes the power loss. At low speeds, this optimum efficiency will coincide with the condition of MTPA control. As described previously, such operation leads to minimal copper losses in the stator windings and also minimum power loss in the semiconductor switches of the inverter.

The relationship between the reference $i_{sd}$ and $i_{sq}$ current components for MTPA control can be derived as

$$
\begin{aligned}
i_{sq} &= I_s \cos(\beta) \\
i_{sd} &= -I_s \sin(\beta)
\end{aligned}
\tag{5.62}
$$

where $\beta$ is the current phase angle as shown in Figure 5.63b. Combining Equation 5.62 and the torque equation in Equation 5.53 results in:

$$
\begin{aligned}
T_e &= \frac{3}{2} p(\psi_m I_s \cos\beta + (L_d - L_q)(-I_s^2)\cos\beta\sin\beta) \\
&= \underbrace{\frac{3}{2} p\psi_m I_s \cos\beta}_{\text{Excitation torque}} + \underbrace{\frac{3}{4} p(L_q - L_d)I_s^2 \sin 2\beta}_{\text{Reluctance torque}}
\end{aligned}
\tag{5.63}
$$

Equation 5.63 shows that the excitation and reluctance torque components vary with the current phase angle as shown in Figure 5.64. For $\beta = 0$ (only $q$-axis current component), the electromagnetic

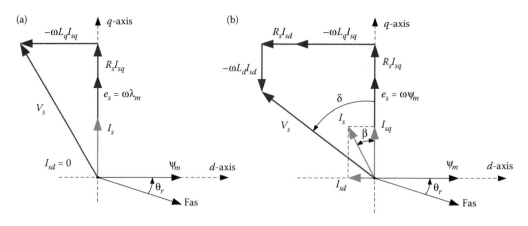

FIGURE 5.63  Vector diagrams for IPM machine for (a) speed lower than base speed and (b) speed higher than base speed.

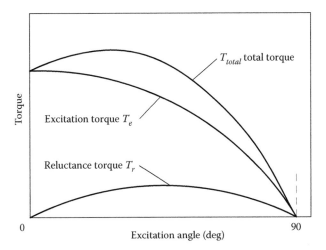

**FIGURE 5.64**  Excitation and reluctance torque components as a function of current phase angle.

torque will be maximum; however, the reluctance torque component will be maximum for $\beta = 45°$. Therefore, the maximum total torque will be achieved within the range $0 < \beta < 45°$.

For IPM motors, $L_q - L_d$ can be large; so, the reluctance torque is not negligible. To operate the IPM motor at high torque and high efficiency, $i_d$ should be determined from Equation 5.62 with $\beta$ corresponding to the maximum torque for a given $I_s$. In addition, fast transient response at high torque operation can be obtained when $\beta$ is controlled.

### 5.6.2  INDUCTION MACHINES

Owing to its simple and robust structure, induction machine has been the workhorse in the industrial electric drive applications for a long time. There are mainly two types of induction machines: In wound-rotor induction machines, the rotor circuit is made up of three-phase windings similar to that of the stator. The rotor windings are then short circuited by using slip rings.

Squirrel-cage induction machines are more commonly used in low-and medium-power applications, including electrified power trains. The rotor bars are inserted by die casting, where melted aluminum is molded in the rotor slots. The bars are then short circuited by using end rings. Figure 5.65 shows the cross-section of a squirrel-cage induction machine.

As its name implies, induction machines generate torque based on the force created by rotor currents, which are induced by the rotating airgap field generated by the stator currents. Stator windings of an induction machine are usually made up of three-phase distributed windings. Details of distributed windings have been discussed in detail in Section 5.4 and a sample three-phase distributed winding configuration has been shown in Figure 5.14. When the stator windings are excited with three-phase currents, a rotating ampere–conductor distribution is maintained, which rotates in synchronous speed as shown in Equation 5.39. The stator ampere–conductor distribution creates a rotating magnetic field in the airgap. As defined by Faraday's law given by Equations 5.15 and 5.16, the time-changing magnetic field induces voltages on rotor conductors. According to Lorentz's force law quantified by Equation 5.13, when the rotor conductors are short circuited, the induced currents generate a force on the rotor and this is the source of the motion in induction machines.

An EMF and, hence, voltage will be induced on the rotor conductors when the time-changing magnetic flux flows through the closed surface of the conductor (see Figure 5.3). If the rotor rotates in synchronous speed, the rotating airgap magnetic field will not be able to flow through the conductor surface and no voltage, current, force, and, hence torque would be induced. Therefore, in an induction machine, the rotational speed of the rotor is slightly lower than synchronous speed. The

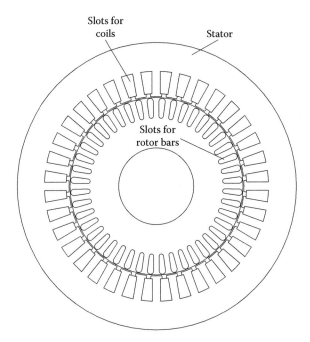

FIGURE 5.65    Cross-section view of a squirrel-cage induction machine.

relative speed between the rotor ($\omega_r$) and the rotating airgap magnetic field ($\omega_s$) is called slip and is defined as

$$s = \frac{\omega_s - \omega_r}{\omega_s} \qquad (5.64)$$

The frequency of the EMFs and currents induced in the rotor is a function of stator frequency and is dependent on the slip:

$$f_r = sf_s \qquad (5.65)$$

It can be observed from Equations 5.64 and 5.65 that when the rotor speed reaches synchronous speed, the slip is zero and the rotor frequency is also zero. When the rotor is at standstill, $s = 1$ the rotor frequency will have its highest value. In motoring mode, slip changes between zero and one. The lower the slip, the higher is the rotor frequency and also the induced currents. This results in higher rotor losses and limits the performance of induction machines. Therefore, in practice, induction machines are designed to operate at small slip, generally <5% at full load.

In the induction motor, the stator magnetic field rotates with synchronous speed. The rotor rotates asynchronously and slower than the stator magnetic field. However, since torque is generated by the interaction of stator and rotor magnetic fields, rotor magnetic field in the airgap rotates synchronously with the stator magnetic field, but it rotates faster relative to the rotor speed. The difference between the speed of the rotor and the rotor magnetic field is dependent on slip.

The operation of induction machines is similar to that of transformers except that the frequency of the induced EMFs on the rotor is different from the stator as defined in Equation 5.65. In addition, unlike a transformer, there is an airgap in the magnetic circuit that increases the level of magnetizing current. As shown in the equivalent circuit in Figure 5.66, the induced voltage in the stator, $E_1$ is coupled with the rotor circuit with the turns ratio, $u$. Here, $V_1$, $I_1$, $r_1$, and $X_1$ represent the stator

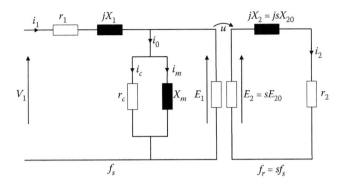

FIGURE 5.66    Single-phase equivalent circuit of an induction machine.

voltage, current, winding resistance, and stator leakage reactance, respectively. When the rotor is stationary, the induction machine behaves like a transformer with an airgap. As the rotor accelerates, the rotor speed gets closer to the synchronous speed; therefore, the frequency of the rotor magnetic field reduces, leading to lower induced voltage. Considering that the slip varies between one (standstill or locked rotor operation) and zero (rotor rotates at synchronous speed), the voltage induced in the rotor is also proportional to the slip:

$$E_2 = sE_{20} \tag{5.66}$$

where $E_{20}$ is the induced rotor voltage at the locked rotor case when $s = 1$ and $f_r = f_s$. Since the rotor leakage inductance, $X_2$ is also frequency dependent; it can be represented as calculated in Equation 5.67.

$$X_2 = 2\pi f_2 L_2 = 2\pi s f_1 L_2 = sX_{20} \tag{5.67}$$

The rotor resistance does not depend on the rotor frequency provided that the skin and proximity effects are neglected. The rotor current can be calculated using the total impedance on the rotor side:

$$i_2 = \frac{sE_{20}}{r_2 + jsX_{20}} = \frac{E_{20}}{\dfrac{r_2}{s} + jX_{20}} \tag{5.68}$$

$$\Rightarrow Z_2 = \frac{r_2}{s} + jX_{20}$$

The term related to the rotor resistance in Equation 5.68 can be divided into two components:

$$\frac{r_2}{s} = r_2 + r_2\left(\frac{1-s}{s}\right) \tag{5.69}$$

The first term represents the rotor copper losses. The second term represents the electromechanical power conversion. To simplify the calculation of motor parameters, rotor side voltage, current, resistance, and leakage reactance can be referred to the stator side by using the turns ratio:

$$E_2' = u^2 E_{20} \quad i_2' = \frac{i_2}{u}$$
$$r_2' = u^2 r_2 \quad X_2' = u^2 X_2 \tag{5.70}$$

Therefore, considering the expressions in Equations 5.69 and 5.70, the equivalent circuit of an induction machine can be updated as shown in Figure 5.67.

In the equivalent circuit, $X_m$ represents the magnetizing reactance and $i_m$ represents the magnetizing current. Owing to the airgap, the magnetizing reactance of an induction machine is significantly higher than that of a transformer. It can also be observed that even though the rotor current is zero, $i_0$ in Figure 5.67 still flows in the machine and maintains magnetization. The parameter $r_c$ represents the iron losses and it is a function of frequency and magnetic flux density. As it was mentioned earlier, induction machines are designed to operate at small slip, usually <5%. Since the rotor frequency will be much less than the stator frequency due to the small slip ratio, rotor iron losses are usually much lower as compared to the stator.

From Figure 5.67, the total airgap power can be represented as

$$P_a = mI_2'^2 \frac{R_s^2}{s} \tag{5.71}$$

where $m$ is the number of phases. Airgap power includes the power used in the electromagnetic energy conversion and the rotor copper losses:

$$P_{conv} = mI_2'^2 R_2' \left(\frac{1-s}{s}\right) \tag{5.72}$$

$$P_{cu\_r} = mI_2'^2 R_2'$$

It can be observed from Equation 5.72 that rotor copper losses in an induction machine increase as the slip increases. This is another factor that signifies why induction machines are designed to operate at low slip. With smaller slip, the portion of the airgap power that is used in electromechanical energy conversion increases, whereas the rotor copper losses decrease.

Using Equation 5.72, the induced torque can be calculated as

$$\tau_{ind} = \frac{mp}{\omega} \frac{R_2'}{s} \frac{V_1^2}{\left(R_1 + \frac{R_2'}{s}\right)^2 + \left(X_1 + X_2'\right)} \tag{5.73}$$

Figure 5.68 shows the typical torque-speed characteristics of an induction machine. When the slip is very small, the rotor runs close to the synchronous speed. This corresponds to no-load conditions, where the rotor frequency, rotor-induced voltage, and, hence, the rotor current are very low. Up to full load torque, there is almost a linear relationship between the load, slip, and, hence, the

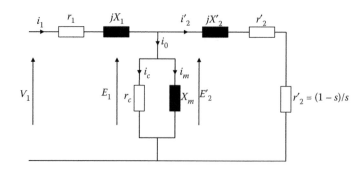

FIGURE 5.67  Single-phase equivalent circuit of an induction machine referred to the stator side.

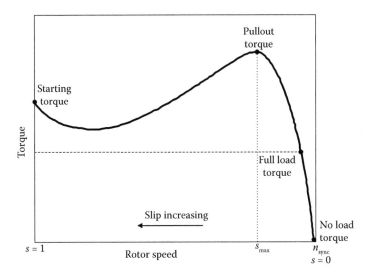

FIGURE 5.68    Typical torque-speed characteristics of an induction machine.

rotor speed. Rotor current increases linearly with speed, whereas rotor speed also reduces linearly. The induction machine operates in this region in steady state. Since the slip and, hence, the rotor frequency is low, the rotor reactance is negligible as compared to rotor resistance. This results in a high-power factor operation.

As the load increases, the slip keeps increasing. At higher slips, up to the pullout torque, the rotor frequency increases and the rotor reactance becomes closer to the rotor resistance. This results in lower power factor. If the load torque keeps increasing and the slip goes beyond its maximum value ($s_{max}$), induced torque starts reducing. This is because the power factor decreases with a higher rate than the increase in rotor current due to the high rotor reactance as a result of higher rotor frequency. Therefore, if the load is increased beyond the pullout torque, the rotor starts to decelerate rapidly and stall.

When the rotor rotates over synchronous frequency, the slip becomes negative as defined in Equation 5.64. In this case, the induced torque also becomes negative and the induction machine operates as a generator as shown in Figure 5.69. However, since the machine itself cannot run over synchronous speed, it has to be accelerated by a prime mover and, this way, it draws power from it.

When working as a generator, induction machine provides active power to the electrical source. However, due to a single source of excitation and, hence, lack of a field excitation on the rotor (this was the case in synchronous machines), it still needs to draw the reactive magnetizing current ($i_m$) in Figure 5.67 from the source.

## 5.6.2.1   Speed Control in Induction Machines

Speed control in induction machines can be achieved either via controlling the stator frequency by varying the synchronous speed or via changing the rotor torque by varying the rotor resistance or the stator voltage. Changing the rotor resistance is possible in wound-rotor induction machines. When additional resistance is added to the rotor circuit, torque-speed profile of the induction machine changes, but this causes extra losses and is not practical in electrified drivetrain applications considering the efficiency constraints. As shown in Equation 5.73, changing the stator voltage also adjusts the torque. However, due to the single source of excitation, the magnetization current in induction machine is dependent on the stator voltage. This can be observed from the equivalent circuit diagram in Figure 5.67. When the stator voltage is reduced, the magnetization current and, hence, the flux reduces. This results in lower-rated and pullout torque.

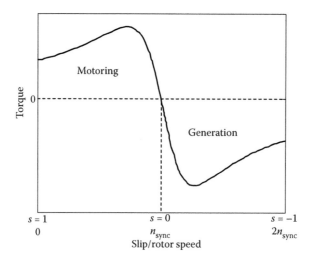

**FIGURE 5.69** Typical torque-speed characteristics of an induction machine considering the over-synchronous-speed operation in the generating mode.

When the stator frequency of an induction motor is changed, synchronous speed varies accordingly as shown in Equation 5.39. As per Equation 5.65, this results in a varying rotor frequency and, hence, the operational speed of the induction machine can be adjusted. The synchronous speed can also be changed by changing the number of poles. However, this requires changing the winding configuration and it is not practical in electrified drivetrain applications due to mechanical constraints and the limitations in terms of the number of poles to be applied.

By using power electronic inverters, stator frequency of an induction motor can be changed and as shown in Figure 5.70, this translates the torque/speed curve along the speed axis. At speeds lower than the base speed, if only the frequency is reduced, this increases the flux, causes the motor to operate in the nonlinear region and, hence, leads to an increase in the magnetization current. To keep the flux constant, the stator voltage should be linearly decreased with the stator frequency. If voltage-to-frequency (V/f) ratio of an induction machine is kept constant, the breakdown torque remains constant and the torque-speed curve is shifted along the speed axis.

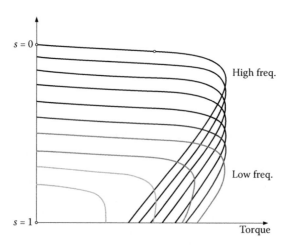

**FIGURE 5.70** Typical torque-speed characteristics of an induction motor for variable-frequency operation.

It can be observed from Figure 5.70 that at lower speeds, the torque-speed curve deviates from its original shape. This is mainly because at low frequencies, the leakage reactance reduces and the rotor resistance dominates. This causes the magnetizing current to drop, leading to a decrease in flux and hence torque. At speeds higher than the base speed, the stator voltage should be kept constant due to the design constraints of the machine. In this case, when the stator frequency is increased, the flux reduces, leading to a reduction in torque. This is similar to the flux-weakening operation in the constant power region of the torque-speed curve given in Figure 5.56.

### 5.6.3 Switched Reluctance Machines

SRM has double saliency in terms of its stator and rotor construction. The mmf is created by energizing the coils concentrated around the salient stator poles. Torque is generated when the flux flowing through the salient rotor poles pulls them toward the stator poles to reduce the reluctance of the magnetic path. SRM configurations are defined by the number of stator and rotor poles. Figure 5.71 shows the cross-section of three-phase 6/4 and four-phase 8/6 SRMs in aligned and unaligned positions, respectively.

SRM has a simple and low cost construction: It has a laminated stator core with concentrated windings around the salient poles. The laminated rotor core has no windings and PMs. This enables high-speed and high-temperature operation on the rotor. As discussed in Section 5.4, salient pole structure enables the phase windings to be isolated from each other. Therefore, the mutual inductance is very low and this makes SRM an inherently fault-tolerant machine. SRM can operate in four quadrants and has the higher constant power speed range as compared to other machines. On the other hand, salient pole structure results in less utilization of the airgap circumference. The magnetic flux density usually concentrates around the salient poles. Owing to a single source of excitation, a small airgap is required to increase the power density. These challenges result in higher radial forces and cause noise and vibration in SRM.

SRMs operate based on the principle that the magnetic flux tends to flow through the magnetic path with the lowest reluctance. When the concentrated coils around the salient stator poles are energized, the magnetic flux closes its path over the salient rotor poles. This generates torque on the rotor pole and pulls it toward the stator pole for alignment. The reluctance at the aligned position is minimum since the magnetic flux travels mostly through the stator and rotor laminations. When the rotor pole moves from an unaligned to an aligned position, the airgap length and reluctance

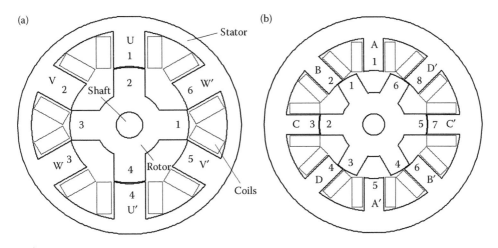

FIGURE 5.71  Cross-section view of SRM: (a) three-phase 6/4 SRM at an aligned position and (b) four-phase 8/6 SRM at an unaligned position.

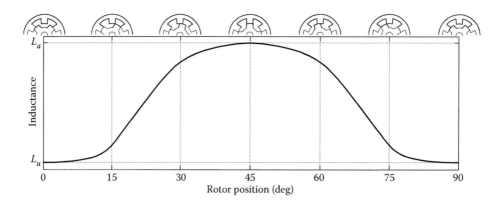

**FIGURE 5.72** Variation of inductance profile in SRM with rotor position.

decreases and, hence, the inductance of the magnetic circuit increases as shown in Figure 5.72. When the rotor pole moves away from the stator pole (from aligned to unaligned position), since the airgap length increases, inductance decreases.

In SRM, aligned position is a stable equilibrium point, where the reluctance is minimum. If a force is applied to move the rotor pole away from the aligned position, the stator flux tends to reduce the reluctance in the magnetic circuit by pulling the rotor pole back and keeping it at the aligned position. Therefore, in the motoring mode of operation, the torque is generated by applying current to the phases sequentially, when the rotor pole moves from unaligned to aligned position for each phase. This corresponds to the increasing inductance profile in Figure 5.72. If a phase is still energized after the alignment, then negative torque will be generated to stop the motion and keep the rotor pole in alignment.

SRM operates in the generating mode, when the inductance profile has the negative slope. This corresponds to the case in Figure 5.72, when the rotor pole moves from an aligned position toward the unaligned position. If the rotor pole is at the aligned position when the phase is energized, an attempt from a prime mover to move the rotor pole away from the alignment will result in an opposing electromagnetic torque (stator pole tends to keep the rotor pole in alignment) and the stored magnetic energy increases. When the rotor pole reaches an unaligned position, all the mechanical energy is converted into magnetic energy and supplied back to the source.

Since the phase coils are isolated from each other, it can be assumed that the flux generated by one phase coil does not link with other phase coils. Therefore, the mutual inductance can be ignored and the phase-equivalent circuit can be represented as shown in Figure 5.73.

The voltage equation for a single phase can be represented as

$$v_s = R_s i_s + \frac{d\psi}{dt} = R_s i_s + \frac{d\psi}{d\theta}\frac{d\theta}{dt} = R_s i_s + \frac{d\psi}{d\theta}\omega_m \tag{5.74}$$

where $v_s$ is the terminal voltage, $i_s$ is the phase current, $\psi$ is the flux linkage, $R_s$ is the phase resistance, $\theta$ is the rotor position, and $\omega_m$ is the angular speed. When operating in the linear region of

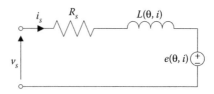

**FIGURE 5.73** Single-phase equivalent circuit of SRM.

the magnetization curve, inductance changes with the rotor position, but it does not change with current:

$$v_s = R_s i_s + \frac{d\psi}{d\theta}\omega_m = R_s i_s + \frac{d(L(\theta)i_s)}{d\theta}\omega_m = R_s i_s + L(\theta)\frac{di_s}{dt} + \omega_m i_s \frac{dL(\theta)}{d\theta} \tag{5.75}$$

where $L(\theta)$ is the phase inductance, which varies with rotor position. The expression in Equation 5.75 represents the equivalent circuit in Figure 5.73 and the right-most term is the back EMF. The instantaneous input power can be calculated as

$$v_s i_s = R_s i_s^2 + L(\theta)i_s \frac{di}{dt} + \omega_m i_s^2 \frac{dL(\theta)}{d\theta} \tag{5.76}$$

If the inductance did not change with rotor position, the term in the middle would represent the rate of change of field energy, as it is the case in a brushed-DC machine:

$$\frac{d}{dt}\left(\frac{1}{2}Li_s^2\right) = Li_s\frac{di_s}{dt} \tag{5.77}$$

However, in SRM, the rate of change of field energy ends up in a different form due to the salient pole construction and single source of excitation:

$$\frac{d}{dt}\left(\frac{1}{2}L(\theta)i_s^2\right) = L(\theta)i_s\frac{di_s}{dt} + \frac{1}{2}i_s^2\frac{dL(\theta)}{dt} = L(\theta)i_s\frac{di_s}{dt} + \frac{1}{2}i_s^2\frac{dL(\theta)}{d\theta}\frac{d\theta}{dt}$$

$$= L(\theta)i_s\frac{di_s}{dt} + \frac{1}{2}i_s^2\frac{dL(\theta)}{d\theta}\omega_m \tag{5.78}$$

The second term in Equation 5.78 shows that the applied mmf in SRM is responsible for both building up of the magnetic field and the torque production. Combining Equations 5.76 and 5.78 results in:

$$v_s i_s = R_s i_s^2 + \frac{d}{dt}\left(\frac{1}{2}L(\theta)i_s^2\right) + \frac{1}{2}\omega_m i_s^2\frac{dL(\theta)}{d\theta} \tag{5.79}$$

The last term represents the airgap power and the instantaneous electromagnetic torque can be calculated as

$$\tau_e = \frac{\frac{1}{2}\omega_m i_s^2\frac{dL(\theta)}{d\theta}}{\omega_m} = \frac{1}{2}i_s^2\frac{dL(\theta)}{d\theta} \tag{5.80}$$

The torque expression in Equation 5.80 is derived considering the linear operation of SRM and has some limitations. In the linear-operating range, the phase inductance is not affected by current; therefore, the flux linkage curves can be represented as a linear function of current as shown in Figure 5.74, where the slope of the lines are the aligned, $L_a$ and unaligned inductances, $L_u$, respectively.

When the rotor is at the unaligned position, if the phase current increases from zero to $i_a$, the flux linkage increases linearly and follows the line OA. In this case, the energy supplied to the magnetic circuit will be

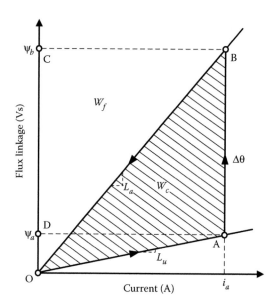

FIGURE 5.74   Flux linkage versus current waveform for SRM operating in the linear region.

$$\text{area}(OAD) = \frac{1}{2}\psi_a i_a = \frac{1}{2}L_u i_a^2 \tag{5.81}$$

When the rotor poles move from an unaligned to an aligned position, the flux linkage changes from $\psi_a$ to $\psi_b$. The electromagnetic energy represented by the shaded area OAB is expressed as the co-energy. It is converted into mechanical energy and validates the torque expression given in Equation 5.80:

$$W_c = \text{area}(OAB) = \frac{1}{2}i_a\psi_b - \frac{1}{2}i_a\psi_a = \frac{1}{2}i_a(\psi_b - \psi_a) = \frac{1}{2}i_a^2(L_a - L_u)$$

$$= \frac{1}{2}i_a^2(L_a - L_u) = \frac{1}{2}i_a^2\Delta L = \tau_e\Delta\theta \Rightarrow \tau_e = \frac{1}{2}i_a^2\frac{\Delta L}{\Delta\theta} \tag{5.82}$$

When the rotor poles move by $\Delta\theta$, the current changes from zero to $i_a$ and the flux linkage changes from $\psi_a$ to $\psi_b$. Therefore, the energy absorbed in the circuit is represented by the area ABCD. This represents the energy stored by the back EMF given in Equation 5.75:

$$\text{area}(ABCD) = i_a(\psi_b - \psi_a) = i_a^2\Delta L = ei_a\Delta t$$

$$\Rightarrow e = i_a\frac{\Delta L}{\Delta t} = i_a\frac{\Delta L}{\Delta\theta}\frac{\Delta\theta}{\Delta t} = i_a\frac{\Delta L}{\Delta\theta}\omega_m \tag{5.83}$$

The total energy supplied to the magnetic circuit is the sum of stored and co-energy:

$$W_t = W_f + W_c = \text{area}(OAD) + \text{area}(ABCD) = \frac{1}{2}L_u i_a^2 + i_a^2\Delta L \tag{5.84}$$

Since the co-energy given in Equation 5.82 is the one converted into electromagnetic torque, the energy conversion ratio can be calculated as

$$K = \frac{W_c}{W_t} = \frac{\frac{1}{2}i_a^2 \Delta L}{\frac{1}{2}i_a^2 L_u + i_a^2 \Delta L} = \frac{\frac{1}{2}(L_a - L_u)}{L_a - \frac{1}{2}L_u} \qquad (5.85)$$

If the saliency ratio is defined as

$$\lambda = \frac{L_a}{L_u} \qquad (5.86)$$

The energy conversion ratio can be derived as

$$K = \frac{\frac{1}{2}(L_a - L_u)}{L_a - \frac{1}{2}L_u} = \frac{\frac{1}{2}(\lambda L_u - L_u)}{\lambda L_u - \frac{1}{2}L_u} = \frac{\lambda - 1}{2\lambda - 1} \qquad (5.87)$$

Equation 5.87 has important conclusions about the operation of SRM in the linear region:

- Since $\lambda > 1$, less than half of the total magnetic energy is converted into mechanical work.
- The rest of the total energy is stored in the magnetic circuit and at the end of the stroke, it is supplied back to the source or dissipated inside the motor.
- Operation in the linear region causes poor use of the converter: the converter is sized for the total energy, but the motor delivers less than half of it.

In practice, SRMs are designed to operate in the nonlinear region, where the co-energy increases for the given total energy. As it can be observed from Figure 5.75, for the same displacement of the rotor pole, saturation limits the maximum flux linkage. Therefore, in the saturated region, the change of peak value of the back EMF with current is limited, whereas peak torque increases, due to higher ratio of the co-energy. This leads to a higher energy conversion ratio, higher power factor, and also better utilization of the power converter.

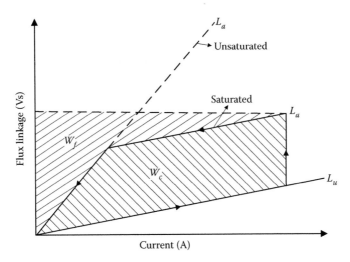

FIGURE 5.75 Flux linkage versus current waveform for SRM operating in the nonlinear region.

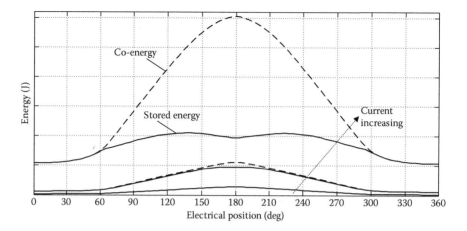

**FIGURE 5.76**  Stored and co-energy waveforms for one stroke with constant current excitation.

Stored and co-energy can be mathematically expressed as

$$W_f = \int i\,d\lambda$$
$$W_c = \int \lambda\,di \tag{5.88}$$

As shown in Figure 5.76, when SRM operates in the linear region, stored and co-energy wave-forms are similar. When the excitation current increases, the motor operates in the nonlinear region. This makes $W_c > W_f$ and, hence, the energy waveforms deviate significantly, especially when the rotor pole is closer to the aligned position. Figure 5.77 shows how to determine the electromagnetic torque from flux linkage–current characteristics. When the rotor pole moves by $\Delta\theta$, the mechanical work that must be done is the area covered by the OAB and hence the co-energy. For constant current excitation, torque can be defined as

$$T_e = \left.\frac{\partial W_c}{\partial \theta}\right|_{i=const} \tag{5.89}$$

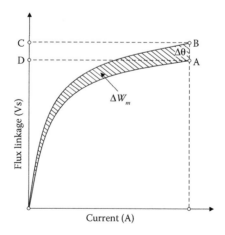

**FIGURE 5.77**  Derivation of instantaneous torque for constant current excitation.

As shown in Figure 5.76, the magnetization curve is a straight line when the motor operates in the linear region. In this case, since the stored and co-energy values are the same, the torque in Equation 5.89 can be calculated as in Equation 5.82.

One of the major challenges in SRM is the acoustic noise and vibration. In electric machines, the electromagnetic forces are produced with the interaction of normal and tangential components in the airgap. Unlike PM or induction machines, the magnetic flux density waveform in a radial flux SRM is not sinusoidally distributed around the airgap, but it is concentrated around the salient poles, usually with a higher peak value. This results in strong radial forces and they create opposing forces on the stator, leading to acoustic noise and vibration.

## 5.7 SPECIFICATIONS FOR TRACTION MOTORS

### 5.7.1 ROLE OF HIGH-VOLTAGE BATTERY AND INVERTER CONTROL METHOD

A typical traction motor drive system is composed of a high-voltage battery, three-phase inverter, three-phase motor, cooling pump, heat exchanger, and motor controller as shown in Figure 5.78. The high-voltage DC battery is highly dependent on temperature; hence, in motor design, the best-and worst-case battery voltage is to be taken into account. High-voltage DC power from the battery is converted into AC power by three-phase inverter. There are many ways to perform this conversion as shown in Figure 5.79. The choice of the inverter control method has a big impact on the motor output torque and efficiency. The choice of the inverter control strategy shows that the same DC bus voltage can produce 27% difference in output torque/power.

### 5.7.2 CHOICE OF MAGNET, EFFECT OF TEMPERATURE, AND DEMAGNETIZATION

Most traction motors are PM machines that have NdFeB PMs in the rotor. These magnets are very sensitive to temperature as discussed in Section 5.5. For example, for a change in magnet temperature from 20°C to 160°C, there can be a 46% drop in the output torque.

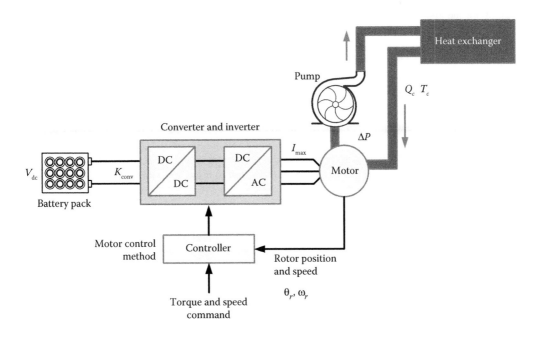

**FIGURE 5.78** Electrical motor drive system.

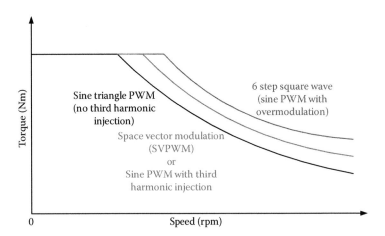

FIGURE 5.79  Motor torque output for various inverter control strategies.

The other point that is directly linked to the rotor temperature and maximum stator current is demagnetization. In the design specification, if the motor is overdesigned for demagnetization condition, then it implies that the magnet used is having high coercivity/thickness and hence a large increase in the motor cost. On the other hand, if the motor is under designed for demagnetization conditions, then during worst-case operating conditions, there is a possibility of magnets getting completely demagnetized, resulting in permanent loss of motor performance.

The stator temperature has a direct impact on the winding resistance and losses. The general equation that relates the resistance chance in conductors as a function of temperature is given in Equation 5.32. PM machines usually have an operating temperature of 100°C and the associated resistance change can be a very significant factor for copper loss and efficiency.

### 5.7.3  COGGING TORQUE AND TORQUE RIPPLE

Cogging torque and torque ripple are important performance indices of noise and vibration as well as smoothness in rotation of the rotor. Torque ripple specified in one application might give better or worse results for another one. This is due to the fact that there is dynamic reduction or amplification in the torque ripple in the driveline and to a large extent depends on the location of the motor in the transmission. Torque ripple data measured at frequencies below resonance are realistic values. Torque ripple data measured at resonant frequencies are often amplified to a very large level. Torque ripple measured at frequencies above resonance is damped out. The same calibration (amplification/reduction) should be applied when writing torque ripple specification. Specification calling for very low torque ripple will require stator or rotor skewing or some other design considerations for torque ripple reduction, which leads to a reduction in performance and efficiency, and increased system cost. On the other hand, having a higher value of torque ripple in the specification can result in noise and vibration issues.

### 5.7.4  MECHANICAL VERSUS ELECTRICAL PERFORMANCE

Structural and electrical performance for a traction motor is inversely related. Hence, it is very important not to over-or underdesign for mechanical performance. Having minimum bridge thickness ensures as shown in Figure 5.34 that leakage flux is reduced, requires less magnet to get the same performance, but can result in a structurally weak rotor (see Figure 5.80). This can lead to failure. On the other hand, if bridge thickness is increased for more structural stability, leakage flux can increase, which leads to the use of more magnets to get the same motor performance. This increases the motor cost.

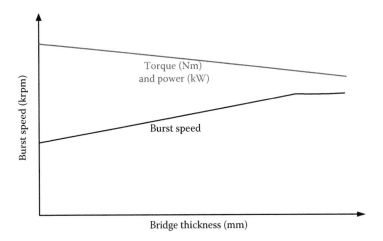

FIGURE 5.80   Mechanical versus electrical performance.

### 5.7.5   MATERIAL AND TRANSMISSION TOLERANCE

Material and transmission tolerances play a vital role in deciding the air gap thickness. Nominal values for air gap for traction motors are between 0.5 and 1 mm. If the tolerances are overspecified, then this can result in designing a motor with larger air gap, which results in increased current/magnet ratio to get the required motor performance. This increases the motor cost. On the other hand, if proper tolerances are not taken into account, then it can result in a motor design with a very small air gap. This can result in motor failure.

Finally, a motor design engineer might be able to design a motor with the lowest motor cost. In this case, the system cost need not be low. On the other hand, if a higher-efficiency motor can be designed with an increased cost to the motor, then this can help with lower battery and cooling cost, which can help to reduce the cost of the system. It is therefore very clear that a system-level approach is needed to write proper motor specifications.

### 5.7.6   DIMENSIONAL DETAILS

In traction applications, the electric motor is usually located in a transmission, which puts a constraint on the maximum motor length and diameter. The motor dimensions along with the maximum RMS current from the inverter, battery voltage, temperature, and cooling conditions, decide the peak and continuous torque, and power requirements of the motor. Usually, the continuous rating is around 50%–60% of the peak rating although this largely depends on how well the motor is cooled.

### 5.7.7   EFFICIENCY REQUIREMENTS AND DC VOLTAGE

The motor has to be efficient in the most frequently used operating range of the vehicle. To meet the miles-per-gallon (mpg) requirements, the peak efficiency target in current generation traction motors is around 96%. Higher battery voltage is usually required if this peak efficiency is desired in high-speed region.

In a motor drive system, maximum DC voltage is defined by the battery, whereas peak RMS current requirement is based on inverter rating. Peak back EMF is specified based on the voltage that the inverter switches can handle, which is also necessary to prevent uncontrolled generation mode fault.

## 5.8 FAULT CONDITIONS IN IPM MACHINES

In a typical traction motor drive system, fault circuitry is included to detect faults and a speed sensor to detect the motor speed. The most common types of motor fault in a hybrid transmission include the open-circuit and short-circuit faults. The motor controller either applies a three-phase short or open circuit to the three-phase inverter when the corresponding fault is detected. Short-circuit fault is applied when the motor speed is less than the transition speed and the fault is detected. Open-circuit fault is applied when the speed of the PM motor is greater than the transition speed and when a fault is detected.

### 5.8.1 UNCONTROLLED GENERATION MODE

A typical IPM machine drive configuration utilizes six switches with antiparallel diodes. The three-phase inverter is used to control the IPM machine to deliver the required torque. For any IPM machine, as the motor speed increases, the machine back EMF increases. A plot of the open-circuit back EMF as a function of speed is shown in Figure 5.81.

The amplitude of the line-to-line back EMF generated in the motor can easily exceed the battery voltage under high-speed operation. As long as the inverter switches are operating in a controlled manner, high-speed operation poses no problem and the machine back EMF is limited to the DC link voltage. If a fault occurs in this high-speed operation, the inverter will be shut down. This means that the gate signals are completely removed from these controlled switches. This can cause a situation where the machine-generated back EMF can be much higher than the DC link voltage. Owing to the presence of antiparallel diodes, the motor phases can conduct current through the diodes and the dc link. This is called uncontrolled generator mode (UCG) fault condition and is represented in Figure 5.82.

### 5.8.2 SHORT-CIRCUIT FAULT

Among the various faults that can occur in a motor, the most dangerous is considered as the three-phase short-circuit fault. In this case, because of the PM flux linkage, the IPM motor works as a brake, limiting the steering movement.

In the synchronous d–q reference frame, the voltages are given by Equation 5.55, where the d-and q-axis flux linkages are defined as

$$\psi_d = \psi_m + L_d i_{sd}$$
$$\psi_q = L_q i_{sq}$$

(5.90)

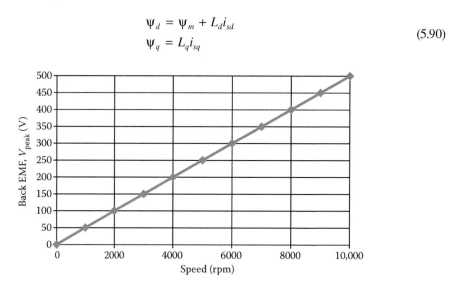

FIGURE 5.81  IPM machine open-circuit back EMF versus speed.

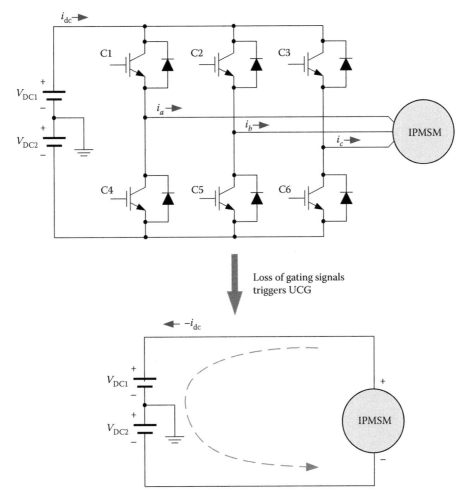

FIGURE 5.82   Circuit operation during UCG mode.

In the case of a three-phase short circuit,

$$v_{sd} = v_{sq} = 0 \tag{5.91}$$

In steady-state short-circuit condition, the currents move from the initial value $(I_{d0}, I_{q0})$ toward the steady-state short-circuit value, which is computed from Equations 5.55 and 5.91. Considering the time derivatives equal to zero, Equation 5.55 becomes

$$0 = R_s I_{d,sc} - \omega\psi_q$$
$$0 = R_s I_{q,sc} - \omega\psi_d \tag{5.92}$$

Substituting Equation 5.90 in Equation 5.92, we get

$$0 = R_s I_{d,sc} - \omega L_q I_{q,sc} \Rightarrow I_{d,sc} = \frac{\omega L_q I_{q,sc}}{R_s}$$

$$0 = R_s I_{q,sc} - \omega(\psi_m + L_d I_{d,sc}) \Rightarrow I_{q,sc} = \frac{\omega(\psi_m + L_d I_{d,sc})}{R_s} \tag{5.93}$$

By substituting the equations together in Equation 5.93, d-and q-axis short-circuit currents can be derived:

$$I_{q,sc} = \frac{\omega(\psi_m + L_d(\omega L_q I_{q,sc}/R_s))}{R_s} = \frac{\omega R_s \psi_m}{(R_s^2 - \omega^2 L_d L_q)}$$

$$I_{d,sc} = \frac{\omega L_q}{R_s}\left(\frac{\omega R_s \psi_m}{R_s^2 - \omega^2 L_d L_q}\right) = \frac{\omega^2 \psi_m L_q}{(R_s^2 - \omega^2 L_d L_q)}$$

(5.94)

The stator short-circuit current can be calculated as

$$I_{sc} = \sqrt{I_{d,sc}^2 + I_{q,sc}^2}$$

(5.95)

Substituting Equation 5.94 in Equation 5.95, we get

$$I_{sc} = \sqrt{\left(\frac{\omega^2 \psi_m L_q}{(R_s^2 - \omega^2 L_d L_q)}\right)^2 + \left(\frac{\omega R_s \psi_m}{(R_s^2 - \omega^2 L_d L_q)}\right)^2}$$

$$\Rightarrow I_{sc} = \sqrt{\left(\frac{\omega^2 \psi_m L_q + \omega R_s \psi_m}{(R_s^2 - \omega^2 L_d L_q)}\right)^2}$$

(5.96)

When the value of $\omega$ becomes very high, $\omega^2 L_d L_q \gg R_s^2$:

$$I_{sc} = \sqrt{\frac{(\omega^2 \psi_m L_q)^2 + (\omega R_s \psi_m)^2}{(-\omega^2 L_d L_q)^2}}$$

$$\Rightarrow I_{sc} = \sqrt{\frac{\psi_m^2}{L_d^2} + \frac{(R_s \psi_m)^2}{(\omega L_d L_q)^2}}$$

(5.97)

When $\omega \xrightarrow{\text{yields}} \infty$, we get

$$\frac{(R_s \psi_m)^2}{(\omega L_d L_q)^2} = 0$$

(5.98)

Therefore, the short-circuit current at very high speed can be calculated as

$$I_{sc,\omega \xrightarrow{\text{yield}} \alpha} = \frac{\psi_m}{L_d}$$

(5.99)

Figure 5.83 shows the motor torque response for short-circuit and UCG fault, and Figure 5.84 shows the motor current response for short-circuit and UCG fault current.

## 5.9 TESTING OF ELECTRIC MACHINES

Electric motors designed for traction applications are expected to run in harsh operating conditions with an extended lifetime. When assembled in the vehicle, traction motors interact with many

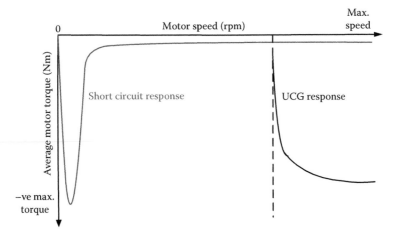

FIGURE 5.83    Motor torque response (short-circuit and UCG fault conditions).

other mechanical components including transmission, shaft, and transaxles and, in case of hybrid electric vehicles, internal combustion engine. Validation of the motor characteristics and performance might be difficult during the complete vehicle testing. For this reason, dynamometer setups are used. Figure 5.85 shows a typical dynamometer setup, where the traction machine under test is mechanically coupled with the dyno machine. By controlling the traction machine inverter and the driver for the dyno machine, and also by measuring many different parameters such as current, voltage, and temperature; various characteristics of the traction machine can be tested, including torque, speed, power, temperature rise, and so on.

   The purpose of the testing is to validate the performance that the customer will experience when driving the vehicle under various drive cycles and operating conditions. Electric traction motors are designed for these specifications and, therefore, the input voltage, ambient temperature, and cooling conditions for the motor should be adequately simulated in the test setup. The experimental testing of the traction machine provides validation on certain characteristics of the motor, which signify the

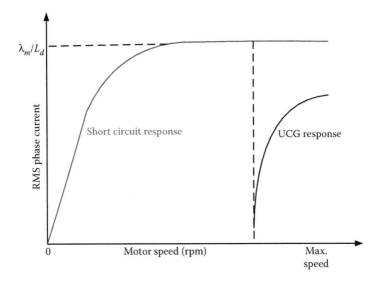

FIGURE 5.84    Motor phase current response (short-circuit and UCG fault conditions).

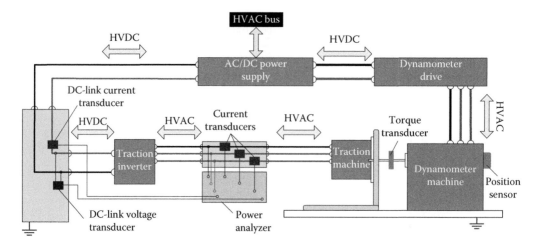

FIGURE 5.85  Typical dynamometer setup for traction motor testing. (Adapted from P. Savagian, S. E. Schulz, and S. Hiti, Method and system for testing electric motors, U.S. Patent 2011/0025447, February 3, 2011.)

operational limits for the specified thermal class. As an example, the maximum temperature inside the machine is limited to 130°C for B-class insulation, whereas for F-and H-class insulations, it is 155°C and 180°C arbitrarily.

When testing the continuous rating, the test results should validate the operation of the traction motor for indefinite time where the maximum temperature is below the limit of the insulation class as shown in Figure 5.86a. In an overload condition, the maximum temperature should be still lower than the maximum temperature of the insulation class, for a certain period of time as shown in Figure 5.86b. This time interval is defined as the overload time to deliver the peak torque. In addition, the dynamic testing also validates some other characteristics including maximum power, maximum speed, and power, and torque density.

During the testing process, the parameters of traction motors are also measured experimentally to validate the design. In PMSMs, static tests are applied to measure the d–q axis inductance, back EMF, and torque profiles [22]. For induction machines, equivalent circuit parameters are identified using open-and short-circuit test [23]. For SRMs, usually, the inductance profile is verified based on either the static characterization of flux linkage or torque [24].

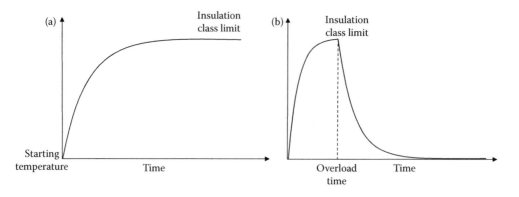

FIGURE 5.86  Temperature rise during continuous and overload conditions. (a) Continuous operation and (b) overload condition.

## 5.10   OTHER MACHINE CONFIGURATIONS

Electric machines operate based on the principles of electromagnetics discussed in Section 5.2 and, based on these principles; various machine configurations can be designed to generate torque. PMSM, induction machine, and SRMs are the most commonly used configurations for traction applications and their operational principles have been discussed in detail in Section 5.6. Here, some other machine topologies will be discussed briefly, emphasizing their advantages and challenges.

### 5.10.1   Synchronous Reluctance Machine

Similar to the SRM, the torque production in synchronous reluctance machine (SyncRM) is based on the varying reluctance profile in the airgap. Synchronous reluctance motor has no magnets or windings on its rotor. As shown in Figure 5.87, the air barriers create the difference between q-and d-axis inductances.

Unlike SRM, SyncRM does not have salient poles. The mmf is created by the distributed windings on the stator. This is an advantage of SyncRM since a similar stator structure and inverter can be used as in induction and PM machines, where the machines are driven with a sinusoidal current excitation [25]. In ideal conditions, this creates a sinusoidal magnetic flux density distribution in the airgap. Owing to its salient poles, SRM is driven with a pulsed current waveform. Therefore, the flux density is concentrated around the poles; resulting in a higher torque density for SRM.

Similar to SRM, since SyncRM also has a single source of excitation, the performance of torque production is still dependent on the nonlinear magnetic properties of the core material, the geometry of the rotor, and, hence, the design of the machine [26]. To increase the efficiency, achieve a higher power factor, and maintain a wider constant power speed range, PMs can be inserted between rotor laminations of SyncRM to assist the torque production. These types of machines are named as PM-assisted SyncRMs and the amount of PMs used is usually smaller than that of IPM machines [27].

### 5.10.2   Transverse Flux PM Machine

As compared to radial flux machines, transverse flux machines have shorter flux paths that result in higher magnetic flux density in the airgap and, hence, higher electromagnetic force [28]. As shown

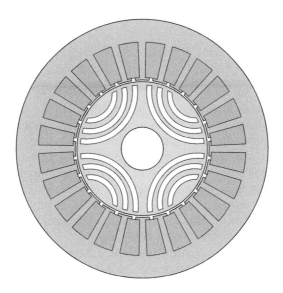

FIGURE 5.87   Synchronous reluctance motor.

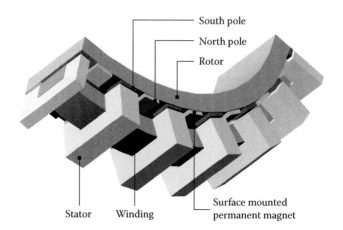

South pole

North pole

Rotor

Stator    Winding

Surface mounted
permanent magnet

**FIGURE 5.88**   Structure of a transverse flux PM machine (single phase).

in Figure 5.88, transverse flux machine is made of different stator cores that enable 3D flux pattern. Each phase winding consists of a solenoidal coil; thus, they are electrically isolated from each other [29].

Transverse flux machines are usually designed with high number of poles. Combined with the high flux density in the airgap, this enables high torque density at low speeds and makes it a possible option for direct drive applications, such as in-wheel drives [30]. On the other hand, transverse flux machine suffers from high leakage flux that reduces the power factor at high-specific torque. Low-power factor usually reduces the power density of the inverter due to the higher volt-ampere requirements. In addition, large number of poles causes a limitation in the high-speed operation of the motor, unless the switching frequency of the inverter can be increased significantly. This would come with an expense of lower inverter efficiency. If the switching frequency of the inverter cannot be increased, the torque ripples of transverse flux machine might be higher during high-speed operation [31].

Owing to its complicated 3D structure, transverse flux machines require 3D FEA approach for design and optimization, which might take significant amount of time. The high implementation cost is also a significant disadvantage in the common use of transverse flux machine in electrified drivetrain applications.

### 5.10.3   Axial Flux PM Machine

As stated by Lorentz's force law given in Equation 5.13, electromagnetic force in electric machines is generated when the magnetic field and the conductor distribution are oriented in different axis. In radial flux machines, which are the most common type in traction applications so far, the magnetic field is oriented in radial direction and the current is in the axial direction. In axial flux machines, the magnetic field is in the axial direction and the conductors are in the radial direction. This can provide a better utilization of space, higher specific power, and improved efficiency.

Figure 5.89 shows the structure of an axial flux PM machine with two rotors and one stator. Axial flux machines provide short axial length and this enables their use in high-torque direct-drive applications, such as in-wheel motor drives [32]. Axial flux machines can be configured by multiple layers of stator–rotor combinations for higher torque output.

The assembly of axial flux motors is rather complicated due to the nature of axial forces. In addition, due to the axial flux orientation, the magnetic core of an axial flux machine cannot be manufactured similar to radial flux machines. Especially for the stator core, spirally wound armature core is used. This is a slow, difficult, and expensive process [33].

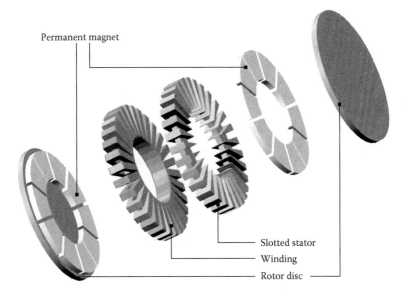

FIGURE 5.89 Axial flux PM machine.

## PROBLEMS

5.1 *Fundamentals of Electromagnetics* The magnetic circuit shown in Figure 5.90a is made up of a linear magnetic material with a relative permeability of 2500. The dimensional parameters for the core are as follows: $g = 0.6$ mm, $w = 18$ mm, $d = 10$ mm, and the

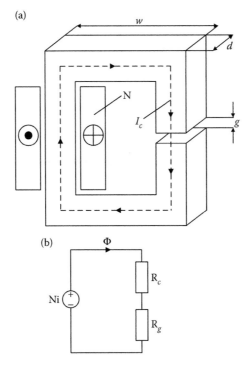

FIGURE 5.90 (a) Magnetic circuit and (b) equivalent circuit.

length of the flux path $l_c = 110$ mm. The coil has 10 turns. If a current of 20 A is applied to the coil, please calculate the magnetic flux density.

5.2 *Losses in Electric Machines and Core Materials* With a sinusoidal magnetic field in the airgap, the magnetization of the core material follows the hysteresis curve as shown in Figure 5.91. The area inside the loop is called the hysteresis loss and it is dependent on the peak value of the magnetic flux density and the frequency.

As described in Section 5.3, due to the nonideal conditions in the machines, such as slot harmonics, space harmonics, and so on, a sinusoidal airgap flux density cannot be usually achieved. The harmonics might distort the airgap flux density waveform as shown in Figure 5.91. This effect is usually called as a minor loop. How do you think minor loops affect the hysteresis curve? What is the frequency and peak flux density of this minor loop?

As discussed in Section 5.3, eddy current losses are due to the induced voltages. If the airgap flux density waveform has harmonics, which will rotate faster than the fundamental, how do you think this affects the eddy current losses? How do you think the harmonics in an electric machine can be reduced?

As shown in Figure 5.11, using laminated steel core reduces the total resistance in the magnetic path of the eddy currents leading to lower core losses. Do you think an electric machine design with a smaller lamination thickness would be effective? Considering the manufacturing process of electric steels shown in Figure 5.35, what would be the challenges reducing the thickness of the laminations?

5.3 *Windings* Figure 5.92 shows two different distributed windings with five phases and 80 slots: one with four poles and four slots per-phase-per-pole and the other with eight poles and two slots per-phase-per-pole. As discussed previously in Figure 5.15, with higher number of slots per-phase-per-pole, the ampere–conductor distribution in the airgap will be closer to a sinusoidal waveform. In Figure 5.92, it can be observed that

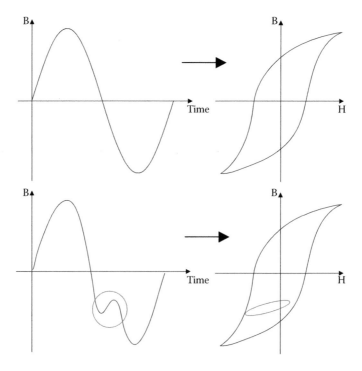

FIGURE 5.91   Magnetic flux density and hysteresis loops for ideal and nonideal conditions.

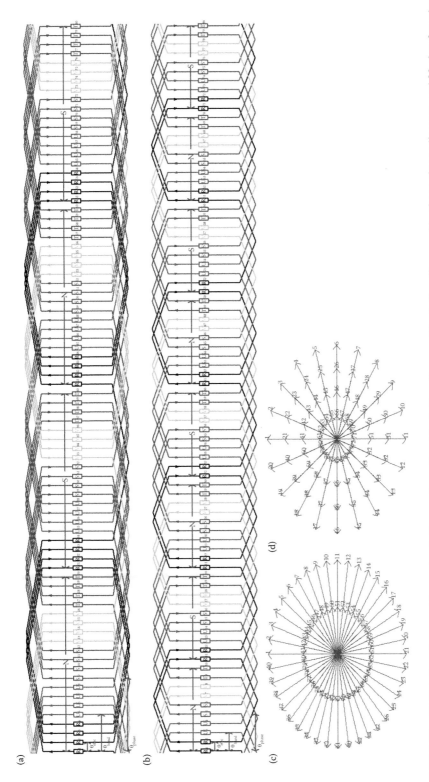

FIGURE 5.92   80-Slot, five-phase, and four-pole winding configuration: (a) winding diagram; (b) winding diagram; (c) voltage phasor diagram, and 80-slot, five-phase, and eight-pole winding configuration: (d) voltage phasor diagram.

with higher number of phases, there are more number of slots with the same current direction under one pole. Therefore, higher number of phases also helps having a more sinusoidal ampere–conductor distribution.

When the winding configurations in Figures 5.15 and 5.92 are compared, what are the challenges in designing and implementing windings with higher number of phases? How do you think the current rating of the winding changes when the number of phases is increased? What are the other challenges with electric machines and motor drives with higher number of phases?

The windings in Figure 5.92 have different number of poles and number of slots per-phase-per-pole. What is the difference between having a four-pole machine and eight-pole machine in terms of torque, speed, and losses? What are the advantages and challenges with the winding design with higher number of slots per-phase-per-pole? For the same number of phases and number of slots, which one is easier to manufacture: two or four slots per-phase-per-pole?

5.4   *Permanent Magnets* Figure 5.93 shows the magnetization curve for a PM at different temperatures. At $T_1$, the remanence of the magnet is 1.2 T. When the temperature increases to $T_2$, the remanence drops to 1.12 T. Initially, the magnet operates at point A. Calculate the PC ($PC_1$) at this point.

When the temperature increases, the operating point shifts to B. What is the magnetic flux density and magnetic field intensity at this point?

When operating at point B, if the PC of the magnetic circuit changes to $PC_2$, what would be the magnetic flux density and field strength at point C? What is the remanent flux density of the new magnetization line? What is the reason for the drop in remanence?

5.5   *Electric Machines* If the base speed of an inverter-driven motor is 2000 rpm at 200 Vdc, what will be the base speed at 400 Vdc? If the peak power of the motor is 50 kW at 200 Vdc, what will be the peak power at 400 Vdc?

Assume there is a motor design (MD1) with a specified current I1 and voltage V1. If the voltage is increased to V2 = 1.5V1, what are the design changes that can be made to optimize the design to get the same performance as MD1? What would the design changes be if the current was increased to I2 = 1.5I1?

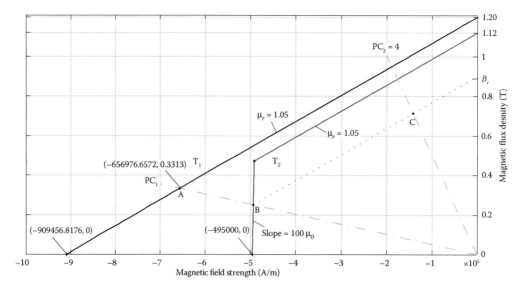

FIGURE 5.93   Typical magnetization curve for a PM and operating points.

If a motor is designed for a certain current and voltage, what happens if the volume of the motor is doubled by increasing the stack length? How are the peak torque and peak power affected?

For a certain motor and slot dimensions, if the number of turns in a motor is doubled, what design changes have to be applied to maintain the same slot fill factor?

In a PM machine, how does magnet remanance affect the back EMF and current for a given motor power? Evaluate the same for winding number of turns.

What are the important losses to be minimized in high-speed region? For a certain design, if the back EMF is reduced in high-speed region, what is the effect on the peak torque and efficiency in the high-speed (constant power) region?

Assume a motor design capable of providing continuous torque of 50 Nm and continuous power of 20 kW at 150°C. If the motor temperature decreases to 75°C, what is the effect on the continuous power and torque?

What is cogging torque? What are the ways to reduce the cogging torque? How does the slot opening affect the cogging torque? What is the effect of skewing on peak torque, average torque, and torque ripple?

## REFERENCES

1. P. Harrop, Electric motors for electric vehicles 2013–2023: Forecasts, Technologies, Players, IDTechEx, 2013.
2. C. Cartensen, *Eddy Currents in Windings of Switched Reluctance Machines*, 1st ed. Aachen, Germany: Shaker Verlag, 2008.
3. K. Yamazaki and A. Abe, Loss investigation of interior permanent-magnet motors considering carrier harmonics and magnet eddy currents, *IEEE Transactions on Industry Applications*, 45, (2), 659–665, March/April 2009.
4. K. Yamazaki, Y. Fukushima, and M. Sato, Loss analysis of permanent magnet motors with concentrated windings—Variation of magnet eddy-current loss due to stator and rotor shapes, *IEEE Transactions on Industry Applications*, 45, (4), 1334–1342, July/August 2009.
5. D. J. Griffiths, *Introduction to Electrodynamics*, Upper Saddle Drive, NJ: Prentice-Hall, Inc., 1999.
6. J. Pyrhonen, T. Jokinen, and V. Hrabovcova, *Design of Rotating Electrical Machines*, West Sussex, United Kingdom: John Wiley & Sons, Ltd., 2008.
7. D. Fleisch, *A Student's Guide to Maxwell's Equations*, Cambridge, UK: Cambridge University Press, 2008.
8. W. G. Hurley and W. H. Wolfle, *Transformers and Inductors for Power Electronics Theory, Design and Applications*, West Sussex, United Kingdom: John Wiley & Sons, 2013.
9. T. A. Lipo, *Introduction to AC Machine Design*, Madison, WI: Wisconsin Power Electronics Center University of Wisconsin, 2011.
10. C. S. Edrington, M. Krishnamurthy, and B. Fahimi, Bipolar switched reluctance machines: A novel solution for automotive applications, *IEEE Transactions on Vehicular Technology*, 54, (3) 795–808, May 2005.
11. M. Krishnamurthy, C. S. Edrington, A. Emadi, P. Asadi, M. Ehsani, and B. Fahimi, Making the case for applications of switched reluctance motor technology in automotive products, *IEEE Transactions on Power Electronics*, 21, (3), 659–675, May 2006.
12. G. J. Li, J. Ojeda, E. Hoang, M. Lecrivain, and M. Gabsi, Comparative studies between classical and mutually coupled switched reluctance motors using thermal-electrodynamic analysis for driving cycles, *IEEE Transactions on Magnetics*, 47, (4), 839–847, April 2011.
13. D. Berry, S. Hawkins, and P. Savagian, Motors for automotive electrification, February 2010. (Online). Available: http://www.sae.org/.
14. K. Rahman, M. Anwar, E. Schulz, E. Kaiser, P. Turnbull, S. Gleason, B. Given, and M. Grimmer, The Voltec 4ET50 Electric Drive System, December 2011. (Online). Available://www.sae.org/.
15. K. Rahman, S. Jurkovic, C. Stancu, J. Morgante, and P. Savagian, Design and performance of electrical propulsion system of extended range electric vehicle (EREV) Chevrolet Voltec, in *Proceedings of Energy Conversion Congress and Exposition*, Raleigh, NC, September 2012, pp. 4152–4159.
16. TDK Corporation, NEOREC series neodymium iron boron magnet datasheet, May, 2011. (Online). Available: http://tdk.co.jp/.

17. S. Constantinides, The demand for rare earth materials in permanent magnets, Arnold Magnetic Technologies. (Online). Retrieved on June 20 2013, http://arnoldmagnetics.com/.

18. S. R. Trout, Material selection of permanent magnets, considering the thermal properties correctly, in *Proceedings of the Electric Manufacturing and Coil Winding Conference*, Cincinnati, OH, October 2001.

19. Hitachi-Metals, Neodymium–iron-boron magnets Neomax, Retrieved on August, 2013. (Online). Available: http://www.hitachi-metals.co.jp/.

20. U.S. Department of Energy, Critical materials strategy, December, 2011. (Online). Available: http://www.energy.gov/.

21. P. Savagian, S. E. Schulz, and S. Hiti, Method and system for testing electric motors, U.S. Patent 2011/0025447, February 3, 2011.

22. J. G. Cintron-Rivera, A. S. Babel, E. E. Montalvo-Ortiz, S. N. Foster, and E. G. Strangas, A simplified characterization method including saturation effects for permanent magnet machines, in *Proceedings of the International Conference on Electric Machines*, Marseille, France, September 2012, pp. 837–843.

23. S. D. Sudhoff, D. C. Aliprantis, B. T. Kuhn, and P. L. Chapman, Experimental characterization procedure for use with an advanced induction machine model, *IEEE Transactions on Energy Conversion*, 18, (1), 48–56, March 2003.

24. J. Zhang and V. Radun, A new method to measure the switched reluctance motor's flux, *IEEE Transactions on Industry Applications*, 42, (5), 1171–1176, September 2006.

25. T. J. E. Miller, A. Hutton, and A. Staton, Design of a synchronous reluctance motor drive, *IEEE Transactions on Industry Applications*, 27, (4), 741–749, July/August 1991.

26. T. A. Lipo, Synchronous reluctance machines—A viable alternative for AC drives, May 1991. (Online). Available: http://lipo.ece.wisc.edu/.

27. A. Vagati, The synchronous reluctance solution: A new alternative in AC drives, in *Proceedings of the IEEE International Conference on Industrial Electronics*, Bologna, Italy, September 1994, pp. 1–13.

28. N. Chaker, I. B. Salah, S. Tounsi, and R. Neji, Design of axial-flux motor for traction application, *Journal of Electromagnetic Analysis and Applications*, 2, 73–83, June 2009.

29. S. C. Oh and A. Emadi, Test and simulation of axial flux—Motor characteristics for hybrid electric vehicles, *IEEE Transactions on Vehicular Technology*, 53, (3), pp. 912–919, May 2004.

30. S. M. Husband and C. G. Hodge, The Rolls-Royce transverse flux motor development, in *Proceedings of the IEEE International Electric Machines and Drives Conference*, Madison, WI, June 2003, pp. 1435–1440.

31. K. Lu, P. O. Rasmussen, and E. Ritchie, Design considerations of permanent magnet transverse flux machines, *IEEE Transactions on Magnetics*, 47, (10), 2804–2807, October 2011.

32. B. E. Hasubek and E. P. Nowicki, Design limitations of reduced magnet material passive rotor transverse flux motors investigated using 3D finite element analysis, in *Proceedings of the IEEE Canadian Conference on Electrical and Computer Engineering*, Halifax, Nova Scotia, May 2000, pp. 365–369.

33. M. Gaertner, P. Seibold, and N. Parspour, Laminated circumferential transverse flux machines—Lamination concept and applicability to electrical vehicles, in *Proceedings of the IEEE International Electric Machines and Drives Conference*, Niagara Falls, ON, May 2011, pp. 831–837.

# 6 Fundamentals of Electric Motor Control

*Nicholas J. Nagel*

## CONTENTS

## 6.1   INTRODUCTION

Motor control in electric vehicles covers a wide range of complexity and integration with system resources. It can be as straightforward as producing motor torque (traction force) proportional to a command input (i.e., foot pedal position), or as complex as a complete integration with an engine and traction control computer in hybrid electric vehicles. Regardless of the level of integration with other vehicle system controllers, traction motor control deals with controlling the torque and/or speed in the electric machine.

Additionally, there are ancillary systems that have been historically powered by the internal combustion engine that are not an option in an all-electric vehicle. These include the power-steering unit and the air-conditioning compressor. These systems are now being powered by electric motors that require torque and/or speed control as well.

While many systems require simple motor controls (i.e., a brush DC motor controlled by a relay), others need tightly regulated torque. Therefore, we will first focus on torque production and control. The logical starting point seems to be with a brush DC motor as this is the least-complicated machine in terms of torque production. Many of the concepts of controlling torque in the brush DC motor can easily be extended to other types of machine. We will then explore control of switched reluctance motors as the torque production and control are not necessarily an extension of the concepts of a brush DC motor. We will complete the chapter by reviewing speed-control techniques for motors.

This chapter assumes that the reader is familiar with basic essentials for electric motor control. These fundamentals include:

- Fundamentals of physics
- Basic control theory
- Fundamentals of electric machines and power electronics

If the reader is not familiar with these topics, references at the end of this chapter will aid in the necessary background.

### 6.1.1   TORQUE PRODUCTION IN BRUSH DC MOTORS

A brush-type DC motor relies on the interaction of armature current and field flux to produce torque. The machine uses commutator bars to maintain the proper orientation of current and flux. Figure 6.1 shows a conceptual diagram of the basic structure of a wound-field DC motor.

Figure 6.2 shows the simplified torque production mechanism in DC motors. It is important to note that torque produced from each conductor is proportional to the sine of the angle between the conductor and the field magneto motive force (MMF). If the angle between the field MMF and armature MMF is maintained at 90°, the machine will produce the maximum torque per ampere (MTPA). The purpose of the commutator is to connect windings with the circuit only when they are near this 90° orientation. This concept of orientating the armature orthogonal to the field MMF is important and is easily extended to AC machines as well. This topic of "field orientation" in AC machines will be discussed later in the chapter.

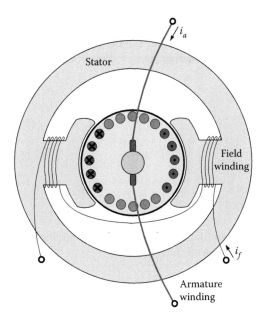

FIGURE 6.1    Conceptual diagram of the basic structure of a wound-field DC machine.

### 6.1.2    DC MOTOR TORQUE CONTROL

To best understand torque control in a DC motor, we will first start with the differential equations that describe it. These equations will then be transformed into the LaPlace domain and finally, a block diagram of the physical system will be formed. From this block diagram, we will explore insightful means of directly controlling torque in the DC motor. The fundamental approach that is used to describe torque control in DC machines will be extended to AC machines further in this chapter. So, it is important that the appropriate level of groundwork is laid down to proceed.

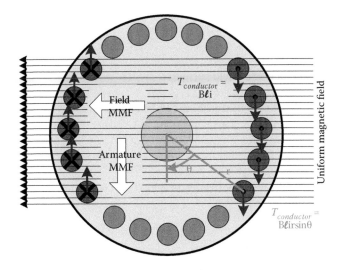

FIGURE 6.2    Simplified torque production mechanism in DC machines.

### 6.1.2.1 Differential Equations of DC Machines

The differential equations of the DC machine can be derived from basic physics. From Faraday's law, the applied voltage in the field winding is

$$v_f = R_f i_f + L_f \frac{di_f}{dt} \tag{6.1}$$

where

$v_f$: Field voltage
$i_f$: Field current
$R_f$: Field winding resistance
$L_f$: Field winding inductance

The armature circuit has a similar equation but has an additional term for the motion of the armature (rotor) conductors through the field flux.

$$v_a = R_a i_a + L_a \frac{di_a}{dt} + e_a \tag{6.2}$$

where

$v_a$: Armature voltage
$i_a$: Armature current
$R_a$: Armature winding resistance
$L_a$: Armature winding inductance
$e_a$: Back electromotive force (EMF)

The back EMF is proportional to the speed of the rotor (how fast armature conductors are moving past the field flux) and the intensity of the field flux. This can be expressed as

$$e_a = K_f i_f \, \omega = K_e \, \omega \tag{6.3}$$

where

$e_a$: Back EMF
$K_f$: Field constant (geometry dependent)
$i_f$: Field current
$\omega$: Rotor speed
$K_e$: Back EMF constant (at a fixed field current)

Finally, the torque production in the machine can be expressed as

$$T_{\mathrm{em}} = K_f i_f i_a = K_t i_a \tag{6.4}$$

where

$T_{\mathrm{em}}$: Electromagnetic torque produced by the motor
$K_f$: Field constant (geometry dependent)
$i_f$: Field current
$i_a$: Armature current
$K_t$: Torque constant (at a fixed field current)

It is important to note that the back EMF constant and the torque constant are physically the same parameter and are a function of field current (field flux) and machine geometry (number of turns,

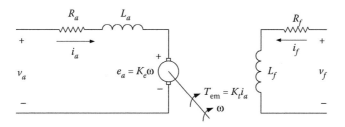

FIGURE 6.3   Equivalent circuit diagram of a wound-field DC machine.

length of wire, rotor radius, etc.). The equivalent circuit diagram of a wound-field DC machine is shown in Figure 6.3.

Equation 6.4 shows that torque production in a DC machine is directly proportional to the field current and the armature current. However, under normal conditions, the machine is operated with a constant field current and the armature current is actively regulated to control the torque. The reason for this is straightforward. The field winding, by design, has a very high inductance. This produces rated flux in the field with relatively low-field currents (and therefore low losses in the field circuit). The armature circuit typically has a significantly lower inductance due to the machine geometry and therefore can have the armature current rapidly changed. A permanent magnet (PM) DC machine does not have direct control over the field circuit and operates in the same way as the wound-field DC machine at a constant field current (constant flux).

In all DC machines, high-performance (high bandwidth) torque control is accomplished by directly regulating the armature current. Regulating field current in a wound-field DC machine does have two advantages though. It allows field weakening (reduction of back EMF constant in Equation 6.3) for higher-than base-speed operation, and it allows the machine to be operated at lower flux levels (which equates to lower magnetic losses) at light loads.

The mechanical equation of motion for the machine is given below. It is simply derived from Newton's second law and states that the acceleration is proportional to the sum of the torques applied to the shaft.

$$J \frac{d\omega}{dt} = T_{em} - b\omega - T_L \tag{6.5}$$

where
   $J$: Rotor inertia
   $b$: Viscous damping
   $T_{em}$: Electromagnetic torque produced by the motor
   $T_L$: Load torque

### 6.1.2.2   LaPlace Representation of DC Machines

The differential equations of the DC machine from the section above are converted into the LaPlace domain by taking the LaPlace transform [1,2]. The equations, assuming zero initial conditions, are defined as follows. The field circuit of Equation 6.1 becomes

$$V_f(s) = R_f I_f(s) + L_f s\, I_f(s) \tag{6.6}$$

where
   $V_f$: Field voltage
   $I_f$: Field current
   $R_f$: Field winding resistance

$L_f$: Field winding inductance

The armature circuit of Equation 6.2 becomes

$$V_a(s) = R_a I_a(s) + L_a s I_a(s) + E_a(s) \tag{6.7}$$

where

$V_a$: Armature voltage
$I_a$: Armature current
$R_a$: Armature winding resistance
$L_a$: Armature winding inductance
$E_a$: Back EMF

Finally, the torque production in the machine of Equation 6.4 becomes

$$J s \omega(s) = T_{em}(s) - b \omega(s) - T_L(s) \tag{6.8}$$

where

$J$: Rotor inertia
$b$: Viscous damping
$T_{em}$: Electromagnetic torque produced by the motor
$T_L$: Load torque

Table 6.1 summarizes the equation of the brush DC machine in both the time domain and LaPlace domain. With the DC machine equations in the LaPlace domain, we can begin to use common control system tools for analysis and ultimately to determine appropriate controls. The initial analysis of the DC machine will assume that the field current is held constant. This same model is used for PM DC machines. Maintaining the DC machine at a constant field current (or assuming a PM DC machine) fixes the back EMF and torque constant of the machine from Equations 6.3 and 6.4.

From Equation 6.7, the armature current is derived from the difference of the applied armature voltage and the back EMF. This is calculated as

$$I_a(s) = \frac{V_a(s) - E_a(s)}{L_a s + R_a} \tag{6.9}$$

From Equation 6.8, the machine speed is derived from the difference of the electromagnetic torque and the load torque. This is calculated as

$$\omega(s) = \frac{T_{em}(s) - T_L(s)}{J s + b} \tag{6.10}$$

---

**TABLE 6.1**

**Time Domain and LaPlace Domain Representation of the Equations of Motion for a Brush DC Motor**

| Equation | Time Domain Representation | LaPlace Domain Representation |
|---|---|---|
| Armature | $v_a(t) = R_a i_a(t) + L_a \dfrac{d i_a(t)}{dt} + e_a(t)$ | $V_a(s) = R_a I a(s) + L_a s I_a(s) + E_a(s)$ |
| Field | $v_f(t) = R_f i_f(t) + L_f \dfrac{d i_f(t)}{dt}$ | $V_f(s) = R_f I_f(s) + L_f s I_f(s)$ |
| Mechanical | $J \dfrac{d\omega(t)}{dt} + b\omega(t) + T_L(t) = T_{em}(t)$ | $J s \omega(s) + b \omega(s) + T_L(s) = T_{em}(s)$ |

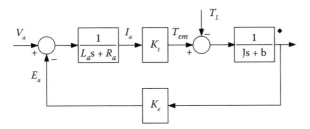

FIGURE 6.4    Block diagram of a DC machine.

The block diagram of a DC machine is shown in Figure 6.4. Note that each of the variables is assumed to be a function of the LaPlace variable "s" and it is therefore omitted in the diagram for simplicity. The left-hand side of the block diagram shows the calculation of the armature current from Equation 6.9. This is a representation of the electrical circuit. The right-hand side of the equation shows the calculation of the machine speed from Equation 6.10. This is a representation of the mechanical circuit. The electrical and mechanical circuits are combined with the electromechanical conversion constants (back EMF and torque) from Equations 6.3 and 6.4.

### 6.1.3   TORQUE CONTROL IN BRUSH DC MOTORS

From the block diagram of a brush DC machine (Figure 6.5) or Equation 6.4, we can see that the electromagnetic torque is linearly proportional to armature current. This condition of course assumes the absence of magnetic saturation in the machine. Given this linear relationship between torque and current, by directly regulating armature current, we directly regulate torque.

The ability to regulate current in the machine directly impacts the ability to control torque. Fast, stable, and high-bandwidth current regulation implies the same is true for torque control. High-performance control of current (torque) offers significant advantages in machine control that will be described later in this chapter. Additionally, most of the concepts of current (torque) regulation in DC machines extend to AC machines as well. High-bandwidth servo motion control loops (which regulate machine speed and/or position) are dependent on high-performance torque regulation.

Directly controlling torque (current) in the motor is, however, not always a requirement. There are many applications in the vehicle that only require modest speed regulation. These applications, such as ancillary pumps, fans, or utility actuation, are cost sensitive. Since servo-grade performance is not a requirement, it does not need the complexity of additional sensors and/or computational complexity of signal-processing algorithms. This will be discussed in the subsequent sections. For now, we will explore high-performance torque (current) control. The next section will discuss current control and it will then be extended to torque control.

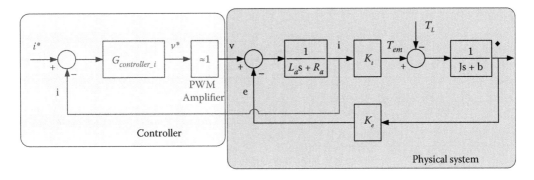

FIGURE 6.5    Block diagram of a DC machine with a current regulator.

### 6.1.3.1  Current Control in Brush DC Motors

Figure 6.5 is the classical block diagram of a brush DC motor from a controls perspective. If the armature current is sensed and fed back to a controller, the block diagram is modified as shown in Figure 6.6. $G_{controller\_i}$ represents the controller transfer function. For now, we will assume that we have an ideal voltage amplifier (i.e., if the controller commands 100 V, the amplifier immediately provides 100 V). In practice, this assumption is not exactly true. Inverter nonlinearities such as voltage drops in devices, inverter deadtime effects, and saturation cause the inverter to deviate from an ideal amplifier. However, these effects can be compensated for (with the exception of saturation), and as an assumption, it is reasonable to ignore these effects for a first-case analysis.

Examination of Figure 6.5 shows the inherent coupling of mechanical and electrical states. Current produces electromagnetic torque in the machine (through $K_t$). The difference between the electromagnetic torque and any load torque accelerates the inertia and causes motion in the machine. This motion (speed) couples back into the electrical side via the back EMF constant ($K_e$). This back EMF opposes the applied voltage used to control current. This coupling of electrical and mechanical states makes analysis and control more complicated.

The transfer function of the current loop (how the current loop responds to a command) can be derived from the block diagram of Figure 6.5 as follows. Substituting in the back EMF in terms of machine speed in Equation 6.9, we get:

$$I_a(s) = \frac{V_a(s) - K_e \omega(s)}{L_a s + R_a} \tag{6.11}$$

Next, substituting in the mechanical state into Equation 6.11, we get:

$$I_a(s) = \frac{1}{L_a s + R_a}\left( V_a(s) - K_e \frac{T_{em}(s)T_L(s)}{J s + b} \right) \tag{6.12}$$

Assuming an ideal voltage regulator (where the applied voltage exactly equals the commanded voltage) and substituting in current times the machine torque constant for the electromagnetic torque term, we get:

$$I_a(s) = \frac{1}{L_a s + R_a}\left( G_{controller\_i}(I_a^*(s) - I_a(s)) - K_e \frac{K_t I_a(s) - T_L(S)}{J s + b} \right) \tag{6.13}$$

Equation 6.13 has current on both sides of the equation. If we solve for current, we get:

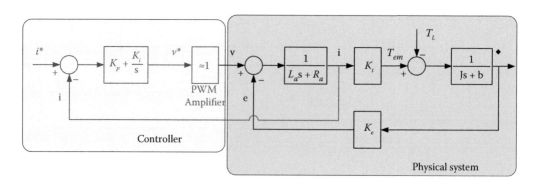

FIGURE 6.6  Block diagram of a DC machine with a PI current regulator.

$$I_a(s) = \frac{(Js + b)G_{controller\_i}}{JL_as^2 + ((G_{controller\_i} + R_a)J + L_ab)s + K_eK_t + (G_{controller\_i} + R_a)b} I_a^*$$

$$(s) + \frac{K_e}{JL_as^2 + ((G_{controller\_i} + R_a)J + L_ab)s + K_eK_t + (G_{controller\_i} + R_a)b} T_L(s)$$

(6.14)

Equation 6.14 shows that the current response is a function of both the commanded current and the load torque on the machine. The load torque affects the velocity that in turn affects the back EMF that in turn couples into the electrical state. This highlights the fact that we want to decouple the electrical and mechanical states so that load torque disturbances do not affect current regulation. This will be discussed below. But there is another very important reason why we want to decouple the electrical and mechanical states in our current controller, and that is for simplicity. This will also be discussed below.

### 6.1.3.2 Standard Proportional Plus Integral Current Control in Brush DC Motors

One of the most pervasive means of current regulation in electric machines is the standard proportional plus integral (PI) control. This is shown in Figure 6.6.

The PI controller has the following structure:

$$G_{controller\_i} = K_p + \frac{K_i}{s}$$

(6.15)

where
$K_p$: Proportional gain
$K_i$: Integral gain

With $G_{controller\_i}$ replaced by the PI controller, Equation 6.14 becomes

$$I_a(s) = \frac{JK_ps^2 + (JK_i + K_pb)s + K_ib}{JL_as^3 + ((K_p + R_a)J + L_ab)s^2 + (JK_i + K_eK_t + (K_p + R_a)b)s + K_ib} I_a^*$$

$$(s) + \frac{K_es}{JL_as^3 + ((K_p + R_a)J + L_ab)s^2 + (JK_i + K_eK_t + (K_p + R_a)b)s + K_ib} T_L(s)$$

(6.16)

Even with a simple PI controller, the transfer function becomes fairly complex and selecting the gains is nontrivial. Recall that Equation 6.16 ignores the dynamics of the voltage regulator as well.

To simplify things, we first confine our analysis to tuning the controller when the rotor is not allowed to move (called locked rotor tuning). This may seem impractical, but we will address this limitation in future sections. Selecting gains in this case is trivial. When the rotor is locked, there is no back EMF that couples into the electrical state. Thus, the locked rotor condition decouples the electrical and mechanical states. The simplified electrical block diagram is shown in Figure 6.7.

The response of the armature current is solely dependent on the commanded current (i.e., no longer depends on load torque). Using the same PI controller, Equation 6.16 reduces to

$$I_a(s) = \frac{K_ps + K_i}{L_as^2 + (K_p + R_a)s + K_i} I_a^*(s)$$

(6.17)

Equation 6.8 is significantly less complicated than Equation 6.7. There is also a significant amount of literature on classical tuning strategies for this case [1]. However, by appropriately forming the transfer function, we can eliminate any tedious techniques and make this a rudimentary problem.

Figure 6.8 shows the simplified block diagram of Figure 6.7 on top and the same diagram with the blocks expressed as time constants on the bottom. By using the classical definition of the electrical

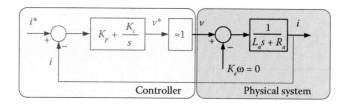

**FIGURE 6.7** Block diagram of a DC machine with a PI current regulator.

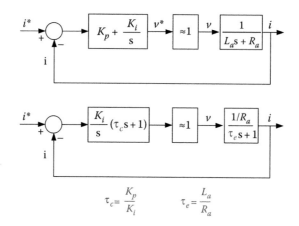

**FIGURE 6.8** Block diagram of a DC machine with a PI current regulator (upper figure) rewritten using electrical and controller time constants (lower figure).

time constant (armature inductance divided by resistance), we can formulate a simple solution. The electrical plant has a pole at the electrical time constant $\tau_e$. Notice that reformulating the PI controller as shown in the lower figure, there is a controller zero at $\tau_c$. The controller zero is simply the ratio of the proportional to integral gains and can be set arbitrarily. By setting the controller zero equal to the electrical pole, there is a pole zero cancelation and the block diagram is further reduced to Figure 6.9. This reduces the block diagram to having an open-loop free integrator and a gain. The closed-loop transfer function for the current loop is simply:

$$\frac{I_a(s)}{I_a^*(s)} = \frac{K_i}{R_a s + K_i} \tag{6.18}$$

### 6.1.3.3 Standard Proportional PI Current Loop Tuning and Response in Brush DC Motors

From Equation 6.18, we see that there is unity steady-state gain (as $s \to 0$). As the commanded frequency increases, the gain rolls off (proportional to $1/s$). The system bandwidth is typically defined

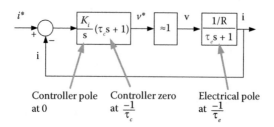

Controller pole    Controller zero    Electrical pole
at 0               at $\frac{-1}{\tau_c}$               at $\frac{-1}{\tau_e}$

**FIGURE 6.9** Block diagram of a DC machine with cancelation of the electrical plant pole by the controller zero.

as the −3 dB crossing of the frequency response magnitude. Solving for the gain that produces −3 dB response, the desired bandwidth is trivial. Equation 6.19 shows the magnitude of the locked rotor current loop transfer function.

$$\left|\frac{I_a(s)}{I_a^*(s)}\right| = \frac{K_i}{\sqrt{(2\pi f\, R_a)^2 + (K_i)^2}} \tag{6.19}$$

The −3 dB point (magnitude ratio of 0.707) occurs when real and imaginary parts are equal in magnitude. The integral gain is solved for as shown in Equation 6.20. Noting that the controller time constant was set equal to the electrical time constant for pole/zero cancelation, the proportional gain is solved for as shown in Equation 6.21.

$$K_i = 2\,\pi f_{bw}\, R_a \tag{6.20}$$

$$K_p = 2\,\pi f_{bw}\, L_a \tag{6.21}$$

where
$f_{bw}$: Desired current loop bandwidth

The response of the system is shown in Figure 6.10 tuned for 1000 Hz. With perfect pole/zero cancelation, the system has a first-order response. The magnitude crosses −3 dB and the phase crosses −45° at the desired 1000 Hz.

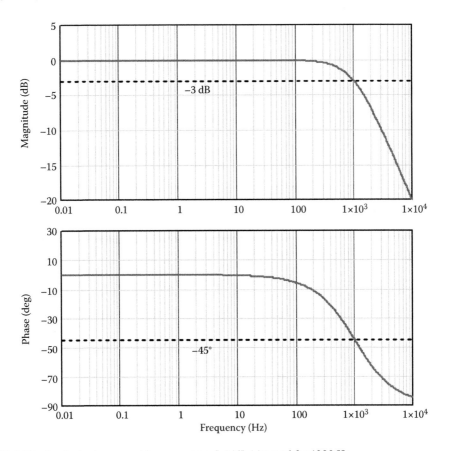

FIGURE 6.10 Locked rotor current loop response $I_a(s)/I_a(s)$ tuned for 1000 Hz.

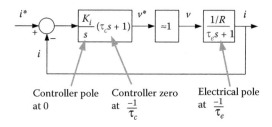

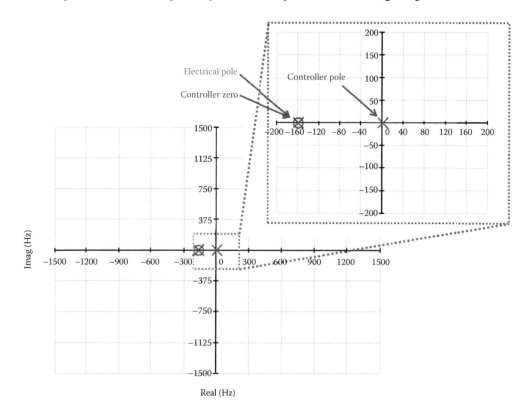

FIGURE 6.11   Block diagram of a DC machine with a PI current regulator highlighting the open-loop system poles and zeros.

Figure 6.11 highlights the open-loop system poles and zeros from the simplified block diagram. Figure 6.12 shows a plot of the poles and zeros on the complex plane. We can see that the controller zero is placed on top of the electrical pole, effectively canceling the dynamics associated with it.

Figure 6.13 shows the root locus plot of the closed-loop system. The controller zero cancels the electrical pole, and the controller pole (beginning at the origin) is moved to the left (until it achieves the desired system bandwidth) by increasing the gain $K_i$. This effectively reduces the system to a simple first order as described by Equation 6.18. The step response of the system is shown in Figure 6.14. The system has a well-behaved first-order response with a rise time of ~1 ms as expected.

### 6.1.3.4   Effects of Mistuned Proportional PI Current Loop Tuning and Response in Brush DC Motors

While it is the goal to place the controller zero of the PI loop on top of (and effectively cancel) the electrical pole, this cannot be perfectly achieved in practice. However, getting the controller zero

FIGURE 6.12   Pole/zero map of locked rotor open-loop system.

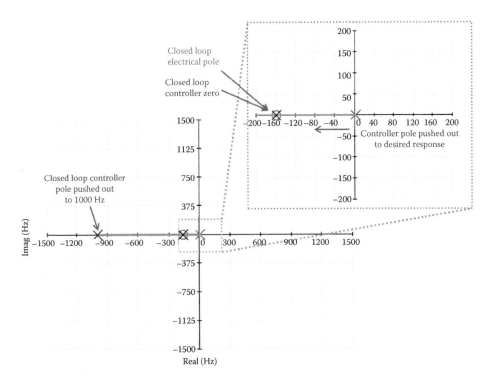

**FIGURE 6.13**   Root locus plot of locked rotor closed-loop system.

close to the electrical pole is usually sufficient for most motor control applications. This section will explore what happens to the system response when the controller zero is greater than or less than the electrical pole.

Figure 6.15 shows the open-loop system poles and zeros when the controller zero is set to 150% of the plant electrical pole. This places the controller zero closer to the origin. Figure 6.16 shows the root locus plot of the system as the gain $K_i$ is increased. Since the controller zero misses to the right, the controller pole migrates toward and ends near the controller zero. The plant electrical pole moves to the left beyond the desired bandwidth of 1000 Hz. The closed-loop system poles end up at −102 and −1557 Hz as shown in Figure 6.16.

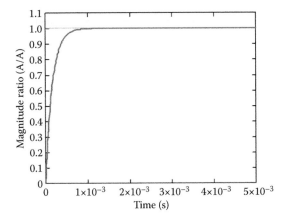

**FIGURE 6.14**   Step response of locked rotor closed-loop system.

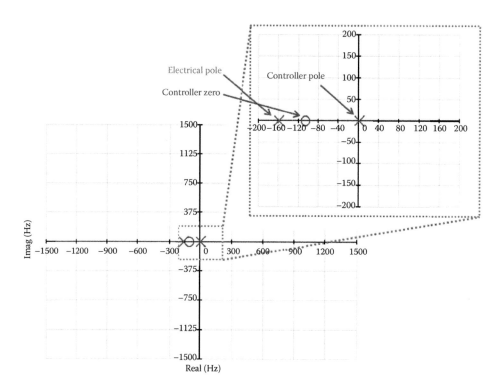

**FIGURE 6.15** Pole/zero map of locked rotor open-loop system with controller zero placed to the right of the plant electrical pole ($\tau_c = 1.5\,\tau_e$).

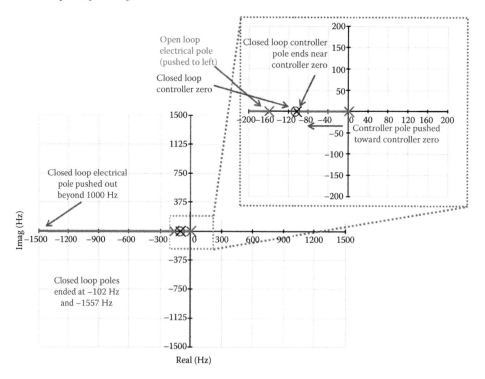

**FIGURE 6.16** Root locus plot of the system with controller zero placed to the right of the plant electrical pole ($\tau_c = 1.5\,\tau_e$).

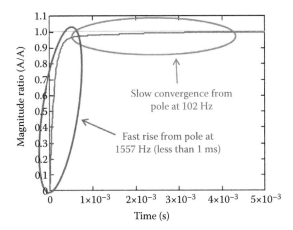

**FIGURE 6.17**    Step response of closed-loop system with controller zero placed to the right of the plant electrical pole ($\tau_c = 1.5\ \tau_e$).

Since there is no longer an exact pole/zero cancelation, the system has closed-loop poles at two frequencies. These two poles give the system response two distinct time constants as can be seen in the step response plot in Figure 6.17. The pole at 102 Hz causes the final system convergence to take approximately 10 ms to reach the steady-state commanded value. The pole at 1557 Hz forces the system to initially converge faster than the original 1 ms (1000 Hz). It is important to note that the response reaches more than 95% of the commanded value in <1 ms which is why this is typically not detrimental for motor control applications. A typical system has outer rate or position loops that are continually modifying the commanded torque value. Achieving 95% or more of the commanded value does not have an impact on these outer loops in all but the most stringent of applications.

Figure 6.18 shows the open-loop system poles and zeros when the controller zero is set to 50% of the plant electrical pole. This places the controller zero further from the origin. Figure 6.19 shows the root locus plot of the system as the gain $K_i$ is increased. Since the controller zero misses to the left, the controller pole and electrical pole migrate toward each other and break off the axis with increasing $K_i$ [1]. These poles migrate toward the controller zero to the left. The closed-loop system poles end up at $-329 \pm j\,225$ Hz as shown in Figure 6.19. The step response of this system is shown in Figure 6.20.

Figure 6.21 shows the frequency response of the locked rotor current loop system. We see that the bandwidth (−3 dB magnitude ratio) is reduced from the desired 1000 Hz requirement. We also observe the slight overshoot (~0.5 dB) in the frequency response function of Figure 6.21. Here again, this response will not affect system performance except in the most stringent motion control applications.

In summary, this section shows that even with significant errors in estimation of the plant electrical pole (errors in estimates of armature resistance and inductance of ±50%), the PI controller is robust. A good deal of attention was paid to the system response in this section as it will be directly applicable in the control of AC machines (induction, PMSM, synchronous reluctance, etc.) as well. The next section will cover a general approach to current regulation with the rotor free to rotate. This is an obvious and important feature to produce useful mechanical energy conversion.

### 6.1.3.5    Back EMF Decoupling of Proportional PI Current Loop in Brush DC Motors

The previous section looked at the current loop response when the back EMF was zero. If the rotor shaft is free to rotate, the back EMF couples the electrical and mechanical states of the machine. One solution to this problem is to decouple the effects of the back EMF by negating it in the controller

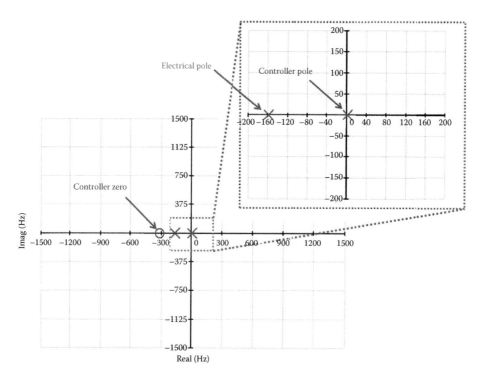

**FIGURE 6.18** Pole/zero map of locked rotor open-loop system with controller zero placed to the left of the plant electrical pole ($\tau_c = 0.5\ \tau_e$).

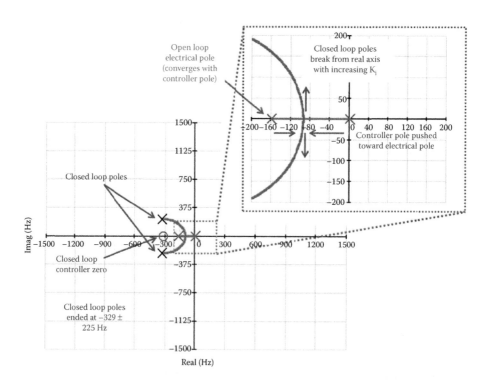

**FIGURE 6.19** Root locus plot of the system with controller zero placed to the left of the plant electrical pole ($\tau_c = 0.5\ \tau_e$).

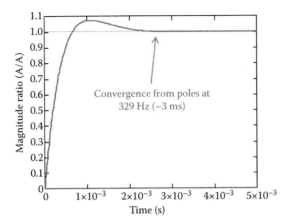

FIGURE 6.20   Step response of the closed-loop system with controller zero placed to the left of the plant electrical pole ($\tau_c = 0.5\,\tau_e$).

[3]. This approach is shown in Figure 6.22. We see that the back EMF term is subtracted from the applied voltage in the physical system. By measuring the rotor speed, we can estimate the back EMF. This is simply the product of the measured speed and an estimate of the back EMF constant (Note: Estimated parameters are expressed with the ^ over the constant). This estimated back EMF is added to the voltage command in an attempt to compensate for the physical back EMF.

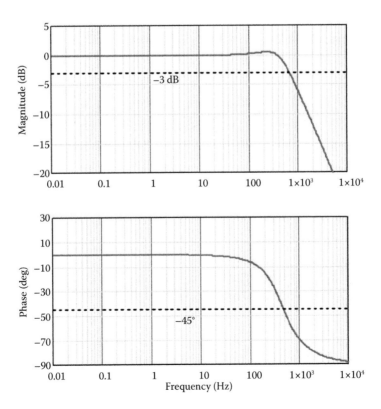

FIGURE 6.21   Locked rotor current loop response ($I_a(s)/I_a^*(s)$) of the system with controller zero placed to the left of the plant electrical pole ($\tau_c = 0.5\,\tau_e$).

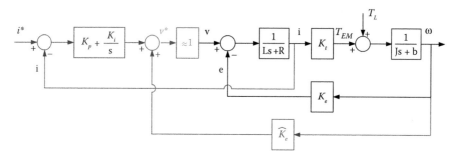

FIGURE 6.22   Back EMF decoupling with PI current loop.

Assuming an ideal voltage regulator, the derivation of the current response is straightforward. Equation 6.16 is modified to include the estimated back EMF term and becomes

$$I_a(s) = \frac{J K_p s^2 + (J K_i + K_p b)s + K_i b}{J L_a s^3 + [(K_p + R_a)J + L_a b]s^2 + [J K_i + (K_e - \hat{K}_e)K_t + (K_p + R_a)b]s + K_i b} I_a^*(s)$$

$$\frac{(K_e - \hat{K}_e)s}{J L_a s^3 + [(K_p + R_a)J + L_a b]s^2 + [J K_i + (K_e - \hat{K}_e)K_t + (K_p + R_a)b]s + K_i b} T_L(s)$$

(6.22)

Figure 6.23 shows the effects of back EMF decoupling on the system response. Without back EMF decoupling, the disturbance from the mechanical system on the electrical system is apparent, beginning below 0.1 Hz. With the addition of back EMF decoupling, the system response is well behaved, even with ±10% error in the estimate of the back EMF constant. The phase response is barely affected in this case.

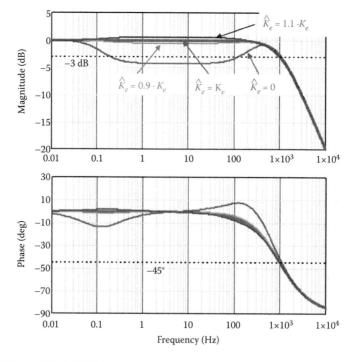

FIGURE 6.23   Effects of back EMF decoupling on system response.

This analysis is only valid when the voltage regulator is assumed to be ideal. Slight errors in voltage amplifier gain affect the system in the same manner as errors in the estimate of the back EMF constant. The main issue is when the voltage amplifier saturates. The faster the motor spins, the higher the back EMF and therefore the larger the commanded voltage. As the motor approaches the no-load speed, the voltage command exceeds the capabilities of the inverter and saturates. It is important to consider that independent of what type of control implemented, there are physical limitations that cannot be exceeded. So, while the reader may think this technique breaks down at high speed, there is no control scheme that does not break down at high speed due to physical limitations.

## 6.2  FUNDAMENTALS OF AC MOTOR CONTROL

The previous sections focused on DC motor control. This section will explore control of AC machines. We will begin this section with the brushless DC (BLDC) machine (trapezoidal back EMF) control, and extend to permanent magnet synchronous machines (PMSMs). PMSMs are often referred to as brushless AC (BLAC) (sinusoidal back EMF) machines and include surface permanent magnet (SPM) and interior permanent magnet (IPM) machines. Control of BLAC machines will then be extended into synchronous reluctance and finally induction machine control. As was the case with the DC machine, we will first explore torque control of these machines through current regulation. The current regulators will be extensions of the DC machine concepts.

BLDC machines are direct extensions of brushed DC machines. The major difference is that the armature circuit is electronically commutated with the inverter bridge as opposed to mechanically commutated with the commutator bars and brushes. We will see there is quite a difference in how BLAC machines are controlled as compared to BLDC machines. The techniques for BLAC machine control are then easily extended to synchronous reluctance and induction machines.

### 6.2.1  FUNDAMENTALS OF BLDC MACHINE TORQUE CONTROL

A typical BLDC machine is constructed inside out of a typical PM brush machine. Instead of the windings on the rotor mechanically commutated and the PMs (field) on the stator, the typical BLDC machine has the PMs on the rotor (still producing the field) and has the armature windings on the stator that are electronically commutated through the inverter bridge. Figure 6.24 shows an equivalent circuit of the BLDC machine and inverter bridge. Unlike a brush DC machine, the BLDC machine cannot simply be connected to a voltage source for operation. The operation is integrally connected to the BLDC controller.

The BLDC machine operates by applying voltage and regulating current in two of the three phases at any one time. The third phase is open circuited during this time. This is shown in Figure 6.25. The machine is commutated between phases every 60 electrical degrees. Each phase is on for 120

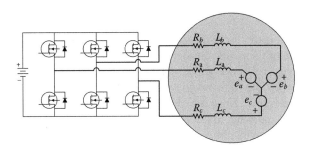

FIGURE 6.24   Equivalent circuit diagram of BLDC with inverter bridge.

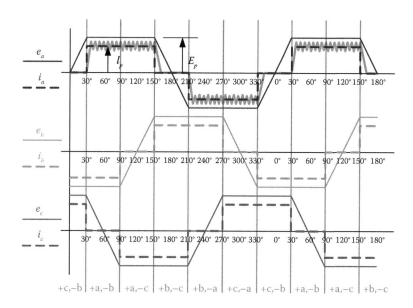

**FIGURE 6.25** Voltage and current waveforms in a BLDC.

electrical degrees. In each 60 electrical degree segment, the machine has the same equivalent circuit as the brush DC machine, namely an equivalent resistance, inductance, and back EMF (which is proportional to speed). Aside from the state machine logic that implements the commutation between phases, current regulation can be achieved as described in the previous sections (although oftentimes, less robust, but more simple cycle-by-cycle current control is used). The commutation state machine logic simply decides which set of power electronic switches to pulse width modulate (PWM) at any given rotor position.

### 6.2.2  FUNDAMENTALS OF BLAC MACHINE TORQUE CONTROL

Unlike BLDC machines, BLAC machines excite all three phases simultaneously. With the proliferation of low-cost, high-performance digital signal processors, the most common approach to BLAC torque (current) control is known as field-oriented control (FOC). FOC transforms quantities (voltages, fluxes, and currents) from their three-phase coordinates (i.e., phase A current) to two-phase orthogonal coordinates in a reference frame that rotates with the rotor. The transformation from three-phase stator variables to two-phase rotor variables is an important step in the control of these machines because it makes the steady-state AC variables appear as DC quantities in the rotor reference frame. Thus, all the analysis of the PI regulator for DC machines is applicable even though the terminal quantities vary sinusoidally.

#### 6.2.2.1  Overview of BLAC Machine Current Control

In BLAC machine control, the three-phase AC currents are measured and converted into two-phase AC currents. Both of them are considered "stationary" frame currents as they are measured in the stator or stationary reference frame. Using the measured rotor flux (rf) position (which rotates synchronously with the rotor magnets), the two-phase currents are transformed into the "synchronous" reference frame. These steady-state quantities appear to be DC terms. These synchronous frame feedback currents are compared to the synchronous frame commands and the error is fed into PI current regulators, one for each of the two phases. The steady-state output is a DC voltage command that gets transformed back into AC by an inverse transformation to the stationary reference frame.

### 6.2.2.2  Definition of Complex Space Vectors in AC Machine Current Control

Field orientation requires mathematical transformations to define quantities in various reference frames. First, the concept of space vectors must be defined. Space vectors, as the name suggests, define an amplitude and an orientation in space. It is important to note that the amplitude is not necessarily a constant but rather a time-varying quantity. The Euler identity listed in Equations 6.23 and 6.24 is an example of a space vector with a constant magnitude of one.

$$e^{j\theta} = \cos(\theta) + j\,\sin(\theta) \tag{6.23}$$

$$e^{-j\theta} = \cos(\theta) - j\,\sin(\theta) \tag{6.24}$$

Space vectors offer a convenient means to describe the spatial location of quantities such as windings in a machine. The spatial location of three-phase windings separated by 120 electrical degrees is easily described with the space vectors $e^{+j0°}$, $e^{+j120°}$, and $e^{-j120°}$. The complex space vector of a machine variable (voltage, current, and flux linkage) is defined in Equation 6.25.

$$\bar{f}_s = \frac{2}{3}(f_{as}e^{+j0°} + f_{bs}e^{+j120°} + f_{cs}e^{-j120°}) \tag{6.25}$$

As an example, the current space vector combines the time-varying current amplitude with the spatial location of the winding. Assuming that the current varies sinusoidally in each of the three phases as given in Equations 6.26 through 6.28, the complex space vector is defined as given in Equation 6.29.

$$i_{as}(t) = I_m \cos(\omega t) \tag{6.26}$$

$$i_{bs}(t) = I_m \cos(\omega t - 120°) \tag{6.27}$$

$$i_{cs}(t) = I_m \cos(\omega t + 120°) \tag{6.28}$$

$$\bar{I}(t) = \frac{2}{3}\left[I_m \cos(\omega t)e^{+j0°} + I_m \cos(\omega t - 120°)e^{+j120°} + I_m \cos(\omega t + 120°)e^{-j120°}\right]$$
$$= I_m \cos(\omega t) + j\,I_m \sin(\omega t) = I_m e^{+j\omega t} \tag{6.29}$$

Equation 6.29 describes a vector of constant amplitude that rotates at an angular rate of the excitation frequency $\omega$. This can be graphically seen in Figure 6.26. The figure represents the vector sum of the amplitudes of each of the phase currents at a given time $t_0$ that corresponds to a spatial location of $\omega t_0$. Also note that the space vector is defined with the constant 2/3 in it. This was chosen so that when summed together, the magnitude of the rotating vector is equivalent to the magnitude of the sinusoidal quantity in any one of the phases. The other common choice for the scale factor is $\sqrt{2/3}$ that maintains a constant power relationship [4].

As seen in Equation 6.25, the space vector is defined with three unit vectors, and each vector is 120 electrical degrees apart. These three unit vectors can be summed to form a space vector anywhere on the plane. However, any two nonparallel vectors can be used to define a space vector on the plane. It is most convenient to define orthogonal (90 electrical degrees apart) unit vectors. Using the complex plane, the orthogonal unit vectors lie on the real and imaginary axes. A second definition of the complex space vector is given in Equation 6.30 with these unit vectors.

$$\bar{f}_{qds} = f_{qs} - j\,f_{ds} \tag{6.30}$$

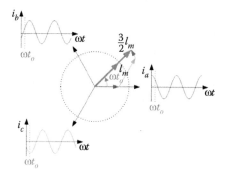

FIGURE 6.26   Space vector at time $t_0$ (position $\omega t_0$).

This is the common "dq" frame. This is one definition and the reader is cautioned that there are multiple definitions used in literature (where the d-axis is aligned with the positive imaginary axis as opposed to the negative imaginary axis, or where the d-axis is aligned with the real axis and the q-axis is aligned with the imaginary axis). How the axes are defined is not important, only that you remain consistent with the definition for both forward and reverse transformations.

A third variable is necessary to define a unique transformation (since there are three variables in the abc frame) when using the dq frame:

$$f_{0s} = \frac{1}{3}[f_{as} + f_{bs} + f_{cs}] \tag{6.31}$$

The first transformation defined converts three-phase variables (separated by 120 electrical degrees) into two-phase dq variables. This is known as the Clarke transformation. The relationship between the abc variables and the dq0 variables is given in Equation 6.32 below. This is the forward transformation from abc variables to dq0 variables. The reverse transformation from dq0 variables to abc variables is given in Equation 6.33.

$$\begin{bmatrix} f_{qs} \\ f_{ds} \\ f_{0s} \end{bmatrix} = \frac{2}{3} \begin{bmatrix} 1 & \dfrac{-1}{2} & \dfrac{-1}{2} \\ 0 & \dfrac{-\sqrt{3}}{2} & \dfrac{\sqrt{3}}{2} \\ \dfrac{1}{2} & \dfrac{1}{2} & \dfrac{1}{2} \end{bmatrix} \begin{bmatrix} f_{as} \\ f_{bs} \\ f_{cs} \end{bmatrix} \tag{6.32}$$

$$\begin{bmatrix} f_{as} \\ f_{bs} \\ f_{cs} \end{bmatrix} = \frac{2}{3} \begin{bmatrix} 1 & 0 & 1 \\ \dfrac{-1}{2} & \dfrac{-\sqrt{3}}{2} & 1 \\ \dfrac{-1}{2} & \dfrac{\sqrt{3}}{2} & 1 \end{bmatrix} \begin{bmatrix} f_{qs} \\ f_{ds} \\ f_{0s} \end{bmatrix} \tag{6.33}$$

The second transformation used is the Park transformation. This transforms variables from the stationary frame to a rotating frame. First, we will define a general notation form for variables so that they are explicitly defined [4]:

$\bar{f}_{qdx}^{y}$: —$f$ is a variable ($v$, $i$, $\lambda$)
    —$x$ is where the variable comes from ($r$—rotor, $s$ —stator)

(actual location of variables)

—$y$ is where variables are referred to ($r$—rotor, $s$ —stator, and $g$—general)

As an example, $v_{qs}^r$ is the q-axis stator voltage referred to the rotor reference frame. $i_{ds}^s$ is the d-axis stator current referred to the stator (or stationary) reference frame.

Rotating to a general reference frame requires only simple trigonometry but is a crucial step in the process of AC machine current regulation. The simple trigonometry is shown in Figure 6.27.

*Stationary to Rotating*

$$\overline{f}_{qdx}^g = e^{-j\theta}\,\overline{f}_{qdx}^s$$

$$\begin{bmatrix} f_{qx}^g \\ f_{dx}^g \end{bmatrix} = \begin{bmatrix} \cos\theta & -\sin\theta \\ \sin\theta & \cos\theta \end{bmatrix} \begin{bmatrix} f_{qx}^s \\ f_{dx}^s \end{bmatrix}$$

*Rotating to Stationary*

$$\overline{f}_{qdx}^s = e^{j\theta}\,\overline{f}_{qdx}^g$$

$$\begin{bmatrix} f_{qx}^s \\ f_{dx}^s \end{bmatrix} = \begin{bmatrix} \cos\theta & \sin\theta \\ -\sin\theta & \cos\theta \end{bmatrix} \begin{bmatrix} f_{qx}^g \\ f_{dx}^g \end{bmatrix}$$

In AC machine control, one convenient reference frame to choose is the rf reference frame. In a surface PM machine, this is simply determined by the location of the PMs. For induction machines, the rf must be determined by inferred measurements or estimation. The reference frame is chosen such that the d-axis is aligned with the rf (rotor PMs) and the q-axis is orthogonal to the rf. The "d" in d-axis stands for "direct" axis as it is directly aligned with the rotor field. The "q" in q-axis stands for "quadrature" and is in quadrature (90 electrical degrees) with the rotor field. Figure 6.28 shows the transformations used in a PM synchronous machine controller. Note that when the q and d axes are chosen as described above, the d-axis aligns with the rf and attempts to add to (with positive d-axis current) or subtract from (with negative d-axis current) the total flux. The q-axis interacts with the rf to produce torque. Thus, the d-axis current is analogous to field current and the q-axis is analogous to armature current in DC machines.

The variables are transformed into the synchronous reference frame (denoted with the "e" superscript for excitation). In this case, the synchronous and rf reference frames are identical

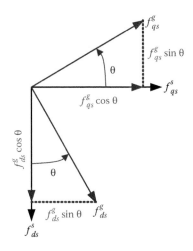

FIGURE 6.27  Transformation from stationary to general rotating reference frame.

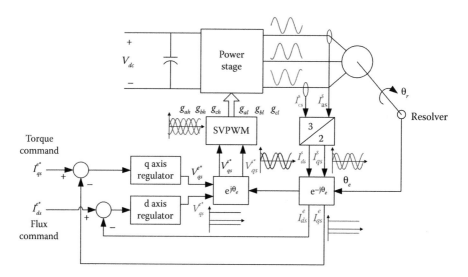

**FIGURE 6.28** Block diagram of PM AC machine controller with transformations.

(which is not the case in induction motor control). The sinusoidal currents, when transformed into the synchronous reference frame, become DC quantities easily controlled as discussed in the previous sections. In fact, the concept of AC machine control is to transform quantities into the synchronous reference frame, regulate currents as if they are DC quantities, and then transform the commanded voltages back to the stationary reference frame where the inverter applies voltages to the machine windings. This is known as synchronous frame current regulation (SFCR).

In surface PM BLAC machines, the d-axis current command is commonly set to zero. This is due to the fact that the PMs produce all the flux linkage necessary for torque production in the machine. Additionally, it becomes difficult to weaken the field in most surface PM machines because of the low values of magnetizing inductance in these machines.

Interior PM machines are different in that they are intentionally designed for field weakening to allow a large constant power-operating regime. In this case, the d-axis (flux-producing axis) is not set to zero but rather regulated based on operating point demand. This will be discussed in the subsequent sections.

### 6.2.2.3 Differential Equations of BLAC Machine

As described in the previous chapters, the differential equations of the BLAC machine in a general reference frame (with PM flux aligned with the d-axis) are

$$v_{qs}^g = r_s i_{qs}^g + \frac{d\lambda_{qs}^g}{dt} + \omega \lambda_{ds}^g \tag{6.34}$$

$$v_{qs}^g = r_s i_{ds}^g + \frac{d\lambda_{ds}^g}{dt} + \omega \lambda_{qs}^g \tag{6.35}$$

$$\lambda_{qs}^g = L_{1s} i_{qs}^g + L_{mq} i_{qs}^g \tag{6.36}$$

$$\lambda_{ds}^g = L_{1s} i_{ds}^g + L_{md} i_{ds}^g + \Lambda_{mf} \tag{6.37}$$

$$T_e = \frac{3}{2}\frac{P}{2}\left[\underbrace{\Lambda_{mf}i_{qs}^g}_{\text{PM torque}} + \underbrace{(L_{md} - L_{mq})i_{ds}^g i_{qs}^g}_{\text{Reluctance torque}}\right] \tag{6.38}$$

where

$v_{qs}^g$: $q$-axis voltage
$v_{ds}^g$: $d$-axis voltage
$i_{qs}^g$: $q$-axis current
$i_{ds}^g$: $d$-axis current
$\lambda_{qs}^g$: $q$-axis flux linkage
$\lambda_{ds}^g$: $d$-axis flux linkage
$T_e$: Electromagnetic torque
$\omega$: Electrical speed of the reference frame
$\Lambda_{mf}$: PM flux linkage
$r_s$: Stator resistance
$L_{mq}$: q-axis mutual inductance
$L_{md}$: d-axis mutual inductance
$L_{ls}$: Leakage inductance
$P$: Number of motor poles

One of the most common techniques used for AC machine control is to transform to the synchronous frame, apply a PI current regulator, and transform back to the stationary reference frame. This takes advantage of the abundance of knowledge and literature on PI regulators. But there is one slight difference that must be noted. Equations 6.34 and 6.35 show the cross-coupling that is introduced between $q$ and $d$ reference frames by the electrical speed of the reference frame. If the stationary frame is used, the axes are fixed at zero speed and there is no cross-coupling ($\omega = 0$). But this would require regulating AC currents as opposed to DC currents as desired. In the synchronous frame, the speed of the frame is equal to the rotor speed ($\omega = \omega_r$); so, cross-coupling is introduced in the controller. This cross-coupling must be accounted for. Additionally, with the axes aligned as described, all the rotor PM flux is aligned with the d-axis and therefore couples into the q-axis via the speed-dependent term. This is the back EMF term and is only present in the q-axis. This must be decoupled as well similar to the approach used in DC machines with back EMF decoupling. Figure 6.29 shows the decoupling strategy used in SFCR systems.

### 6.2.2.3.1 Surface PM versus Interior PM Structure

As the name denotes, surface PM (SPM) machines have their magnets on the surface of the rotor, and interior PM (IPM) machines have their magnets buried in the rotor iron. This difference in physical construction results in changes to the magnetic structure. Figure 6.30 shows the rotor structure

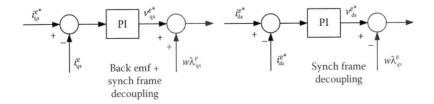

FIGURE 6.29   Reference frame and back EMF decoupling in AC synchronous frame controllers.

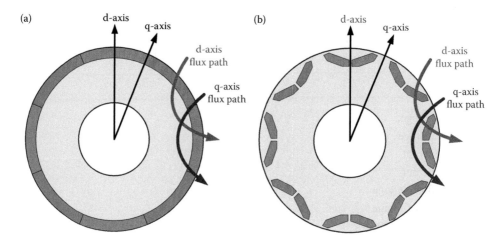

(a)   d-axis
      q-axis
            d-axis
            flux path
            q-axis
            flux path

(b)   d-axis
      q-axis
            d-axis
            flux path
            q-axis
            flux path

FIGURE 6.30   Rotor structure of a surface PM machine (a) and interior PM machine (b).

of both a surface PM machine and an interior PM machine of an eight-pole motor. Note that these represent typical configurations, but they are only two of numerous rotor designs. Fundamentally, the electromagnetic structure is similar to this regardless of configuration. Figure 6.30 shows the d (direct) and q (quadrature) axes of an eight-pole machine as well as the path of flux in each of these axes. The path of the d-axis flux is directly through the magnets.

Figure 6.30 shows that for the case of the SPM, the reluctance path for the *d*-and *q*-axes is the same. Both axes have a flux path that flows from the stator magnetic steel, through a small air gap, through the PMs, and then through rotor magnetic steel before following a similar path back to the stator magnetic steel. The reluctance of both flux paths is dominated by the reluctance of the surface magnet that is typically orders of magnitude greater than the air gap and has the same relative permeability of air. The PMs "look" like a large air gap to the magnetic circuit. This large effective air gap makes the reluctance relatively large in SPM machines. The large reluctance means that the stator inductance (which is inversely proportional to reluctance) is relatively low.

The low inductance of SPM machines often makes current control difficult. If high switching frequencies are not used, the current ripple in the machine windings can be large, leading to high switching losses and excessive noise in the current feedback circuit. Additionally, current control sample rates must be sufficiently fast to ensure that currents do not exceed inverter ratings before the controller reacts.

Since the reluctance of both paths is identical (or nearly identical based on the design), the *d*-axis and *q*-axis inductances are equal (or nearly equal). Even when the inductances are not equal (whether it is due to slight geometry differences or magnetic saturation due to the magnets in the *d*-axis path), the difference between them is typically negligible in SPM machines. With a negligible difference in *d*-axis and *q*-axis inductance, the reluctance portion of the torque in Equation 6.38 becomes negligible. Since the reluctance torque portion is nearly zero, SPM machines normally command zero *d*-axis current and apply all of the current in the *q*-axis (torque-producing axis). In addition to the reluctance torque being negligible, there is another reason that the *d*-axis current is commonly commanded to zero. As stated above, it becomes difficult to weaken the field in most surface PM machines (there are exceptions to this [5]) because of the low values of magnetizing inductance in these machines. With zero *d*-axis current, the torque in Equation 6.38 simplifies to Equation 6.39.

$$T_e = \frac{3}{2}\frac{P}{2}[\Lambda_{mf} i_{qs}^r].$$                                        (6.39)

Figure 6.30 shows that for the case of the IPM, the reluctance path for the *d*-axis and q-axis is not the same. The d-axis flux path crosses two sets of magnets in the rotor that presents significantly higher reluctance than the *q*-axis flux path that travels only through magnetic steel in the rotor. Therefore, the d-axis inductance is, by design, significantly lower than the *q*-axis inductance. Figure 6.31 highlights the resultant torque as the sum of the PM torque (first term in Equation 6.38) and the reluctance torque (second term in Equation 6.38).

With interior PM machines, the d-axis current command (flux command) can be positive or negative. A negative command attempts to weaken the resultant *d*-axis flux that decreases the effective back EMF constant to allow operation at higher speeds. This reduction in *d*-axis flux also reduces the effective torque constant (first term in Equation 6.38), but this is offset by an increase in torque due to the reluctance term (second term in Equation 6.38). In fact, this reluctance term is used to increase the overall torque per ampere of the total stator current. IPMs are usually operated with the MTPA control up to the inverter current-rating limit. The maximum currents are constrained by Equation 6.40.

$$i_{qs}^2 + i_{ds}^2 = I_{max}^2 \tag{6.40}$$

The inverter also has a voltage constraint given by Equation 6.41.

$$v_{qs}^2 + v_{ds}^2 = V_{max}^2 \tag{6.41}$$

Considering steady-state operation of the IPM, the voltage equations of Equations 6.34 and 6.35 simplify to Equations 6.42 and 6.43, respectively.

$$v_{qs} = r_s i_{qs} + \omega \lambda_{ds} \tag{6.42}$$

$$v_{ds} = r_s i_{qs} - \omega \lambda_{qs} \tag{6.43}$$

If the IR (current × resistance (resistive voltage drop)) is neglected and the flux linkages are expressed in terms of currents, Equations 6.42 and 6.43 reduce further to Equations 6.44 and 6.45.

$$v_{qs} = \omega (L_{ds} i_{ds} + \Lambda_{mf}) \tag{6.44}$$

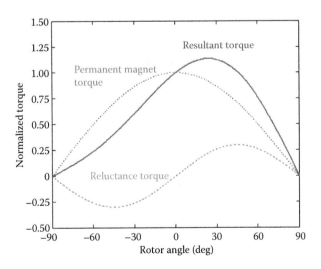

**FIGURE 6.31** Torque as a function of rotor angle for an interior PM machine with fixed stator current.

$$v_{qs} = \omega L_{qs} i_{qs} \qquad (6.45)$$

where

$L_{qs}$: q-axis inductance $= L_{mq} + L_{ls}$
$L_{ds}$: d-axis inductance $= L_{md} + L_{ls}$

The inverter voltage limit in Equation 6.41 can now be expressed in terms of the d-axis and q-axis currents as shown in Equation 6.46.

$$(L_{ds}i_{ds} + \Lambda_{mf})^2 + (L_{qs}i_{qs})^2 = \left(\frac{V_{max}}{\omega}\right)^2 = \lambda_{max}(\omega) \qquad (6.46)$$

The equation for maximum inverter current (Equation 6.40) describes a circle in the dq current plane. The equation for maximum voltage (Equation 6.46) describes an ellipse in the dq current plane. Figure 6.32 shows the map of the dq current plane with the maximum current circle and maximum voltage ellipses. Notice that the maximum voltage is a function of speed as shown in Equation 6.46. Figure 6.32 plots three constant speed constraints, but there is actually an infinite number of ellipses on the plane that vary in size inversely with speed. Figure 6.32 also highlights the constant torque lines for three different amplitudes of torque (both positive and negative). Finally, Figure 6.32 shows the MTPA line. The equation for this line is given by Equation 6.47 shown below [6]:

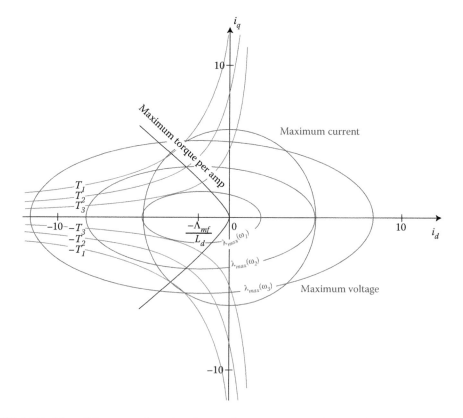

**FIGURE 6.32** Map of dq current plane showing maximum current, maximum voltage, constant torque values, and maximum torque per amp.

$$i_d(i_q) = \frac{-\Lambda_{mf}}{2(L_q - L_d)} - \sqrt{\frac{\Lambda_{mf}^2}{4(L_q - L_d)^2} + i_{qs}^2} \qquad (6.47)$$

Control of the IPM machine with MTPA utilizes look-up tables or inverse functions to derive $d$-axis and $q$-axis currents from the torque command. The analysis above is obviously idealized. The nonlinearities of saturation and inverter deadtime can be accounted for in the look-up tables or inverse functions. The determination of the parameters in the look-up table or the inverse functions used is done by finite-element analysis or measured data.

Synchronous reluctance machines (SynRM) are similar to IPM machines. There are no PMs to produce torque; so, only the second portion of Equation 6.38 exists (the reluctance torque-producing term). The torque equation is given in Equation 6.48.

$$T_e = \frac{3}{2}\frac{P}{2}(L_{md} - L_{mq})i_{ds}^r i_{qs}^r \qquad (6.48)$$

For the ideal case of SynRMs, the MTPA occurs when the d-axis and q-axis currents are equal. Here again, look-up tables or inverse functions can be used for MTPA, including the effects of saturation and other nonlinearities.

### 6.2.2.4 Overview of AC Induction Machine Current Control

As with all forms of machine control, there are several techniques of control for AC induction machines (ACIMs), and each technique is with multiple variants. This section will describe two forms of FOC that are suitable for traction control in electric vehicles. These two forms, indirect and direct FOC have a tremendous amount of literature published and this section will only serve as a brief overview.

ACIM control is very similar to BLAC machine control. The currents are transformed into a reference frame aligned with the rf, a PI controller determines commanded voltage, and these commanded voltages transformed back to the stationary reference frame to be applied by the inverter to the machine windings. The major difference is that the rf is not synchronous with the terminal excitation frequency and must be determined (either directly or indirectly).

### 6.2.2.5 Differential Equations of ACIM

As described in the previous chapters, the differential equations of the ACIM in a general reference frame are (neglecting zero-sequence terms)

For stator:

$$v_{qs}^g = r_s i_{qs}^g + \frac{d\lambda_{qs}^g}{dt} + \omega\lambda_{ds}^g \qquad (6.49)$$

$$v_{ds}^g = r_s i_{ds}^g + \frac{d\lambda_{ds}^g}{dt} - \omega\lambda_{qs}^g \qquad (6.50)$$

$$\lambda_{qs}^g = L_{1s} i_{qs}^g + L_m(i_{qs}^g + i_{qr}^g) \qquad (6.51)$$

$$\lambda_{ds}^g = L_{1s} i_{ds}^g + L_m(i_{ds}^g + i_{dr}^g) \qquad (6.52)$$

For rotor:

$$v_{qr}^g = r_r i_{qr}^g + \frac{d\lambda_{qr}^g}{dt} + (\omega - \omega_r)\lambda_{dr}^g \tag{6.53}$$

$$v_{dr}^g = r_r i_{dr}^g + \frac{d\lambda_{dr}^g}{dt} - (\omega - \omega_r)\lambda_{qr}^g \tag{6.54}$$

$$\lambda_{qr}^g = L_{1r} i_{qr}^g + L_m(i_{qs}^g + i_{qr}^g) \tag{6.55}$$

$$\lambda_{dr}^g = L_{1r} i_{dr}^g + L_m(i_{ds}^g + i_{dr}^g) \tag{6.56}$$

$$T_e = \frac{3}{2}\frac{P}{2} L_m(i_{qs}^g i_{dr}^g - i_{qr}^g i_{ds}^g) \tag{6.57}$$

where
  $v_{qs}^g$: q-axis stator voltage
  $v_{ds}^g$: d-axis stator voltage
  $i_{qs}^g$: q-axis stator current
  $i_{ds}^g$: d-axis stator current
  $\lambda_{qs}^g$: q-axis stator flux linkage
  $\lambda_{ds}^g$: d-axis stator flux linkage
  $v_{qr}^g$: q-axis rotor voltage
  $v_{dr}^g$: d-axis rotor voltage
  $i_{qr}^g$: q-axis rotor current
  $i_{dr}^g$: d-axis rotor current
  $\lambda_{qr}^g$: q-axis rf linkage
  $\lambda_{dr}^g$: d-axis rf linkage
  $T_e$: Electromagnetic torque
  $\omega_e$: Excitation frequency
  $\omega_r$: Rotor frequency (electrical rotor speed)
  $r_s$: Stator resistance
  $L_m$: Mutual inductance
  $L_{ls}$: Stator leakage inductance
  $L_{lr}$: Rotor leakage inductance
  $P$: Number of motor poles

Using the complex notation defined in Equation 6.30, it is often more convenient to describe the machine as a set of complex vector equations as opposed to the scalar equations described above. This makes comparison of methods more straightforward with the use of the compact notation. It should be noted that the scalar equations can easily be derived from the complex vector equations and vice versa. Equations 6.58 and 6.59 replace Equations 6.49 through 6.56. Here, the "p" operator replaces the common LaPlace operator "s" so that the "s" is not confused with slip in the induction machine. The "p" operator represents the time rate of change of the variable it is operating on.

$$\overline{v}_{qds}^g = r_s \overline{i}_{qds}^g + L_s(p + j\omega)\overline{i}_{qds}^g + L_m(p + j\omega)\overline{i}_{qdr}^g \tag{6.58}$$

$$\bar{v}_{qdr}^g = r_r \bar{i}_{qdr}^g + L_r[p + j(\omega - \omega_r)]\bar{i}_{qdr}^g + L_m[p + j(\omega - \omega_r)]\bar{i}_{qds}^g \tag{6.59}$$

$$T_e = \frac{3}{2}\frac{P}{2}L_m \text{Imag}\{\bar{i}_{qds}^g \cdot \bar{i}_{qdr}^{g\,*}\} \tag{6.60}$$

where

$\bar{v}_{qds}^g$: Complex vector stator voltage
$\bar{v}_{qdr}^g$: Complex vector rotor voltage
$\bar{i}_{qds}^g$:Complex vector stator current
$\bar{i}_{qdr}^g$:Complex vector rotor current
$T_e$: Electromagnetic torque
$\omega$:  Reference frame electrical frequency
$\omega_r$:  Rotor frequency (electrical rotor speed)
$r_s$:  Stator resistance
$r_r$:  Rotor resistance
$L_m$:  Mutual inductance
$L_s = L_m + L_{ls}$: stator inductance
$L_r = L_m + L_{lr}$: rotor inductance
$p$:  Differential operator
$j$:  Imaginary number
$P$:  Number of motor poles
$*$:  Complex conjugate

### 6.2.2.6   Indirect Field Orientation of ACIMs

ACIM control often uses the same FOC techniques discussed above, but transforms into the rf reference frame as opposed to the synchronous reference frame. Indirect field-oriented control (IFOC) indirectly determines the location of the rf based on the relationship of the slip (excitation speed minus the rotor speed). This is given in Equation 6.61.

$$\omega_e - \omega_r = s\omega_e = \frac{r_r}{L_r}\frac{i_{qs}^g}{i_{ds}^g} \tag{6.61}$$

where

$\omega_e$:  Excitation frequency
$\omega_r$:  Rotor frequency (electrical rotor speed)
$s$:  Slip
$s\,\omega_e$: Slip frequency
$r_r$:  Rotor resistance
$L_r$:  Rotor inductance
$i_{qs}^g$:  $q$-axis stator current
$i_{ds}^g$:  $d$-axis stator current

This slip frequency is calculated from the machine parameters (rotor resistance, rotor inductance, and mutual inductance) as well as the q-and d-axes stator currents. Therefore, the commanded value of slip frequency is calculated from the commanded values of $q$-and $d$-axes stator currents. This commanded slip frequency is integrated to get the commanded value of slip position. The commanded slip position is added to the rotor position to determine the rf reference frame as shown in Figure 6.33. Once the rf reference frame is determined, the same synchronous frame PI current regulators are used as discussed above.

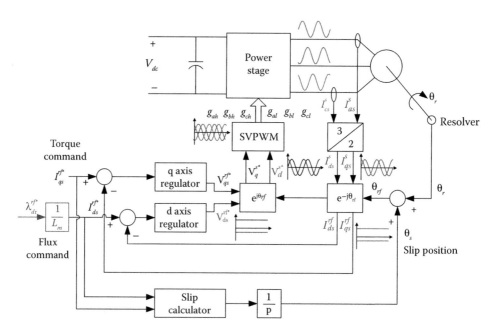

**FIGURE 6.33**  Block diagram of ACIM controller with transformations in the rotor flux reference frame.

### 6.2.2.7  Direct Field Orientation of ACIMs

In direct field orientation control (DFOC), the rf angle is "directly" measured (or estimated) rather than calculating it using the slip relationship. Originally, flux sensors were embedded in the stator windings to measure the mutual air gap flux. The mutual air gap flux is the sum of the stator and rotor currents times the mutual inductance as shown in Equation 6.62. Note that these variables are referenced to the stationary frame as that is where the measurements take place. From this, the complex rotor current vector can be derived as listed in Equation 6.63. The rf is then calculated as the air gap mutual flux plus the rotor currents multiplied by the rotor leakage inductance as given in Equation 6.64.

$$\bar{\lambda}_{qdm}^{s} = L_m(\bar{i}_{qds}^{s} + \bar{i}_{qdr}^{s}) \tag{6.62}$$

$$\bar{i}_{qdr}^{s} = \frac{\bar{\lambda}_{qdm}^{s}}{L_m} - \bar{i}_{qds}^{s} \tag{6.63}$$

$$\bar{\lambda}_{qdr}^{s} = \bar{\lambda}_{qdm}^{s} + L_{lr}\bar{i}_{qdr}^{s} = \bar{\lambda}_{qdm}^{s} + \frac{L_{lr}}{L_m}\bar{\lambda}_{qdm}^{s} - L_{lr}\bar{i}_{qds}^{s} = \frac{L_r}{L_m}\bar{\lambda}_{qdm}^{s} - L_{lr}\bar{i}_{qds}^{s} \tag{6.64}$$

This approach had two major disadvantages. First, the air gap sensors had reliability problems. Second, the calculation of rf linkage is based on leakage inductance that substantially varies with load.

Most modern methods estimate the rf linkage from measured terminal variables. The terminal voltage minus the IR drop is integrated (the 1/p operator is used for integration) as shown in Equation 6.65. Note again that the variables are referenced to the stationary frame where the terminal quantities are measured. The rf linkage is the stator current times the mutual inductance plus the rotor term times the self-inductance. This is shown in Equation 6.66. The rotor current is solved from Equation 6.66 and given in Equation 6.67. From this, the rf linkage is now solved for in

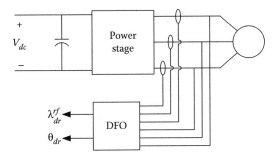

FIGURE 6.34   Direct field orientation measurements and calculations.

terms of the stator flux and stator current as given in Equation 6.68. Thus, the rf is derived from the terminal voltage and current only.

$$\bar{\lambda}_{qds}^{s} = \frac{1}{p}(\bar{v}\lambda_{qds}^{s} - r_s \bar{i}_{qds}^{s}) \tag{6.65}$$

$$\bar{\lambda}_{qdr}^{s} = L_m \bar{i}_{qds}^{s} + L_r \bar{i}_{qdr}^{s} \tag{6.66}$$

$$\bar{i}_{qdr}^{s} = \frac{\bar{\lambda}_{qds}^{s}}{L_m} - \frac{L_s}{L_m}\bar{i}_{qds}^{s} \tag{6.67}$$

$$\bar{\lambda}_{qdr}^{s} = \frac{L_r}{L_m}(\lambda_{qds}^{s} - L_s'\bar{i}_{qds}^{s}) \tag{6.68}$$

where

$L_s' = L_s - \dfrac{L_m^2}{L_r}$: Stator transient inductance

Once the complex rf linkage vector is calculated, the magnitude is determined by the square root of the sum of the squares of the d-and q-axes. The rf angle (used for transformation in the SFCR) is determined from the inverse tangent of the ratio of the q-axis to d-axis values. This is shown in Figure 6.34.

The main disadvantage of this approach is with low-speed operation. At low speed, the fundamental frequency approaches DC and the applied voltage is low. Both of these conditions lead to integration errors. As the frequency approaches zero, integration of a constant continues to grow without bounds. So, any offsets in measurements will produce ever-increasing errors. The second issue is with low values of stator voltage. At low values of stator voltage and high load (high stator currents), the IR drop becomes an increasing portion of the integration term. Slight errors in resistance estimation or changes with temperature are magnified under these conditions. There has been a lot of work to overcome both issues with DFOC [7].

## 6.3   SWITCHED RELUCTANCE MACHINE CONTROL

There have been significant developments in switched reluctance machines (SRMs) in the past several decades. SRMs have a significant advantage over PMs and induction machines in that there are no magnets or windings on the rotor. The structure of the SRM is extremely robust. It is constructed

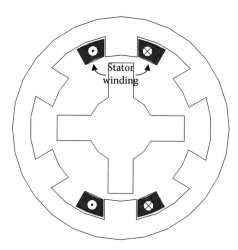

**FIGURE 6.35**  Construction of a typical 6/4 switched reluctance machine showing the winding of one phase.

of stacked laminations for both the stator and rotor and works on the principle of the reluctance force that attempts to align the magnetic pole of the rotor to the stator when the stator is energized. A typical 6/4 SRM construction is shown in Figure 6.35.

### 6.3.1  Torque Production in SRMs

Torque is produced in SRMs from the reluctance force on the rotor that attempts to align with the stator poles. While there is a significant amount of literature on torque production and control in SRMs, this section will focus on the basics.

#### 6.3.1.1  Magnetic Characteristics of SRMs

Most machines are pushed up to saturation levels in the magnetic steel in which they are made. They are pushed up to saturation but not significantly beyond because there are diminishing returns for doing so and typically losses increase considerably faster than torque (or mechanical output power) increase when operating above saturation. The SRM is very different in that it is often operated well above saturation in the magnetic steel. It might seem odd that we want to operate the machine that is heavily saturated, but in fact, it is advantageous to do so to aid in the electromechanical energy conversion process.

Since the machine is operated well into magnetic saturation, the characteristic of the machine is highly nonlinear. This is one of the main challenges of controlling these machines. We must first determine what the flux linkage versus rotor position versus current curves are for a given SRM. This can be accomplished by finite-element modeling (FEM) or by measurements. One method of measuring these curves is to lock the machine at various rotor angles and calculate the flux linkage versus current at each of these positions. This is typically done by applying bus voltage to one phase until a desired current is reached. The phase voltage and currents are measured versus time and recorded. The flux linkage versus time is calculated by integrating the phase voltage minus the IR drop and then plotted versus current. Figure 6.36 shows the measured current versus time and the corresponding flux linkage versus current at various rotor positions.

#### 6.3.1.2  Energy/Co-Energy in SRMs

Torque produced in SRMs is derived from a change in energy or co-energy with respect to rotor position. Stored energy in a magnetic field is a familiar concept from basic physics. This section will discuss energy and a new term, co-energy, in SRMs. The stored energy and co-energy in one phase

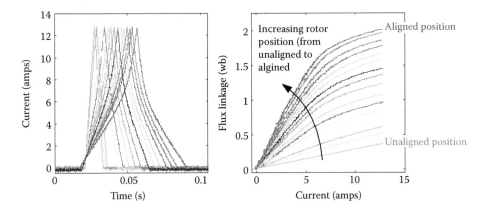

**FIGURE 6.36** Measured current versus time and flux linkage versus current for one phase of an SRM for one electrical revolution.

of an SRM is graphically shown in Figure 6.37. The equation for stored energy is given in Equation 6.69. Co-energy is the portion under the curve of flux linkage versus current at a given position. Co-energy is mathematically expressed in Equation 6.70.

$$W_f(\lambda_o, \theta_o) = \int_0^\lambda i(\lambda, \theta_o) d\lambda \tag{6.69}$$

where
  $W_f$: Stored field energy
  $\lambda$: Flux linkage
  $\theta$: Rotor position

$$W_{co}(i, \theta) = i\,\lambda(i, \theta) - W_f(i, \theta) \tag{6.70}$$

where
  $W_{co}$: Co-energy
  $i$: Stator current
  $\theta$: Rotor position

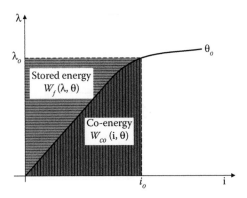

**FIGURE 6.37** Plot of flux linkage versus current showing the stored energy and co-energy in the magnetic field for a given level of flux linkage at a given position.

Stored energy is expressed in terms of flux linkage and rotor position whereas co-energy is expressed in terms of current and rotor position. Either one of these terms can be used to define torque produced by one phase of an SRM.

### 6.3.1.3 Torque Equations in SRMs

The equation for torque produced in an SRM using flux linkage and rotor position as state variables is given in Equation 6.71 and using current and rotor position as state variables is given in Equation 6.72. In both cases, torque is expressed as a change in the quantity (stored energy or co-energy) with rotor position at a fixed value of flux linkage or current, respectively.

$$T(\lambda, \theta_r) = \left. \frac{-\partial W_f(\lambda, \theta_r)}{\partial \theta_r} \right|_{\lambda = \text{const.}} \tag{6.71}$$

$$T(i, \theta_r) = \left. \frac{\partial W_{co}(i, \theta)}{\partial \theta_r} \right|_{i = \text{const.}} \tag{6.72}$$

The point at $\lambda_0$, $i_0$ represents the total electrical energy input into the system. A graphical example of Equation 6.71 is shown in Figure 6.38. This highlights the change in stored field energy with a change in rotor position. The instantaneous torque is this change in stored energy with an infinitesimal change in rotor position.

### 6.3.1.4 Torque Equations in Unsaturated SRMs

As previously stated, it is typical to run an SRM heavily into saturation. However, at light loads, the machine is not saturated. Additionally, it is more convenient to look at the basics of machine control with the machine unsaturated. This simplifies the analysis and the concepts can be extended to the saturated machine case. Figure 6.39 shows the flux linkage versus current showing stored energy and co-energy with and without magnetic saturation.

Figure 6.39 also highlights why SRMs are typically operated heavily into saturation. In the unsaturated case, half of the input electrical energy goes into stored field energy (as represented by the triangular area). This is equivalent to a power factor equal to 0.5 and therefore increases the volt-ampere ratings of the power electronics for a given output power. The further the machine is pushed into saturation, the lower the percentage of stored field energy to input electrical energy, thus increasing the effective power factor.

For unsaturated operation of the machine, the co-energy and corresponding torque are given by Equations 6.73 and 6.74.

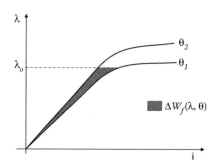

**FIGURE 6.38** Plot of flux linkage versus current showing torque as the differential stored energy as a function of position for a fixed flux linkage.

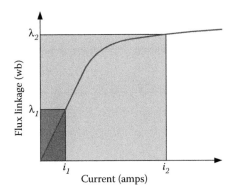

FIGURE 6.39    Plot of flux linkage versus current showing stored energy and co-energy without saturation ($\lambda_1$, $i_1$) and with saturation ($\lambda_2$, $i_2$).

$$W_{co}(i, \theta_r) = \frac{1}{2}L(\theta_r)i^2 \tag{6.73}$$

$$T(i, \theta_r) = \frac{1}{2}i^2\frac{dL(\theta_r)}{d\theta_r} \tag{6.74}$$

Equation 6.74 highlights one of the difficulties in controlling torque in an SRM. Even with the simplification, the machine is operated without saturation, and torque is still a nonlinear function of current. Equation 6.74 also shows that torque production is independent of the direction of current because it is proportional to the current squared. The sign of torque (positive or negative) is dependent on the change of inductance with rotor position.

A further simplification in SRM control is that the inductance linearly changes with position. An idealized phase inductance versus rotor position plot is shown in Figure 6.40 for a 6/4 SRM. Figure 6.41 shows torque production in an idealized SRM for both the motoring case (positive torque) and generating case (negative torque). Note again that the sign of the torque is only dependent on the change of inductance with rotor position.

The standard inverter bridge for a three-phase 6/4 SRM is shown in Figure 6.42. Figure 6.43 shows the operation of one phase of the inverter. With both switches in the phase turned on, the bus voltage is applied across the winding. Note that the phase winding is placed in between the two switches; so, shoot through conditions that exist in AC inverter bridges that do not occur in SRMs. When both switches are opened, the current in the winding continues to flow through the

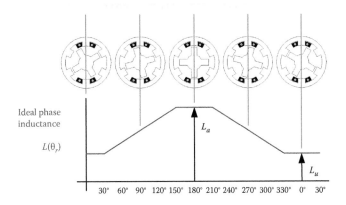

FIGURE 6.40    Plot of an idealized phase inductance versus rotor position in a 6/4 SRM.

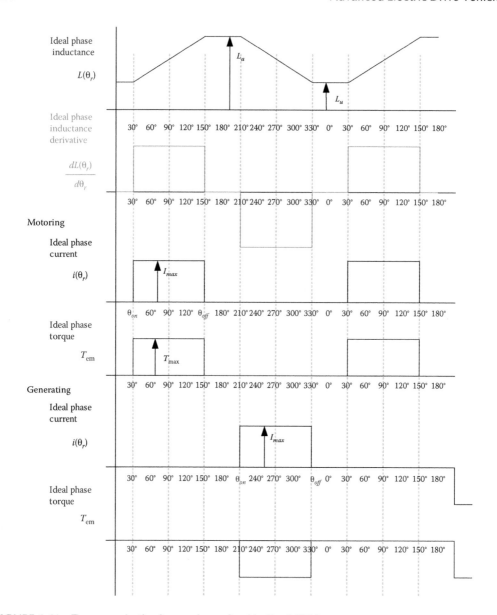

**FIGURE 6.41**  Torque production in one phase of an idealized SRM.

reverse-blocking diodes until it reaches zero. During this time, negative bus voltage is applied to the winding, driving the current to zero as quickly as possible. There are many different topologies of SRM inverters that have been implemented, but ultimately, they have the same goal in terms of regulating current in the machine to regulate torque. This chapter only focuses on the standard two-switch, two-diode-per-phase implementation.

Typically, a simple hysteresis-type current controller is used to regulate current. The hysteresis current controller is shown in Figure 6.44. The output of the hysteresis controller is a gate command to the inverter stage. If the command is 1, the two switches are turned on and +Vdc is applied across the phase winding. If the command is 0, the two switches are turned off and −Vdc is applied across the windings (until the current reaches zero). The output is summarized in Equation 6.75.

$$S = 1 \quad \text{if } i_{err} > \delta$$

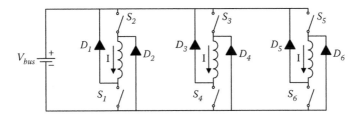

FIGURE 6.42    Standard drive topology of a three-phase SRM.

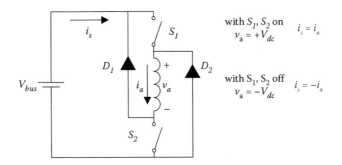

FIGURE 6.43    Operation of one phase of an SRM drive.

$$S = 0 \quad \text{if } i_{err} < \delta$$

$$S \text{ remains unchanged} \quad \text{if } -\delta < i_{err} < \delta \qquad (6.75)$$

where
    $S$: Switch state ($S_1$ and $S_2$ in Figure 6.43)

The hysteresis current controller is used to control the commanded current value. Since torque is a nonlinear function of current, it is common for look-up tables to be used to map torque command to current command since no closed-form solution exists. A high-level block diagram of the SRM controller is shown in Figure 6.45.

As the speed of the SRM increases, the time required to bring the phase current to the commanded value or from the commanded value to zero cannot be neglected. The phase inductance limits the rate of rise of current into the phase winding. The higher the speed, the more advance angle is required to get the current to the commanded level when the change of inductance is positive (for motoring operation). Figure 6.46 shows low-speed and high-speed motoring operation for a typical SRM. At high speed, the current is no longer pulse width modulated. This is known as single-pulse operation as the switches are turned on once during each phase's excitation. Note the adjustment of $\theta_{on}$ (turn-on angle) and $\theta_{off}$ (turn-off angle) as the speed increases. These angles are

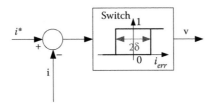

FIGURE 6.44    Hysteresis current controller.

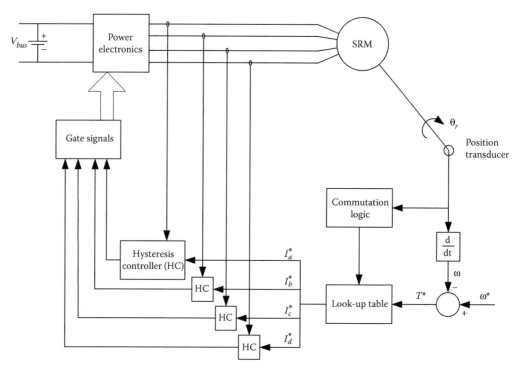

**FIGURE 6.45** High-level block diagram of a four-phase SRM controller.

often optimized and stored in a look-up table as well. The look-up table thus takes in torque command as an input and outputs turn-on time, turn-off time, and current command as a function of speed. This can be a highly nonlinear function and must be optimized for each new SRM.

One additional drawback of SRMs is the significant torque ripple associated with their operation. There has been a considerable amount of research that aims to minimize this torque ripple by either machine design (modifying the pole geometry) or by shaping the phase currents to achieve a more constant torque throughout the rotation of the rotor. Each specific application determines the need for torque ripple requirements. Modern internal combustion engines have significant torque ripple associated with the firing of cylinders. This torque ripple has been managed by simple flywheels since the advent of the internal combustion engine.

## 6.4 SPEED CONTROL IN MACHINES

Speed control in electric machines is straightforward for applications requiring modest specifications, and more difficult in cases that require tightly regulated speed control. For simple cases with modest bandwidth requirements, simple proportional or proportional plus integral plus derivative (PID) control is used. As an example, a radiator fan might be required to run at a constant speed in an on/off arrangement (on when additional cooling is required, off when it is not), or it may be used in a variable-speed application where the fan speed is proportional to the cooling needs of the system. In either case, the bandwidth and accuracy requirements are modest. The fan speed can be regulated to ±5% and has bandwidth requirements in the low (0.1–2) Hz range.

Higher accuracy or higher bandwidth speed loop requirements dictate tighter control. This tighter control needs more accurate sensors (or accurate estimation techniques) as well as a means to reject disturbances quickly. Most of the previous sections in this chapter focused on regulating torque in electric machines. High bandwidth torque regulation allows us to tightly regulate speed control. Assuming the bandwidth of the torque regulator is sufficiently faster than the bandwidth of

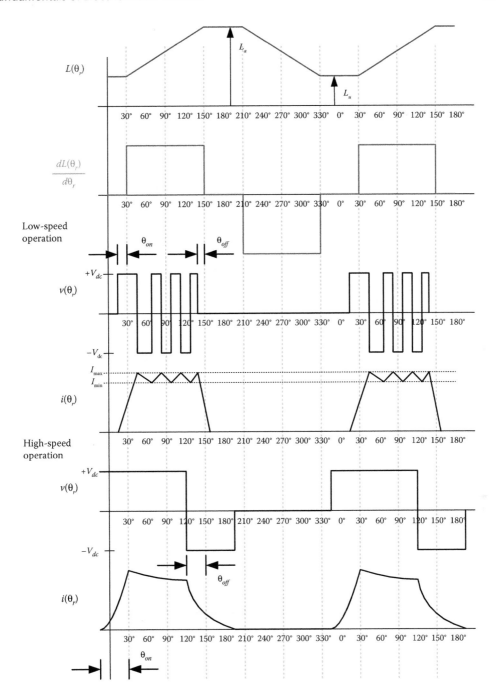

**FIGURE 6.46** Torque production in one phase of a typical SRM with idealized inductance variation.

the speed loop that we are trying to achieve, the model of the system can be greatly simplified. A model of a simple DC machine with high bandwidth torque control can be simplified as shown in Figure 6.47.

The goal of the speed loop then becomes to derive a commanded torque that will regulate the speed to a commanded value. Separating out the control loops into a torque loop and a speed loop greatly simplifies the problem of controlling speed.

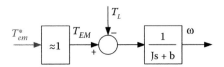

**FIGURE 6.47**   Simplified DC machine block diagram with ideal torque regulator.

### 6.4.1  CLASSICAL METHODS OF SPEED CONTROL

Before looking at speed control with a cascaded torque loop, let us first look at the simple example of a brush DC machine where only the armature voltage is controlled. It is important to note that the machine speed can be controlled by increasing or decreasing the armature voltage. A current (torque) loop is not required; however, it is often used to ensure that the currents do not exceed rated device values. Speed control of a DC machine without a current loop is shown in Figure 6.48, where $G_{controller\_\omega}$ must be defined.

The block diagram is simplified (neglecting the load torque) by determining the transfer function of speed output to a given voltage input. This relationship is found through algebraic manipulation and is given in Equation 6.76. The block diagram then reduces to Figure 6.49.

$$\frac{\omega(s)}{V_a(s)} = \frac{K_t}{JL_a s^2 + (JR_a + L_a b)s + R_a b + K_e K_t} \tag{6.76}$$

While Figure 6.49 has a simplified DC machine model (ignores load torque), it does not provide an intuitive method of determining what the controller structure should be. If a PID controller is

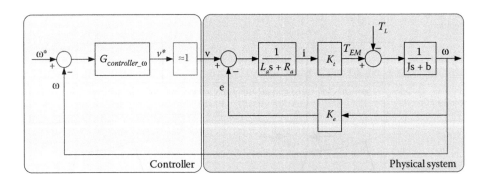

**FIGURE 6.48**   Speed control of a brush DC machine without a current (torque) loop.

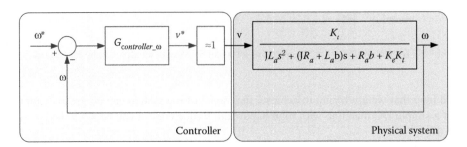

**FIGURE 6.49**   Speed control of a brush DC machine without a current (torque) loop with a simplified DC machine model.

used, the values for each term (proportional, integral, and derivative) must be determined or tuned in the application.

### 6.4.1.1 Classical Control: Proportional Speed Loop

Since a current feedback device is used in almost all modern motor controllers for device and machine protection, this feedback is available for current (torque) loop control. Closing the current (torque) loop allows the separation of torque and speed loops. Returning to Figure 6.47, we can close a simple proportional speed loop around this simple model as shown in Figure 6.50.

The transfer function for speed (as a function of input speed command and load torque) is shown in Equation 6.77. Neglecting load torque, the speed loop transfer function (speed response over speed command) is given in Equation 6.78. Even neglecting load torque, there are steady-state speed errors. In the steady state ($s = 0$), the ratio of speed response to speed command is not unity (implying that the actual speed equals the commanded speed). The steady-state error is given in Equation 6.79.

$$\omega(s) = \frac{K_{pv}\omega^*}{Js + (K_{pv} + b)} - \frac{T_L}{Js + (K_{pv} + b)} \tag{6.77}$$

$$\frac{\omega(s)}{\omega^*(s)} = \frac{K_{pv}}{Js + (K_{pv} + b)} \tag{6.78}$$

$$\omega_{err\_ss} = 1 - \frac{\omega(s)}{\omega^*(s)}\bigg|_{s=0} = 1 - \frac{K_{pv}}{K_{pv} + b} - \frac{b}{K_{pv} + b} \tag{6.79}$$

By making the proportional gain $K_{pv}$ arbitrarily big, we can make the error arbitrarily small. However, there are practical limits on how large the proportional gain can be increased. Setting the gain too large makes the system responsive to noise. Additionally, the system is not truly a first-order equation. The idealized assumption made in the torque loop eventually breaks down at higher frequencies, and nonlinearities in the system ultimately lead to instabilities with excessive controller gains.

### 6.4.1.2 Classical Control: Proportional PI Speed Loop

In the case of the proportional-only controller, the damping term leads to steady-state errors as shown in Equation 6.79. Now, we can explore using a PI controller for speed control. The same approach that was used in current regulation can be used here. Figure 6.51 shows the DC machine with an ideal torque regulator with a PI speed loop controller. The inertia and damping terms can be reformed to a mechanical time constant as was done for the electrical system above. The same plant pole, controller zero cancelation technique will be used as was done for the current loop above.

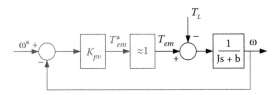

FIGURE 6.50 Simplified DC machine block diagram with an ideal torque regulator with a proportional speed loop.

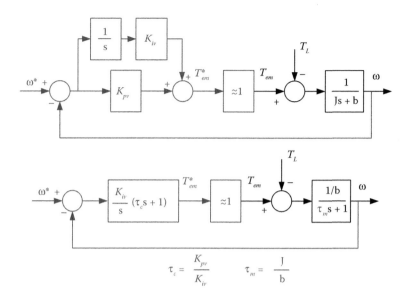

FIGURE 6.51   Block diagram of a DC machine with an ideal torque regulator with a PI speed loop (upper figure) rewritten using mechanical and controller time constants (lower figure).

With ideal pole/zero cancelation, the block diagram reduces to Figure 6.52. The transfer function for this system is given in Equation 6.80. The steady-state error for this system is zero.

$$\frac{\omega(s)}{\omega^*(s)} = \frac{K_{iv}}{bs + K_{iv}} \tag{6.80}$$

### 6.4.1.3   State Feedback Control

Returning to the simplified model in Equation 6.76 where a current (torque) loop is used, physical intuition can be applied to determine what the controller structure should be. The first thing to note is that the viscous-damping term can be separated from the mechanical block ($J\,s + b$) in Figure 6.47. This is shown in Figure 6.53. This simple change shows the physical state feedback nature of viscous damping and forms the beginning of a methodology of control [3]. For example, if an estimate of viscous damping is fed back into the torque command, the viscous-damping term can be decoupled. This is shown in Figure 6.54a where the "^" denotes an estimated variable.

Figure 6.54b shows the system with the viscous damping decoupled with a proportional speed loop. Since the proportional gain takes speed error and generates a torque command, it has units of torque per rate of speed (Nm/rad/s in SI units). This is the same units as the physical-damping

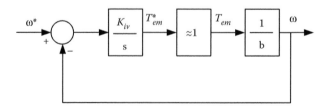

FIGURE 6.52   Block diagram of a DC machine with an ideal torque regulator and with a PI speed loop assuming ideal pole zero cancelation.

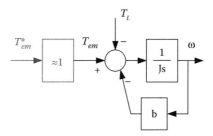

FIGURE 6.53   Simplified DC machine block diagram with ideal torque regulator with viscous damping term shown as physical state feedback.

coefficient and is denoted as $b_a$ for "active damping." The transfer function for the speed loop is given in Equation 6.81. It is clear from Equation 6.81 that the steady-state error of this system is zero if the estimated value of viscous damping is equal to the actual value. This zero steady-state error occurs without an integrator in the controller that simplifies the system and eliminates overshoot. Additionally, if the physical damping is decoupled, it is easy to tune the value of active damping required for the system. One very attractive feature is this value can be measured in the lab with the appropriate equipment (a torque and speed transducer).

$$\frac{\omega(s)}{\omega^*(s)} = \frac{b_a}{Js + b_a + (b - \hat{b})} \tag{6.81}$$

Typically, viscous damping is caused by grease in bearings that produce drag torque as a function of speed. In reality, this damping coefficient has a slight dependence on load and a strong dependence on temperature (especially at cold temperatures). The total friction can be a nonlinear function of speed as well. No matter how this function varies (as a function of load, temperature, speed, or any combination thereof), if it can be quantified, it can be decoupled. Figure 6.55 shows the friction torque as a nonlinear function of speed. Each of the coefficients can be mapped out in terms of any other dependence (i.e., temperature). By accurately estimating each of these coefficients, the nonlinear relationship is decoupled in the same manner that the simple linear relationship was.

With the state feedback approach, it is easy and intuitive to decouple physical system parameters that adversely affect the system response and augment the parameters that are desired. This technique is also extended to applications in which position must be controlled.

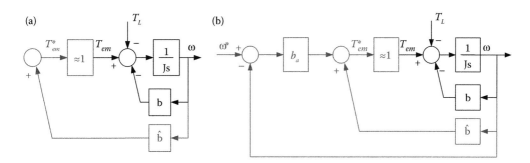

FIGURE 6.54   Simplified DC machine block diagram with ideal torque regulator with viscous damping state feedback decoupling (a) and with speed loop added (b).

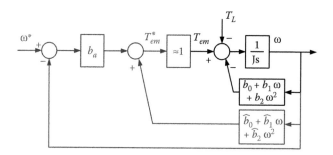

FIGURE 6.55   Simplified DC machine block diagram with ideal torque regulator with nonlinear friction and nonlinear friction decoupling.

## 6.5   CONCLUSION

This chapter presented an overview of electric machine control. Electric machine control is a very broad and very deep topic and is difficult to quantify in one chapter. In an attempt to narrow the scope, we first focused on torque control in electric machines. One method was highlighted in this chapter, specifically the control of electromagnetic torque by tightly regulating current in the machine. It should be noted that there are other techniques for directly controlling the torque in an electric machine without regulating torque [8]. While this in itself is a broad topic, it was felt that we would lose focus by presenting too many topics. The advantage of focusing on current regulation in machine control is that many of the same techniques can be extended to speed control. Also, by focusing on torque control, we attempt to get the machine to behave as an ideal torque modulator (i.e., the actual value of torque is equal to the commanded value of torque within a specified performance bandwidth). This allows us to decouple the mechanical and electrical interactions of electric machines. It also allows us to keep the same control strategies regardless of whether the machine is an induction motor or a PM motor. This abstraction in the torque loop also simplifies the speed loop analysis.

The reader is strongly encouraged to additional topics not discussed here. Additional topics of interest in machine control in electric vehicle applications include position sensorless control and other estimation techniques. These become especially important in electric vehicles where cost reduction and reliability make these techniques extremely attractive.

## PROBLEMS

6.1 Figure P6.1 shows the physical system of a DC machine without any control. From this diagram:
   a. Derive the transfer function for speed output as a function of voltage input $\omega(s)/V(s)$
      *Hint*: Neglect the load torque input $T_L(s)$.

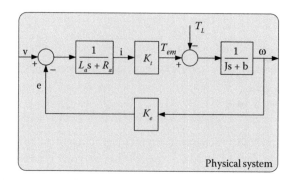

FIGURE P6.1   Block diagram of a DC machine physical system.

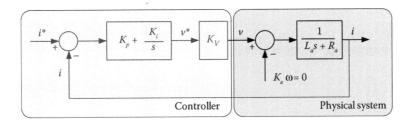

FIGURE P6.2    Block diagram of a locked rotor DC machine with current regulator.

  b.  Derive the transfer function for speed output as a function of load torque input
      $\omega(s)/T_L(s)$. *Hint*: Neglect the voltage input $V(s)$.
6.2  Figure P6.2 shows a locked rotor DC machine with a current regulator. In this case, the
     voltage amplifier is not assumed to be unity but rather has a gain of $K_V$. For this problem:
  a.  Derive the transfer function for current output as a function of current command input
      $I(s)/I^*(s)$
  b.  Determine the values of controller gains $K_p$ and $K_i$ in terms of the armature inductance
      $L_a$, armature resistance $R_a$, voltage amplifier gain $K_V$, and desired bandwidth $f_d$.
6.3  A PM DC machine has the following parameters:

| | |
|---|---|
| $V_a = 500$ V | Armature voltage |
| $R_a = 60$ mΩ | Armature resistance |
| $L_a = 60$ μH | Armature inductance |
| $P_r = 150$ Hp | Rated power |
| $\omega_{nl} = 1800$ rpm | No-load speed |
| $\omega_r = 1750$ rpm | Rated speed (at rated power) |

For this problem:
  a.  Determine the back EMF constant $K_e$.
  b.  Determine the rated torque and rated current.
  c.  Given a PI controller that is used to regulate current, find the proportional in integral
      gains necessary to achieve a 1000 Hz bandwidth.
6.4  Given the following currents in an AC machine:

$$i_{as}(t) = I_m \cos(\omega t)$$

$$i_{bs}(t) = I_m \cos\left(\omega t - \frac{2\pi}{3}\right)$$

$$i_{cs}(t) = I_m \cos\left(\omega t + \frac{2\pi}{3}\right)$$

With no zero sequence term, that is, $i_{as}(t) + i_{bs}(t) + i_{cs}(t) = 0$
For this problem:
  a.  Find $i_{qs}^s(t)$ and $i_{ds}^s(t)$.
  b.  Using $\theta = \omega t$ as, the rotary transformation variable, and the transforms of Figure 6.27,
      find $i_{qs}^e(t)$ and $i_{ds}^e(t)$.
  c.  Plot $i_{qs}^s(t)$, $i_{ds}^s(t)$, $i_{qs}^e(t)$, and $i_{ds}^e(t)$ by two electrical cycles on the same graph.
6.5  For PMSMs:
  a.  Describe the major differences between surface PM and interior PM machines.
  b.  Describe torque control in terms of q-axis and d-axis currents.
  c.  Describe what typical speed versus torque curves look like for surface PM and interior
      PM machines.

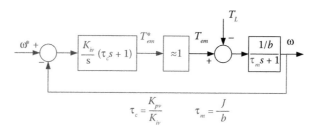

FIGURE P6.3  Block diagram of a speed loop with an ideal torque regulator.

6.6 The figure below shows the equivalent circuit of a separately excited DC machine. Torque is proportional to the product of the armature current and the field current. Explain why torque is controlled by regulating the armature current as opposed to the field current. Also explain when the field current would be adjusted.

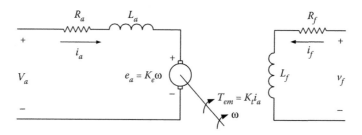

6.7 Figure P6.3 shows an ideal torque control drive (dynamics are sufficiently fast so that they can be neglected) in pole/zero form. The motor viscous damping and inertia are

$$b = 1.5 \times 10^{-5} \frac{\text{Nm}}{\text{rad/s}} \text{ armature voltage}$$

$$J = 5 \times 10^{-5} \text{ kg} \cdot \text{m}^2 \text{ armature voltage}$$

For this problem, determine the values of $K_{pv}$ and $K_{iv}$ to achieve a 50 Hz speed loop bandwidth.

6.8 Describe the difference between indirect and direct rf orientation of induction machines.

6.9 What are the two benefits of SRMs over more conventional motors? What are the two drawbacks?

## REFERENCES

1. Kuo, B. C., *Automatic Control Systems*, 5th edition, Prentice-Hall, Inc., Englewood Cliffs, NJ, 1987.
2. Anand, D. K., *Introduction to Control Systems*, 2nd edition, Pregamon Press, Oxford, England, 1984.
3. Lorenz, R. D., ME 746, Dynamics of controlled systems: A physical systems-based methodology for nonlinear, multivariable, control systems design, Course notes, University of Wisconsin-Madison, 1993.
4. Novotny, D. W., Lipo, T. A., *Vector Control and Dynamics of AC Drives*, Clarendon Press, Oxford, England, 1996.
5. Jahns, T. M., Kliman, G. B., Neumann, T. W., Interior permanent-magnet synchronous motors for adjustable-speed drives, *IEEE Transactions on Industry Applications*, 1986, 738–747.
6. Peng, H., Chang-yun, M., Hong-qiang, L., Cheng, Z., Maximum-torque-per-ampere control of interior permanent magnet synchronous machine applied for hybrid electric vehicles, *2011 International Conference on Control, Automation and Systems Engineering (CASE)*, Singapore, 2011, pp. 1–3.

7. Jansen, P. L., Lorenz, R. D., Novotny, D. W., Observer-based direct field orientation: Analysis and comparison of alternative methods, *IEEE Transactions on Industry Applications,* 1994, 30, 945–953.
8. Takahashi, I., Noguchi, T., A new quick-response and high-efficiency control strategy of an induction motor, *IEEE Transactions on Industry Applications,* 1986, IA-22, 820–827.

## FURTHER READING

1. Slemon, G. R., *Electric Machines and Drives,* Addison-Wesley Publishing Company, Inc., Reading, MA, 1992.
2. Schmitz, N. L., Novotny, D. W., *Introductory Electromechanics,* The Ronald Press Company, New York, NY, 1965.
3. Miller, T. J. E., *Switched Reluctance Motors and Their Control,* Clarendon Press, New York, NY, 1993.
4. Liwschitz-Garik, M., Weil, R. T., *D-C and A-C Machines Based on Fundamental Laws,* D. Van Nostrand Company, Inc., New York, NY, 1952.
5. Elis, G., *Control System Design Guide,* 3rd edition, Elsevier Academic Press, San Diego, CA, 2004.
6. De Donker, R., Pulle, D. W. J., Veltman, A., *Advanced Electrical Drives Analysis, Modeling, and Control,* Springer, Dordrecht, Netherlands, 2011.
7. Bolognani, S., Sgarbossa, L., Zordan, M., Self-tuning of MTPA current vector generation scheme in IPM synchronous motor drives, *2007 European Conference on Power Electronics,* Aalborg, Denmark, 2007, pp. 1–10.
8. Meyer, M., Bocker, J., Optimum control for interior permanent magnet synchronous motors (IPMSM) in constant torque and flux weakening range, *Twelfth International Power Electronics and Motion Control Conference,* Portoroz, Slovenia, 2006, pp. 282–286.
9. Bilewski, M., Fratta, A., Giordano, L., Vagati, L., Control of high-performance interior permanent magnet synchronous drives, *IEEE Transactions on Industry Applications,* 1993, 29, 328–337.
10. Shi, Y., Sun, K., Huang, L., Li, Y., Control strategy of high performance IPMSM drive in wide speed range, *Thirty-Seventh Annual Conference on IEEE Industrial Electronics Society,* Melbourne, Australia, 2011, pp. 1783–1788.
11. Limsuwan, N., Shibukawa, Y., Reigosa, D., Lorenz, R. D., Novel design of flux-intensifying interior permanent magnet synchronous machine suitable for power conversion and self-sensing control at very low speed, *2010 IEEE Energy Conversion Congress and Exposition,* 2010, 555–562.
12. De Donker, R. W., Novotny, D. W., The universal field oriented controller, *IEEE Transactions on Industry Applications,* 30, 1994, 92–100.
13. De Donker, R. W., Profumo, F., Pastorelli, M., Ferraris, P., Comparison of universal field oriented (UFO) controllers in different reference frames, *IEEE Transactions on Power Electronics,* 1995, 10, 205–213.
14. Briz, F., Degner, M. W., Lorenz, R. D., Dynamic analysis of current regulators for AC motors using complex vectors, *Thirty-Third Industry Applications Annual Meeting,* St. Louis, MO, 1998, pp. 1253–1260.
15. Zhang, Y., Shao, K., Li, L., Li, X., Theory and simulation of vector control based on rotor position orientation of rare-earth permanent magnet motor, *Sixth International Conference on Electric Machines and Systems,* Beijing, China, 2003, pp. 530–533.
16. Finch, J. W., Atkinson, D. J., Acarnley, P. P., Scalar to vector: General principles of modern induction motor control, *Fourth International Conference on Power Electronics and Variables-Speed Drives,* 1991, pp. 364–369.
17. Zhong, H. Y., Messinger, H. P., Rashad, M. H., A new microcomputer-based direct torque control system for three-phase induction motor, *IEEE Transactions on Industry Applications,* 1991, 27, 294–298.
18. Ayaz, M., Yildiz, A. B., Control of a switched reluctance motor containing a linear model, *Fourteenth Mediterranean Conference on Control and Automation,* Ancona, Italy, 2006, pp. 1–6.
19. Herrera, E. B., Guerrero, G., Dur'an, M. A., Astorga, C., Medina, M. A., Oiberio, G., Switched reluctance motor control in two-quadrants with electric vehicle applications, *World Automation Congress,* 2012, pp. 1–6.

# 7 Fundamentals of Electric Energy Storage Systems

*Pawel P. Malysz, Lucia Gauchia, and Hong H. Yang*

## CONTENTS

## 7.1   INTRODUCTION

Electrical energy storage systems (ESS) have a history that dates back to at least 1745 when Musschenbroek and Cunaeus were able to store charge in a glass filled with water to produce an electric shock. A simplified approach consisting of metal foil wrapped around the inside and outside of the jar led to the development of the Leyden Jar, essentially a capacitor that stores energy electrostatically. In 1748, Benjamin Franklin coined the term "battery" to describe an array of charged glass plates. A few decades later in 1786, Galvani's famous experiments involving twitching frog legs and dissimilar metals lead him to believe that bioelectricity was responsible for the apparent electricity. Volta was not satisfied with Galvani's explanation and showed in 1799 that combining dissimilar metals that are separated by a salt/acidic solution can generate electricity. A series of these rudimentary electrochemical cells became known as the Volta pile, it became one of the first commercially available batteries.

In the early days of the automotive industry, vehicles powered by electrical energy storage using batteries competed with designs based on internal combustion (IC) engines. The latter became dominant and for the better part of the twentieth century, the use of automotive batteries was limited to basically providing starting energy and powering electronics and lights. Now, in the twenty-first century, the increased electrification of automobiles is driven by goals of increasing vehicle efficiency, ensuring long-term sustainability of the automobile sector, and minimizing negative environmental impacts. Many electrified vehicles have been developed and are commercially available. Table 7.1 lists a variety of vehicles that have employed significant electrical energy storage. The list is dominated by electrochemical-based storage due to its technological maturity.

### TABLE 7.1
### Battery Types Used in Selected Electrified Vehicles

| Company | Country | Vehicle Model | Battery Type |
|---|---|---|---|
| GM | United States | Chevy-Volt, Spark | Li-ion |
| | | Saturn Vue Hybrid | NiMH |
| Ford | United States | Escape, Fusion, MKZ HEV | NiMH |
| | | Escape PHEV, and Focus EV | Li-ion |
| Toyota | Japan | Prius, Lexus | NiMH |
| | | Scion iQ EV, RAV4 EV | Li-ion |
| Honda | Japan | Civic, Insight | NiMH |
| | | Fit EV | Li-ion |
| Hyundai | South Korea | Sonata | Li polymer |
| Chrysler/Fiat | United States | Fiat 500e | Li-ion |
| BMW | Germany | X6 | NiMH |
| | | Mini E, ActiveE | Li-ion |
| BYD | China | E6 | Li-ion |
| Daimler Benz | Germany | ML450, S400 | NiMH |
| | | Smart EV | Li-ion |
| Mitsubishi | Japan | iMiEV | Li-ion |
| Nissan | Japan | Altima | NiMH |
| | | Leaf EV | Li-ion |
| Tesla | United States | Roadster, Model S | Li-ion |
| Think | Norway | Think EV | Li-ion/sodium–Ni–Cl |
| Iveco | Italy | Electric Daily | Sodium–Ni–Cl |

*Source:*   Modified from Young, K., C. Wang, and K. Strunz. Electric vehicle battery technologies. *Electric Vehicle Integration into Modern Power Networks.* Springer, New York, 2013: 15–56.

There is a large amount of research and development effort to improve electrical ESS for vehicles to make them practical alternatives to vehicles powered only by IC engines. One aspect has focused on only improving electrochemical-based storage through improved chemistries, new materials, or improved pack/cell designs. A variety of chemistries have been explored. Another approach considers adding electrostatic-based storage to complement electrochemical batteries. This is being explored at both the pack and cell levels, for example, hybrid packs employing separate battery and ultracapacitor cells, and/or hybrid cells that exhibit both battery and ultracapacitor characteristics.

This chapter begins by discussing energy storage requirements for electrified vehicles. A section on electrochemical-based storage then follows, where four different types are described: lead-acid, nickel metal hybrid, lithium-ion, and sodium nickel chloride. A discussion on ultracapacitor cells is given in Section 7.4 where two subtypes are described: electric double-layer capacitors (EDLCs) and ultracapacitors with pseudocapacitance. Characteristic terminology and performance parameters are presented in Section 7.5. This is followed by modeling in Section 7.6. Section 7.7 presents time/frequency domain testing and measurement approaches used to characterize ESS. Pertinent aspects and approaches to packs, management systems, and cell balancing are highlighted in the Section 7.8. Section 7.9 concerns battery state and parameter estimation.

## 7.2  ENERGY STORAGE REQUIREMENTS FOR ELECTRIFIED VEHICLES

With hybrid electric vehicles (HEVs), plug-in hybrid electric vehicles (PHEVs), and electric-only vehicles (EVs) approaching the point of entering the market in mass-production volumes, the search has intensified for advanced energy storage technologies that offer the ESS with increased energy density, power density, durability, safety, and affordability.

Despite advancements in recent years of lithium-based batteries for electrified vehicle application, this battery technology remains expensive and the realistic electric-only driving range with a fully charged battery is still very limited when compared to conventional IC engine-powered vehicles. Research continues into a wide range of alternative energy storage technologies that appear to offer varying degrees of promise in performance and ability to meet market requirements for electrified vehicles.

Critical measures of an ESS performance are

1. Safety: Ensure safe operation of ESS, no risk of thermal runaway, or exothermic behavior in the event of a crash or short circuit.
2. Cycle life: The number of full charge/discharge cycles until the ESS reaches end-of-life (EOL) condition; the definition of EOL condition varies with the usage of the ESS. One commonly used EOL condition is battery's remaining capacity at 80% of its beginning-of-life (BOL) capacity.
3. Calendar life: The calendar longevity of the ESS when it is at storage (month or year).
4. Energy density: The amount of energy each kilogram or liter of ESS contains (Wh/kg and Wh/L).
5. Power density: The amount of power each kilogram or liter of ESS delivers per second (W/kg and W/L).
6. Charge acceptance capacity: The amount of energy the ESS absorbs per second (W/kg and W/L).
7. Cost: $ per kWh.

The ultimate goal for ESS performance would be to offer similar energy and power densities to petroleum fuels used in conventional vehicles, with a comparable cost of an IC engine. However, this is not feasible with the current technology and a compromise has to be made. A number of organizations have set targets for future ESS requirements. As of 2013, the U.S. Advanced Battery Consortium

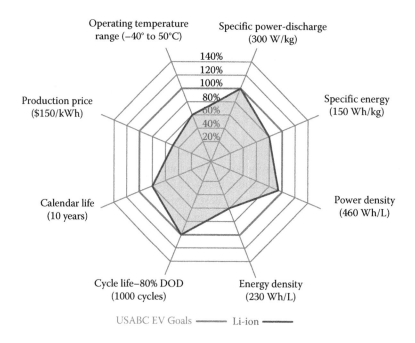

**FIGURE 7.1**  Performance of current Li-ion batteries.

(USABC) has set a 15-year life target for an ESS, with different requirements for HEV and PHEV applications. For HEVs, the target is a 300 W/L power density combined with >25 kW power delivery for 10 s at <$20 per kW. For PHEV applications, the target is 3.4 kWh usable energy or 16 km all-electric range, discharge of 45 kW for 10 s at a cost of <$500 per kWh. These goals are intended to bring ESS costs in line with conventional IC engines. Although the ESS technology can advance at least 5% per year in the near future as industry expected, the USABC's target is hard to reach before 2020.

The spider web in Figure 7.1 shows the major ESS performance metrics, which compares the performance of state-of-the-art Li-ion technology with USABC ESS targets for EVs. Li-ion technology suffers from a number of trade-offs between power and energy density, cycle life, safety, temperature, and a number of other factors discussed here. With the improvement in engineering capabilities and reduced Li-ion battery cost, a combination of ultracapacitors and Li-ion batteries could achieve a very-high-power density, high-energy density, longer cycle life, and cold-temperature performance.

### 7.2.1  ENERGY DENSITY AND SPECIFIC ENERGY

Energy density is the amount of energy stored per unit volume for an energy storage device. Specific energy is the amount of energy stored per unit mass. Electrical energy requirements for HEVs differ from those for pure electric vehicles and PHEVs; these requirements affect the design of ESS. For pure electric vehicles, a large amount of energy must be stored to transport the vehicle over an acceptable range and therefore, a high-energy density is required. However, the total energy storage is limited by weight, size, and cost of the ESS. Therefore, a key target in the development of ESS for electric vehicles is to maximize energy density and specific energy.

The trends in energy density of advanced batteries have shown a considerable improvement over time. In the past, EVs used lead-acid batteries with a low specific energy of <50 Wh/kg and energy density <90 Wh/L. Li-ion cells today can be manufactured with energy densities as high as 175 Wh/L and specific energy of 144 Wh/kg, with a targeted value of 200 Wh/kg. The basic energy storage capability of base materials determines the theoretical energy density and it is this that will limit the density of future technologies. For example, the theoretical maximum for Li-ion is over 300 Wh/kg, but these maximums are not realistic because the cells are not 100% efficient for a number of reasons. 700 Wh/kg energy density is needed to bring ESS closer to liquid fuels. This will require new ESS cell materials or novel energy storage solutions. Therefore, long-term targets for cost reduction and energy density improvement will require a technological breakthrough in battery cell chemistry.

### 7.2.2 POWER DENSITY AND SPECIFIC POWER

Power density is the amount and rate at which that energy can be delivered per unit volume, whereas specific power is per unit mass. For mild and full HEVs, where the main source of energy is petroleum fuel, the electrical energy requirement is limited. The ESS provides a power boost for rapid acceleration and energy recapture via regenerative braking; therefore, the ESS requires a higher power density over a short period. For this reason, ultracapacitors perform particularly well in low-energy, frequent stop–start cycles, compared to electrochemical cells.

The main differences between an ESS device optimized for use in a PHEV and EV (high-energy density) and the one optimized for use in a HEV (high-power density) are the size of the ESS and the relative quantities of active materials. A high-power lithium cell may have 1.3 kW/kg specific power and a specific energy of 70 Wh/kg, about half the specific energy of the high-energy lithium cell. Figure 7.2 shows the power-to-energy ratio (P/E) requirement for different HEV, PHEV, and EV applications. It should be noted that battery packs have lower energy and power densities than individual cells due to the packaging factor. Typically, a packaging factor of 0.6–0.8 is applied to cell power and energy density to calculate these densities for packs; therefore, only 0.6–0.8 kWh/kg energy density can be reached for a pack formed with cells with 1 kWh/kg energy density.

### 7.2.3 CYCLE LIFE AND CALENDAR LIFE

Cycle life is a measure of the longevity of the ESS based on the number of charge and discharge cycles it can achieve. The number of cycles illustrates how often an ESS device can be charged and discharged repeatedly before an EOL condition is reached, for example, lower limit of the capacity, or maximum limit of cell impedance. EOL capacity is often set at 80% of the nominal capacity at BOL. Depending on the application, the ESS will be cycled through a number of charge and discharge cycles by a specific amount known as depth of discharge (DOD). For EV applications,

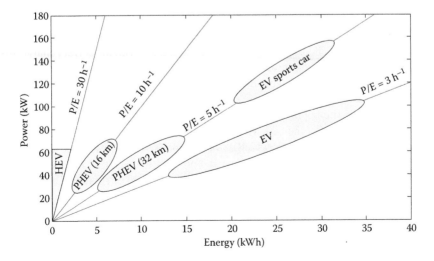

FIGURE 7.2   Requirements for automotive battery systems.

a high DOD is required, to maximize the range of the electrified vehicle. For HEV applications, the DOD will be low, but a much higher number of cycles will be needed due to higher cycling of energy throughput.

Calendar life is a separate measure, based on the calendar longevity of the ESS. The ESS can deteriorate as a result of chemical side reactions that proceed not only during charging and discharging, but also during storage. ESS design, storage temperature, and the charge state affect the shelf life and the charge retention of an ESS.

Lead-acid batteries may be limited to only 500 deep discharge cycles, limiting their suitability for an EV. For conventional vehicles, the cycle life of a 12-V battery will be measured in terms of the number of vehicle starts. Typically, this could be over 30,000 cycles before replacement is necessary. For a microhybrid application with engine stop–start capability, the cycle life will be considerably higher, of the order of 150,000 starts. In these applications, the batteries are only used to start the vehicle and therefore, DOD is relatively low, usually <10%. But for microhybrid application, the power requirement is high and there are a greater number of charges and discharges that the battery has to capture energy during regenerative braking. Advanced lead-acid batteries such as valve-regulated lead-acid batteries offer improved performance for deep cycle and deep discharge applications. On larger vehicles, these could be coupled with ultracapacitors or replaced by more expensive Li-ion technologies. For PHEVs, the DOD will be as high as 80%, and for EVs, it will be >90%. For PHEVs, the trade-off between microcycles (power/acceleration) and deep cycles (energy/range) creates a challenge for Li-ion-based batteries. According to the USABC, Li-ion can achieve over 300,000 50 Wh pulse cycles satisfactorily. However, whether or not this can be achieved over deep cycles remains to be seen. The USABC's 15-year target, at 330 days per year is 5000 deep cycles, a more modest 10 years is still 3300 cycles. This still remains a challenge for many cell chemistries.

### 7.2.4 OPERATING TEMPERATURE

Typically, the operating temperature range for an automotive ESS is from about −40°C to about 60°C, while the specific impact on performance varies by application. Most ESS technologies, especially the standard Li-ion, will likely suffer a decrease in cycle life at extreme high temperatures and/or lower power capabilities at extreme low temperatures. Li-ion is typically more susceptible than lead-acid to performance degradation due to extreme temperatures, which requires Li-ion battery systems to be built with a heating mechanism for cold-weather operation and cooling for extreme heat, which adds cost, weight, and additional system complexity.

### 7.2.5 SAFETY

It is of crucial importance to choose the right ESS in combination with the correct charge, discharge, and storage conditions to assure optimum, reliable, and safe operation. Li-ion batteries, unlike other ESS devices, typically have a flammable electrolyte kept in pressure. A too low end-of-discharge voltage, a too high end-of-charge voltage, or a too high charge or discharge rate can not only affect the lifetime and the cycle life but it can also amount to abuse of the equipment resulting in possible venting, rupture, or explosion of the cell. While charging at temperatures below 0°C, the anode of the cells gets plated with pure lithium, which can compromise the safety of the whole pack. Short circuiting a battery will cause the cell to overheat and possibly to catch fire. Adjacent cells may then overheat and fail, possibly causing the entire battery pack to ignite or rupture. To avoid this, a cell and/or a battery should include protective devices to avoid:

- Application of too high charge or discharge rates
- Improper charge or discharge voltage or voltage reversal
- Short circuiting
- Charging or discharging at too high or too low temperatures

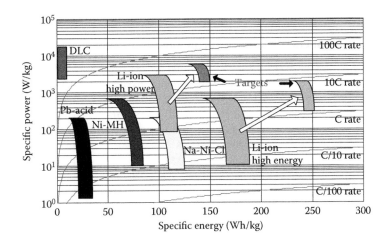

FIGURE 7.3    Comparison of different electric storage systems via Ragone plot.

To ensure that the right operating conditions are used, a battery management system (BMS) is used to monitor the cell voltage, current, and temperature conditions, and to ensure proper operation condition of the battery system.

### 7.2.6   OUTLOOK

The Ragone plot is commonly used for performance comparison of various energy-storing devices. On such a chart, the values of energy density (in Wh/kg) versus power density (in W/kg) are plotted. Both axes are logarithmic, which allows comparing the performance of very different devices. The Ragone chart in Figure 7.3 shows the current state of the art and the comparison of different ESS technologies. Li-ion, which is currently the leading ESS solution for electrified vehicle application, still has a long-reaching target to meet USABC performance requirements. Detailed USABC long-term targets are summarized in Table 7.2. Continuous search of new ESS technologies with high potential will not stop in the foreseeable future. A comparison of current ESS technologies is given in Table 7.3.

TABLE 7.2

**USABC Goals for Advanced Batteries for EVs**

| Parameter of System (Units) | Minimum Long-Term Goal | Desired Long-Term Goal |
| --- | --- | --- |
| Power density (W/L) | 460 | 600 |
| Specific power discharge 80% DOD/30 s (W/kg) | 300 | 400 |
| Specific power—Regen 20% DOD/10 s (W/kg) | 150 | 200 |
| Energy density—C/3 discharge rate (Wh/L) | 230 | 300 |
| Specific energy—C/3 discharge rate (Wh/kg) | 150 | 200 |
| Specific power/specific-energy ratio | 2 | 2 |
| Total pack size (kWh) | 40 | 40 |
| Life (years) | 10 | 10 |
| Cycle life—80% DOD (cycles) | 1000 | 1000 |
| Power and capacity degradation (%) | 20 | 20 |
| Selling price ($/kWh) | 150 | 100 |
| Operating temperature (°C) | −40 to 50 | −40 to 85 |
| Normal recharge time (h) | 6 | 3 |
| High rate charge (min) | 30 (20%–70% SOC) | 15 (40%–80% SOC) |
| Continuous discharge in 1 h (% of capacity) | 75 | 75 |

**TABLE 7.3**

**Comparison of Electrical Energy Storage Technologies**

| | Lead Acid | NiMH | Li-ion | Na–Ni–Cl | EDLC | Hybrid UC |
|---|---|---|---|---|---|---|
| Specific energy (Wh/kg) | 30–50 | 60–120 | 100–265 | 100–120 | 2.5–15 | 2.84–120 |
| Energy density (Wh/L) | 50–80 | 140–300 | 250–730 | 150–180 | 10–30 | 5.6–140 |
| Specific power (W/kg) | 75–300 | 250–1000 | 250–340 | 150–200 | 500–5000 | 2300–14,000 |
| Power density (W/L) | 10–400 | 80–300 | 100–210 | 220–300 | 100,000 | 2500–27,000 |
| Round-trip efficiency (%) | 70–80 | 60–70 | 85–98 | 85–90 | 90–98 | 95–99 |
| Self-discharge (%/day) | 0.033–0.3 | 25–30 | 0.1–0.3 | 15 | 20–40 | 0.1–12.5 |
| Cycle lifetime (cycles) | 100–2000 | 500–1000 | 400–1200 | 2500 | 10,000–100,000 | 5000–200,000 |
| Power capacity Cost ($/kW) | 175–600 | 150–1500 | 175–4000 | 150–300 | 100–360 | 50–320 |
| Energy capacity Cost ($/kWh) | 150–400 | 150–1500 | 500–2500 | 100–200 | 300–94,000 | 600–50,000 |

*Source:* Augmented from Bradbury, K. *Energy Storage Technology Review.* Duke University, Durham, NC, 2010: 1–34.

## 7.3 ELECTROCHEMICAL CELLS

### 7.3.1 BASIC PHYSICS AND ELECTROCHEMISTRY

Electrochemical cells operate by converting electrical energy into chemical energy through a pair of reduction–oxidation (redox) reactions. In redox reactions, electrons are transferred. Redox reactions in electrochemical cells take place with a reduction reaction at one electrode and an oxidation reaction at the other. This is possible by the transferal of electrons from the electrode suffering oxidation to the one being reduced. The particularity in batteries is that these reactions take place in separate places in the battery, forcing the electrons to travel from one electrode to the other. If the load to be powered is located in the electrical path the electrons are moving through, then power can be used to generate work. The following terminology is used to define redox reactions:

| | |
|---|---|
| Reduction | A gain of electrons or a decrease in the oxidation state. |
| | An example reduction reaction: $A^{n+} + ne^- \rightarrow A$ |
| Oxidation | A loss of electrons or increase in the oxidation state. |
| | An example of an oxidation reaction: $A^{n-} \rightarrow A + ne^-$ |
| Reducing agent | Also known as a reducer, it is a substance that donates electrons to another species, and as such, it is said to have been oxidized. |
| Oxidizing agent | Also known as an oxidizer, it is a substance that accepts electrons from another species, and as such, it is said to have been reduced. |

In typical chemical redox reactions, electrons are transferred between molecules. An electrochemical reaction is one that is either facilitated through an externally applied voltage, for example, electrolysis, or one that generates an electric voltage. Electrochemistry deals with cases where redox reactions are separated by space or time and connected by an external circuit. Rather than redox occurring simultaneously at a common location, electrons flow externally to enable redox to occur at different places and at different times.

Electrochemical cells can be classified as galvanic cells or electrolytic cells. A galvanic cell is one that spontaneous redox reactions occur and produce positive cell voltages. Nonrechargeable batteries can be referred to as galvanic cells. Electrolytic cells involve nonspontaneous redox reactions that are driven by an externally applied voltage. Rechargeable (secondary) batteries act as galvanic cells when discharged and as electrolytic cells when charged.

An electrochemical cell contains the following three basic parts: positive electrode, negative electrode, and electrolyte. The electrodes are typically referred to as the anode and cathode, and the exact definition depends on whether the cell is charging or discharging. The following definitions are used to classify the electrodes during different operation conditions:

*Cathode*—Electrode where electric current flows out or electrons flow in
*Anode*—Electrode where electric current flows in or electrons flow out

Given the above definitions during discharge, the positive electrode is the cathode and the negative electrode is the anode; during cell-charging operation, the reverse definitions apply. The positive and negative electrodes are surrounded by the electrolyte. It is electrically insulating and ion conducting. Liquid electrolytes typically have dissolved compounds that are ionized in an aqueous solution such as water. To increase power and energy density, electrodes are typically placed as close as possible to each other; to prevent short circuit, a separator is also a part of the cell. The design of the separator is cell-type specific; a key property of it is that it allows ion flow through it. The flow of current between the positive and negative electrodes occurs via electrons externally and ions internally. Two types of ions can be found in the electrolyte:

*Anion*—An ion with net negative charge, that is, more electrons than protons
*Cation*—An ion with net positive charge, that is, more protons than electrons

The operation of the cell during discharge is described as follows. Electrons flow from the negative electrode through the load to the positive electrode. This causes the positive electrode to become slightly less positive and the negative electrode to become slightly more positive resulting in a charge imbalance within the cell. To compensate for this charge imbalance, ions flow inside the cell, effectively completing the current flow. Any anions in the electrolyte drift through the electrolyte toward the negative electrode to donate electrons in an oxidation reaction. Simultaneously, any cations drift toward the positive electrode to accept electrons in a reduction reaction. This process is depicted in Figure 7.4. During charging operation of the cell, the reverse processes occur where electrons are forced into the negative electrode and any anions in the electrolyte drift toward the positive electrode and any cations drift toward the negative electrode.

Since the mass transport of the ions does not occur instantaneously within the cell, it is expected that when the electrodes are disconnected from the external circuit, ions will continue to drift toward their respective electrodes to internally balance the charge within the cell. This is one of the major processes that is responsible for a measurable voltage relaxation at the cell terminals. The ions in the electrolyte are also always subjected to spatial diffusion effects. This is another dominant mechanism that causes a dynamic voltage response to step current inputs. The maximum drift velocity of the ions in the electrolyte is limited; therefore, the cell has a nonzero internal resistance.

The amount of energy that can be stored in an electrochemical cell is limited by the amount of active chemical species in the electrolyte that can be stored in the cell. The power capability of the cell is determined by the surface area of the electrode/electrolyte interface. Therefore, a given cell of a finite size has a limited storage capacity and power rate determined by the internal packaging design of the cell.

The voltage across the terminals of the cell is dependent on the chemical potential voltages of the two reactions occurring at each electrode or half-cell. It is approximately equal to

$$E^o = E^o_{pos} - E^o_{neg} \tag{7.1}$$

where the superscript "*o*" refers to standard conditions of gas pressures at 1 atm, temperature of 25°C, and 1 M (molar) concentrations. The voltage values $E^o_{pos}$ and $E^o_{neg}$ are reported as chemical potential voltages to a reference standard hydrogen electrode (SHE). This reference is the reaction

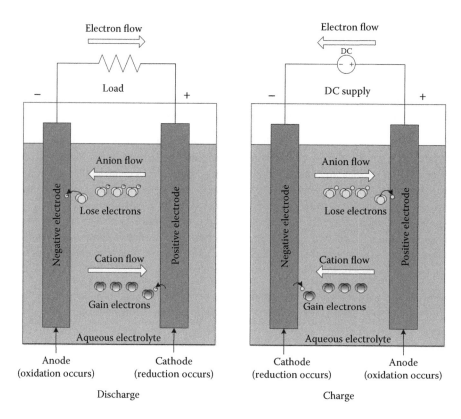

**FIGURE 7.4** Discharge and charge of an electrochemical cell. (Adapted from Linden, D., and T.B. Reddy. *Handbook of Batteries*. New York, 2002.)

$2H^+_{(aq)} + 2e^- \rightarrow H_{2(g)}$, where "(aq)" signifies the ion is dissolved in an aqueous/water solution and "(g)" denotes the gaseous state; this reaction has, by definition, the reference potential voltage of 0 V. Values for $E^o_{pos}$ and $E^o_{neg}$ have been empirically measured for many reduction reactions at standard conditions and are reported in so-called standard reduction tables [1]. To obtain the voltage for an oxidation reaction, one can reverse the chemical equation and multiply the tabular reference voltage by −1.

The conditions in and around the cell are rarely at standard conditions. For example, the concentrations of active species in the electrode/electrolyte, for example, the dissolved cations/anions in the electrolyte or absorbed ions in electrodes, vary as the cell is charged or discharged. Therefore, it is expected the cell terminal voltage is a function of the concentrations of the reactants and products available for the two half-cell reactions at each electrode. A deviation of voltage from standard conditions can be given by the Nernst equation

$$E = E^o - \frac{RT}{nF}\ln\left(\frac{[C]^c[D]^d}{[A]^a[B]^b}\right) \tag{7.2}$$

where $R$ is the gas constant (8.314 J K$^{-1}$ mol$^{-1}$), $T$ is temperature, $n$ is the number of moles of electrons transferred, $F$ is Faraday's constant (96,485 C mol$^{-1}$), and the square brackets indicate species concentrations of the overall chemical reaction

$$aA + bB \rightarrow cC + dD \tag{7.3}$$

The state of charge (SOC) of the cell is directly related to the concentrations described above such that at equilibrium, the open-circuit-measured cell potential is typically a monotonically increasing function of SOC. Besides the deviations described by the Nernst equations, other effects such as hysteresis also affect the measured open-circuit potential. An empirical model that lumps all these effects will be described later in this chapter.

## 7.3.2 LEAD ACID

A very common and mature battery type particularly used for automotive starting, lighting, and ignition (SLI) applications is lead-acid batteries. The reactions that drive this cell are as follows:

$$\text{Positive electrode } PbO_2 + SO_4^{2-} + 4H^+ + 2e^- \underset{\text{Charge}}{\overset{\text{Discharge}}{\rightleftharpoons}} PbSO_4 + 2H_2O$$

$$\text{Negative electrode } Pb + SO_4^{2-} \underset{\text{Charge}}{\overset{\text{Discharge}}{\rightleftharpoons}} PbSO_4 + 2e^-$$

$$\text{Net cell reaction } Pb + PbO_2 + 2H_2SO_4 \underset{\text{Charge}}{\overset{\text{Discharge}}{\rightleftharpoons}} 2PbSO_4 + 2H_2O$$

$$E^o = 2.04 \text{ V (discharge)}$$

The positive electrode is lead oxide and the negative electrode is lead. As the cell is discharged, lead sulfate is formed at both electrodes. Flooded lead-acid batteries use liquid water with dissolved sulfuric acid as the electrolyte. As the cell is discharged, the concentration of the acid decreases. The cell in its theoretical fully charged and discharged states is shown in Figure 7.5. There are both anions ($SO_4^{2-}$) and cations ($H^+$) to facilitate current and ion flow inside the cell. Since the densities of water and sulfuric acid differ, these ions are also subject to convection flow as well as diffusion and drift from an electric field.

The construction of a cell is typically in the form of multiple plates in the arrangement depicted in Figure 7.6 to increase cell power output. A porous separator allows electrolyte and ion flow and prevents the electrodes from short circuiting. The electrode plates typically contain spongy portions to further increase surface area and power capability. Energy capacity of the cell can be increased by increasing plate thickness.

Overcharging of flooded lead-acid batteries generates oxygen gas and hydrogen gas via electrolysis and results in a water loss. This can be compensated for via periodic water replacement maintenance. Special vents also need to be designed into the cell to prevent gas buildup. Lead-acid batteries also typically last longer and are less prone to freezing when they are stored at high SOC levels. Common faults of the battery/cell include corrosion at the cell terminals and plate cracking.

Fully charged                    Fully discharged

FIGURE 7.5   Charged and discharged states of a lead-acid electrochemical cell.

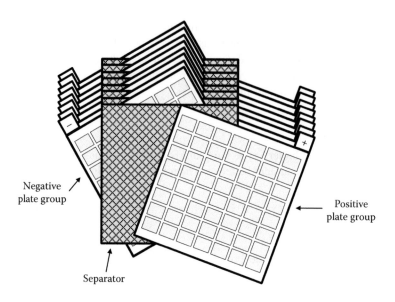

**FIGURE 7.6**    Plate construction of lead-acid battery cell.

A dominant aging mechanism is sulfation, the crystallization of lead sulfate, which prevents the ions from dissolving in the electrolyte and participating in current flow within the cell. At the end of cell life, lead-acid batteries are routinely recycled.

Other variants include sealed lead acid (SLA) and valve-regulated lead acid (VRLA). A minimum amount of electrolyte is used. Absorbed gas mat (AGM) designs immobilize the electrolyte by absorbing it into a porous glass microfiber. Alternately, gel electrolytes can be employed. Both designs enable the cell to be arbitrarily oriented. A key design feature, as illustrated in Figure 7.7, is their recombinant nature of oxygen and hydrogen gas into water inside the cell; this reduces water loss and vents less hydrogen and oxygen gas during charging. The reduced water loss and sealed packing design result in these batteries to be typically referred to as maintenance-free lead-acid batteries. Compared to flooded leadacid, SLA/VRLA have minimal or no leakage in the event of cell puncture, they use less volume, have higher cycle life, and operate at a slightly higher floating voltage of 2.25 V compared to 2.17–2.22 V. However, SLA/VRLA technology is less mature, and potential for thermal runaway exists.

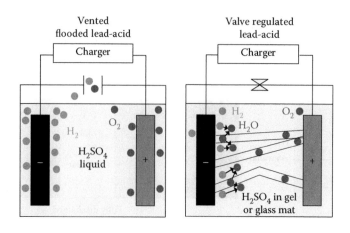

**FIGURE 7.7**    Comparison between flooded lead acid and VRLA.

### 7.3.3 NICKEL–METAL HYDRIDE

Nickel–metal hydride (NiMH)-based cells are also relatively mature. They have largely replaced the older but similar nickel cadmium (NiCd)-based technology due to the toxicity of cadmium, its memory effect, and reduced cost for NiMH. The metal M is typically an intermetallic compound of the form $AB_5$ or $AB_2$ [2]. In the former case, A is a combination of La, Ca, Pr, and Nd; B is a combination of Ni, Co, Mn, and Al. For $AB_2$ cases, A is a combination of Ti, V, and Zr; B is a combination of Ni, Co, Cr, Mn, Al, and Sn. The electrolyte is typically an aqueous solution of 30 wt% KOH. The reactions that govern this cell are

$$\text{Positive electrode } NiOOH + H_2O + e^- \underset{\text{Charge}}{\overset{\text{Discharge}}{\rightleftharpoons}} Ni(OH)_2 + OH^-$$

$$\text{Negative electrode } MH + OH^- \underset{\text{Charge}}{\overset{\text{Discharge}}{\rightleftharpoons}} M + H_2O + e^-$$

$$\text{Net cell reaction } MH + NiOOH \underset{\text{Charge}}{\overset{\text{Discharge}}{\rightleftharpoons}} M + Ni(OH)_2$$

$$E^o = 1.35 \text{ V (discharge)}$$

From an electrochemistry point of view, the negative electrode is hydrogen; however, it is absorbed in the metal alloy, for example, $LaNi_5H_6$. Internal ion current flow in the electrolyte is due to only the anion $OH^-$. The cell operation is depicted in Figure 7.8. Proton movement ($H^+$) occurs at the negative electrode. Nickel is connected to each electrode to act as current collectors for electron flow.

NiMH can suffer from pole reversal when over discharged permanently damaging the cells. This occurs in multicell arrangements when other cells drive the polarity reversal. Modern cells contain catalysts to recombine hydrogen and oxygen gas to form water during low (trickle) overcharge currents and in the process generate heat. However, this is less effective at larger currents; hence for

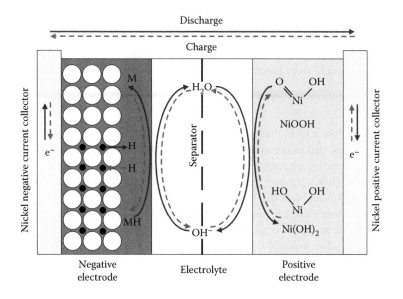

**FIGURE 7.8** NiMH cell operation.

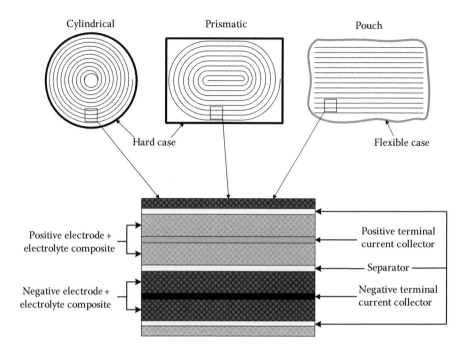

FIGURE 7.9    Cross sections of example cell-packaging styles.

safety, vents are still part of the cell design to prevent gas buildup; internal-pressure switches to disconnect the cell can also be present. Compared to other cell types, NiMH has a relatively high self-discharge rate ranging from 5% to 20% per day. The dominant aging occurs at the MH electrode due to the following reasons: repeated mechanical stress from expansion/contraction during cycling/usage causing particle breakdown, corrosion of MH resulting in growth of a resistive layer, gas production due to overdischarge and overcharge, and self-discharge consumption of the active electrode material and water from the electrolyte.

Common form factors for the cells are cylindrical and prismatic styles with hard cases. A spirally wound layering approach as depicted in Figure 7.9 is common to increase power capability. A stacked approach similar to lead acid is also possible for prismatic design to utilize space better. The composite layers are typically spongy structure-like electrodes embedded/filled with an electrolyte. This also maximizes surface area to increase power output. Separator layers are needed; for NiMH, a common material is made from grafted polyethylene/polypropylene nonwoven fabric.

### 7.3.4   Lithium Ion

Cells that $Li^+$ cations flow within the cell are referred to as Li-ion. The technology is still heavily researched with designs still being optimized. There are many variations consisting of different electrode materials. Positive electrode materials include lithium cobalt oxide ($LiCoO_2$), lithium manganese oxide ($LiMnO_4$), lithium iron phosphate ($LiFePO_4$), lithium nickel manganese cobalt oxide ($LiNi_\alpha Mn_y Co_\gamma O_2$), and lithium nickel cobalt aluminum oxide ($LiNi_\alpha Co_y Al_\gamma O_2$); the latter two are blended variations that are referred to as NMC and NCA, respectively. The most common negative electrode material is graphite intercalated with lithium ($LiC_6$), although lithium titanate ($Li_4 Ti_5 O_{12}$) is also available. The nominal cell voltage is dependent on the materials chosen [3].

Given the high reactivity of lithium with water, nonaqueous or aprotic electrolytes are used. Liquid electrolytes consist of lithium salts ($LiPF_6$, $LiBF_4$, and $LiCoO_4$) dissolved in an organic solvent such as dimethyl carbonate, diethyl carbonate, and ethylene carbonate. Solid electrolytes are also possible such as polyoxyethylene, the so-called dry Li-polymer cells that can be made using solid electrolyte polymers. A separator material is also present that can be porous polypropylene, polyethylene, or composite polypropylene/polyethylene films. Compared to other cell types, Li-ion has advantages of high voltage, low self-discharge, and high efficiencies.

The cell-operating principles of the different chemistries are similar; $LiCoO_2$ will be described here as an example. The reactions governing this cell chemistry are as follows:

$$\text{Positive electrode } CoO_2 + Li^+ + e^- \underset{\text{Charge}}{\overset{\text{Discharge}}{\rightleftharpoons}} LiCoO_2$$

$$\text{Negative electrode } LiC_6 \underset{\text{Charge}}{\overset{\text{Discharge}}{\rightleftharpoons}} Li^+ + C_6 + e^-$$

$$\text{Net cell reaction } LiC_6 + CoO_2 \underset{\text{Charge}}{\overset{\text{Discharge}}{\rightleftharpoons}} LiCoO_2 + C_6,$$

$$E^o = 3.7 \text{ V (discharge)}$$

The lithium cation flow is depicted in Figure 7.10 for the charge and discharge modes where an intercalation process is shown. The subscripts $x,y$ are fractional values between 0 and 1 indicating the proportion of lithium ions intercalated within each electrode. For real cells, the value of $y$ in the positive electrode is typically between 0.5 and 1. Owing to the reactivity of lithium with air, copper, and aluminum electrodes are used as the cell terminals.

A key feature of Li-ion cells is that upon initial charge, a surface film or surface electrolyte interface (SEI) layer is formed between the electrodes and the electrolyte. It effectively protects against a perpetual reaction of intercalated lithium with the electrolyte. The growth of this layer is one mechanism for capacity loss and impedance growth of the cell as it ages. Mechanical stress as a result of negative electrode expansion/contraction during battery cycling is another aging and failure mechanism.

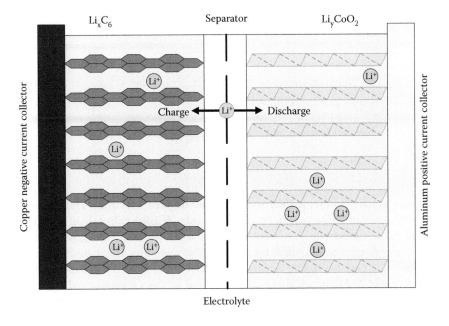

FIGURE 7.10  Li-ion cell operation.

Potentially catastrophic failure can occur when the cell is overcharged or overdischarged. Overcharging the cell can lead to cell swelling and pressure buildup within the cell; moreover, terminal cell voltages of >5.2 V cause side reactions that are detrimental to the cell. Overdischarging can potentially internally short circuit the cell and lead to excessive overheating possibly leading to fire. Safety features such as vents, thermal interrupts, and external short-circuit interrupting are the common built-in cell features. Regardless of these safety features, terminal cell voltage and thermal management are key issues in the application of Li-ion technology in a battery pack.

Thermal runway is a well-known catastrophic failure mode of Li-ion cells. It typically arises when there is an internal short circuit or separator breakdown within the cell. An exothermic reaction then results between the positive and negative electrode materials, which causes cell overheating and excessive gas release. The electrolytes tend to be flammable exacerbating the fire hazard. Among the cell chemistries, $LiCoO_2$ is most sensitive to the potential of thermal runway, particularly in mechanical failure modes involving accidental cell puncture and rupture.

Li-ion cells come in a variety of form factors including cylindrical, prismatic, and pouch styles, as depicted in Figure 7.6. Cell packing can be similar to NiMH. Pouch styles utilize a flexible case and are common for solid electrolyte Li-polymer cells since electrolyte leakage is less of a concern. The pouch-style form factors tend to utilize space better and can comparatively have better energy/power densities than hard case cells. The terminal connection mechanism between pouch cells that utilize stacked layers and hard case cells that are spirally wound is also different.

### 7.3.5  SODIUM NICKEL CHLORIDE

An alternative cell chemistry type that is finding its way into some automotive and transportation applications is the sodium nickel chloride cell, commercially known as ZEBRA batteries. They were invented over 25 years ago within the Zeolite Battery Research Africa project by Coetzer's research group in South Africa. It is considered as an improvement over its older derivative of sodium sulfur batteries since the replacement of sulfur eliminated much of its safety concerns. Both types are referred to as molten salt batteries since their high operating temperature results in molten electrodes. The typical operating temperature of ZEBRA cells is between 270°C and 350°C. The electrochemical reactions of this cell are given as follows:

$$\text{Positive electrode} \quad NiCl_2 + 2Na^+ + 2e^- \underset{\text{Charge}}{\overset{\text{Discharge}}{\rightleftharpoons}} Ni + 2NaCl$$

$$\text{Negative electrode} \quad 2Na \underset{\text{Charge}}{\overset{\text{Discharge}}{\rightleftharpoons}} 2Na^+ + 2e^-$$

$$\text{Net cell reaction} \quad NiCl_2 + 2Na \underset{\text{Charge}}{\overset{\text{Discharge}}{\rightleftharpoons}} Ni + 2NaCl$$

$$E^o = 2.58 \text{ V (discharge)}$$

At the operating temperature, molten sodium is the negative electrode. Unlike sodium sulfur cells that employ a molten sodium sulfur positive electrode, ZEBRA cells utilize a solid porous-like nickel chloride positive electrode. Two electrolytes are used, one is $NaAlCl_4$, which has a melting point of 157°C; therefore, it is a liquid at the operating temperature. This liquid electrolyte permeates through the solid porous positive electrode. A second solid electrolyte known as β-alumina is an isomorphic form of $Al_2O_3$, which is a hard ceramic that acts as a separator and also prevents molten sodium from reacting with the liquid electrolyte. The cells are packed in upright rectangular box shapes with the height being the longest dimension, as depicted in Figure 7.11. A square horizontal cross section in Figure 7.11c depicts a clover-leaf geometry design to increase the surface area and

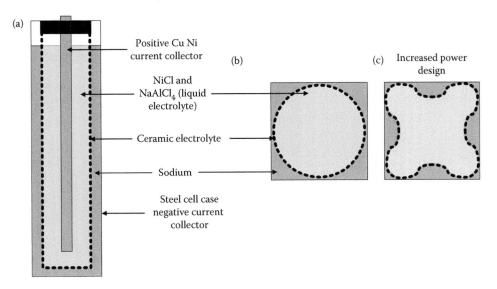

**FIGURE 7.11** Cross section of sodium nickel chloride battery cell: (a) vertical cross section (side view), (b) horizontal cross section (top-down view), and (c) alternate design.

power capability of the cell. A steel case encloses the cell and acts as a negative terminal current collector. A solid copper nickel alloy is used as the positive terminal current collector. Alternative positive electrode compositions where iron partially replaces nickel are also possible. This lowers the voltage of the cell but overall improves power and energy capabilities [4].

The necessity to maintain a high temperature even when the cell is in a stand-by state generates heat loss that effectively acts as a self-discharge loss. While the cell is in operation, these losses are minimal due to self-heating of the cell. A special pack design and insulation is employed to deal with the thermal considerations of this cell, and as a result, they are fairly insensitive to ambient operating temperatures. A key safety and fault-tolerant feature of this cell is that in the event of solid electrolyte cracking, a mild exothermic reaction between negative electrode sodium and the liquid electrolyte generates solid aluminum that effectively shorts the cell with resistance compared to that of an intact cell. Therefore, a pack can tolerate 5%–10% of the cell failing and can still be operational.

## 7.4  ULTRACAPACITOR CELLS

### 7.4.1  Basic Physics

Capacitors store energy electrostatically in an electric field. A conventional capacitor, see Figure 7.12, can be made from two oppositely charged plates separated by a dielectric material. Traditional dielectric materials are ceramics, polymer films, or aluminum oxide. Charge movement occurs via the alignment of molecular dipoles in the dielectric material. The capacitance for this conventional construction is inversely proportional to the separation distance, and directly proportional to the surface area of the electrode plates and the dielectric constant, for example, $C = \varepsilon A/d$.

Ultracapacitors (or supercapacitors) are ESS that employ electrostatic-based storage and electrochemical-based mechanisms; however, their materials and disposition endow them with a much higher capacitance than conventional designs. This higher capacitance allows them to present higher energy densities (6 Wh/kg) than conventional capacitors and very high power densities (14,000 W/kg). Ultracapacitors can be classified into different groups depending on whether the energy storage mechanism at each electrode is non-Faradaic or Faradaic.

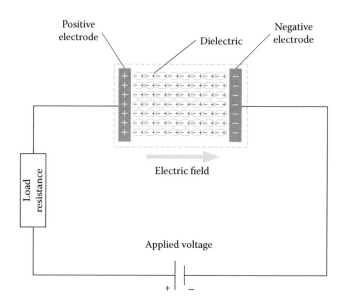

FIGURE 7.12   Conventional capacitor diagram.

| Non-Faradaic electrode | The energy mechanism is entirely electrostatic, where oppositely charged ions attract and move toward each other; however, no charge or electrons are transferred between these ions. |
|---|---|
| Faradaic electrode | Charge transfer at the electrode can occur via electrons. The energy storage mechanism is partially electrochemical. |
| Pseudocapacitance | An observed capacitance that arises from charge transfer at a Faradaic electrode. The charge transfer can arise via redox reactions, electrosorption, or an intercalation process. |

A common ultracapacitor where both electrodes are non-Faradaic is the EDLC. Another commercially available ultracapacitor type are hybrid capacitors (HCs); they contain one Faradaic electrode and one non-Faradaic electrode. An ultracapacitor that has a Faradaic electrode is said to exhibit pseudocapacitance.

### 7.4.2   Electric Double-Layer Capacitors

Compared to conventional capacitors, an EDLC has higher electrode surface area (2000 m²/g), thinner and porous electrodes (smaller than 150 μm), and electrode micropores in the Angstrom order of magnitude. Unlike conventional capacitors that employ a solid dielectric material, the material between both electrodes in an EDLC is an electrolyte, and it can be either aqueous or organic based. Both the electrolyte solvent and any dissolved ions can be absorbed into the micropore structures. A separator is also present to separate the electrodes. Each part is depicted in Figure 7.13. An example of a microstructure is shown in Figure 7.14.

Energy is stored in an EDLC as charge separation, where the electrolyte ions diffuse toward the electrodes. This ion diffusion process in the electrolyte is similar to the one in electrochemical-based storage described in Section 7.3. The internal separation of charge by the movement of ions generates an internal electric field. Most of the energy is stored in the interface between the electrode and electrolyte, the so-called Helmholtz layer. It is sandwiched by two layers, one is the accumulated electrolyte ions, and the other is attracted charge in the electrode. The Helmholtz layer is typically in the order of one atom in thickness, which yields a large capacitance at the electrolyte/electrode interface. There are two Helmholtz layers in an EDLC, one at each positive/ negative electrode. The mobility of the ions is effectively exploited to reduce as much as possible

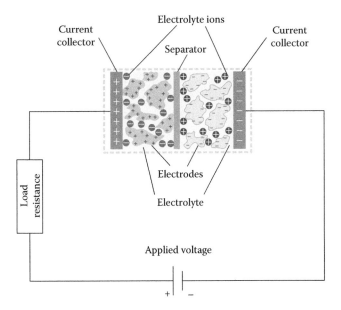

FIGURE 7.13   EDLC diagram.

the charge separation distance needed to increase capacitance. The amount of charge stored is further increased due to the high electrode surface and small micropore diameter. The energy storage mechanism for an EDLC is completely non-Faradaic. These ultracapacitors normally have carbon-based electrodes made from activated carbon or are constructed from carbon nanotubes/nanostructures. The electrolyte composition can vary from aqueous ($H_2SO_4$ or KOH) to organics such as acetonitrile. The main difference in terms of operation is in the equivalent series resistance (ESR) and pore size needed. Normally, aqueous-based electrolytes need smaller pore size and present a lower ESR. Organic electrolytes present a higher breakdown voltage. When voltage is applied, the electrolyte ions are separated, each being attracted by the electrode with opposite charge. This non-Faradaic energy storage is highly reversible, allowing EDLCs to reach million cycle lifetimes. The packaging styles can be similar to the cell packaging shown in Figure 7.9 to further increase power and energy density. The conceptual difference between EDLC and electro-chemical storage is shown in Figure 7.15.

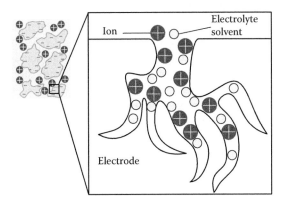

FIGURE 7.14   Micropore structure in EDLC.

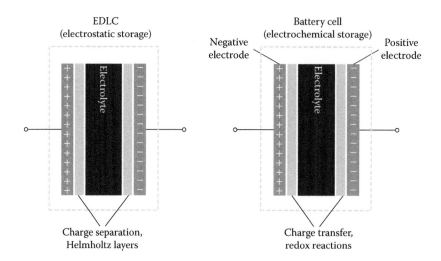

FIGURE 7.15    Comparison between EDLC and electrochemical cell.

### 7.4.3    ULTRACAPACITORS WITH PSEUDOCAPACITANCE

Pseudocapacitance can arise at an electrode that exhibits charge transfer. One mechanism for charge transfer is through redox reactions similar to electrochemical cells via ions in the electrolyte. In electrochemical cells, these redox reactions tend to exhibit a slow response as a result of phase changes during the redox reaction. Ultracapacitors with pseudocapacitance typically employ fast sequences of reversible redox reactions with no phase changes. The other processes that can facilitate charge transfer are electrosorption and intercalation, where in both cases, the ions or atoms once absorbed in the electrode cling to the atomic structure after electron transfer without making or breaking chemical bonds. Both mechanisms interact with the lattice structure of the electrode, and electrosorption is differentiated by deposition of hydrogen or metal adatoms in the surface lattice sites. An electrode can exhibit both electrostatic energy storage and electrochemical energy storage, for example, a Helmholtz layer that contains leaky Faradaic current. In this case, the total capacitance at the electrode/electrolyte interface is composed of traditional electrostatic capacitance and the pseudocapacitance. Electrode materials can be either polymers or metal oxides. Polymer electrodes present a low ESR and high capacitance, but a lower stability. Metal oxides present low ESR but at a higher cost. Pseudocapacitance can offer higher energy densities but can pose additional stress in the cell and potentially limit the cycle life.

Ultracapacitors that exhibit pseudocapacitance are frequently referred to as pseudocapacitors. Different variations are possible depending on where and how much pseudocapacitance is in the cell. Symmetric pseudocapacitors contain two Faradaic electrodes and in principle are conceptually identical to the electrochemical cell process depicted in Figure 7.4. Symmetric pseudocapacitors employing the fast electron transfer mechanisms for pseudocapacitance are yet to attain commercial success; traditional electrochemical battery cells dominate energy storage cells containing two Faradaic electrodes.

An asymmetric pseudocapacitor, for example, HC or hybrid ultracapacitor (HUC), depicted in Figure 7.16, presents both Faradaic and non-Faradaic electrodes to combine each of its strengths. One electrode can be similar to an EDLC electrode employing electrostatic storage, and the other can be an electrode employing pseudocapacitance and charge transfer. A battery-type HC design is also possible, where one electrode employs traditional electrochemical redox reactions; the other electrode is a non-Faradaic electrode. HCs that have lithium ions in the electrolyte are also referred to as Li-ion capacitors. Various types of HC designs are possible; the first uses a positive Faradaic electrode as in Figure 7.16, where metal oxides can be used to enable ion intercalation, for example,

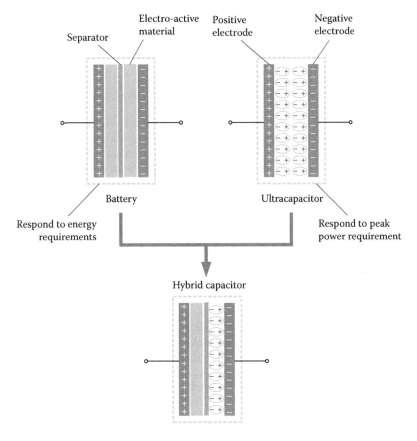

**FIGURE 7.16** Hybrid capacitor.

Nesscap pseudocapacitor. The second type employs a negative Faradaic electrode, where carbon has been used by JSR Micro as the electrode material. The two aforementioned approaches fully replace a non-Faradaic electrode with a Faradaic one. A partial replacement is also possible, for example, the so-called ultrabattery technology based on lead acid uses a composite negative electrode composed of a lead Faradaic electrode and a non-Faradaic carbon electrode.

## 7.5 CHARACTERISTIC TERMINOLOGY AND PERFORMANCE PARAMETERS

| | |
|---|---|
| Specific energy | Energy per unit mass, typically expressed as Wh/kg |
| Energy density | Energy per unit volume, typically expressed as Wh/L |
| Specific power | Power per unit mass, typically expressed as W/kg |
| Power density | Power per unit volume, typically expressed as W/L |
| Round-trip efficiency | Expressed as percentage of fraction of energy output during discharge compared to input energy during charge |
| Self-discharge | Energy loss per unit time, typically %/day |
| Cycle lifetime | Expected number of useful cycles, dependent on definition of cycle |
| Power capacity cost | Monetary cost per unit power, for example, $/kW |
| Energy capacity cost | Monetary cost per unit energy, for example, $/kWh |
| SOC | Either expressed as a percentage 0%–100% or number between 0 and 1. Indicates the recommended extremes of cell states, for example, active-material/ion concentrations in an electrochemical cell, which represent a fully charged or fully discharged cell condition. Typical definitions of SOC involve integrating current flow and dividing by the rated capacity. |

| | |
|---|---|
| DOD | Indicates the ratio or percentage to the total capacity that has been discharged, for example, DOD = 1 − SOC or DOD% = 100 − SOC%. |
| Ampere hour (Ah) capacity | A common unit manufacturers use to describe the total capacity of a cell under some predescribed conditions, for example, 20°C at 1/20 C discharge rate. A related metric is watt-hour (Wh) obtained by multiplying rated Ah by rated cell voltage. |
| Discharge capacity | The total capacity that can be extracted from a fully charged cell from SOC = 1 to SOC = 0. Denoted as $Ah_d$ or $Wh_d$. |
| Charge capacity | The total capacity that can be charged into a cell from SOC = 0 to SOC = 1. Denoted as $Ah_c$ or $Wh_c$. |
| Coulombic efficiency | Also called Faradaic efficiency for an electrochemical cell, it describes the efficiency in which charge (e.g., electrons) is transferred in a system to facilitate the main electrochemical reaction. Participation in side reactions generating heat is an example of Faradaic loss. Separate efficiencies for charging and discharging are possible. The metric is typically measured stoichiometrically by comparing transferred charge to amounts of active material. |
| C-rate | A charge or discharge rate equal to the rate that can fully charge or discharge the cell in 1 h. |
| Min/max cutoff voltage | Manufacturer-specified voltages limit the cell. Current limitations need to be applied to avoid exceeding these limits. |
| Calendar life | The expected life span of the cell under storage conditions. Can be sensitive to factors such as ambient temperature and SOC. |
| Cell voltage reversal | Weak cells forcibly operated at negative voltages can result in a possibly permanent, polarity reversal at the cell terminals. Can shorten cell life or lead to complete failure. |
| Remaining useful life (RUL) | Typically, a unit of time representing the remaining life of the cell before it no longer satisfies application requirements of power output and available capacity. It is heavily dependent on operating conditions such as environmental factors and usage history. |
| State of health (SOH) | A dimensionless metric used to define the general health or usability of the cell usually for the purpose of estimating RUL. Impedance-based metrics, for example, $SOH_r$, are typically defined based on ratios of the aged cell impedance and fresh cell impedance. Capacity-based SOH, for example, $SOH_c$, is based on ratios of aged cell-degraded capacity and fresh cell capacity. Many variations of SOH equations are possible. |
| BMS | The BMS is a combination of battery pack hardware and software with measurement, control, and communication capabilities to report the status/states of the multiple cells and perform low-level protection features such as cell balancing, thermal management, and optimal cell control. |
| Battery state estimation (BSE) | A collection of algorithms and software residing in the BMS to perform key functions such as estimating SOC, SOH, RUL, capacity, and maximum available power output. |
| Open-circuit voltage (OCV) | An internal voltage of the cell that is measured across the cell terminals in zero-current/disconnected rested equilibrium conditions. Its value is dependent on cell SOC and hysteresis effects. |
| Electrochemical impedance spectroscopy (EIS) | A frequency domain-testing procedure employing small signal input perturbations at common cell-operating points. |
| Potentiostatic | A term to describe an EIS testing condition where the average input signal is a constant voltage (CV) and output is the current response. |
| Galvanostatic | A term to describe an EIS testing condition where the average input signal is a constant current (CC) and output is the voltage response. |
| ESR | Typically related to the series resistance to represent losses in inductors/capacitors. |
| ECM | Equivalent circuit model |

## 7.6 MODELING

### 7.6.1 ELECTROCHEMICAL CELL—EQUIVALENT CIRCUIT MODEL

There are a variety of ECM approaches and topologies for electrochemical cells. They typically contain common circuit elements to mimic the experimental responses observed from cell characterization data. In this section, a specific ECM with $n$ RC elements, depicted in Figure 7.17, is presented that mimics effects such as transient response, hysteresis, nonlinear OCV, asymmetric internal resistance, and thermal dependence. It can be readily modeled in circuit simulation software using standard packages such as PLECS and/or Matlab Simscape/SimPowerSystems.

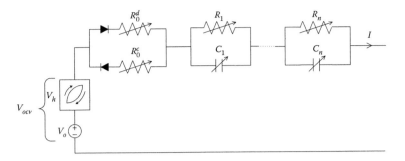

FIGURE 7.17   Equivalent circuit model of battery cell.

The SOC is modeled as an integrator with the following dynamics:

$$\frac{dSOC(t)}{dt} = \frac{-1}{CAP}[\eta_c I^-(t) + \eta_d^{-1} I^+(t)] - \rho_{sd} \tag{7.4}$$

$$I^-(t) = \min(I(t),0), \quad I^+(t) = \max(I(t),0) \tag{7.5}$$

where positive current $I$ indicates discharging, $\eta_c, \eta_d$ are charging and discharging efficiencies, $CAP$ is cell capacity, and $\rho_{sd}$ is the self-discharge rate. The SOC impacts the values of other elements in the ECM since they are modeled as functions of SOC and temperature, that is,

$$\begin{aligned}
V_o &= g_{V_o}(SOC,T) \\
R_i &= g_{R_i}(SOC,T) \quad i = 0\ldots n \\
C_j &= g_{C_j}(SOC,T), \quad j = 1\ldots n
\end{aligned} \tag{7.6}$$

The OCV of the cell is decomposed into two components, $V_{OCV} = V_o + V_h$, as depicted in Figure 7.18. One is dependent on SOC, and the other is dependent on hysteresis. The so-called combined model can be used to represent the former component, which is based on the equation

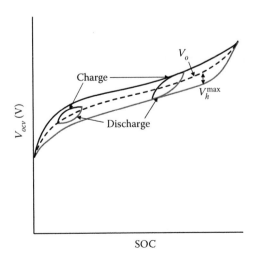

FIGURE 7.18   Components of open-circuit voltage and hysteresis loops.

$$V_o = k_0 - k_1/SOC + k_2 SOC + k_3 \log SOC - k_4 \log(1 - SOC) \tag{7.7}$$

where the coefficients $k_i$ are empirically found from experimental OCV measurement data. The above is a combination of the OCV forms from the Shepherd, Unnewehrl, and Nernst models, each of which contain OCV–SOC monotonic dependent terms [5].

A simple method to model hysteresis is a zero-state hysteresis approach, that is,

$$V_h = -V_h^{\max} \operatorname{sgn}(I) \tag{7.8}$$

where $V_{\max}^h$ represents the maximum hysteresis voltage, and it can also be modeled as a function of SOC and temperature, for example, $V_h^{\max} = g_{V_h^{\max}}(SOC, T)$. This approach is simple; however, it has been shown to poorly mimic cell voltage response [6] since it neglects hysteresis-specific transient behavior such as the inner hysteresis loops depicted in Figure 7.18. An improvement can be made via the introduction of a hysteresis state within the cell, and this approach is referred to as the one-state hysteresis model [6], which has the first-order dynamics

$$V_h = V_h^{\max} \cdot \upsilon_h$$
$$\frac{d\upsilon_h}{dt} = -\gamma I \cdot (\operatorname{sgn}(I)\upsilon_h + 1) \tag{7.9}$$

where $\upsilon_h$ is a hysteresis state with values between −1 and 1, and $\gamma$ is a slew-type rate.

The RC pairs follow the first-order dynamics

$$\frac{dV_j}{dt} = \frac{-V_j}{R_j C_j} + \frac{I}{C_j} \tag{7.10}$$

where $V_j$ is the voltage across the $j$th RC pair. The voltage of the ECM can be summarized as

$$V = V_h + V_o - I^+ R_0^d - I^- R_0^c - \sum_{j=1}^{n} V_j \tag{7.11}$$

It is noted here that the model order $n$ depends on the application and cell. Common choices are between 1 and 3. For modeling purposes, higher orders may be desirable, whereas for online estimation filter design, a lower-order filter may be acceptable. Cells phenomena such as diffusion is best approximated with one or two RC pairs with long time constants.

Discretizing the ECM dynamics enables the use of many offline parameterization and online estimation algorithms. A zero-order hold (ZOH) approach is taken where the sample time is chosen to be small enough such that the parameter variations during this interval are assumed to be constant; therefore, without loss of generality, the functional dependence on SOC and temperature will be dropped for neatness of presentation. The discrete-time dynamics of Equation 7.1 become

$$SOC_k = SOC_{k-1} - \frac{\Delta t}{CAP}[\eta_c I_{k-1}^- + \eta_d^{-1} I_{k-1}^+)] - \rho_{sd}\Delta t \tag{7.12}$$

where the subscript $k$ indicates a sample, $\Delta t$ is sample time, and Equation 7.2 still applies. A discrete-time expression for the expression of $V_{o_k}$ is identical to Equation 7.7 with the exception of added subscripts $k$. Hysteresis discrete-time dynamics are derived as

$$
V_{h_k} = \begin{cases} -V_h^{\max}\,\mathrm{sgn}\,(I_k) & \text{Zero-state model} \\ V_h^{\max} \cdot \upsilon_{h_k} & \text{One-state model} \end{cases} \tag{7.13}
$$

$$
\upsilon_{h_k} = -\mathrm{sgn}(I_{k-1}) + (\upsilon_{h_{k-1}} + \mathrm{sgn}(I_{k-1}))e^{-\gamma|I_{k-1}|\Delta t} \tag{7.14}
$$

where Equation 7.14 is derived by first discretizing the charging and discharging case separately and then forming the above general expression. For the RC elements, their discrete-time dynamics become

$$
V_{j_k} = e^{-\frac{\Delta t}{R_j C_j}} V_{j_{k-1}} + R_j \left( 1 - e^{-\frac{\Delta t}{R_j C_j}} \right) I_{k-1} \tag{7.15}
$$

Finally, the discrete-time voltage output is simply

$$
V_k = V_{h_k} + V_{o_k} - I_k^+ R_0^d - I_k^- R_0^c - \sum_{j=1}^{n} V_{j_k} \tag{7.16}
$$

## 7.6.2  Electrochemical Cell—Enhanced Self-Correcting Model

An alternative so-called enhanced self-correcting (ESC) model directly describes the discrete-time voltage dynamics via a selected number of internal filter states [6]. The development of this model is performed in the discrete-time domain. It can be considered as a black-box approach to model the transient dynamics since the parameters of this model are solely based on empirical measurement data. Instead of $n$ RC voltage states, there are $n$ ESC voltage filter states that constitute a portion of the cell terminal voltage. The voltage response of the ESC model in the discrete-time domain is as follows:

$$
V_k = V_{h_k} + V_{o_k} - I_k^+ R^d - I_k^- R^c + \underbrace{\sum_{j=1}^{n} \beta_j V_{ESC,j_k}}_{V_{ESC_k}} \tag{7.17}
$$

where $V_{h_k}, V_{o_k}, I_k^+$, and $I_k^-$ are modeled as Equations 7.13, 7.7, and 7.5. The discrete-time SOC integrator (7.12) is also employed. The resistances $R^c$ and $R^d$ represent the overall steady-state charging and discharging resistances. $V_{ESC,j_k}$ represents the $j$th internal ESC state with $\beta_j$ as its empirically found coefficient. The latter terms equal $V_{ESC_k}$, the linear combination of internal states. The ESC states satisfy two properties: (1) each internal state is stable, and (2) the value of $V_{ESC_k}$ is zero under steady-state current. These conditions provide the desirable voltage transient and relaxation effects. The internal states are modeled as

$$
V_{ESC,j_k} = \alpha_j V_{ESC,j_{k-1}} + I_{k-1} \tag{7.18}
$$

where $\alpha_j$ specifies a filter pole, and choosing its value to be $-1 < \alpha_j < 1$ ensures stability of state $V_{ESC,j_k}$. The ESC voltage can be summarized as the following discrete-time filter state-space dynamics:

$$
\begin{bmatrix} V_{ESC,1_k} \\ \vdots \\ V_{ESC,n_k} \end{bmatrix} = \overbrace{\begin{bmatrix} \alpha_1 & 0 & \cdots & 0 \\ 0 & \alpha_2 & \cdots & 0 \\ \vdots & \vdots & \ddots & \vdots \\ 0 & 0 & \cdots & \alpha_n \end{bmatrix}}^{A_{ESC}} \begin{bmatrix} V_{ESC,1_{k-1}} \\ \vdots \\ V_{ESC,n_{k-1}} \end{bmatrix} + \overbrace{\begin{bmatrix} 1 \\ \vdots \\ 1 \end{bmatrix}}^{B_{ESC}} I_{k-1}
$$

$$
V_{ESC_k} = \underbrace{\begin{bmatrix} \beta_1 \cdots \beta_n \end{bmatrix}}_{C_{ESC}} \begin{bmatrix} V_{ESC,1_k} \\ \vdots \\ V_{ESC,n_k} \end{bmatrix}
$$

(7.19)

The corresponding discrete-time $z$-transfer function of the above is

$$
G(z) = \mathbf{C}_{ESC} (\mathbf{I}Z - \mathbf{A}_{ESC})^{-1} \mathbf{B}_{ESC}
$$

(7.20)

To satisfy the second zero-gain steady-state condition on $V_{ESC}$, $G(z) = 0$ for $z = 1$ must be satisfied. This leads to the following constraint on the ESC model parameters:

$$
\sum_{j=1}^{n} \frac{\beta_j}{1 - \alpha_j} = 0
$$

(7.21)

The above constraint can be embedded into an optimization-based fitting routine when performing ESC model fitting. It is also noted here that the typical filter orders for the ESC model are $n = 2$ and $n = 4$ [6].

### 7.6.3 Ultracapacitor Cell

Compared to electrochemical cells, ultracapacitor cells exhibit less nonlinear behavior and can be modeled fairly well using equivalent circuit-based models. The simplest approach is to model the cell with an ideal capacitor element and an ESR to represent losses; this approach is depicted in Figure 7.19.

For an EDLC, the model can be decomposed to consider the different dominant parts of the ultracapacitor model in Figure 7.15 to obtain the equivalent circuit model in Figure 7.20. Ideal capacitors $C_{HL1}$, $C_{HL2}$ can model the two capacitive Helmholtz dual layers at each electrode/electrolyte interface. In between these layers, a resistor $R_{se}$ can model the ionic resistance due to the separator and movement through the electrolyte. An additional two resistances $R_{e1}$, $R_{e2}$ can represent the electrical resistances at the electrode and cell terminal.

Transient and high-frequency responses can be modeled by adding inductor and RC elements, as depicted in Figure 7.21. This approach can be amenable to HCs since RC dynamics can be used to model transient behavior from Faradaic electrodes. Additional RC pairs can be added if necessary.

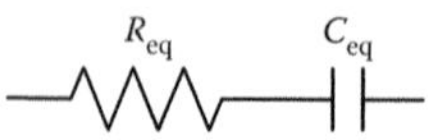

FIGURE 7.19  Series resistance ultracapacitor equivalent circuit model.

FIGURE 7.20  EDLC equivalent circuit model.

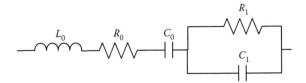

FIGURE 7.21 Ultracapacitor equivalent circuit model with inductor and RC dynamics.

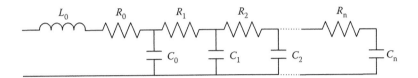

FIGURE 7.22 Multistage ladder model of ultracapacitor.

A multistage model depicted in Figure 7.22 is another common approach. It mimics the distributed physical nature of the ultracapacitor fairly well, for example, the large surface is at the porous electrode/electrolyte interface in an EDLC. This model has also been shown to fit empirical experimental data fairly well over a large frequency range. The number of stages or model order can be chosen based on the number of dominant time constants observed from experimental data. In an EDLC, effects such as ionic diffusion in the electrolyte and ion movement through macro/micro pores can occur at different timescales. An order selection of three or higher is commonplace.

## 7.7 TESTING PROCEDURES

### 7.7.1 TIME DOMAIN

In this section, common time-domain tests used to characterize a cell are described. The example responses shown are typical to that of electrochemical cells; however, the same procedures can be applied to ultracapacitor cells. Tests are performed in controlled environment conditions such as constant atmosphere pressure and temperature via the use of thermal environmental chambers.

A common test on a fully charged cell involves a constant discharge to a minimum cutoff voltage, otherwise known as a C-rate discharge capacity test. The typical voltage responses are shown in Figure 7.23. From this figure, the internal resistance of the cell contributes to two effects observed; the first is an initial voltage drop that increases with higher discharge currents, and the second is underutilization of the total cell capacity that reduces the usable capacity at higher currents. The cell impedance is typically temperature dependent; therefore, the voltage response for C-rate discharge tests and the usable capacity are also temperature dependent.

One expects that the total capacity can be independent of temperature and current rates. For electrochemical cells, the amount of available active chemical material dictates capacity; for ultracapacitors, the surface area of charge storage dictates capacity. Both are practically constant across a wide range of normal operating conditions, and any cell capacity degradation occurs over long time periods and/or high number of usage cycles. A total capacity test can be performed with the use of CC and CV charge–discharge segments. An example of voltage (black) and current (blue) responses is shown in Figure 7.24. A CC charge is performed until the maximum cutoff voltage is reached. This is followed by a CV charge until the charging current diminishes to some small threshold, usually of magnitude C/100. This point is defined as 100% SOC. Then a CC discharge is performed until the lower voltage threshold is reached. This is followed by a CV discharge to the same (C/100) threshold point, that is, 0% SOC. Current integration then defines a

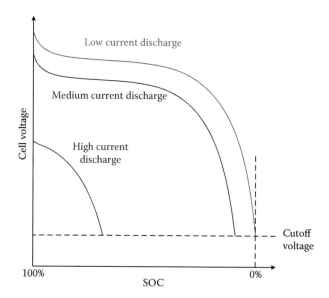

FIGURE 7.23    Cell voltage response at different discharge currents (C-rates).

total capacity estimate, and the shaded area in Figure 7.24 indicates the total discharge capacity, that is, $Ah_d$. Repeating CC–CV charge segments, a similar total charge capacity estimate ($Ah_c$) can be obtained by current integration from 0% to 100% SOC. Usually, multiple CC–CV charge–discharge cycles are performed and the results are averaged. Any inefficiency in the cell will result in a total discharge capacity to be less than total charge capacity. These two total capacity estimates can be used to determine the charging and discharging efficiencies described in Section 7.6. The values of these efficiencies depend on whether cell capacity is defined as total charge capacity or total discharge capacity.

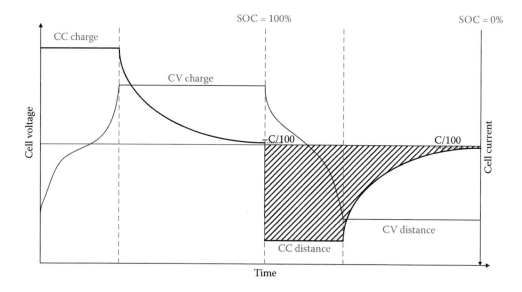

FIGURE 7.24    CC and CV charge/discharge profiles for a total capacity test measurement. The shaded area is a total discharge capacity measurement.

The CC–CV charge/discharge procedure described above defines a pair of extreme OCV–SOC points namely at SOC = 0% and SOC = 100%. The CC portions of this test can be modified to find intermediate points and construct the OCV versus SOC curve. Two curves are needed as depicted in Figure 7.18, one for charge and the other for discharge. The OCV–SOC test procedure is summarized as follows:

- Cell charge that ends with a CV charge segment to bring SOC to 100%. A 1–2 h rest time then follows.
- Repeated low rate, for example, C/10, discharge pulses, and resting periods. Each discharge pulse is of constant time interval that decreases the SOC by 5%–10%. The subsequent rest time is usually 1–2 h to allow the cell to reach an equilibrium state. At the end of the rest period, the measured voltage is taken as an OCV measurement.
- At low SOC, the lower voltage cutoff will be reached during the last discharge pulse; once this occurs, a CV discharge segment is performed to reach SOC 0%. A 1–2 h rest time then follows. This ends the discharge OCV–SOC test, and the charge OCV–SOC test follows next.
- Repeated low-rate charge pulses are followed by resting periods. This part follows the same guidelines as the discharge pulses.
- At high OCV, the upper voltage limit will be reached during the last charge pulse, and a CV charge segment is performed to reach SOC 100%. A 1–2 h rest time then follows. This ends the charge OCV–SOC test.

Typical voltage and current profiles are depicted in Figure 7.25. A capacity test is done beforehand; therefore, intermediate SOC points are found by current integration of the charge/discharge pulses. The OCV–SOC test is done over a wide operating range, typically a cell whose OCV–SOC curve is insensitive to changes in temperature and aging effects are desired since they make calibration and cell state/parameter estimation easier.

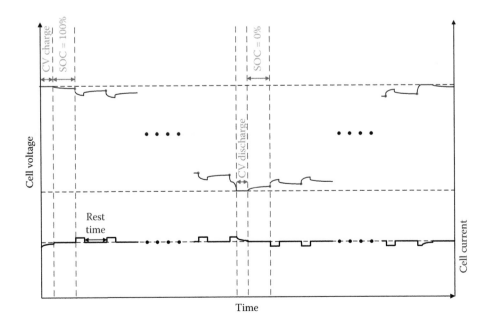

FIGURE 7.25  OCV–SOC test for discharge and charge. The red-circled voltage points are OCV measurements.

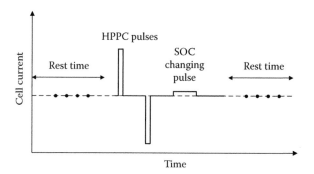

FIGURE 7.26   Typical current pulses for HPPC tests.

A similar procedure that tests cell impedance and power capabilities is the so-called high (or hybrid) pulse power characterization (HPPC) test. It adds additional high-current pulses, for example, 10 C, for short durations, for example, 10 s, to the current profile for an OCV–SOC test. Usually, pairs of equal magnitude charge and discharge pulses are added at each different SOC operating point, as shown in Figure 7.26. Rest periods following the HPPC charge and discharge pulses may or may not be used. At the extreme SOC operating points, only one of the two HPPC pulses can be used to avoid overcharge/discharge and to respect cell voltage limits; alternatively, a pulse employing CV charge/discharge can be used. Variations to the HPPC pulse times, current magnitudes, sequence, and temporal spacing are possible. For example, having multiple HPPC current pulses with different magnitudes can enable characterization of current-dependent ECM parameters and also enable better characterization of any hysteresis dynamics.

One of the main purposes of HPPC testing is to enable ECM parameter fitting; these parameters are used for modeling and peak power estimation. A typical voltage response of a HPPC pulse is shown in Figure 7.27a. An ohmic resistance response can be readily calculated from this curve, given the instantaneous voltage and current changes. Dynamics effects are also present and include changes in OCV (from a change in SOC), hysteresis, and dynamic resistances such as the RC pairs

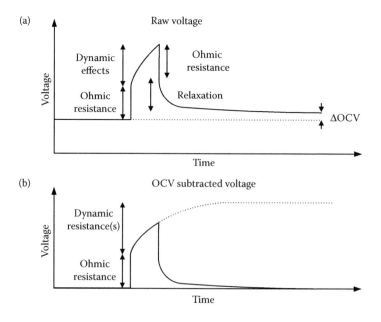

FIGURE 7.27   (a) Raw and (b) processed voltage response for cell characterization.

in the cell ECM of Figure 7.17. Curve fitting can be applied to the voltage response curve to obtain model parameters.

Assuming the ECM model of Figure 7.17, it is desirable to process the voltage response to focus parameterization of the linear ECM elements. The OCV response is subtracted from the measured voltage to produce the response shown in Figure 7.27b. The OCV response can be found by using the bracketing resting voltage points (OCV endpoints) and current integration; special testing/processing considerations may need to be applied to correctly handle any cell hysteresis and lack of sufficient rest/relaxation times. The ideal OCV-subtracted response of $n$ RC pairs can be derived as

$$
V_{OS}(t) = \begin{cases} -\left( R_0 + \displaystyle\sum_{j=1}^{n} R_j \left( 1 - e^{\frac{-t}{\tau_j}} \right) \right) i_{pulse} & t < t_{pulse} \\[2em] -\displaystyle\sum_{j=1}^{n} R_j \left( 1 - e^{\frac{-t_{pulse}}{\tau_j}} \right) i_{pulse} e^{\frac{-(t - t_{pulse})}{\tau_j}} & t \geq t_{pulse} \end{cases} \tag{7.22}
$$

where $t_{pulse}$ is the pulse time duration and $i_{pulse}$ is the pulse current. The above equation assumes zero-state initial conditions, that is, sufficient prior rest time. Equation 7.22 or its discrete-time analogs can be used for parameter fitting.

Finally, it is noted that many other time-domain tests are available for modeling and characterization purposes. The so-called dynamic stress test (DST) [7] is composed of a sequence of CC charge/discharges steps at various C-rates. It is designed to operate the cell at a wide SOC range. For automotive applications, current profiles derived from expected drive-cycle response are used to excite the cell under more realistic conditions. These drive-cycle tests can be used for both model parameterization and validation. The nonlinearity of electric energy storage devices, particularly electrochemical cells, warrants extensive testing to examine and model their response over the application lifetime. Calendar-aging tests involve testing the effect of ambient environmental conditions over long time durations on a rested cell at different SOC levels. Only periodic capacity/characterization tests are performed. Drive-cycle aging tests are a form of accelerated testing that repeatedly cycle the cell with drive-cycle usage profiles. The repeated drive-cycle current profiles can be separated with CV charge/discharge segments to maintain a desired SOC operating range. Periodic HPPC and capacity tests are also applied. Energy throughput of the cell is typically recorded and used as a metric for cell age.

## 7.7.2 FREQUENCY DOMAIN

EIS is a frequency domain-testing procedure that employs small amplitude signal perturbations at different cell-operating points, as depicted in Figure 7.28. Each operation point is defined by the SOC, the operating current, temperature, and whether it is a charging or discharging process. If the signal imposed is voltage, the test is potentiostatic, whereas if it is current controlled, it is called galvanostatic. This signal must be of small amplitude to respect the initial hypothesis of linearity for each test at the operating point. This will in practice mean that the SOC, temperature, and amplitude must be kept in a tight interval to guarantee that results can be related to a specific operating point. This is one of the most important conditions to fulfill during frequency domain tests, as electrical storage devices typically present a great deal of nonlinear behavior, and the results for one operating point should not overlap other operating points.

The signal used to carry out the frequency domain tests can be sinusoidal, delta function, white noise, or pseudorandom binary signals. In cell testing, the approach chosen by most is to use sinusoidal signals; this is convenient for equivalent circuit model fitting, capacity fading, SOC, or SOH

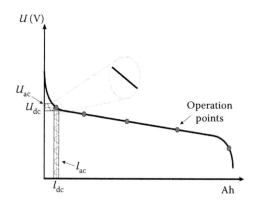

**FIGURE 7.28**  Battery operation points during EIS tests.

estimation. Its main advantage is the simplicity of the waveform used, but it presents some disadvantages due to the high number of measurements needed to be done to guarantee a reasonable accuracy. This can be a considerable drawback when working at low frequencies (1 mHz or less) due to the length for each test, which can be above several hours. This issue can severely compromise keeping a tight operation condition, especially for the cell temperature and SOC.

The signal imposed, a small AC ripple that is a fraction of the DC current flowing, can have a fixed or variable frequency. The most usual approach is having a frequency sweep between low frequencies (a few mHz) and high frequencies (a few kHz). This frequency sweep allows to obtain the cell impedance spectra in a wide range, being able to observe different behaviors. An example of frequency response result for a Li-ion cell is shown in Figure 7.29; the distinctive areas in the Nyquist plot are matched to their dominant equivalent circuit-modeling elements. It is a common convention to invert the vertical imaginary axis as depicted in Figure 7.29.

At the mid-frequency range, the response exhibits behavior that can be described using RC circuit elements. These elements produce semicircular behavior that is depicted in Figure 7.30. The ratio between the different time constants changes the relative blending between multiple RC pairs.

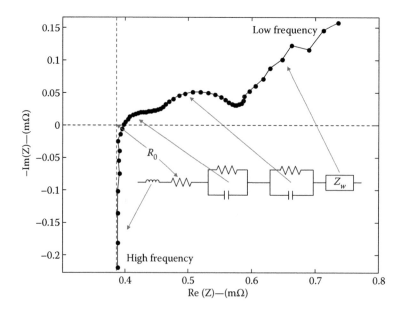

**FIGURE 7.29**  Example of an Nyquist plot for a Li-ion battery.

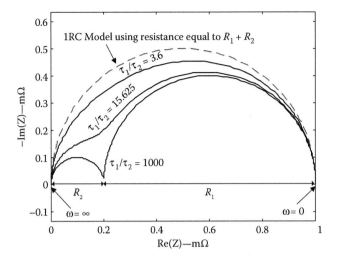

FIGURE 7.30 Nyquist plot for 1RC/2RC model elements with different time constants $\tau_1 = R_1C_1$ and $\tau_2 = R_2C_2$. Here, $R_1 = 0.8$ m$\Omega$ and $R_2 = 0.2$ m$\Omega$.

At higher frequencies, inductive behavior is usually evident, which can sometimes be a by-product of the wiring used to connect the cells. Capacitive behavior is typically evident at low frequencies.

The frequency response can exhibit sloped behavior that cannot be represented by simple circuit elements. These sloped responses can be described by the so-called constant phase elements (CPEs); they are represented by the formula

$$CFE(\omega) = A(j\omega)^a = \begin{cases} L_a(j\omega)^a & a > 0 \\ R & a = 0 \\ \dfrac{1}{C_a(j\omega)^a} & a < 0 \end{cases} \tag{7.23}$$

where "$a$" is a number between $-1$ and $1$, and "$A$" is a coefficient that scales the circuit element response. Inductor/capacitor elements have a constant phase lead/lag of 90°, and the CPE is a generalization of these standard elements. Nyquist plots of different CPEs are depicted in Figure 7.31. The usage of CPEs in the frequency domain can model nonstandard inductance, corrosion observed at low frequencies, and the so-called Warburg impedance effects.

At low frequencies, a sloped behavior is commonly observed; this portion is commonly referred to as Warburg impedance, and its simplest form is given by

$$Z_w(\omega) = \frac{R_w}{\sqrt{\omega}}(1 - j) \tag{7.24}$$

where $R_w$ is a resistance-like parameter. Warburg impedance typically describes transport effects such as semi-infinite diffusion in an electrochemical cell or capacitive charging with a porous electrode; the latter case can involve no diffusion. At very low frequencies, the shape of the frequency response can either converge toward the real axis or tend toward the negative imaginary axis; in these cases, modified Warburg impedance terms expressed as Equation 7.25 or 7.26 can be employed. The different representations of Warburg impedance are shown in Figure 7.32.

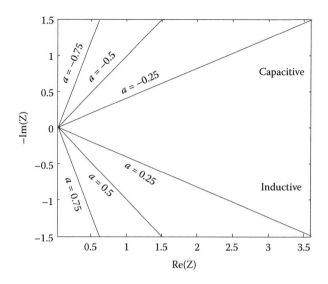

FIGURE 7.31   Nyquist plots for different values of "$a$" for each CFE $A(j\omega)^a$.

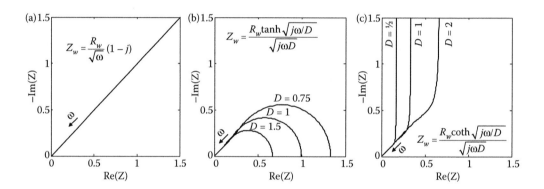

FIGURE 7.32   Nyquist plots for different representations of Warburg element [17]: (a) CFE model, (b) tanh-based model, and (c) coth-based model. A unit resistance was used in the plots.

$$Z_w(\omega) = \frac{R_w \tan h \sqrt{j\omega/D}}{\sqrt{j\omega D}} \tag{7.25}$$

$$Z_w(\omega) = \frac{R_w \coth \sqrt{j\omega/D}}{\sqrt{j\omega D}} \tag{7.26}$$

## 7.8   PACKS AND MANAGEMENT SYSTEMS

### 7.8.1   FUNCTIONS AND DESIGN CONSIDERATIONS

The power and energy requirements for a vehicle energy storage pack exceed what a single electrochemical cell or ultracapacitor cell can provide. Therefore, ESS packs are designed with a collection of many cells. Typically, multiple cells of approximately a dozen are grouped in a module, and then multiple modules are grouped to form a pack. This approach allows modularity in the design. Many cell arrangements and topologies are possible for the design of the energy module/pack. The common wiring-only types are depicted in Figure 7.33, where to increase output voltage, cells are connected in series, and to increase current output, parallelization is employed. Strings that are connected in series can be prone to failure/showdown by the weakest cell in the chain. For a series-only configuration, Figure 7.33a, open-circuit failures will cause total failure of the pack; complete pack shutdown will also occur if the weakest cell needs to be shut down for safety reasons. Series–parallel configurations, Figure 7.33b, are somewhat less prone to this since other parallel cell strings can potentially meet the power/energy requirements, if only at least for some time duration. The most complex arrangement is a matrix topology shown in Figure 7.33c, where parallel cell groups are placed in series. This topology can, in principle, bypass a single open-circuit cell failure and utilize the remaining working cells. The manufacture of this topology can be prohibitive due to the extra wiring involved. Moreover, uneven current circulations can arise when parallel cell arrangement is employed.

Ultimately, individual cell management may be required to ensure optimal operation and safety of the pack; therefore, more complex topologies are needed. This can be achieved through cell-balancing topologies to be described in the next section. These topologies can be designed to not only perform cell equalization, but to also protect against overcharge/discharge, short circuits, and mechanical cell defects/failure.

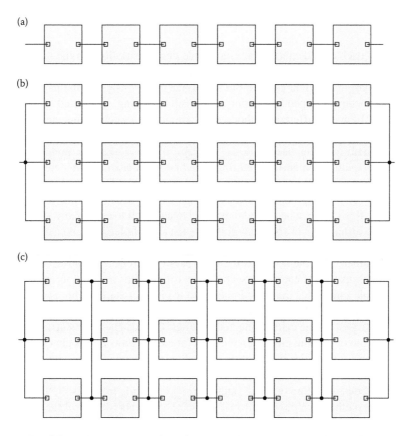

FIGURE 7.33  Possible cell arrangement for a battery pack/module: (a) series configuration, (b) series–parallel configuration, and (c) matrix configuration.

An energy management system (EMS) is a key component of the energy pack. For systems that utilize electrochemical cells only, the EMS is the BMS. The EMS performs critical reporting, sensing, communication, and control functions of the energy pack. The key reporting and sensing functions are

- Cell measurements—for example, voltage, current, temperature, and stress/strain
- Pack measurements—for example, DC link voltage and current, and ambient temperature
- Fault detection—for example, open/short-circuit failures, sensing failures, and cooling system failures
- SOC, SOH, impedance, and capacity estimation
- Maximum power output capability
- Available pack energy
- Data logging, for example, usage history

The key communication and control functions for the EMS/BMS are

- Cell balancing
- Controlling the main contactor switches
- Thermal management—for example, active heating/cooling systems
- Controlling any cell-protecting circuitry
- Communication handshaking with power-train controller
- Communication with power-electronic components—for example, DC/DC converters, charger, and auxiliary power unit

### 7.8.2  CELL BALANCING

Within the energy pack cell, deviations are inevitable due to variations in the manufacturing processes and unequal aging/usage of the cells. Balancing SOC among the cells is needed to avoid overcharging/discharging deviant cells and to maximize energy usage of the pack. A wide variety of topologies have been developed for the purpose of cell balancing. A selected few are presented in this section to highlight the different methodologies employed for cells connected in series.

The common terminology to describe the balancing methods includes the use of terms active and passive. The latter has been used to describe one of the two scenarios, both of which dissipate energy to perform equalization during charging. The first is a noncontrollable topology where cells are connected in series and they can tolerate a slight overcharge. The overcharged cells naturally dissipate energy usually through the generation of heat. During this time, this allows other cells to catch up in charge. Resistors can be placed in parallel to each cell to exploit any voltage/SOC differences among cells; differences in voltage will dictate the rate of bleeding and cell charging leading the different cells to naturally converge to a balanced state. Lead-acid and some NiMH batteries can be used in this manner. This method uses the term passive to reflect the absence of control in the topology and the dissipation of energy. A second scenario that has used the term passive balancing refers to controllable topologies that bleed energy to avoid overcharge. This is typically done via shunting resistors and controllable switches to enable energy bleeding. An example is depicted in Figure 7.34 for the case of three cells connected in series. Confusion arises since such examples are active in the sense of control, but passive in the sense of energy flow. In this chapter, the term active balancing will refer to topologies that contain controllable elements that are actively controlled by the BMS. Since this defines nearly all topologies to be active, three main categories are described here: shunting methods, shuttling methods, and energy-converting methods [8,9].

Active shunting methods employ controllable bypass elements to control the flow of charge to a cell. A common, simple, and low-cost approach is the dissipative shunting method depicted in Figure 7.34. The size of the resistor controls the rate of balancing. It functions to protect only against overcharge by bleeding charge from high SOC cells while lower SOC cells catch up in charge. The creation of heat in this topology is a major source of inefficiency. This may complicate thermal

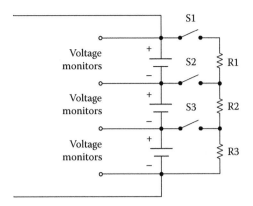

FIGURE 7.34   Dissipative shunting topology.

management of the pack and can result in uneven temperature distribution within the pack, further aggravating imbalances between cells.

A complete shunting topology [8] depicted in Figure 7.35 can balance the cells in both charging and discharging modes by allowing bypass of individual cells that are at the extremes of SOC or voltage. Efficiency improvements arise from eliminating deliberate energy dissipation. For example, in the normal charging state, switches S1–S6 are open; if the first cell is at the maximum, then S4 is closed to bypass charging of that cell. In normal discharging mode, switches S1–S3 are closed and S4–S6 are all open; if the first cell is at the minimum, then S1 is opened to prevent further discharge of the first cell. A variable and potentially wide pack voltage range can result from the disconnection of multiple cells, which may necessitate additional converter power electronics at the pack output to regulate this voltage. The additional components and control complexity increase the cost of this topology.

The second category of active balancing are shuttling methods [8,9]; they use external energy-storing elements, for example, capacitors, to move charge between cells within a pack. They can be used in both charging and discharging modes. An example for $n = 3$ cells is depicted in Figure 7.36 employing $n - 1$ capacitors and $n$ three-terminal switches. In this example, charge

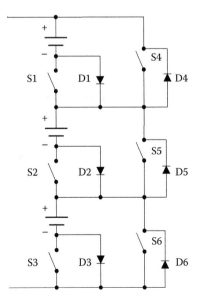

FIGURE 7.35   Complete shunting topology.

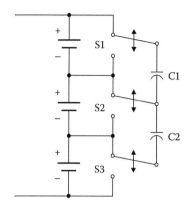

FIGURE 7.36   Switched capacitor-shuttling topology.

is shuttled between nearby cells depending on the switch configuration and cell voltage conditions. For example, consider the case when the first cell is at a higher SOC/voltage state than
the second cell, C1 would be connected to be in parallel with this cell to drain it and charge
the capacitor. Then C1 would be connected in parallel to the second lower SOC/voltage cell to
transfer charge to it. The same strategy can be employed for pack charging and discharging.
The size of the capacitor affects the speed both in terms of voltage change time constants and
maximum amount of charge transfer shuttling. Equalization speed can be low in cases where
charge needs to be transferred from faraway cells. Alternative shuttling topologies that change
the arrangement and number of capacitors have also been developed to improve balancing control and speed [9].

The third category of balancing methodologies include energy converter methods. They can
feature multiple and possibly isolated converters such as buck and/or boost, Cûk, flyback, or quasi-
resonant [9]. They are characterized by their increased balancing control capability and relatively
high complexity. An example employing isolated bidirectional step-up (boost) converters with isolated grounds is depicted in Figure 7.37. A single-cell-to-pack philosophy is used here where the

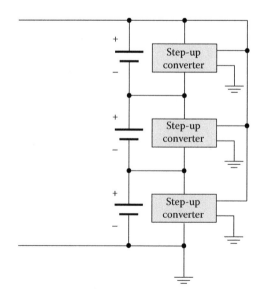

FIGURE 7.37   Step-up energy converter topology.

converters can control the rate and direction of balancing of each cell by controlling the current flows through the converters. More complex control is needed for this topology since low-level control is needed at each converter and a higher level of control is employed for balancing. The design of the converter is also dependent on the number of cells in series since it needs to accommodate a potentially large discrepancy between cell and pack voltage. Besides the advantage of enhanced balancing control, this design is also fairly modular.

## 7.9 STATE AND PARAMETER ESTIMATION

### 7.9.1 ESTIMATION ALGORITHMS

In this section, common discrete-time estimation algorithms used for online energy storage state/parameter estimation are briefly reviewed. Specifically, recursive least squares (RLS), Kalman filter (KF), and extended Kalman filter (EKF) are presented. A brief overview of each algorithm and its mechanics is given. The focus is on the implementation of the algorithm rather than the theory of its derivations.

RLS is an algorithm that can estimate parameters using a recursive implementation of weighted least-squares fitting [10]. It assumes the following regressed form:

$$y_k = \mathbf{c}_k^T \mathbf{x}_k + v_k \tag{7.27}$$

where $y_k$ is a measurement, $\mathbf{c}_k$ is a regression column vector composed of known/measurable quantities, $v_k$ is a noise vector, and $\mathbf{x}_k$ is a vector to be estimated. RLS aims to find recursively in time an estimate of $\mathbf{x}_k$ that minimizes the following sum of squared errors:

$$E(k) = \sum_{i=1}^{k} \lambda^{k-i} (y_i - \mathbf{c}_i^T \mathbf{x}_k)^2 \tag{7.28}$$

where $0 < \lambda \leq 1$ is a forgetting factor that diminishes the influence of old data; it is typically chosen close to one. The algorithm also recursively updates an estimate of the covariance matrix; it is denoted as $\mathbf{P}_k$. The algorithm requires initial estimates $\mathbf{x}_0$ and $\mathbf{P}_0$. The mechanics of the algorithm are illustrated in Figure 7.38 and Table 7.4.

The second algorithm presented is the KF. It fundamentally differs from RLS in that it embeds a model into the filter and uses it to predict an intermediate estimate. The KF also generalizes to consider a multidimensional output vector. Therefore, a discrete-time state-space model of the following form is assumed:

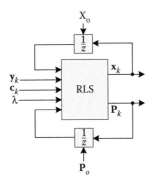

**FIGURE 7.38** RLS block diagram.

**TABLE 7.4**

**RLS Algorithm**

1. Kalman gain calculation          $\mathbf{k}_k = \mathbf{P}_{k-1}\mathbf{c}_k(\lambda + \mathbf{c}_k^T\mathbf{P}_{k-1}\mathbf{c}_k)^{-1}$

2. Covariance matrix update

$\mathbf{P}_k = \dfrac{1}{\lambda}(\mathbf{I} - \mathbf{k}_k\mathbf{c}_k^T)\mathbf{P}_{k-1}$

3. Estimate update

$\mathbf{x}_k = \mathbf{x}_{k-1} + \mathbf{k}_k(y_k - \mathbf{c}_k^T\mathbf{x}_{k-1})$

$$\mathbf{x}_k = \mathbf{A}_{k-1}\mathbf{x}_{k-1} + \mathbf{B}_{k-1}\mathbf{u}_{k-1} + \mathbf{w}_{k-1} \tag{7.29}$$

$$\mathbf{y}_k = \mathbf{C}_k\mathbf{x}_k + \mathbf{D}_k\mathbf{u}_k + \mathbf{v}_k \tag{7.30}$$

where $\mathbf{x}_k$ is to be an estimated system state vector at time $k$, and $\mathbf{u}_k$ is a known input vector, $\mathbf{y}_k$ is a measurement vector, and $\mathbf{w}_k$ and $\mathbf{v}_k$ represent the process noise and measurement noise, respectively.

The KF is an optimal estimator [11] under the assumption of normally distributed, zero mean, and independent process and measurement noise, that is, with probability distributions

$$P(\mathbf{w}) \sim N(\mathbf{0}, \mathbf{Q}), \quad P(\mathbf{v}) \sim N(\mathbf{0}, \mathbf{R}) \tag{7.31}$$

where $\mathbf{Q}$ is the process noise covariance matrix and $\mathbf{R}$ is the measurement noise covariance matrix. The mechanics of the KF are illustrated in Figure 7.39 and Table 7.5.

In the time update steps 1 and 2, *a priori* state and *a priori* covariance are predicted from time step $k-1$ to $k$. During the measurement update steps 3–5, the Kalman gain matrix $\mathbf{K}_k$ is used to obtain *a posteriori* covariance matrix and state estimate. Note that in the last step, the predicted output based on the predicted state estimate is used. It is noted for single-dimension measurement estimation, that the latter half of the KF is nearly identical to RLS with forgetting factor equal to one. The matrices $\mathbf{Q}_k$ and $\mathbf{R}_k$ are used to tune the filter, and $\mathbf{x}_{0|0}$ and $\mathbf{P}_{0|0}$ are used to initialize it.

The third estimation algorithm handles nonlinear systems of the form

$$\mathbf{x}_k = f(\mathbf{x}_{k-1}, \mathbf{u}_{k-1}) + \mathbf{w}_{k-1} \tag{7.32}$$

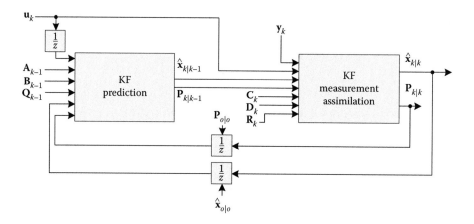

**FIGURE 7.39**   KF block diagram.

**TABLE 7.5**

**KF Algorithm**

1. *A priori* covariance update $\quad\quad\quad\quad\mathbf{P}_{k|k-1} = \mathbf{A}_{k-1}\mathbf{P}_{k-1|k-1}\mathbf{A}_{k-1}^T + \mathbf{Q}_{k-1}$

2. *A priori* estimate prediction $\quad\quad\quad\mathbf{x}_k = \mathbf{A}_{k-1}\mathbf{x}_{k-1} + \mathbf{B}_{k-1}\mathbf{u}_{k-1}$

3. Kalman gain calculation $\quad\quad\quad\quad\mathbf{K}_k = \mathbf{P}_{k|k-1}\mathbf{C}_k^T(\mathbf{C}_k\mathbf{P}_{k|k-1}\mathbf{C}_k^T + \mathbf{R}_k)^{-1}$

4. *A posteriori* covariance update $\quad\quad\mathbf{P}_{k|k} = (\mathbf{I} - \mathbf{K}_k\mathbf{C}_k)\mathbf{P}_{k|k-1}$

5. *A posteriori* estimation correction $\quad\mathbf{x}_{k|k} = \mathbf{x}_{k|k-1} + \mathbf{K}_k(\mathbf{y}_k - \mathbf{C}_k\mathbf{x}_{k|k-1} - \mathbf{D}_k\mathbf{u}_k)$

$$\mathbf{y}_k = g(\mathbf{x}_k, \mathbf{u}_k) + \mathbf{v}_k \tag{7.33}$$

where $f()$ and $g()$ are nonlinear functions, and $\mathbf{w}_k$ and $\mathbf{v}_k$ are process and measurement noise vectors. For such nonlinear systems, a modified KF that linearizes around the state estimates can be used. It is referred to as the EKF [12] and for nonlinear systems, it becomes a suboptimal filter.

For EKF, the nonlinear functions of $f()$ and $g()$ are used for prediction; however, covariance matrix updates and Kalman gain calculation employ linearized Jacobian matrices of $f()$ and $g()$ as follows:

$$\mathbf{A}_k = \left.\frac{\partial f(\mathbf{x}, \mathbf{u}_k)}{\partial \mathbf{x}}\right|_{\mathbf{x}=\mathbf{x}_{k|k}}, \quad \mathbf{C}_k = \left.\frac{\partial g(\mathbf{x}, \mathbf{u}_k)}{\partial \mathbf{x}}\right|_{\mathbf{x}=\mathbf{x}_{k|k-1}} \tag{7.34}$$

In effect, the nonlinear functions are approximated as first-order Taylor series expansion around the most recent estimates. Normal noise distributions are assumed around these points. This assumption may not be valid, due to, for example, nonlinear mapping of process noise, which can result in a bias in the state estimation. Nonetheless, the EKF is still widely used and can be an effective estimator for nonlinear systems. The mechanics of the EKF are summarized in Figure 7.40 and Table 7.6.

There are other estimation algorithms that have been used by researchers for BSE. Plett has employed a sigma-point Kalman filter (SPKF) [13]; it uses a collection of the so-called sigma points to approximate the mean and covariance of the state vector. SPKF can deal with bias better than EKF. Multiscale dual EKF estimation has been employed by Xiong et al. to deal with estimation of states/parameters that evolves at very different timescales [14]. Another approach employing the so-called smooth variable structure filter (SVSF) has been used by Farag et al. for SOC estimation

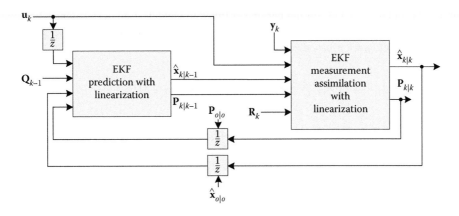

FIGURE 7.40 EKF block diagram.

**TABLE 7.6**

**EKF Algorithm**

| | |
|---|---|
| 1. *A priori* linearization | $\mathbf{A}_{k-1} = \dfrac{\partial f(\mathbf{x}, \mathbf{u}_{k-1})}{\partial \mathbf{x}}\bigg|_{\mathbf{x} = \mathbf{x}_{k-1|k-1}}$ |
| 2. *A priori* covariance update | $\mathbf{P}_{k|k-1} = \mathbf{A}_{k-1}\mathbf{P}_{k-1|k-1}\mathbf{A}_{k-1}^T + \mathbf{Q}_{k-1}$ |
| 3. *A priori* estimate prediction | $\mathbf{x}_k = f(\mathbf{x}_{k-1}, \mathbf{u}_{k-1})$ |
| 4. *A posteriori* linearization | $\mathbf{C}_k = \dfrac{\partial g(\mathbf{x}, \mathbf{u}_k)}{\partial \mathbf{x}}\bigg|_{\mathbf{x} = \mathbf{x}_{k|k-1}}$ |
| 5. Kalman gain matrix calculation | $\mathbf{K}_k = \mathbf{P}_{k|k-1}\mathbf{C}_k^T(\mathbf{C}_k\mathbf{P}_{k|k-1}\mathbf{C}_k^T + \mathbf{R}_k)^{-1}$ |
| 6. *A posteriori* covariance update | $\mathbf{P}_{k|k} = (\mathbf{I} - \mathbf{K}_k\mathbf{C}_k)\mathbf{P}_{k|k-1}$ |
| 7. *A posteriori* estimate correction | $\mathbf{x}_{k|k} = \mathbf{x}_{k|k-1} + \mathbf{K}_k(\mathbf{y}_k - g(\mathbf{x}_{k|k-1}, \mathbf{u}_k))$ |

[5]; SVSF does not employ covariance matrix updates and uses a different gain correction matrix instead of the Kalman gain matrix.

### 7.9.2 ONLINE SOC AND IMPEDANCE ESTIMATION

In this section, it is described how the algorithms in the previous section can be used for online estimation of cell SOC and impedance. A 1RC ECM from Section 7.6.1 will be used as an illustrative example; for simplicity, a symmetric ohmic resistance is used, and both hysteresis effects and self-discharge are ignored; therefore, $R_0 = R_0^c = R_0^d$, $V_{OCV} = V_o$, and $\rho_{sd} = 0$. The approaches in this section assume that some of the cell model parameters are known or are modeled to sufficient accuracy, for example, inefficiencies, capacity, and the known OCV–SOC curve.

If an ideal current sensor is used, Equation 7.12 is sufficient for online SOC measurement provided the initial SOC is known; the latter can be obtained from a rested voltage measurement when $V \approx V_o$. Owing to the presence of sensor noise and unavoidable sensor bias, current-only SOC estimation leads to estimation drift. Filtering approaches that employ both current and voltage measurements can be used to prevent drift. The approach by Verbrugge and Koch [15] updates SOC according to

$$SOC_k = w_{SOC}SOC_k^C + (1 - w_{SOC})SOC_k^V \tag{7.35}$$

where $SOC_k^C$ is an estimate based on current integration, $SOC_k^V$ is an estimate that employs voltage measurements, and $w_{SOC}$ is a weighting parameter tuned to yield desired performance. For example, using $w_{SOC} = 0$ is acceptable when the cell is in a rested state; otherwise, values close to one are used. The current integration-based estimate is calculated using

$$SOC_k^C = SOC_{k-1} - \frac{\Delta t}{CAP}\left[\eta_c I_{k-1}^- + \eta_d^{-1}I_{k-1}^+\right] \tag{7.36}$$

where Equation 7.18 defines charging and discharging currents; note $I_k = I_k^+ + I_k^-$.

The voltage-based estimate is found by using the inverse of the OCV–SOC curve with an estimate of $V_o$, that is, $SOC_k^V = f_{V_o}^{-1}(V_o, T)$; a look-up-table approach can be employed. Owing to cell ohmic and RC resistance, the cell voltage measurement cannot be directly used; therefore, $V_o$ must be estimated.

RLS can be employed to estimate both $V_o$ and impedance. To derive the regression equation, the discrete-time state-space model is first presented; simplifying Equations 7.15 and 7.16 leads to

$$V_{1_k} = e^{-\frac{\Delta t}{R_1 C_1}} V_{1_{k-1}} + R_1 \left( 1 - e^{-\frac{\Delta t}{R_1 C_1}} \right) I_{k-1} \tag{7.37}$$

$$v_k = v_{O_k} - R_0 I_k - V_{1_k} \tag{7.38}$$

The above two equations can be manipulated to eliminate RC voltage $V_1$ and yield the regressed form $y_k = V_k = \mathbf{c}_k^T \mathbf{x}_k$ where

$$\mathbf{c}_k^T = \begin{bmatrix} v_k - I_k & I_{k-1} & 1 \end{bmatrix} \tag{7.39}$$

$$\mathbf{x}_k^T = \left[ e^{-\frac{\Delta t}{R_1 C_1}}, R_0 \ R_a \left( e^{-\frac{\Delta t}{R_1 C_1}} - 1 \right) + R_0 \ e^{-\frac{\Delta t}{R_1 C_1}}, \left( 1 - e^{-\frac{\Delta t}{R_1 C_1}} \right) V_{o_k} \right] \tag{7.40}$$

The last equation can be rearranged to yield impedance and OCV estimates

$$
\begin{aligned}
V_{o_k} &= \frac{\mathbf{x}_k(4)}{1 - \mathbf{x}_k(1)} \\
R_{0_k} &= \mathbf{x}_k(2) \\
R_{1_k} &= \frac{\mathbf{x}_k(3) - \mathbf{x}_k(2)\mathbf{x}_k(1)}{\mathbf{x}_k(1) - 1}
\end{aligned}
\tag{7.41}
$$

Alternatively, the KF can be used to estimate $V_o$, the resistances, and even the RC voltage state $V_{1_k}$; the estimation vector becomes $\mathbf{x}_k^T = \begin{bmatrix} V_1 & V_o & R_o & R_1 \end{bmatrix}$. An assumption will be made that the time-constant-related parameter $\theta = e^{-\Delta t/(R_1 C_1)}$ is known and inputted to the filter. Considering dynamics (7.37) and (7.38), and using

$$V_{o_{k+1}} = V_{o_k}, R_{0_{k+1}} = R_{0_k}, R_{1_{k+1}} = R_{1_k} \tag{7.42}$$

results in KF state-space matrices

$$\mathbf{A}_k = \begin{bmatrix} \theta & 0 & 0 & (1-\theta)I_k \\ 0 & 1 & 0 & 0 \\ 0 & 0 & 1 & 0 \\ 0 & 0 & 0 & 1 \end{bmatrix}, \quad \mathbf{B}_k = \varnothing \tag{7.43}$$

$$\mathbf{C}_k = \begin{bmatrix} -1 & 1 & -I_k & 0 \end{bmatrix}, \quad \mathbf{D}_k = \varnothing \tag{7.44}$$

It is noted here that both the KF and RLS filters presented here inherently assume a model with constant OCV; therefore, even with proper filter tuning, the estimation response of $V_o$ is slow compared to the typical drive-cycle excitation dynamics. They cannot solely be used for SOC estimation; therefore, they should be combined with the current integration methods as in Equation 7.35. For the same reasons, the estimated resistances represent more average values rather than instantaneous resistances. The RLS filter is somewhat less sensitive to this since it employs a forgetting factor; however, decreasing $\lambda$ generally comes at a cost of increased estimate noise.

A third approach employing EKF can be used to estimate all states and parameters simultaneously while employing a current integration SOC model as in Equation 7.12. The nonlinear

OCV–SOC mapping is also embedded into the filter. Many variations of this filter are present in the literature; two examples can be found in [6,14]. These types of filters do not need to use Equation 7.35 since they automatically weight current integration and voltage-based estimations in an arguably optimal manner. The estimation vector is defined as $\mathbf{x}_k^T = \begin{bmatrix} V_1 & SOC & R_0 & R_1 & \theta \end{bmatrix}$, where $\theta = e^{-\Delta t / R_1 C_1}$. The discrete-time nonlinear model equations become

$$V_{1_k} = \theta_k V_{1_{k-1}} + R_1(1 - \theta_k)I_{k-1} \tag{7.45}$$

$$SOC_k = SOC_{k-1} - \frac{\Delta t}{CAP}\eta I_{k-1} \tag{7.46}$$

$$V_k = g_{V_o}(SOC) - R_0 I_k - V_{1_k} \tag{7.47}$$

where $\eta = \eta_c$ for charge and $\eta = \eta_d^{-1}$ for discharge. Equation 7.44 is nonlinear due to the bilinear term $R_1\theta_k$, and the output equation (7.46) is also nonlinear since it contains the OCV–SOC mapping. The discrete-time models $\theta_{k+1} = \theta_k$, $R_{0_{k+1}} = R_{0_k}$ and $R_{1_{k+1}} = R_{1_k}$ are also employed. Using these and Equations 7.45 through 7.47, the linearized EKF matrices (7.34) become

$$\mathbf{A}_k = \begin{bmatrix} \theta_k & 0 & 0 & (1-\theta_k)I_k & V_{1_k} - R_{1_k}I_k \\ 0 & 1 & 0 & 0 & 0 \\ 0 & 0 & 1 & 0 & 0 \\ 0 & 0 & 0 & 1 & 0 \\ 0 & 0 & 0 & 0 & 1 \end{bmatrix} \tag{7.48}$$

$$\mathbf{C}_k = \begin{bmatrix} -1 & \dfrac{\partial g_{V_o}(SOC)}{\partial SOC}\bigg|_{SOC=SOC_{k|k-1}} & -I_k & 0 & 0 \end{bmatrix} \tag{7.49}$$

Recall that $SOC_{k|k-1}$ is the predicted SOC, that is, the right-hand side of Equation 7.46. It is noted that EKF methods perform best when the initial estimates are close to the actual values; this improves the stability of the filter and strengthens the validity of the first-order linearization. Ad hoc approaches such as limiting the estimates to a specific range can improve the performance of this filter. Moreover, simplifying this filter to take the parameter $\theta$ as an input still requires the use of EKF since the nonlinear OCV–SOC mapping is still present.

A remark concerning observability is warranted; observe that some state-space matrix elements contain current measurements; under static conditions of CC, one can show that the entire estimation vector is not observable. For example, a time-invariant observability matrix [16] can be shown to be rank deficient. Sufficient excitation is needed particularly to estimate the dynamic RC-related parameters. This condition can be readily satisfied in vehicle applications where the drive-cycle operating conditions span a wide range of currents and voltages over a relatively large frequency spectrum.

## EXERCISES

7.1   Calculate theoretical Wh/kg of (a) lead acid, (b) Ni–MH, (c) $LiCoO_2$, and (d) Na–Ni–Cl.

7.2   Use 1RC discrete-time ECM dynamics to derive equations for maximum current given voltage limits $V_{max}$ and $V_{min}$. (a) Assume constant model parameters and (b) comment on difficulties encountered when variations of model parameters are considered.

7.3    Perform OCV–SOC curve fitting using the combined model on sample data. Consider using an optimization approach and extending the model.

7.4    Build a cell ECM in a circuit simulation software and generate sample output from HPPC-style pulses to observe effects of relaxation and hysteresis.

7.5    Extend the ECM model to consider asymmetric charge/discharge behavior in the RC elements, consider different approaches, and comment on them.

7.6    Research an active balancing topology not described in this chapter and compare it to the ones presented.

7.7    Research $LiFePO_4$- and $LiMn_2O_4$-based Li-ion cells, then compare them with each other, and the $LiCoO_2$-based cell described in this chapter.

7.8    Research another equivalent circuit model and then derive its continuous time and discrete-time dynamics.

7.8    Rederive the ECM discrete-time dynamics using a linear interpolation hold for input current instead of a zero-order hold.

7.10   Derive the 2RC ECM regression equation that can be used for RLS.

7.11   Compare estimation performance of RLS and KF/EKF on a fixed parameter 1RC ECM.

7.12   Research three other energy storage technologies not described in this chapter and comment on their suitability for automotive applications.

## REFERENCES

1.  Linden, D., and T.B. Reddy. *Handbook of Batteries*. McGraw-Hill professional, New York, NY. 2002.
2.  Chandra, D., W.-M. Chien, and A. Talekar. Metal hydrides for NiMH battery applications. *Materials Material* 6, 2011: 48–53.
3.  Tao, H., Z. Feng, H. Liu, X. Kan, and P. Chen. Reality and future of rechargeable lithium batteries. *Open Material Science Journal* 5, 2011: 204–214.
4.  Soloveichik, G.L. Battery technologies for large-scale stationary energy storage. *Annual Review of Chemical and Biomolecular Engineering* 2, 2011: 503–527.
5.  Farag, M.S., Ahmed, R., Gadsden, S. A., Habibi, S. R., and Tjong, J. A comparative study of Li-ion battery models and nonlinear dual estimation strategies. *Transportation Electrification Conference and Expo (ITEC), 2012 IEEE*, Dearborn, MI. IEEE, 2012.
6.  Plett, G.L. Extended Kalman filtering for battery management systems of LiPB-based HEV battery packs: Part 2. Modeling and identification. *Journal of Power Sources* 134(2), 2004: 262–276.
7.  *Electric Vehicle Battery Test Procedures Manual, Rev 2*, 1996. http://avt.inel.gov/battery/pdf/usabc_manual_rev2.pdf
8.  Cao, J., N. Schofield, and A. Emadi. Battery balancing methods: A comprehensive review. *Vehicle Power and Propulsion Conference, 2008. VPPC'08. IEEE*, Harbin, Hei Longjiang, China. IEEE, 2008.
9.  Daowd, M., Omar, N., Van Den Bossche, P., and Van Mierlo, J. Passive and active battery balancing comparison based on MATLAB simulation. *Vehicle Power and Propulsion Conference (VPPC), 2011 IEEE*, Chicago, IL. IEEE, 2011.
10. Haykin, S. *Adaptive Filter Theory*, Chapter 9, 4th ed., Prentice-Hall, Upper Saddle River, NJ, 2002.
11. Kalman, R.E., A new approach to linear filtering and prediction problems. *Journal of Basic Engineering* 82(1), 1960: 35–45.
12. Bishop, G., and G. Welch. An introduction to the Kalman filter. *Proceedings of SIGGRAPH, Course* 8, Los Angeles, CA. 2001: 27599–3175.
13. Plett, G.L. Sigma-point Kalman filtering for battery management systems of LiPB-based HEV battery packs: Part 2: Simultaneous state and parameter estimation. *Journal of Power Sources* 161(2), 2006: 1369–1384.
14. Xiong, R., Sun, F., Chen, Z., and He, H. A data-driven multi-scale extended Kalman filtering based parameter and state estimation approach of lithium-ion polymer battery in electric vehicles. *Applied Energy* 113, 2014: 463–476.
15. Verbrugge, M., and B. Koch. Generalized recursive algorithm for adaptive multiparameter regression application to lead acid, nickel metal hydride, and lithium-ion batteries. *Journal of the Electrochemical Society* 153.1, 2006: A187–A201.

16. Chen, C.-T. *Linear System Theory and Design*. Oxford University Press, Inc., New York, NY. 1998.
17. Buller, S., M. Thele, R.W.A.A. De Doncker, and E. Karden. Impedance-based simulation models of supercapacitors and Li-ion batteries for power electronic applications. *IEEE Transactions on Industry Applications* 41(3), 2005: 742–747.
18. Bradbury, K. *Energy Storage Technology Review*. Duke University, Durham, NC, 2010: 1–34.
19. Young, K., C. Wang, and K. Strunz. Electric vehicle battery technologies. *Electric Vehicle Integration into Modern Power Networks*. Springer, New York, 2013: 15–56.

# 8 Hybrid Energy Storage Systems

*Omer C. Onar and Alireza Khaligh*

## CONTENTS

In this chapter, hybrid energy storage systems (ESS) are presented for the decoupled power and energy components of the plug-in electric vehicles (PEVs[*]). A stand-alone battery system with an energy capacity sized to propel the vehicle some distance at a moderate speed on a single charge in an all-electric mode may not be sufficient to satisfy peak demand periods and transient load variations in PEVs. In such a case, the battery would need to be oversized to supply the extra power needed to overcome these limitations, thus increasing the weight, volume, and cost as well as the number and depth of charge/discharge cycles. All these factors lead to concern over battery lifetime, which is one of the strongest barriers currently preventing rapid commercialization of PEVs. Alternatively, an ultra-capacitor (UC) bank can supply or recapture large bursts of power at high C-rates.

---

[*] PEV includes all-electric vehicles (also known as battery electric vehicles) and plug-in electric vehicles (PHEVs).

Battery/UC hybrid operation provides an improved solution over the stand-alone battery design in terms of improved power management and control flexibility. Moreover, the voltage of the battery pack can be selected to be lower than the UC bank, which will result in cost and size reduction of the battery. Furthermore, since the battery is not prone to supply peak and sharp power variations, the stress on the battery is reduced and the battery lifetime can be increased. Utilizing UCs tends to result in a more effective capturing of braking energy, especially in sudden/hard braking conditions, and this would further increase the fuel economy as larger energy transients are able to flow or be recaptured more easily.

## 8.1 COMBINED BATTERY AND UC TOPOLOGIES

To provide more efficient propulsion without sacrificing the performance or increasing fuel consumption, more than one energy storage device, each with different power/energy characteristics, can be used in PEVs. In such a system, proper power budgeting based on the specific characteristics of energy sources should result in higher efficiency, longer life and reduced wear on energy sources, and an overall reduction in size and cost. The combination of energy sources should be able to store, supply, and recapture high-power pulses in a typical or worst-case drive cycle, as well as supply the steady demands of the car. A hybrid topology composed of a high-power-density component such as an UC and a high-energy-density component such as a rechargeable battery offers a compromise of both [1,2].

The energy storage devices in electric vehicles (EVs) should be able to meet the demands that the vehicle may encounter under any condition. Rechargeable chemical batteries are the most traditional energy storage sources for EVs. However, since the source needs to supply the peak power demands of the traction motor during transient and rapid accelerations, and since the current technologies do not provide a battery with sufficiently high-power densities, the size and cost of the battery pack significantly increase if it is required to supply all the load demands.

A PEV traction battery may be sized to successfully meet the energy capacity needs for a given single-charge travel distance requirement, but since the present generation of highly energy-dense lithium-ion battery technology has a relatively low-power density, this single power source may not be capable of sourcing or sinking large, short bursts of acceleration or regenerative braking energy. Moreover, battery longevity is directly related to both the depth of discharge and quantity of micro or macro charge/discharge cycles, that is, the short, powerful charge/discharge cycles associated with sharp acceleration and hard regenerative braking. Battery C-rate is defined as the parameter, which expresses battery discharge intensity [3], and when designing a battery-powered system, low C-rates will tend to increase battery life span and thus, instantaneous charge/discharge pulses or fast-fluctuating currents should be avoided. The problems associated with cycling batteries at high C-rates include decreased capacity, excessive heating (which would require additional cooling), and increased DC resistance (DCR); with capacity and DCR being those metrics used to define battery performance and therefore end of life.

Without a secondary ESS, the battery pack must supply all vehicle power demands that may result in an oversized system with a massive energy density to compensate for power-density shortcomings. This would result in increased cost and size of the battery pack or, if a smaller pack was used, it would shorten the battery lifetime, causing potential thermal runaway problems [4]. For this reason, UCs are proposed, because of their higher specific power and cycling efficiency [5], to relieve the battery bank of peak power transfer stresses. Combining these two energy sources, a hybrid system composed of batteries and UCs can not only meet both energy and power requirements of the drive train better, but also provide the flexibility of using smaller batteries with less peak-output power [6,7]. Owing to their very low internal resistances, UCs have very small time constants and can deliver high-power charge and discharge pulses [8] for relatively short durations. The manufacturer performance rating for certain UCs states a 20% decrease of initial capacitance and doubling of internal resistance over a period of 1,000,000 cycles [9]. The curves showing whether this wear and tear is linear or exponential over time are not provided, but with 1,000,000 cycle rating, it is likely that a system will not see serious degradation for a substantial length of time or energy throughput.

By employing UCs with batteries in a proper, efficient, and cost-effective manner, the peak current capacity of the overall topology can be increased. Therefore, the hybrid topology could benefit from an intermediate storage of high power in a buffer stage designed to deliver or receive current for the highest peaks, thus reducing both the number and the depth of discharge cycles of the batteries [7–11].

Different topologies for combining hybrid energy sources have been studied in the literature [7–15]. A direct parallel connection of two sources [12], a bidirectional DC/DC converter between the battery and UC bank [3–13], and two DC/DC converters sharing the same output [14] are among the conventional options.

### 8.1.1  Topology 1: Passive Parallel Configuration

Combining two power sources passively in parallel is the simplest method as the output voltages of both sources are automatically equalized by virtue of being directly connected to the same bus. The passive parallel connection topology is shown in Figure 8.1 with the bidirectional converter interfacing the common ESS bus to the DC link and motor drive. Since the converter is operated to provide constant input voltage to the DC link/motor drive, when the mechanical load on the machine increases, motor current will tend to increase, decreasing the DC link voltage. This will cause more power to be drawn from the ESS through the bidirectional converter to return the DC link to its nominal voltage. On the other hand, whenever braking occurs, the motor drive operates as a generator and captures the braking energy back to the DC link. Therefore, DC link voltage increases during braking or a reduction in mechanical load, and the bidirectional converter is operated in reverse (from DC link to ESS bus) to regulate the DC link voltage.

This topology provides simplicity and cost-effectiveness for hybrid ESSs. In this connection topology, it is expected that the UC will act faster than the battery due to its lower time constant. Therefore, it is anticipated that the UC will provide the transients and fast power variations while the battery supplies a relatively slow-varying current due to its slower dynamics [12].

However, in this topology, directly after current is drawn from the UC, since they are directly connected, the battery will supply a similar current profile for voltage equalization, since there is no active battery current waveform shaper, limiter, or controller. Battery current is, therefore, not effectively controlled [16] and this presents a drawback to the passive parallel connection topology. In addition to this drawback, the nominal battery and UC voltages must be sized to match, which puts an extra restriction on system configuration.

### 8.1.2  Topology 2: UC/Battery Configuration

The most common UC/battery configuration connects the UC terminals to the DC link through a bidirectional DC/DC converter [17]. In this topology, the battery is directly connected to the DC link and a bidirectional interface is used for UC connection, as shown in Figure 8.2. Figure 8.3

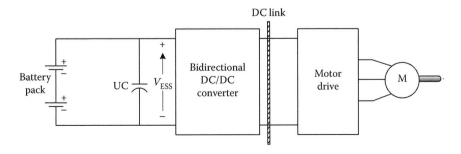

FIGURE 8.1    Passive parallel connection topology.

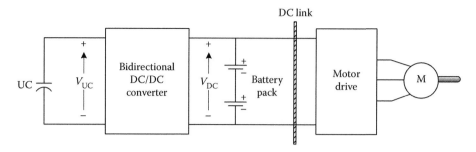

FIGURE 8.2   UC/battery configuration.

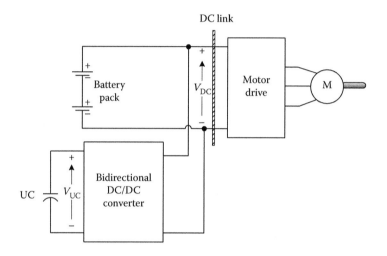

FIGURE 8.3   Another representation of the same UC/battery configuration.

shows another common view of the same connection topology. In this case, the power contribution from the UC can be effectively controlled [18], and the bidirectional DC/DC interface also helps to efficiently and more completely recapture braking energy. Moreover, an UC voltage can be selected that is different from the nominal DC link voltage, allowing the UC energy capacity to be increased or decreased regardless of the system DC voltage, since UC energy capacity varies by the square of its voltage. Since the battery is directly connected to the DC link, input voltage to the motor drive is relatively constant and further DC link voltage regulation is not required. This provides simplicity of control and voltage control loops may be eliminated.

The disadvantage of this topology is that the braking energy captured by the battery is not directly controlled. Braking energy recovered by the battery depends on the power level, battery state of charge (SoC), and the amount of energy captured by the UC. The other drawback is that the bidirectional DC/DC converter must operate properly even with low UC voltages and therefore higher current values; so, current ratings of the switches and other power electronics should be chosen appropriately.

### 8.1.3   TOPOLOGY 3: BATTERY/UC CONFIGURATION

For the battery/UC topology shown in Figure 8.4, the positions of the devices are simply switched as compared to the UC/battery configuration [19,20].

The main advantage of this topology is that the battery voltage can be maintained at a lower level. During braking, the UC is recharged directly from the DC link and some portion of the

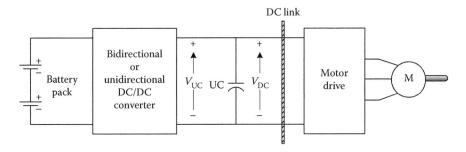

FIGURE 8.4   Battery/UC topology.

braking energy, appropriately current limited, can be transferred to the battery. Since the UC is directly connected to the DC link, it acts as a low-pass filter and takes care of fast load transients. However, the battery pack should be controlled in a way such that it continually maintains an appropriate voltage across the UC and DC link. The control strategy for the battery pack may be designed such that it supplies the average and slow-changing load variations, while the UC supplies the rest, acting as a buffer with faster dynamics. If the UC is not sized large enough or charged continually, the DC link voltage will be allowed to fluctuate over a wide range, and in this case, the motor drive inverter should be capable of operating over a large input voltage range.

In this topology, for simplicity and cost-effectiveness, the battery pack converter can be unidirectional. Since the overall system is that of a plug-in hybrid, the battery pack can be configured to only receive a charge from an on-board generator or an external source, and UC can be the only device responsible for capturing braking energy. This scheme would provide a significant amount of simplicity for power budgeting during braking.

### 8.1.4   TOPOLOGY 4: CASCADED CONVERTERS CONFIGURATION

Alternatively, one energy storage device can be cascaded to the motor drive through a DC/DC converter and the other cascaded through the first and a second DC/DC converter [21,22]. The cascaded converters configuration is presented in Figure 8.5.

In this configuration, both battery and UC voltages can be decoupled from the system voltage and from each other. It is preferred that the battery's converter controls the battery output current and therefore the stress on the battery. If the UC is undersized or the battery converter is not appropriately controlled, the UC voltage can vary substantially. At low UC voltages, the input current to the UC converter can be very high, leading to higher conduction losses and the need for high-current-rated switches. Additionally, the UC DC/DC converter must provide stable operation over a wide voltage input range.

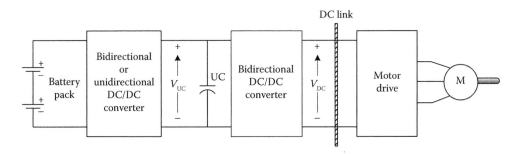

FIGURE 8.5   Cascaded converters configuration.

The major disadvantage of the cascaded converter topology is that additional losses may be encountered at the battery power flow path since there are two cascaded converters between the battery and DC link.

The battery converter can again be a unidirectional converter for control and configuration simplicity. In the case of a boost converter for the battery, the battery's power contribution can be easily controlled by current control mode [16]. The UC converter can be controlled for DC link regulation and the battery can be controlled such that it supplies a smoother current profile during the operation.

As in topology 3, the positions of the battery and UC can be switched, leading to another cascaded converters topology. However, in this case, the power contribution from the UC could possibly result in a more fluctuation voltage applied to the battery terminals. On the other hand, since the battery is the DC link side energy storage device, DC link voltage regulation could be easily accomplished by virtue of a nearly constant battery voltage.

### 8.1.5 TOPOLOGY 5: MULTIPLE PARALLEL-CONNECTED CONVERTERS CONFIGURATION

In this topology, each energy storage device has its own bidirectional DC/DC converter for interfacing with the DC link [11,22–24] and the outputs of each converter are all held in parallel. The block diagram of the multiple parallel-connected converters topology is given in Figure 8.6.

Although this topology is called a multiple-input converter, this is not a "true" multiple-input converter since each energy storage device has an individual converter and its contributions to the DC link are paralleled. This topology offers the highest flexibility and provides better functionality than that of the cascaded converters topology. The voltages of the battery and UC are decoupled from each other as well as the DC link voltage. Since power controls and power flow paths from the energy storage devices are totally decoupled, this topology is superior from stability, efficiency, and control simplicity points of view. The reliability is also improved since one source can keep operating through the failure of another.

As discussed in Reference 24, the battery can be operated in current control mode, supplying the load variations averaged and smoothened over a period of time. Meanwhile, the UC can be operated in voltage control mode, maintaining a nearly constant voltage across the DC link. Therefore, the UC will supply fast load variations and transients during both rapid acceleration and sudden braking conditions. This is due to the fact that these load activities directly affect the DC link voltage, and as long as DC link voltage is regulated fast enough, all load demands will be satisfied.

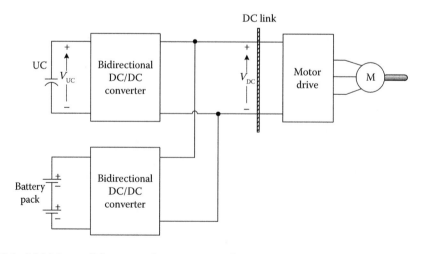

FIGURE 8.6 Multiple parallel-connected converters topology.

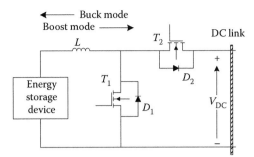

FIGURE 8.7    Bidirectional DC/DC converter.

The bidirectional DC/DC converters discussed for topologies 2 through 4 could be typical two-quadrant converters, able to operate in boost mode in one direction and buck mode in the other, as shown in Figure 8.7. This topology is also called a half-bridge bidirectional DC/DC converter.

As shown in Figure 8.7, for power flow from an energy storage device to the DC link, inductor $L$, switch $T_1$, and diode $D_2$ form a boost converter. To accommodate power flow from the DC link to an energy storage device, switch $T_2$, diode $D_1$, and inductor $L$ form a buck converter. Of course, other types of bidirectional converters could also be used, and some of these are presented in the following sections.

### 8.1.6    TOPOLOGY 6: MULTIPLE DUAL-ACTIVE-BRIDGE CONVERTERS CONFIGURATION

It is well known that conventional buck–boost converters can step the source voltage up or down at the cost of having a negative voltage output. Therefore, an inverting transformer is usually employed to obtain a positive output voltage. Although the transformer adds cost and volume to the system, it may be advantageous when there are two input sources for both isolation and coupling. Two buck–boost-type DC/DC converters for the UC and battery can be combined through the magnetic coupling of a transformer reactor [25]. However, neither conventional buck–boost nor buck–boost with transformer topologies is suitable for vehicle propulsion systems since they are not capable of bidirectional operation. On the other hand, dual-active-bridge DC/DC converters [26] can be employed for the combined operation of batteries and UCs. Although transformers typically add cost and volume to a system, the transformer in the dual-active-bridge converter operates at a very high frequency and it may therefore be very small and cheap. Having a transformer in the converter topology may be advantageous when there are two or more input sources since they can be combined through the magnetic coupling of the transformer reactor. The dual-active-bridge converter with two input sources is presented in Figure 8.8. Although this topology completely isolates the input sources from the DC link, it requires a greater number of switches at increased cost. If only isolation is required, the number of switches can be reduced by employing half-bridge inverters/rectifiers instead of full-bridge versions. This dual-active-bridge topology with half-bridge converters would cut the number of switches into half, and is given in Figure 8.9.

### 8.1.7    TOPOLOGY 7: DUAL-SOURCE BIDIRECTIONAL CONVERTERS CONFIGURATION

In the multiple-converters configuration, as discussed earlier, each individual converter shares the same output; hence, the combination of converters occurs at the output. Instead of paralleling the converters' outputs at the DC link, the combination can be applied at the input, as in dual-source bidirectional converters [27,28]. The dual-source bidirectional converters configuration is presented in Figure 8.10.

Although this topology is very similar to topology 5, there is one less switch in this dual-source input case. For the UC, inductor $L_1$, switch $T_1$, and diode $D_2$ form a boost converter when transferring

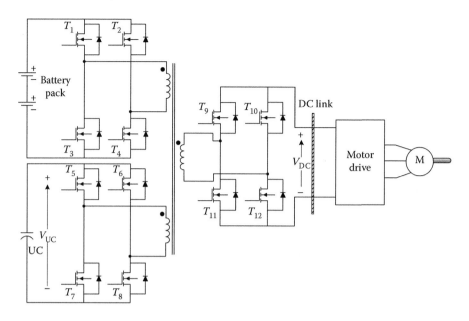

**FIGURE 8.8**   Dual-active-bridge converter topology with full-bridge converters.

power from battery to the DC link, and for the battery, inductor $L_2$, switch $T_3$, and $D_1$–$D_2$ path form a boost converter when transferring power from UC to the DC link. During regenerative braking, the interface should be operated in buck mode. Switch $T_2$, diode $D_1$, and inductor $L_1$ form a buck converter from the DC link to the UC. On the other hand, some regenerative braking energy can be conveyed to the UC applying a pulse width modulation (PWM) signal to $T_1$. In this case, switches $T_2$ and $T_1$, diode $D_3$, and inductor $L_2$ will form a buck converter from DC link to the UC. By applying an appropriate duty cycle to $T_1$ and $T_2$, braking energy can be properly shared.

Although one switch is eliminated as compared to the multiple-converters configuration, a complicated control system is the main drawback of this configuration.

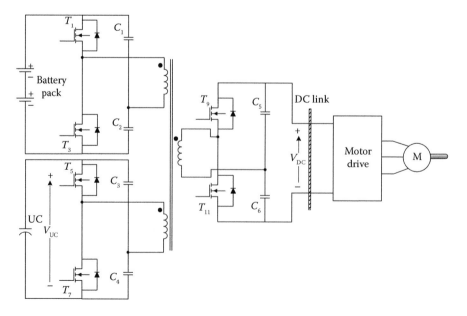

**FIGURE 8.9**   Dual-active-bridge converter topology with half-bridge converters.

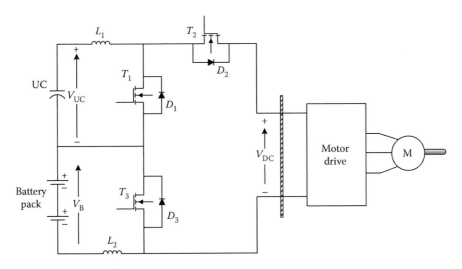

**FIGURE 8.10**   Dual-source bidirectional converters topology.

### 8.1.8   Topology 8: Multiple-Input Converter Configuration

In the previously discussed topologies, battery and UC energy storage devices were employed through their individual DC/DC converters. Unlike these configurations, the multiple-input DC/DC converter has the flexibility of adding a number of inputs at the added cost of one switch and one diode (two switches and two diodes in the case of bidirectional operation). The multiple-input DC/DC converter topology with battery and UC input sources is given in Figure 8.11.

In this converter topology, the inputs share the same converter inductor and are connected through bidirectional switches [28–31]. This converter is capable of operating in buck, boost, and buck–boost modes for power flow in both directions. The assumption of a continuous inductor current requires that at least one input switch or diode is conducting at all times. The respective input diode is ON if all the input switches are OFF. If more than one switch is turned ON at the same time, the inductor voltage equals to the highest input voltage [32].

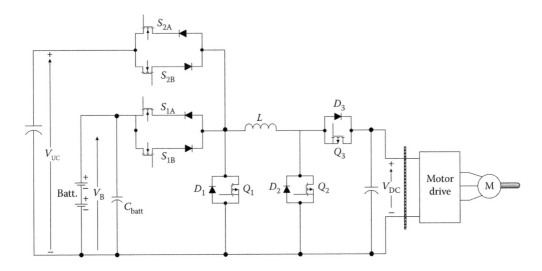

**FIGURE 8.11**   Multiple-input DC/DC converter topology.

Under acceleration conditions, both sources deliver power to the DC link. Since the UC voltage varies in a wider range than that of the battery, battery voltage can be selected higher than the UC for simpler operation. Since battery voltage is higher than that of the UC, $S_{2A}$ is turned ON in boost mode. Switch $Q_2$ can be switched to control the inductor current, and power flow from battery to the UC can be controlled by switch $S_{1A}$. Diode $D_3$ conducts when $Q_2$ is turned OFF. During deceleration, the braking energy is transferred from the DC link to the energy storage devices and the converter is operated in buck mode. Since the UC voltage is less than the battery voltage, $S_{1B}$ is always turned ON. Switch $Q_3$ can be used to control the inductor current. Power sharing between inputs is accomplished by controlling $S_{2B}$. Diode $D_2$ cannot conduct until switch $Q_3$ is turned OFF.

The main advantage of this topology is that only one inductor is required for the whole converter even if more inputs are connected, which can decrease the volume and weight of the converter significantly as compared to multiinductor or transformer topologies. Conversely, power budgeting in both boost and buck modes is very challenging and requires advanced control design.

### 8.1.9   TOPOLOGY 9: MULTIPLE MODES SINGLE-CONVERTER CONFIGURATION

In this design, only one bidirectional converter is required, with the UC voltage selected to be higher than the battery voltage. The UC is directly connected to the DC link to supply peak power demands, and the battery is connected to the DC bus through a diode. The bidirectional DC/DC converter is connected between the battery and UC, as shown in Figure 8.12, to transfer power between them [18]. This converter is controlled to maintain a higher voltage across the UC than the battery, and therefore, the diode is reverse biased for the majority of operations.

This converter has four different modes of operation: low-power mode, high-power mode, regenerative braking mode, and acceleration mode.

In low-power mode, it is assumed that the total power demand is less than the power capacity of the bidirectional converter. In this mode, since the UC/DC link voltage is higher than the battery voltage, the diode $D_B$ is reverse biased. Since the power demand is lower than the capacity of the bidirectional converter, there is no power flow from the battery to the DC link. The battery supplies only power to the UC to keep its voltage at some predetermined higher level.

Whenever the power demand of the vehicle is greater than the converter power capacity, the system operates in high-power mode. In this mode, the UC voltage cannot be maintained at that high value since the power from the battery to the UC is less than the power from UC to the DC link. In this case, diode $D_B$ is forward biased and the battery also directly supplies power to the DC link along with the UC.

During regenerative braking mode, since the UC is directly connected to the DC link, it is recharged by virtue of its position in the circuit, while diode $D_B$ blocks the DC link power to prevent recharging of the battery directly. Some portion of the recaptured braking energy can be transferred to the battery through the bidirectional converter. Therefore, this mode provides controlled recharging for the battery; that is, whenever UC is fully charged but there is still regenerative energy available, the rest of the energy can be transferred to the battery as long as the regenerative current does

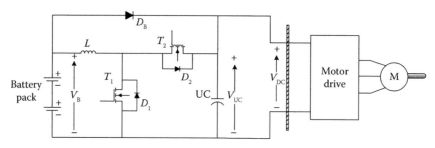

FIGURE 8.12   Single bidirectional converter topology.

not exceed the maximum battery-charging current. In the latter case, mechanical brakes could be utilized to keep the battery current below the maximum limit.

When the vehicle first starts to accelerate, the voltage across the UC is higher than that of the battery, and is equal to that of the DC link. Power demand on the vehicle is high, and UC voltage therefore drops. During the acceleration mode, the UC discharges through the DC link and the battery supplies power to the DC link through the bidirectional converter. Whenever the DC link voltage decreases to the level of the battery voltage, $D_B$ becomes forward biased and the system switches to the high-power mode.

The advantage of this topology is that it requires only one converter. However, although power is shared among battery and UC during different modes, the battery current is not effectively controlled, especially in the regenerative mode with its potentially sharp transients.

### 8.1.10 TOPOLOGY 10: INTERLEAVED CONVERTER CONFIGURATION

The combination of battery and UC can also be achieved by using interleaved converters. The interleaved converter configuration is composed of a number of switching converters connected in parallel, as shown in Figure 8.13.

When low-current ripples or very tight tolerances are required, interleaved converters tend to be preferred. Interleaved converters offer very lower inductor current ripple than regular bidirectional converters, and the overall efficiency for a given power requirement is greater since each interleaved architecture has smaller power rating and smaller overall loss. Interleaved converters also have faster transient response to load changes [33–35].

As shown in Figure 8.13, the battery is interfaced to the UC terminals through the interleaved converters with UC directly connected to the DC link. Alternatively, UC and battery positions can be reversed, as shown by the dashed lines in Figure 8.13. In addition to these two configurations, the interleaved converters can be employed within topologies 2 through 5, presented in the earlier sections.

### 8.1.11 TOPOLOGY 11: SWITCHED CAPACITOR CONVERTER CONFIGURATION

Another bidirectional interface that combines battery and UC operation in PEVs, the switched capacitor converter (SCC), can also be employed [33]. An SCC is basically a combination of switches and capacitors, and by different combinations of switches and capacitors, these converters can produce an output voltage that is higher or lower than the input voltage. In addition, reverse polarity at the

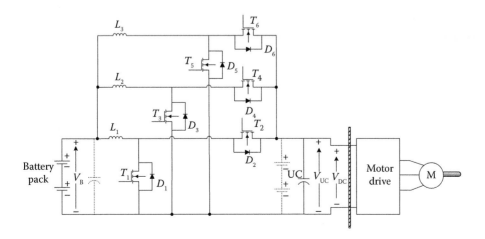

**FIGURE 8.13**    Parallel interleaved three-stage bidirectional converter.

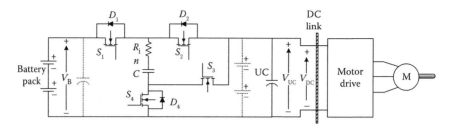

**FIGURE 8.14** SCC configuration.

output can be provided if necessary. The capacitor can be charged or discharged through various paths formed by the controlled switches. Four switches, three diodes, and one switched capacitor can be used for a typical SCC. SCC can have a large voltage conversion ratio with very high efficiency, and therefore, they appear to be well suited for automotive applications [36–39].

An example of a battery/UC combination through SCC is provided in Figure 8.14.

On the basis of the circuit configuration shown in Figure 8.14, battery energy can be delivered to the load side by buck mode operation and the battery can be recharged by boost mode operation. In buck mode, switches $S_1$ and $S_4$ are turned ON until capacitor $C$ is charged to some desired voltage level less than that of the battery, at which point, $C$ is disconnected from the battery terminal by turning switches $S_1$ and $S_4$ OFF and connected to the load by turning switch $S_2$ ON to transfer its stored energy through $S_2$ and diode $D_4$. In boost mode, $C$ can be charged from the load side through $D_2$ and $S_4$. After this stage, the $S_3$ and $D_1$ become the operating switches and the energy in $C$ is discharged to the battery side. This control strategy offers control simplicity, continuous input current waveform in both modes of operation, and low source current ripple [33].

As shown in Figure 8.14, the battery is interfaced to the UC terminals through the SCC and the UC is directly connected to the DC link. Alternatively, UC and battery positions can be reversed, as shown by the dashed lines in Figure 8.14. In addition to these two configurations, the SCC can be employed within topologies 2 through 5, as a bidirectional converter, presented in the earlier sections.

### 8.1.12 TOPOLOGY 12: COUPLED INDUCTOR-BASED HYBRIDIZATION ARCHITECTURE

The system layout of this hybrid battery/UC architecture is demonstrated in Figure 8.15 [40]. The converter is composed of four switches $T_1–T_4$ with their internal diodes $D_1–D_4$, battery and UC, a DC link capacitor $C_1$, a capacitor in parallel with the battery ($C_2$), and an integrated magnetic structure with self-inductances $L_1$, $L_2$, and the mutual inductance $M$. The use of coupled inductors for the battery/UC combination can further reduce the bidirectional converters' most bulky and expensive components.

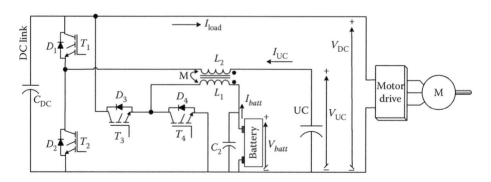

**FIGURE 8.15** Coupled inductor-based battery/UC hybridization architecture.

This converter has five main operational modes: Mode-(1) Plug-in AC/DC charging of the energy storage devices with buck mode of operation from DC link voltage to the battery and UC. Mode-(2) Plug-in DC/AC discharging of the energy storage devices with boost mode of operation from the battery and UC to the DC link. Mode-(3) Boost mode of operation of the battery and UC to the DC link for acceleration, idling, or cruising during driving. Mode-(4) Buck mode of operation from DC link to the battery and UC for regenerative braking during driving. Mode-(5) Boost mode of operation and buck mode of operation when needed, that is, if the UC's SoC drops to the minimum allowed UC SoC. In plug-in charging, the power flow direction is identical to that of the regenerative braking mode. Therefore, Modes 1 and 4 are similar and the DC link acts as a common DC bus in these modes. Correlatively, when discharging energy storage devices in plug-in mode, power is transferred from ESS to the DC link with the same power flow direction in the acceleration mode. Therefore, Modes 2 and 3 are identical. Consequently, the operational modes can be generalized and the number of modes can be reduced to three.

In buck mode of operation for battery and UC, the vehicle can either be grid connected for charging or in regenerative braking condition. Switch $T_3$, diode $D_4$, and inductor $L_1$ form a buck converter from DC link to the battery. When $T_3$ is turned ON, current from DC link passes through $T_3$ and $L_1$ while energizing the inductor. When $T_3$ is turned OFF, $D_4$ is ON and the output current is freewheeled through $D_4$ and the inductance, decreasing the average current transferred to the battery. The current flow path of this operation mode is presented by blue lines in Figure 8.3. Similarly, the DC link voltage is stepped down to recharge the UC through the buck converter that is made up of switch $T_1$, inductor $L_2$, and diode $D_2$. When $T_1$ is turned ON, power flows through $T_1$ and $L_2$. When it is turned OFF, the energy stored in the inductor is freewheeled through $D_2$. In boost mode of operation for battery and UC discharging, power is delivered from storage devices to the DC link, that is, plug-in discharging when grid connected or acceleration, cruising, and idling conditions during driving. The battery voltage can be boosted to the DC link by the inductor $L_1$, switch $T_4$, and diode $D_3$ that form a step-up converter. When $T_4$ is turned ON, the battery terminals are shorted through $T_4$ and $L_1$, while energizing the inductor. When $T_4$ is turned OFF, $D_3$ turns ON and the energy is stored in the inductor, and the battery is transferred through $D_4$ to the DC link. Similar to the battery discharging, inductor $L_2$, switch $T_2$, and diode $D_1$ make up a boost converter from the UC to the DC link. When $T_2$ is turned ON, the inductor $L_2$ is energized and the UC discharges. When $T_2$ is turned OFF, $D_1$ turns ON and the energy stored in the inductor and the UC power flows through $D_1$ to the DC link.

The final operation mode is utilized when the battery is in boost mode and UC is in buck mode. It is very unlikely that the UC SoC drops below the minimum allowed value, since UC gains energy transiently when the load level decreases or a regenerative braking occurs. Therefore, this mode of operation is included as a fail-safe measure. If by any chance, the UC SoC drops to its minimum limit, the battery supplies power to the DC link while the UC retrieves power from the DC link. In this way, both the load demand is met and the UC is recharged by the battery. Whenever a regenerative braking occurs, the UC captures the available braking energy and this mode gets terminated. This mode will naturally end when the UC's SoC increases above the minimum limit. Table 8.1 summarizes all the operation modes of the proposed system.

TABLE 8.1

**Operation Modes of the Proposed Battery/UC Converter**

| Mode | Source | Load | Operation |
|---|---|---|---|
| 1 and 4 | Grid and regenerative braking | Battery and UC | Buck for battery and buck for UC |
| 2 and 3 | Battery and UC | Grid and propulsion | Boost for battery and boost for UC |
| 5 | Battery | Propulsion and UC | Boost for battery and buck for UC |

## 8.2   OTHER ENERGY STORAGE DEVICES AND SYSTEMS: FLYWHEELS, COMPRESSED AIR STORAGE SYSTEMS, AND SUPERCONDUCTING MAGNETIC STORAGE SYSTEMS

As an alternative to the battery/UC hybrid ESS, flywheels, compressed air storage systems, and superconducting magnetic storage systems can be incorporated within plug-in hybrid EVs. Although these ESS are still being investigated and currently, there are only research-level applications, they may be applicable to the plug-in hybrid EVs when they are commercialized due to the characteristics discussed in the following sections.

### 8.2.1   Flywheel ESS

Unlike batteries and UCs, a flywheel stores energy in a kinetic device rather than in an electrochemical or electrostatic device. A flywheel designed for an energy storage application is basically a large rotating disk with a very high moment of inertia, designed to spin at very high speeds—from 20,000 to 50,000 rpm. Stored kinetic energy can be converted into electricity, or electricity can be converted into kinetic energy. A motor/generator is typically coupled to the flywheel, or the flywheel rotor itself is used as the motor/generator rotor, to convert kinetic energy into electric energy or electric energy into kinetic energy. In motoring mode, the electric machine increases the stored kinetic energy within the flywheel by simply increasing its speed, and in generating mode, the electric machine's shaft is mechanically driven by the flywheel, sapping energy [41]. Whenever power is drawn through the generator by the electrical system of the vehicle, it extracts energy from the flywheel by decreasing its rotational speed, and whenever excess needs to be dumped into the storage medium, its speed is increased. To increase the energy storage capacity in a flywheel system, the moment of inertia (a physical characteristic related to mass and geometry), or the maximum rated speed of rotation may be increased. Owing to their smaller size and the fact that stored energy increases as the square of rotational speed, high-speed flywheels are more preferable for automotive applications than the low-speed flywheels. However, high-speed flywheels must be isolated in a vacuum to reduce the windage and ventilation losses, and mechanical bearings—the other mode of energy losses in a flywheel system—should be replaced with contactless magnetic bearings so that the system floats on a "cushion" of electromagnetic force. These bearings are still being researched as practical replacements for mechanical bearings as a means to reduce the friction losses [42].

As is the case with other energy storage devices, safety is a concern that must be addressed when using flywheels. Since flywheels are high-speed devices, containment vessels are used in case of mechanical rotor failures, and flywheels are inherently designed to fail in some manner other than by flying apart. The other drawbacks are that they are relatively large, and heavy systems and rotational energy losses limit the long-term storage capabilities. Although size can be reduced as the speed is increased (with a maximum of about 1,00,000 rpm), this option would also increase the rotational losses and self-discharge (slowing down).

The advantages of flywheel storage devices are that they have a long rated life (typically 20 years [43]), can deliver large amounts of energy in a very short time, and are free of the deep discharge and high cycle count issues that tend to plague batteries. The power that can be delivered to/from a flywheel is limited only by the electric machine that is mechanically connected to it. Flywheels have been considered for large vehicles such as buses, trucks, and high-speed rail locomotives where the battery costs are inherently high [41]. A schematic example of the implementation of a flywheel ESS is shown in Figure 8.16.

In Figure 8.16, the electrical energy storage devices are shown in a generalized form that can be the representation of any of the hybridization topologies discussed earlier, and with the addition of a flywheel, it may be possible to include only one of these electrical storage systems into a design. Since the vehicle provides a common DC link, the flywheel can be connected to this DC link through a bidirectional DC/AC converter and a high-speed electric machine. Whenever there

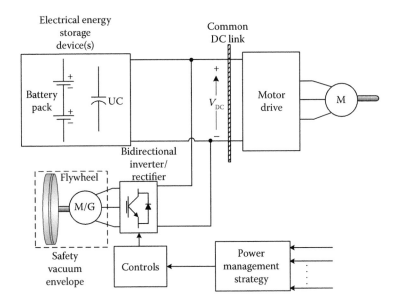

FIGURE 8.16   Hybrid flywheel ESS.

is a surplus of braking or buffering energy, it can be delivered to the flywheel by operating the machine in motoring mode and the converter in inverter mode, and whenever the energy stored in the flywheel is needed for propulsion, the machine is operated in generator mode and the converter is operated in rectifier mode [43–46]. Permanent magnet synchronous machines are typically preferred to drive flywheels due to their high-speed operation capability and control and drive simplicity [47]. The inputs to the power management strategy can be simply flywheel speed, torque demand on the vehicle, engine power, and power from electrical energy storage devices. From these variables, the power from/to flywheel can be controlled accordingly.

As an alternative to this electrical connection of flywheels through a power electronic converter and an electric machine, flywheels can also be mechanically coupled to the traction drive. However, in this mechanical case, they cannot be actively controlled to deliver or capture a certain amount of power; instead, they would only help torque-ripple cancelation as a passive flywheel [48].

### 8.2.2   Compressed Air-Pumped Hydraulics-Based Storage System

Compressed air or pump storage devices may also be promising candidates for future energy storage solutions in PEVs. These storage alternatives have a much longer lifetime than batteries, typically 75 years for pumped hydro and 40 years for compressed air cases [49], and have comparable efficiency ratings of around 75%–80%. In addition to the high efficiency and long lifetime, they are more environmentally friendly as they do not produce problematic waste materials. Compressed air and pump storage device sizes vary greatly and can be implemented in both ultra-large facilities [50] and small applications using mechanical/hydraulic conversion with the liquid piston technique [49]. A simple implementation of a compressed air storage system for PEVs is given in Figure 8.17.

The compressed air-based potential energy storage devices can be interfaced to the common DC link through a motor/generator and a bidirectional DC/AC–AC/DC converter. Electrochemical storage devices such as batteries and UCs can still be employed within the vehicular power system, but since the compressed air storage system offers high-energy capacity and density, it could replace the batteries entirely. However, a storage device with higher power density and faster response, that is, UCs, is still needed for situations involving hard braking and quick acceleration. During the energy

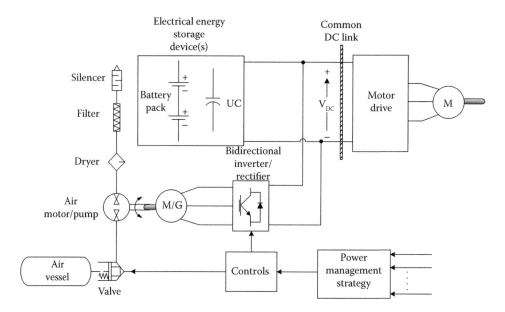

**FIGURE 8.17**   Compressed air storage system implementation.

storage process, the converter is operated in inverter mode and the electric machine is operated as a motor. The motor drives a pneumatic air pump that operates as a compressor and fills the vessel with pressurized air.

Whenever power is needed from the compressed air storage system, the converter is operated in rectifier mode and the electric machine is operated as a generator. In this mode, the pneumatic machine is driven directly by the compressed air that expands and is released from the air vessel. The pneumatic machine then drives a generator and power is supplied to the DC link via the power electronic converter. The power electronic converter and the valve of the air vessel are controlled based on the mode of operation and the amount of power that is to be delivered to/from compressed air storage system.

The main drawback of the pneumatic storage system is its low efficiency when compared to hydraulics-based storage systems. This is mainly due to the inefficiency of the compressor motor/pump. Therefore, oil-hydraulics/pneumatics-based ESS can be implemented to achieve higher efficiency and energy-density levels [49,51,52]. Hydraulic motors have exceptionally high-energy conversion efficiency and can achieve higher energy densities in storage applications due to their high-pressure ratings. These systems employ piston vessels or a high-pressure bladder where nitrogen is compressed by injecting high-pressure fluid (oil) in the body or in the shell using a piston or membrane as a gas/liquid separation medium. Combining gas pressurizing (pneumatics) and fluid compression (hydraulics), one produces a hydro-pneumatic storage system [49]. The operating principle of the hydro-pneumatic storage system for a PEV is provided in Figure 8.18.

The hydro-pneumatic system operating principle is very similar to that of the compressed air storage system. The major difference is that the hydraulic pump/motor, to compress the nitrogen gas, pumps oil rather than directly pressurizing the air. The most significant challenge with the compressed liquid/air systems is the relatively large size and the high number of components involved. This increases the cost and reduces the overall efficiency in vehicular applications.

### 8.2.3   Superconducting Magnetic ESS

Superconducting magnetic energy storage systems (SMES) can store electric energy in the form of magnetic energy. SMES are capable of transferring large amounts of power quickly in both charge

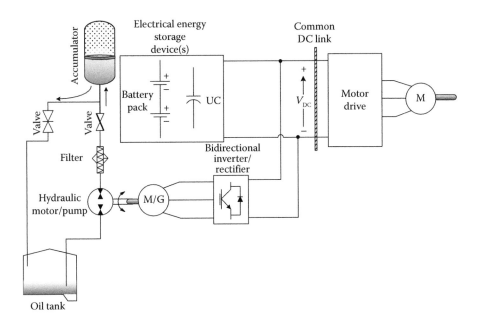

FIGURE 8.18    Hydro-pneumatic ESS implementation.

and discharge directions, and their efficiency is very high with a charge–discharge efficiency of over 95% [53]. Moreover, a relatively small magnet can be employed. Although the power density is high but the energy density is not very high, the battery energy storage devices cannot be completely eliminated in most of the cases. However, SMES can offer a suitable replacement for UCs.

As in battery/UC combined operation, battery/SMES combinations also provide high-power and high-energy density without any rotating parts [54–56]. Since batteries are still the main energy storage device, the size and cost of the SMES can be kept relatively low and this hybrid ESS can be made applicable to plug-in vehicles. Longer life and higher efficiency are the other advantages of hybrid battery/SMES system. As stated in Reference 57, when subjected to fast transients and repeated charge–discharge pulses, batteries present technical problems such as reduced efficiency, degradation, and overheating. Although UCs can mitigate these problems by providing energy during fast, high-demand transients, their energy density is very low and they are not a practical choice as the sole energy storage device for a PEV. With proper design, SMES may replace both batteries and UCs [57]. However, SMES devices present the major drawback of needing to be operated at very low temperatures for their coils to act as a superconducting material and reduce ohmic losses. For refrigeration and device containment, a cryogenic system must be constructed, the complexity and refrigeration power of which reduces the overall system efficiency and ease of implementation, and is a serious drawback to the commercial viability of vehicle-scale SMES systems. This refrigeration energy can be obtained through a closed-cycle system, that is, cryocooling, or cooling can be provided through evaporation of a suitable cryogenic liquid such as helium, nitrogen, or neon [57], and for vehicular applications, these are serious challenges that prevent the use of SMES. The use of high-temperature superconductors would make the SMES cost-effective and more efficient due to the reductions in refrigeration needs, but high-temperature superconducting material developments are still in the research and development stages [58,59].

To achieve the best possible performance with least cost, several factors should be considered for the design of SMES systems with coil configuration, structure, operating temperature, and energy and power capabilities being some of these key factors [58,60]. Energy/mass ratio, Lorentz forces (the force due to the electromagnetic field), and stray magnetic field are the parameters between which a compromise must be made for a stable, reliable, and economic SMES system. The SMES

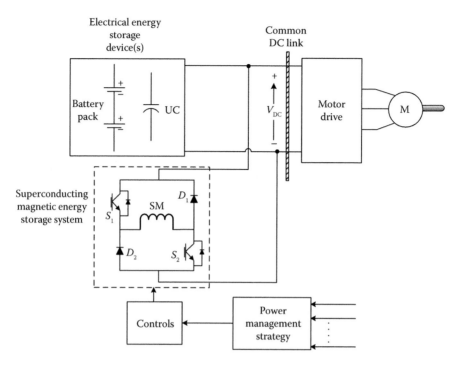

FIGURE 8.19   SMES implementation for a PEV.

coil can be toroidal or solenoidal; the solenoid coil type is preferred due to its simplicity and manufacturing cost-effectiveness, although toroids may be more suitable for small-scale applications [53]. Inductance wiring and the rating of the power electronic converter limit the maximum power (voltage and current) that can be drawn or injected from/to SMES system.

A typical implementation of an SMES to PEVs is shown in Figure 8.19. Although the battery and UC are shown in the system, UCs can be eliminated as mentioned earlier. In this configuration, a power electronic converter is used for SMES utilization.

During charging, switches $S_1$ and $S_2$ are turned ON allowing positive current to flow and increasing the voltage of the superconducting magnet, storing energy. By keeping $S_1$ turned ON and $S_2$ OFF, the energy stored in the magnet will circulate through $S_1$–SM–$D_1$, shorting the SM. Since it is composed of a superconducting material, energy can be stored by circulating the current through this path with only the circuit losses being those caused by the internal resistances of the switch and diode. During discharging mode, both switches, $S_1$ and $S_2$, are turned OFF and the diodes become forward biased such that stored energy is transferred to the DC link. During discharge, it is possible to stop the flow of energy and switch back to the energy storage mode by keeping $S_1$ OFF and turning $S_2$ ON. In this case, the current of the superconducting magnet will circulate through the path SM–$S_2$–$D_2$ and no energy transfer will occur since the SM is shorted through its own terminals.

The other drawback to the practical implementation of an SMES system is that, although the coil is superconducting, the switches are not ideal; therefore, the charge/discharge current will gradually decrease due to semiconductor losses in energy-storing mode. By using switches with low internal resistances and applying soft switching or switching loss recovery techniques, higher efficiencies can be achieved, but parasitic losses can never be completely eliminated.

## 8.3   CONCLUSIONS

In this chapter, different topologies offering the combined operation of several ESS have been reviewed. In total, 12 possible hybridization topologies are described for the combined operation of batteries and UCs. The advantages and drawbacks of the passive parallel connection, UC/battery, battery/UC, cascaded converters, parallel converters, multiinput converters, dual-active-bridge converters, dual-source converters, interleaved converters, and SCC have all been highlighted. In addition, kinetic, potential, and magnetic ESS such as flywheels, compressed air/pumped hydraulics, and SMES devices have been described along with possible implementation scenarios, including advantages and disadvantages for plug-in hybrid EVs. Although there is no commercially manufactured plug-in hybrid vehicle powered by batteries and UCs together in the market so far, the hybridization of these energy storage devices has shown to be academically and analytically very beneficial in terms of battery life, vehicle performance, and fuel economy. However, the hybridization of energy storage devices is a challenging, multivariable problem requiring appropriate sizing and control of power-sharing strategies. Furthermore, nontraditional forms of ESS can be promising candidates for plug-in hybrid EVs due their longer lifetime, efficiency, and high specific power and energy densities, and further research and development of these technologies may produce some unforeseen ideal combination of energy density, power availability, efficiency, and easy if implemented in the future.

## 8.4  SIMULATIONS AND ANALYSES OF HYBRID ESS TOPOLOGIES FOR PEVs

In this section, three examples of hybridization topologies for the combined operation of batteries and UCs have been modeled and simulated. First, the passive parallel configuration topology is simulated; second, battery/UC cascaded and connected converters topology; and third, the parallel connected multiconverters configuration, making a case for the effectiveness and feasibility of each topology through the discussion that follows.

For the simulations, a portion of the urban dynamometer-driving schedule (UDDS) is used for the time interval $t = [690, 760]$. This driving cycle period of 80 s includes acceleration, braking, and idling conditions for the vehicle. For the analysis, a plug-in version of Toyota Prius has been used, and the battery parameters of the Toyota Prius plug-in are given in Table 8.2 [61,62].

For the UC, a BMOD0165 UC module manufactured by Maxwell has been chosen, the parameters of which are given in Table 8.3 [63].

Since one of the test topologies calls for a passive parallel connection, the UC voltage should be chosen such that it is close to that of the battery. Therefore, seven BMOD0165 modules are

### TABLE 8.2
### Toyota Prius PEV Battery Parameters

| Parameter | Value |
| --- | --- |
| Battery type | Lithium ion |
| Rated voltage | 345.6 V |
| Rated energy capacity | 5.2 kWh |
| Rated Coulomb capacity | 15.04 Ah |
| Internal resistance | 0.56104 Ω |

### TABLE 8.3
### Maxwell BMOD0165 UC Parameters

| Parameter | Value |
| --- | --- |
| Nominal capacitance | 165 F |
| Rated voltage | 48.6 V |
| Equivalent series resistance | 6.3 mΩ |
| Peak current | 1970 A |

connected in series, resulting in 23.57 F capacitance, 340.2 V rated terminal voltage, and 44.1 mΩ of internal series resistance.

### 8.4.1 SIMULATION AND ANALYSIS OF PASSIVE PARALLEL CONFIGURATION

In this configuration, the battery and UC are connected directly in parallel without any interfacing converter in between, and the common battery/UC terminals are connected to the DC link through a bidirectional converter. The power demand for the vehicle has been obtained through powertrain system analysis toolkit (PSAT) simulations considering a typical mid-size sedan vehicle configured as a PEV. Since the motor drive voltage is almost constant, the power demand of the vehicle can be divided by the DC link voltage to obtain the motor drive current, and the motor drive and load demand variation have therefore been modeled and implemented as a controlled current source.

During the simulation, the reference DC link voltage was selected as 400 V and the bidirectional converter was controlled through a double-loop voltage and current controller. A proportional–integral (PI) controller was used in the voltage loop, while a peak current mode controller was used in the current loop, as shown in Figure 8.20.

The load current for the $t = [690, 760]$ time interval varies, as shown in Figure 8.21. As shown, this load current includes positive and negative current variations, simulating acceleration, and braking conditions. On the basis of this load current variation, the bidirectional converter is controlled such that it maintains a constant DC link voltage while supplying power from sources during acceleration, and recharging them during braking. The battery and UC current variations are given in Figures 8.22 and 8.23, respectively.

As shown in Figures 8.22 and 8.23, the battery inherently supplies a smoother current profile in comparison to the UC due to its slower dynamics. However, since there is no interface that controls the battery current, its current has some fluctuations that could likely be eliminated by other connection

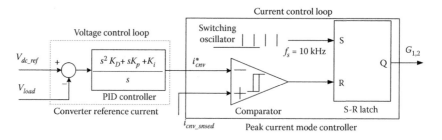

FIGURE 8.20    Control system for the passive parallel connection topology.

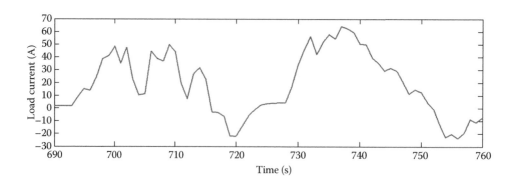

FIGURE 8.21    Load current variation.

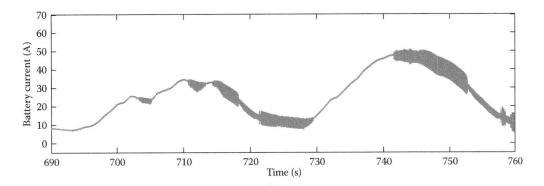

FIGURE 8.22 Battery current variation in passive parallel topology.

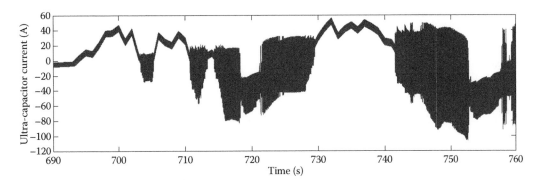

FIGURE 8.23 UC current variation in passive parallel topology.

topologies. Owing to the voltage balance between the battery and UC, battery current varies automatically to maintain similar terminal voltage with the UC at all times. If the UC voltage was higher than the battery voltage due to some large braking energy recovery, the battery current would reverse direction, but here, only the UC receives power from the application of regenerative braking.

The SoC variations of the battery and UC are given in Figures 8.24 and 8.25, respectively.

The initial SoCs for both battery and UC were selected as 90%. Since the battery voltage is higher than the UC, the battery is always discharging as explained for the current variations. However, the SoC of the UC is sometimes increasing as it is recharged during braking conditions, that is, the negative current variations of the UC.

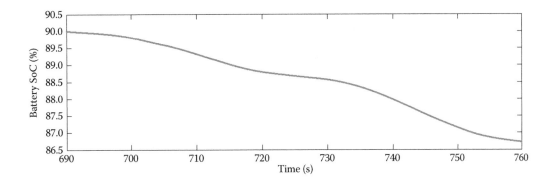

FIGURE 8.24 SoC of the battery for passive parallel topology.

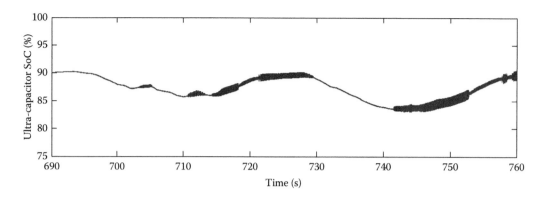

FIGURE 8.25    SoC of the UC for passive parallel topology.

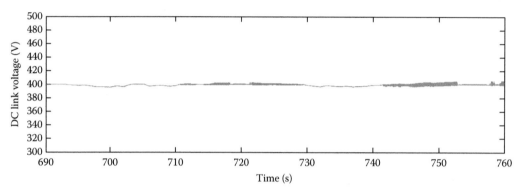

FIGURE 8.26    DC link (load bus) voltage variation for passive parallel topology.

Finally, the DC link voltage variation, to which the motor drive inverter is connected, is given in Figure 8.26. As observed from Figure 8.26, the DC link voltage varies steadily around the 400-V reference set point. During high-power demands and operation mode changes of the bidirectional converter, the voltage fluctuations increase. For this topology and control strategy, the maximum voltage seen at the DC link is 405.3 V with a minimum of 395.2 V, and therefore, a maximum amplitude of the voltage fluctuation of 2.5% over the simulation period.

### 8.4.2    SIMULATIONS AND ANALYSIS OF CASCADED CONVERTERS TOPOLOGY

In this configuration, the battery is connected to the UC through a bidirectional converter and the UC is connected to the DC link through another bidirectional converter; therefore, the battery, converter 1, UC, and converter 2 are all in cascade connection. The same drive cycle over the same time interval was used for load modeling in this topology as in the previous simulation, with the DC link voltage reference kept at 400 V. For the UC controls, a double-loop controller is employed for DC link voltage regulation, and for the battery controls, only a peak current mode controller is used. The reference current for the battery can be obtained as

$$I_{batt}^* = \frac{V_{load} \times I_{load}}{V_{batt}} G_{LP}(s) \tag{8.1}$$

where $I_{batt}^*$ is the battery reference current, $V_{load}$ and $I_{load}$ are the instantaneously measured DC link voltage and current, and $V_{batt}$ is the battery terminal voltage, which is nearly constant during the whole drive cycle. The transfer function represented by $G_{LP}(s)$ is a low-pass Bessel filter that is applied to eliminate any spikes and fast transients from the battery reference current. These fast transients come

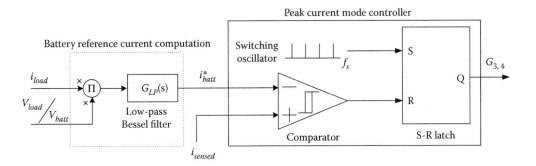

FIGURE 8.27    Battery current controller.

inherently from the variation in instantaneously measured load current, and by employing this filter, the battery current can be smoother and the stress on the battery is reduced since there is an additional converter regulating the battery current. The battery current controller is depicted in Figure 8.27.

The load current drawn from the DC link varies in the same manner as shown in Figure 8.21, and vehicle specifications and battery and UC parameters are the same as in the previous example. On the basis of this load current variation, the UC's bidirectional converter is controlled such that it maintains a constant DC link voltage. The bidirectional converter connected to the battery is controlled so that the battery supplies the average load demand to the converter's output. Whenever the DC link sees a reference voltage >400 V, both converters are controlled to change their modes of operation from boost to buck so that the braking energy can be recovered back into the storage devices. The battery and UC current variations are given in Figures 8.28 and 8.29, respectively.

From Figures 8.28 and 8.29, the battery current ripples are reduced due to the control strategy employed. Moreover, the power contribution is greater as compared to the previous topology since the battery current is actively controlled, allowing it to slowly supply the actual load demands. A benefit of this configuration is that, at any time, a limitation can be placed on the maximum allowable battery current to reduce the battery contribution and allow the UC to supply more power to the DC link to maintain the 400-V regulation. In this topology, the current ripple of the UC is greater than in the simpler passive parallel connection, but since it can successfully supply these current variations without seeing a shorter life span, this is not an issue for the UC.

The SoC variations of the battery and UC are given in Figures 8.30 and 8.31, respectively.

In this configuration, the battery is utilized in a manner similar to the passive parallel case. Therefore, the SoC usage window for the UC is smaller since it continually receives charge from the battery. However, since the battery contributes more, its SoC decreases more quickly in comparison to the passive parallel case.

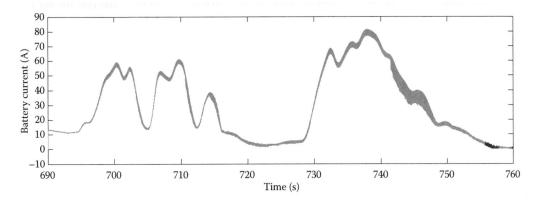

FIGURE 8.28    Battery current variation in cascaded converters topology.

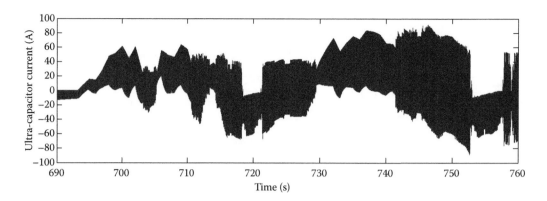

**FIGURE 8.29**  UC current variation in cascaded converters topology.

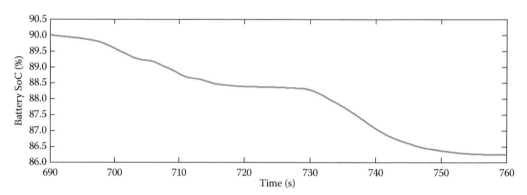

**FIGURE 8.30**  SoC of the battery for cascaded converters topology.

The DC link voltage variation for the cascaded converters configuration is represented in Figure 8.32. As seen from the figure, the DC link voltage varies around the 400-V reference set point, and during high-power demands and operation mode changes of the bidirectional converters, voltage fluctuations become more apparent. For this topology and control strategy, the DC link voltage reaches a maximum of 405.0 V and a minimum of 395.3 V. Therefore, the maximum amplitude of the voltage fluctuation has been calculated as 2.4% over the simulation period.

Since this configuration employs an individual DC/DC converter for the battery, it has the built-in flexibility of tuning and manipulating battery current controls. Therefore, a rate limiter and a saturation limiter can be implemented within the battery current control loop: the rate limiter will

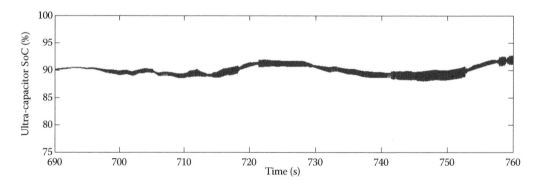

**FIGURE 8.31**  SoC of the UC for cascaded converters topology.

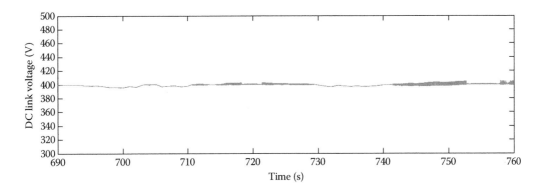

FIGURE 8.32   DC link (load bus) voltage variation for the cascaded converters topology.

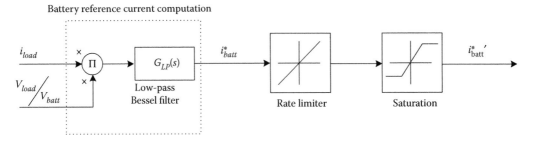

FIGURE 8.33   Battery reference current manipulation.

limit the slope of the battery reference current, while the saturation limiter will limit the battery current magnitude. The implementation of rate and saturation limiters into the battery controller is shown in Figure 8.33.

The rate limiter applied here has a rising slew rate of +0.1 and a falling slew rate of −0.1 placed on the rising and falling rates of the battery current. At the same time, the saturation block limits the maximum battery reference current by +50 A and negative battery reference current by −50 A to ensure the further reduction of battery stress and maximum battery charge and discharge current. In this case, the current variations of the battery and UC are recorded as given in Figures 8.34 and 8.35.

The battery current given in Figure 8.34 resulted from implementation of the rate and saturation limiters within the battery current control loop. This modification improves the battery current

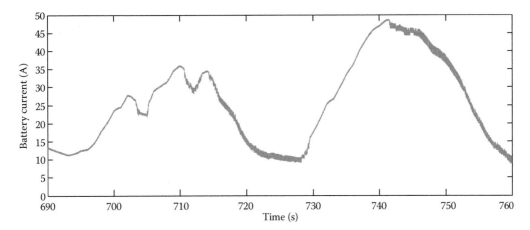

FIGURE 8.34   Battery current variation with modified controls.

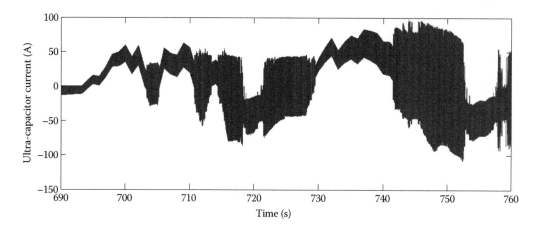

FIGURE 8.35 UC current variation with modified controls of battery.

waveform by eliminating the natural high slew rates of the load current (see Figure 8.28 vs. Figure 8.34). Moreover, maximum charge and discharge current rates can be defined and battery protection can be realized. In this case, the UC tends to vary faster in time and larger in amplitude (see Figure 8.29 vs. Figure 8.35), but again, the selected UC should be capable of supplying this type of current demand. Since the battery usage is reduced and more power is supplied from the UC, the modified current controller affects the SoC variations, as shown in Figures 8.36 and 8.37.

Figures 8.36 and 8.37 show that the battery SoC remains higher (see and compare Figure 8.30) while UC SoC drops more drastically (see and compare Figure 8.31) since the battery response to power-throughput demands is reduced and the UC must deliver more power to the DC link to regulate its voltage during transients.

While using any of the topologies discussed, whenever the UC SoC falls below a certain point, the battery controller should bring it above a certain point while supplying the load demands at the same time. A typical lower limit for the UC can be selected as 20%. Although a deep discharge does not tend to be a problem for UCs, such a limitation would prevent the associated DC/DC converter from operating in extreme voltage conversion ratios. Moreover, a fully discharged UC would draw an excessive high current at the initial charging if the charge current is not appropriately controlled.

The DC link voltage for this topology with a modification allowing for current limiting of battery is presented in Figure 8.38. Since the UC supplies more power to maintain constant DC link

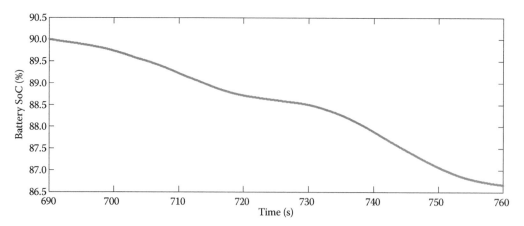

FIGURE 8.36 SoC of the battery for cascaded converters topology with modified controls.

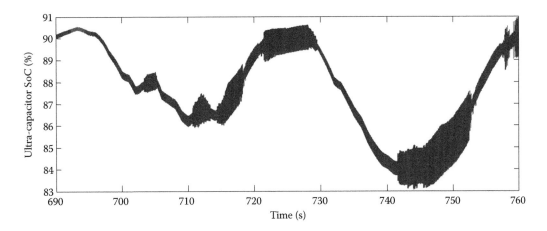

FIGURE 8.37    SoC of the UC with modified controls of the battery.

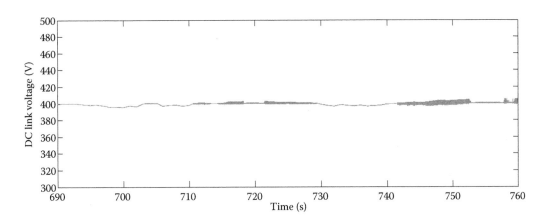

FIGURE 8.38    DC link voltage variation after modified battery current controls.

voltage, the resulting DC link voltage sees slightly more voltage ripple in comparison to the previous configurations. The maximum DC link voltage for this simulation was 405.2 V with a minimum of 395.2 V, and therefore a max/min ripple percentage of 2.5%.

### 8.4.3    SIMULATION AND ANALYSIS OF PARALLEL-CONNECTED MULTICONVERTERS TOPOLOGY

In this configuration, the battery is connected to the DC link through a bidirectional converter and the UC is connected to the same DC link through another bidirectional converter. The battery and UC are therefore connected to the common DC link in parallel through their individual converters as shown earlier. The same drive cycle was used for load modeling over the same time interval of the previous simulations, the DC link voltage reference was kept the same, and the same strategies were applied for the battery and UC control loops.

The battery and UC current variations are given in Figures 8.39 and 8.40, respectively.

Owing to the battery current control strategy used here and to the parallel-connected individual battery DC/DC converter, the battery current has been further smoothed with reduced current ripples. Although the battery current is limited to be within [−50, +50] A, the battery current stays less than the maximum limit due to the Bessel reference current filter and rising–falling slew rate limiter controller. The only trade-off for having less distortion with the battery current is having

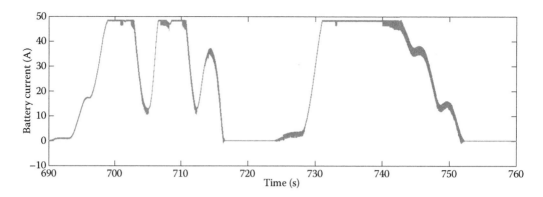

FIGURE 8.39   Battery current with parallel converters topology.

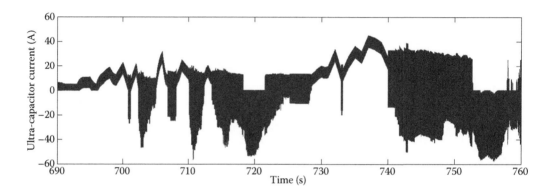

FIGURE 8.40   UC current with parallel converters topology.

huge fluctuations with the UC current. However, the UC is capable of supplying these types of current profiles without sacrificing lifetime and performance.

The SoC variations of the battery and UC are recorded as shown in Figures 8.41 and 8.42, respectively.

From Figures 8.41 and 8.42, it can be observed that the battery is utilized less and maintains a higher SoC at the end of the drive cycle. Since the UC makes a greater contribution, another mode

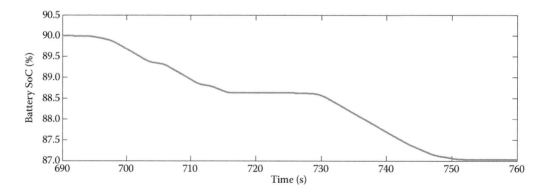

FIGURE 8.41   SoC of the battery with parallel converters topology.

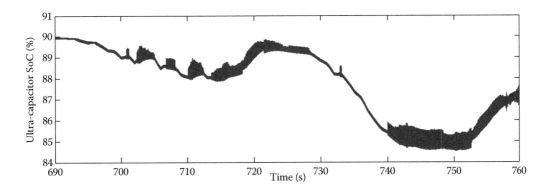

**FIGURE 8.42**    SoC of the UC with parallel converters topology.

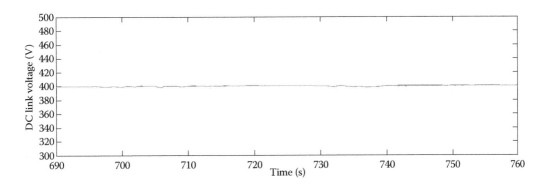

**FIGURE 8.43**    DC link voltage variation with parallel converters topology.

of operation could be employed such that the battery recharges the UC whenever the UC SoC drops below a certain lower limit. The last result for this topology is shown in Figure 8.43 that is the DC link voltage variation.

Since both UC and its individual parallel converter are controlled to maintain a constant DC link voltage, the DC link voltage has a much smaller voltage ripple than in the previous configurations. In this case, the DC link voltage sees a maximum of 400.7 V and a minimum of 397.6 V, resulting in a maximum ripple percentage of 0.8%.

### 8.4.4    CONCLUSIONS

This section presented simulations of battery/UC-based hybrid ESS including passive parallel connection, cascaded converters configuration, and the parallel-connected converters configuration. The results of the analysis of these three topologies have been consolidated for comparison in Table 8.4. For some comparison criteria, these topologies have been graded by the authors on a points scale with 1 indicating the best, 2 indicating the better, and 3 indicating the average.

As presented in Table 8.4, the control system is the simplest for the passive parallel topology since there is only one converter current to be regulated. Control of the cascaded converters is more complex as there are two converter currents to be controlled, and the addition of current and slew rate limiters into the cascaded converters controller is obviously still more complicated. Parallel converters also have a similarly high level of complexity but with a bit more freedom in control of current magnitude and direction.

TABLE 8.4

TABLE 8.4

**Comparisons of Hybrid ESS Configurations**

| Criteria | Passive Parallel | Cascaded Converters | Cascaded (Manipulated Controls) | Parallel Converters |
|---|---|---|---|---|
| Control simplicity | 1 | 2 | 3 | 3 |
| Structure complexity | 1 | 2 | 2 | 2 |
| Number of converters | 1 | 2 | 2 | 2 |
| Number of inductors | 1 | 2 | 2 | 2 |
| Total inductor mass | 2 | 3 | 3 | 2 |
| Number of transducers | 5 | 6 | 6 | 6 |
| Cycle-end battery SoC | 86.72% | 86.24% | 86.66% | 87.03% |
| Cycle-end UC SoC | 89.90% | 91.91% | 90.45% | 87.10% |
| Maximum battery current ripple | ~7 A | ~9 A | ~1.7 A | ~1.8 A |
| Cycle-based topology efficiency | 95.24% | 90.34% | 90.72% | 95.25% |
| Maximum DC link voltage variation percentage | 2.52% | 2.42% | 2.51% | 0.77% |

The passive parallel configuration also has the most basic structure. The other converters have a similar level of structural complexity as they clearly have a larger number of converters and therefore switches, inductors, bus bars, and so on. The total inductor mass of the passive parallel topology and of the cascaded converters topology is higher than that of the parallel converters topology. In the passive parallel case, 100% of the UC and battery current must pass through a single inductor, requiring high-current-rating wiring of the inductor, and in the cascaded converter, the battery converter carries only the battery current, but the UC converter carries the sum of both the battery and UC currents. However, in the case of parallel converters, although two inductors are required, their sizes are relatively smaller as compared to inductors of the other topologies since each converter carries the current of one source and not of the two sources.

When the topologies are compared in terms of cycle-end battery SoC, the parallel converters are best because of the battery current profile. However, in this case, the UC is utilized more that results in less end of cycle SoC. In the cascaded converters case, the battery sustainably recharges the UC; that is, the battery power is transferred to the UC continually; therefore, the UC's cycle-end SoC is more. The highest battery current ripple occurs when using either the cascaded converter or the passive parallel converters topology since battery current is not effectively controlled and limited in these topologies. The cascaded converters with manipulated controls and the parallel converters inherently provide less battery current ripples and therefore prolong battery life.

The cycle-based energy efficiencies are calculated by numerically integrating the battery power, UC power, and load power over the drive cycle to obtain the total energy flow from each source to the load. Once the energy levels are obtained, the output and input energy relationship defines the cycle-based efficiency. In this case, the cascaded converter topology was the least efficient since there are two cascaded converters and one of them should carry all of the current (again, battery current must pass through two converters). In the passive parallel case, there is only one converter that improves the efficiency, but the most efficient topology is that of the parallel converters, since each of the energy storage devices has its own converter, and power from a single device must never pass through multiple converters. The parallel converters topology is also the best topology in terms of DC link voltage variation due to the fact that one of the converters is always utilized to independently regulate the DC link voltage.

## PROBLEMS

An UC module has the following specifications:

| Parameter | Value | Unit |
| --- | --- | --- |
| Rated capacitance | 63 | (F) |
| Maximum $ESR_{DC}$ | 18 | (mΩ) |
| Rated voltage | 125 | (V) |
| Absolute maximum voltage | 136 | (V) |
| Maximum continuous current at 45°C | 240 | (A) |
| Maximum peak current for 1 s, nonrepetitive | 1800 | (A) |
| Mass | 60.5 | (kg) |

1. Calculate the stored energy of the UC module.
2. Calculate the specific energy of the UC module.
3. Calculate the maximum continuous power and the specific power (watts per kilogram) of the UC module.
4. Verify that the maximum peak current for 1 s is 1800 A as given in the datasheet.
5. Calculate the amount of energy that the UC releases when it is discharged from 125 to 100 V.
6. Assume that this UC module is being discharged with 100 A from initially charged condition. Calculate the module voltage for 10 and 50 s after the discharge starts.
7. When this module is discharging at 100 A in constant current discharge mode, how long does it take to discharge the UC from 125 to 45 V?
8. Calculate the discharge current if the UC voltage is reducing from 125 to 75 V in 300 s.
9. Calculate the power loss if the module is discharged at 10 A of constant current.
10. Calculate how long does it take to discharge the UC from 125 to 5 V at a constant power discharge rate of 1000 W?

   A battery pack has the following specifications:

| Parameter | Value | Unit |
| --- | --- | --- |
| Nominal voltage | 360 | (V) |
| Total stored energy | 24 (21, total usable) | (kWh) |
| Maximum continuous output power | 100 | (kW) |
| Weight | 293.93 | (kg) |

11. If this battery pack is hybridized with the UC module of which parameters were earlier expressed, calculate the power density of the hybrid ESS and state what percent of power density has been increased as compared to the battery-alone case according to the IEC definition.
12. Calculate the energy density for the same hybrid configuration.

## REFERENCES

1. S. Williamson, A. Khaligh, and A. Emadi, Impact of energy storage devices on drive train efficiency and performance of heavy-duty HEVs, *IEEE Vehicle Power and Propulsion Conference (VPPC)*, Chicago, September 2005.
2. A. Khaligh, A. M. Rahimi, Y. J. Lee, J. Cao, A. Emadi, S. D. Andrews, C. Robinson, and C. Finnerty, Digital control of an isolated active hybrid fuel cell/Li-ion battery power supply, *IEEE Transactions on Vehicular Technology*, 56, 3709–3721, November 2007.

3. P. Tiehua, J. Zang, and E. Darcy, Cycling test of commercial nickel–metal hydride (Ni–MH) cells, *Battery Conference on Applications and Advances*, Long Beach, CA, pp. 393–397, January 1998.

4. S. Lukic, S. Wirasingha, F. Rodriguez, J. Cao, and A. Emadi, Power management of an ultra-capacitor/battery hybrid storage system in HEV, *IEEE Vehicle Power and Propulsion Conference* (VPPC), Windsor, United Kingdom, pp. 1–6, September 2006.

5. A. Emadi, M. Ehsani, and J. M. Miller, *Vehicular Electric Power Systems: Land, Sea, Air, and Space Vehicles*, New York: Marcel Dekker, 2003.

6. J. P. Zheng, T. R. Jow, and M. S. Ding, Hybrid power sources for pulsed current applications, *IEEE Transactions of Aerospace Electronic Systems*, 1(1), 288–292, January 2001.

7. A. Emadi, S. S. Williamson, and A. Khaligh, Power electronics intensive solutions for advanced electric, hybrid electric, and fuel cell vehicular power systems, *IEEE Transactions on Power Electronics*, 21(3), 567–577, May 2006.

8. P. A. Flatherty, Multi-stage hybrid drives for traction applications, in *Proceedings of the Joint Rail Conference*, pp. 171–175, March 2005, Pueblo, Colorado.

9. Maxwell® Technologies, BMOD0063 P125 B33 Ultra-capacitor datasheet, HTM Heavy Transportation Series, available online at: http://www.maxwell.com/ultracapacitors/datasheets/DATASHEET_BMOD 0063_1014696.pdf

10. R. A. Dougal, S. Liu, and R. E. White, Power and life extension of battery-ultra-capacitor hybrids, *IEEE Transactions on Components and Packaging Technologies*, 25(1), 120–131, March 2002.

11. L. Solero, A. Lidozzi, and J. A. Pomilio, Design of multiple-input power converter for hybrid vehicles, *IEEE Transactions on Power Electronics*, 20(5), 1007–1016, September 2005.

12. S. Kim and S. H. Choi, Development of fuel cell hybrid vehicle by using ultra-capacitors as a secondary power source, *2005 SAE World Congress*, Detroit, Michigan, April 2005.

13. J. M. Miller and M. Everett, An assessment of ultra-capacitors as power cache in Toyota THS-11, GM-Allision AHS-2 and Ford FHS hybrid propulsion systems, *IEEE 20th Applied Power Electronics Conference and Exposition*, Austin, TX, 1, pp. 481–490, March 2005.

14. A. Napoli, F. Crescimbini, F. Capponi, and L. Solero, Control strategy for multiple input DC–DC converters for hybrid vehicles propulsion systems, *IEEE Power Electronics Specialists Conference*, L'Aquila, Italy, pp. 1685–1690, June 2002.

15. S. Liu and R. A. Dougal, Design and analysis of a current-mode controlled battery/ultracapacitor hybrid, in *Proceedings of the IEEE Industry Applications Society Annual Meeting*, pp. 1140–1145, Seattle, WA, October 2004.

16. O. Onar and A. Khaligh, Dynamic modeling and control of a cascaded active battery/ultra-capacitor based vehicular power system, in *Proceedings of the IEEE Vehicle Power and Propulsion Conference (VPPC)*, Harbin, China, pp. 1–4, September 2008.

17. M. Ortuzar, J. Moreno, and J. Dixon, Ultracapacitor-based auxiliary energy system for an electric vehicle: Implementation and evaluation, *IEEE Transactions on Industrial Electronics*, 54(4), 2147–2156, August 2007.

18. J. Cao and A. Emadi, A new battery/ultra-capacitor hybrid energy storage system for electric, hybrid, and plug-in hybrid electric vehicles, in *Proceedings of the IEEE Vehicle Power and Propulsion Conference (VPPC)*, Dearborn, MI, pp. 941–946, 2009.

19. L. Gao, R. A. Dougal, and S. Liu, Power enhancement of an actively controlled battery/ultracapacitor hybrid, *IEEE Transactions on Power Electronics*, 20(1), 236–243, January 2005.

20. W. Lhomme, P. Delarue, P. Barrade, A. Buoscayrol, and A. Rufer, Design and control of a supercapacitor storage system for traction applications, in *Proceedings of the IEEE Industry Application Conference*, Kowloon, Hong Kong, 3, pp. 2013–2020, October 2005.

21. Z. Jiang and R. A. Dougal, A compact digitally controlled fuel cell/battery hybrid power source, *IEEE Transactions on Industrial Electronics*, 53(4), 1094–1104, June 2006.

22. S. M. Lukic, S. G. Wirashanga, F. Rodriugez, C. Jian, and A. Emadi, Power management of an ultracapacitor/battery hybrid energy storage system in an HEV, in *Proceedings of the IEEE Vehicle Power and Propulsion Conference*, Windsor, United Kingdom, pp. 1–6, 2006.

23. S. M. Lukic, J. Cao, R. C. Bansal, F. Rodriguez, and A. Emadi, Energy storage systems for automotive applications, *IEEE Transactions on Industrial Electronics*, 55(6), 2258–2267, June 2008.

24. Z. Li, O. Onar, A. Khaligh, and E. Schaltz, Power management, design, and simulations of a battery/ultra-capacitor hybrid system for small electric vehicles, in *Proceedings of the SAE (Society of Automotive Engineers) World Congress*, Detroit, MI, USA, April 2009.

25. H. Matsuo, L. Wenzhong, F. Kurokawa, T. Shigemizu, and N. Watanabe, Characterization of the multiple-input DC–DC converter, *IEEE Transactions on Industrial Electronics*, 51(3), 625–631, June 2004.

26. M. H. Kheraluwala, R. W. Gascoine, D. M. Divan, and B. Bauman, Performance characterization of a high power dual active bridge DC/DC converter, in *Proceedings of the IEEE Industry Applications Society Annual Meeting*, 2, Seattle, WA, pp. 1267–1273, 1990.

27. M. Marchesoni and C. Vacca, New DC–DC converter for energy storage system interfacing in fuel cell hybrid electric vehicles, *IEEE Transactions on Power Electronics*, 22(1), 301–308, January 2007.

28. M. C. Kisacikoglu, M. Uzunoglu, and M. S. Alam, Fuzzy logic control of a fuel cell/battery/ultra-capacitor hybrid vehicular power system, in *Proceedings of the Vehicle Power and Propulsion Conference (VPPC)*, Arlington, TX, pp. 591–596, 2007.

29. B. G. Dobbs and P. L. Chapman, A multiple-input DC–DC converter topology, *IEEE Power Electronics Letters*, 1(1), 6–9, March 2003.

30. H.-J. Chiu, H.-M. Huang, L.-W. Lin, and M.-H. Tseng, A multiple-input DC/DC converter for renewable energy systems, in *Proceedings of the IEEE Industrial Conference on Industrial Technology*, Hong Kong, pp. 1304–1308, 2005.

31. Y.-M. Chen, Y.-C. Liu, and F.-Y. Wu, Multi-input DC/DC converter with ripple-free input currents, in *Proceedings of the IEEE Power Electronics Specialists Conference*, Cairns. Qld, Australia, 2, pp. 796–802, 2002.

32. Z. Li, O. Onar, A. Khaligh, and E. Schaltz, Design and control of a multiple input DC/DC converter for battery/ultra-capacitor based electric vehicle power system, in *Proceedings of the IEEE 24th Annual Conference on Applied Power Electronics and Exposition (APEC)*, Washington DC, pp. 591–596, February 2009.

33. Z. Amjadi and S. S. Williamson, A novel control technique for a switched-capacitor–converter-based hybrid electric vehicle energy storage system, *IEEE Transactions on Industrial Electronics*, 57(3), 926–934, March 2010.

34. S. Dwari and L. Parsa, A novel high efficiency high power interleaved coupled-inductor boost DC–DC converter for hybrid and fuel cell electric vehicle, in *Proceedings of the IEEE Vehicle Power and Propulsion Conference*, Arlington, TX, pp. 399–404, September 2007.

35. M. B. Camara, F. Gustin, H. Gualous, and A. Berthon, Supercapacitors and battery power management for hybrid vehicle applications using multi boost and full bridge converters, in *Proceedings of the IEEE Europe Conference of Power Electronics Applications*, Aalborg, Denmark, pp. 1–9, September 2007.

36. A. Ioinovici, H. S. H. Chung, M. S. Makowski, and C. K. Tse, Comments on unified analysis of switched-capacitor resonant converters', *IEEE Transactions on Industrial Electronics*, 54(1), 684–685, February 2007.

37. Y. Berkovich, B. Axelrod, S. Tapuchi, and A. Ioinovici, A family of four-quadrant, PWM DC–DC converters, in *Proceedings of the IEEE Power Electronics Specialists Conference*, Orlando, FL, pp. 1878–1883, June 2007.

38. H. S. Chung and A. Ioinovici, Development of a general switched-capacitor DC/DC converter with bi-directional power flow, in *Proceedings of the IEEE International Symposium on Circuits and Systems*, Geneva, Italy, 3, pp. 499–502, May 2003.

39. O. C. Mak, Y. C. Wong, and A. Ioinovici, Step-up DC power supply based on a switched-capacitor circuit, *IEEE Transactions on Industrial Electronics*, 42(1), 90–97, February 1995.

40. O. C. Onar and A. Khaligh, A novel integrated magnetic structure based DC/DC converter for hybrid battery/ultracapacitor energy storage systems, *IEEE Transactions on Smart Grid*, 3(1), 296–307, March 2012.

41. R. Hebner, J. Beno, and A. Walls, Flywheel batteries come around again, *IEEE Spectrum*, 39(4), 46–51, April 2002.

42. T. M. Mulcahy, J. R. Hull, K. L. Uherka, R. C. Niemann, R. G. Abboud, J. P. Juna, and J. A. Lockwood, Flywheel energy storage advances using HTS bearings, *IEEE Transactions on Applied Superconducting*, 9(2), 297–300, June 1999.

43. A. Jaafar, C. R. Akli, B. Sareni, X. Roboam, and A. Jeunesse, Sizing and energy management of a hybrid locomotive based on flywheel and accumulators, *IEEE Transactions on Vehicular Technology*, 58(8), 3947–3958, October 2009.

44. O. Briat, J. M. Vinassa, W. Lajnef, S. Azzopardi, and E. Woirgard, Principle, design, and experimental validation of a flywheel-battery hybrid source for heavy-duty electric vehicles, *IET Electric Power Applications*, 1(5), 665–674, 2007.

45. S. Talebi, B. Nikbakhtian, and H. Toliyat, A novel algorithm for designing the PID controllers of high-speed flywheels for traction applications, in *Proceedings of the Vehicle Power and Propulsion Conference (VPPC)*, Arlington, TX, pp. 574–579, 2007.

46. S. Shen and F. E. Veldpaus, Analysis and control of a flywheel hybrid vehicular powertrain, *IEEE Transactions on Control Systems*, 12(5), 645–660, September 2004.

47. J. G. Oliviera, A. Larsson, and H. Bernhoff, Controlling a permanent-magnet motor using PWM converter in flywheel energy storage systems, in *Proceedings of the IEEE Industrial Electronics Conference (IECON)*, Orlando, FL, pp. 3364–3369, 2008.

48. R. I. Davis and R. D. Lorenz, Engine torque ripple cancellation with an integrated starter alternator in a hybrid electric vehicle: Implementation and control, *IEEE Transactions on Industry Applications*, 39(6), 1765–1773, November/December 2003.

49. S. Lemofouet and A. Rufer, A hybrid energy storage system based on compressed air and supercapacitors with maximum efficiency point tracking (MEPT), *IEEE Transactions on Industrial Electronics*, 53(4), 1105–1115, August 2006.

50. J. Lehmann, Air storage gas turbine power plants, a major distribution for energy storage, in *Proceedings of the International Conference on Energy Storage*, United Kingdom, pp. 327–336, April 1981.

51. A. Rufer and S. Lemofouet, Energetic performance of a hybrid energy storage system based on compressed air and super capacitors, in *Proceedings of the International Symposium on Power Electronics, Electrical Drives, Automation, and Motion (SPEEDAM)*, Taormina, Italy, pp. 469–474, 2006.

52. A. Rufer and S. Lemofouet, Efficiency consideration and measurements of a hybrid energy storage system based on compressed air and super capacitors, in *Proceedings of the International Power Electronics and Motion Control Conference (EPE-PEMC)*, Portoroz, Slovenia, pp. 2077–2081, 2006.

53. P. F. Ribeiro, B. K. Johnson, M. L. Crow, A. Arsoy, and Y. Liu, Energy storage systems for advanced power applications, *Proceedings of the IEEE*, 89(12), 1744–1756, 2001.

54. M. H. Ali, B. Wu, and R. A. Dougal, An overview of SMES applications in power and energy systems, *IEEE Transactions on Sustainable Energy*, 1(1), 38–47, April 2010.

55. T. Ise, M. Kita, and A. Taguchi, A hybrid energy storage with a SMES and secondary battery, *IEEE Transactions on Applied Superconductivity*, 15(2), 1915–1918, June 2005.

56. H. Zhang, J. Ren, Y. Zhong, and J. Chen, Design and test of controller in power conditioning system for superconducting magnetic energy storage, in *Proceedings of the International Conference on Power Electronics (ICPE)*, Daegu, South Korea, pp. 966–972, 2001.

57. L. Trevisani, A. Morandi, F. Negrini, P. L. Ribani, and M. Fabbri, Cryogenic fuel-cooled SMES for hybrid vehicle application, *IEEE Transactions on Applied Superconductivity*, 19(3), 2008–2011, June 2009.

58. R. F. Giese, Progress toward high temperature superconducting magnetic energy storage (SMES) systems—A second look, *Technical Report by Argonne National Laboratory*, 1998.

59. A. P. Malozemoff, J. Maguire, B. Gamble, and S. Kalsi, Power applications of high-temperature superconductors: Status and perspective, *IEEE Transactions on Applied Superconductivity*, 12(1), 778–781, March 2002.

60. C. A. Luongo, Superconducting storage systems, *IEEE Transactions on Magnetics*, 32(4), 2214–2223, 1996.

61. Toyota officially launches plug-in Prius program, retail sales in 2011, Autobloggreen, Available online: http://green.autoblog.com/2009/12/14/toyota-officially-launches-plug-in-prius-program-retail-sales-i/

62. Y. Tanaka, Prius plug-in hybrid vehicle overview, *Technical Report by Toyota Passenger Vehicle Development Center*, Available online: http://www.toyota.co.jp/en/tech/environment/conference09/pdf/phv_overview_en.pdf, December 2009.

63. Maxwell Technologies BMODO165-48.6 V ultra-capacitors' data sheet, Available online: http://www.maxwell.com/ultracapacitors/datasheets/DATASHEET_48V_series_1009365.pdf

# 9 Low-Voltage Electrical Systems for Nonpropulsion Loads

*Ruoyu Hou, Pierre Magne, and Berker Bilgin*

## CONTENTS

## 9.1 INTRODUCTION

In conventional vehicles, the traction power is supplied by the internal combustion engine. In order to provide power to the vehicle electrical loads, a low-voltage system is utilized, which includes a belt-driven alternator, low-voltage battery, and various electrical loads. When the engine is running, it provides torque to the alternator, which then provides electrical energy to the 12 V battery. In conventional vehicles, claw–pole synchronous generators are utilized, due to their low-cost structure and reliable operation. However, claw–pole alternators usually have low efficiency because of the high leakage flux. Depending on the charging current of the low-voltage battery and the load requirements in the vehicle electrical system, the field current of the claw–pole alternator is controlled by a regulator to keep the system voltage constant. In light-duty vehicles, battery voltage is usually 12 V, and when the vehicle is running, system voltage is approximately 13.5 V in summer time and 14.5 V in winter time. With the engine stopped, only low-voltage battery supplies power to the electrical loads. The battery also behaves like a buffer in the electrical systems and stores energy.

With the improvements in vehicle technology, safety requirements, and increasing customer demands, many electric and electronic loads have been added to the vehicle electrical system. In conventional vehicles, the electrical system has to supply enough power to the entire vehicle network, provided that the quality of the voltage is high enough to ensure the functional safety of electronic loads, especially control units.

In electrified vehicles, similar low-voltage electric and electronic loads still exist. However, traction system voltage is usually much higher than the vehicular electrical system voltage. As an example, in 2010 Toyota Prius, battery voltage is 201.6 V (168 NiMH battery cells at 1.2 V) and the voltage supplied to the inverters is varied between 225 and 650 V by a 27 kW boost converter. The vehicular system voltage is 12 V. In all-electric vehicles, the voltage levels are similar. In belt-driven starter generator applications, the traction power is usually supplied by lead-acid batteries and the voltage is set around 48 V. This is mainly because the precautions required for high-voltage systems do not need to be applied in a 48 V system, since high-voltage standards are applied over 60 V DC. The traction motors, generator, or starter alternators in electrified powertrain applications are designed for the above-mentioned voltage levels and they cannot be utilized to supply power directly to the vehicular loads. For this reason, power converters are required, which convert high voltage from the traction battery to a lower voltage in order to supply power to the vehicular electrical and electronic loads, and also to charge the low-voltage battery. This power converter is usually called as auxiliary power module (APM).

Depending on the road and weather conditions, many electric loads are on and off when the vehicle is being driven or stopped. Therefore, APM can draw power from the high-voltage battery anytime throughout the drive cycle and it might affect the state-of-charge (SOC) of the high-voltage battery. In hybrid electric vehicles, if the SOC of the high-voltage battery is low, the engine turns on and charges the battery through the generator. This increases vehicle's emissions and fuel consumption. In all-electric vehicles, lower SOC reduces the range of the vehicle. Therefore, efficiency of the APM is very important to maintain a higher vehicle performance.

## 9.2 LOW-VOLTAGE ELECTRICAL LOADS

The low-voltage system in a vehicle constitutes many different loads. These can be categorized as lighting, air conditioning, wiper and window systems, electronic, and accessory loads. As shown in Figure 9.1, air conditioning loads draw most of the power from the electrical system. These include radiator fan, blowers, and seat heaters. In conventional vehicles, cabin heating is usually maintained from the waste heat of the engine. In hybrid electric vehicles, the engine waste heat can still be utilized; however, in all-electric vehicles, the entire cabin heating should be provided by an electrical heating system.

Lighting loads consume around 24% of the total power in a vehicular electrical system. They are composed of many different loads, including headlights, fog lamps, park lamps, flashers, turn signals, and so on. Among these, back lights, headlamps, and fog lamps draw most of the power. In a typical vehicle, wiper and window system-related loads draw around 10.30% of the total power. Electronic loads include the control units and displays. Power outlets, CD player, and Bluetooth are

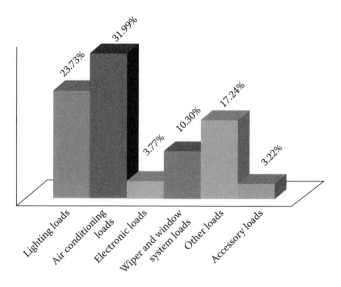

FIGURE 9.1 Typical low-voltage loads in a vehicle electric system.

some of the accessory loads. Electric power steering and motor engaging park brake are some other loads for the low-voltage electric power system.

Most of the loads in the low-voltage electrical system are resistive loads. The resistance that is seen from the supply side changes with the current drawn by the load. The fans, pumps, wipers, and power windows all have electric motors, which are usually controlled by their corresponding control system. The power drawn by the fans is dependent on the fan speed. Ambient temperature and the cabin temperature set by the driver determines the coolant flow rate, and, hence, the electrical power drawn by the coolant pump. In some circumstances, many of these loads can operate together. However, the status of the vehicle and the driving conditions usually determine the activation of the loads.

In a typical vehicle, low-voltage electrical system should be sized at around 3 kW. This is the maximum power that the APM should supply. In vehicles where additional luxury loads are requested, such as power sunroof, active suspension system, or entertainment systems, the power level of the APM could be higher.

## 9.3   REQUIREMENTS OF AUXILIARY POWER MODULE

APM draws power from the high-voltage battery and powers the loads in the low-voltage system. In an electrified powertrain, the size of the high-voltage battery determines the range and the emissions of the vehicle. The more current the APM draws, the higher the drop in the SOC of the high-voltage battery. This might have a significant effect on the vehicle performance. Therefore, the most important requirement for APM is its efficiency. With a higher efficiency, APM draws less power from the high-voltage battery, and the battery charge can be utilized more to power the drivetrain. In practice, the efficiency of APM is expected to be higher than 95% in the medium and heavy load conditions. The reliability of APM is also very important since it powers all microprocessors in the vehicle and, thus, keeps the vehicle awake.

As the APM creates an electrical conversion between the high-voltage/power system and the low-voltage/power system of the vehicle, a galvanic isolation must be used for safety reasons. This ensures that a failure within the high-voltage system will not affect the low-voltage system and shut down the vehicle. The opposite is also true; galvanic isolation would protect the high-voltage system from a failure happening on the low-voltage system, which is directly accessible to the driver and passengers within the vehicle.

The other important requirement for APM is the quality of the output voltage. Especially electronic loads, such as the control units, radio, and the CD player, are very sensitive to the ripple content of the voltage supplied by the APM. For this reason, the output voltage ripple of APM should be quite low, which might require designing output filters. As such, a filter is generally bulky in comparison to the converter; it brings challenges in defining the switching frequency, which strongly affects the filtering requirements, but also losses, as well as the output capacitance and inductance of the converter.

The SOC of the high-voltage battery varies depending on the traction power requested from the high-voltage battery. The terminal voltage and, hence, the input of the APM changes in this case. Therefore, APM is required to operate in a certain input voltage range and provide the output voltage specifications for the entire input voltage range.

Finally, APM should be designed to operate in various temperature conditions. In automotive system, the operating temperature usually varies between −40°C and 85°C, so that the vehicle can operate in different climatic regions around the world. For a power converter with high efficiency requirements, the ambient temperature is very important when defining the size of the cooling system. As an example, the resistance of the transformer and inductor windings and the conduction losses of the power semiconductor switches are dependent on temperature. Therefore, the designer should design the thermal management system for the given specifications, which ensures that the required efficiency can be maintained in various ambient conditions.

## 9.4 CONVERTER TOPOLOGIES FOR AUXILIARY POWER MODULE

In a typical electrified powertrain architecture shown in Figure 9.2, the APM is required to deliver power from high-voltage (HV) DC bus to 12 V loads. The converter must incorporate galvanic isolation to protect the low-voltage (LV) electronic system from the potentially hazardous high voltage [1,2]. This requirement restricts the available topologies to those containing a transformer [3]. In the following, possible candidates for the APM are introduced and discussed.

### 9.4.1 FLYBACK CONVERTER

As shown in Figure 9.3, flyback converter has a single switch and it employs its transformer's magnetizing inductance for energy storage. However, the magnetic flux in the flyback transformer has a DC component. Because of this, the size of the transformer core increases as the power requirements increase [4]. Especially, in high-input-voltage applications where high conversion ratio is required, the voltage stress on the flyback converter switch can be a limiting factor in the design. The switch voltage stress in a flyback converter can be represented as

$$V_{in} + V_o \times \left( \frac{N_1}{N_2} \right)$$

where $V_{in}$ is the input voltage, $V_o$ is the output voltage, $N_1$ and $N_2$ represent the number of turns of the primary and the secondary windings, respectively. The output voltage of the flyback converter can be expressed as

$$V_o = V_{in} \left( \frac{D}{1 - D} \right) \left( \frac{N_2}{N_1} \right)$$

where $D$ is the duty cycle. For 300 V input voltage, if the converter operates at 50% duty cycle, the transformer turns ratio $(N_1/N_2)$ would be 25:1 to achieve 12 V at the output. In this case, the switch voltage stress ends up at 600 V. Considering the voltage overshoots due to the stray inductance in

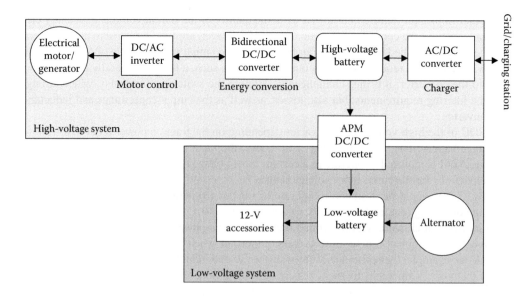

FIGURE 9.2 Typical electrified powertrain system with low-voltage network.

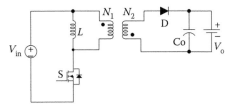

FIGURE 9.3   Flyback converter.

the circuit, the switch should be rated higher than this value. This increases the cost and reduces the power density. Indeed, switches enabling high switching frequency (several kHz) are required to keep the transformer size reasonable. Hence, MOSFET is usually the preferred choice. However, for a 600 V voltage stress, most of the current MOSFETs available in the market might not be capable of handling that high voltage, and the ones rated for these values are usually more expensive than insulated gate bipolar transistor (IGBT) for the same power rating. IGBTs can handle higher voltages, but they usually might not be capable of operating at high switching frequencies. In either case, there is a restriction to achieve high power density and reasonable cost of the converter at the same time.

### 9.4.2   FORWARD CONVERTER

Compared to the flyback converter, forward converter does not need to store the energy in the transformer. The energy is transferred from the source to the load while the switch is closed. As shown in Figure 9.4, a third winding is applied to provide a path for the magnetization current when the switch is open in order to reduce the magnetizing current to zero before the start of each switching period. This provides a smaller transformer size for the forward converter [4]. However, the transformer in a forward converter still employs DC flux similar to the flyback converter.

The semiconductor switch in the forward converter is still exposed to high-voltage stress, which can be represented as

$$V_{in} \times \left(1 + \frac{N_1}{N_3}\right)$$

where $N_3$ is the number of turns of the third winding. Since the magnetizing current must be zero before the start of the next switching period, the following condition must be followed in forward converter:

$$D\left(1 + \frac{N_3}{N_1}\right) < 1$$

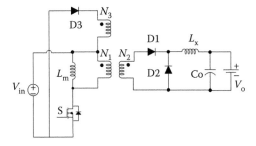

FIGURE 9.4   Forward converter.

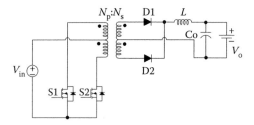

FIGURE 9.5   Push–pull converter.

Thus, $N_3$ must be smaller than $N_1$. For the same operating conditions with the flyback converter ($V_{in} = 300$ V, $D = 50\%$), the switch voltage stress in the forward converter will be greater than 600 V.

### 9.4.3   Push-Pull Converter

Figure 9.5 shows the typical circuit diagram of the push-pull converter. In steady state, the input and output voltage relationship can be represented as

$$V_o = 2V_{in}\left(\frac{N_s}{N_P}\right)D$$

where $D$ is the duty cycle for each switch.

Compared to flyback and forward converters, the number of semiconductor switches is higher in push-pull converter. The voltage stress on the switches is also twice the input voltage. However, unlike flyback and forward converters, the transformer of the push-pull converter has AC flux. Therefore, the transformer does not need to store energy, yielding a relatively smaller transformer core, which can be designed in a smaller volume. This results in better potential power density than flyback and forward converters.

Flyback, forward, and push-pull converters all provide galvanic isolation using a transformer. It is also possible to design the converter by selecting different topologies for the primary and the secondary sides of the transformer. Depending on the operational requirements of the APM, various topologies can be used on both sides, which will also affect the design of the transformer.

### 9.4.4   Topologies for the Primary Side

In general, full-bridge and half-bridge topologies can be utilized for the primary side. Figure 9.6 shows the circuit diagram of these two topologies. For high-power applications, full-bridge converter is usually applied as it is relatively simple and robust, and it offers good power density and efficiency. The switch voltage stress is equal to the input voltage, which leads to a flexible switch

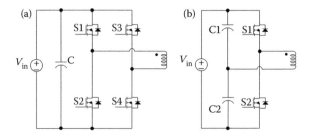

FIGURE 9.6   Primary-side topology candidates: (a) full bridge and (b) half bridge.

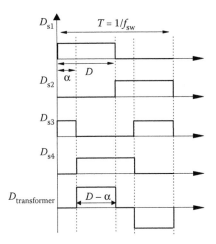

FIGURE 9.7 Phase-shift full-bridge control scheme.

selection for the APM. In addition, zero voltage switching (ZVS) technique can be implemented on the full bridge by employing phase-shift control in order to reduce the switching loss [5], as shown in Figure 9.7, where $D$ is the duty cycle for each switch and $\alpha$ is the phase-shift angle between S1 and S4. In the 2004 model of Toyota Prius, an isolated APM topology has been used with a full-bridge converter on the primary side [6].

Compared to the full-bridge converter, half-bridge converter only needs two switches instead of four. However, these two switches are required to carry two times as much current as compared to the full-bridge converter. Meanwhile, the voltage stress for these two switches still equals the input voltage. Thus, the switch requirements for the half-bridge topology are higher than the full-bridge topology, which restricts its feasibility in high-current applications. In addition, half bridge requires two input capacitors instead of one for the full bridge.

### 9.4.5 Topologies for the Secondary Side

Owing to the low output voltage and high current requirements, conduction losses dominate on the secondary side. For a 3 kW application, an output voltage of 12 V results in an output current around 250 A. This yields large conduction loss and strongly affects the efficiency of the secondary side converter [8]. Hence, it is critical to select the most suitable topology to maximize the converter efficiency for high-current operations. This point is especially important because power requested in modern vehicles is continuously increasing. This results in higher current rating on the secondary side. As a result, topologies proposing better capabilities in handling higher current are appropriate for the secondary side in APM converters.

Figure 9.8 shows the center-tapped rectifier and current doubler rectifier topologies, which can be used as the secondary-side topology in a unidirectional APM. The main waveforms for these topologies are shown in Figure 9.9.

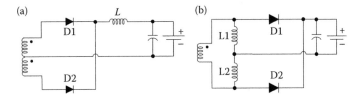

FIGURE 9.8 Secondary-side topology candidates: (a) center-tapped rectifier and (b) current doubler rectifier.

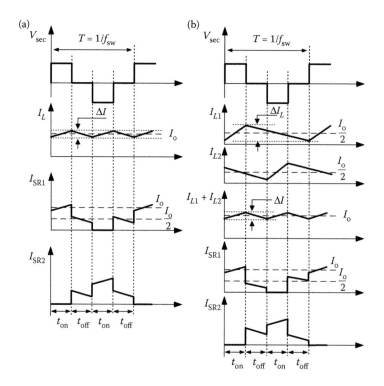

**FIGURE 9.9** Main waveforms of (a) center tapped and (b) current doubler rectifier. (From P. Alou et al. In *Proceedings of Applied Power Electronics Conference and Exposition*, Dallas, TX, Mar. 2006.)

From the inductor aspect, since current doubler has two switches and two inductors, each inductor operates at the same switching frequency as the semiconductor device. Center-tapped rectifier obtains two switches with one inductor; therefore, the inductor current ripple oscillates at twice the switching frequency of the switches.

From the transformer aspect, the current doubler might be more attractive than the center-tapped rectifier. One of the drawbacks of the center-tapped rectifier is that its transformer winding has double winding. The secondary side in the current doubler rectifier has a single winding. This decreases the utilization factor of the transformer in the center-tapped rectifier. Owing to the single secondary winding, it is possible to parallel more coils in the current doubler rectifier for the same window area, enabling lower resistance for high-current operation.

### 9.4.6 SYNCHRONOUS RECTIFICATION

High-current requirement on the secondary side usually results in high conduction losses. The conduction loss of diode rectifiers contributes significantly to the overall power loss due to the high voltage drop. A typical PN-junction power diode voltage drop is 1.2 V and even Schottky barrier diode (SBD) still has 0.6 V voltage drop [9]. For a 12 V output APM application, this becomes a significant portion of the voltage drop (10%) and penalizes the efficiency. MOSFET presents lower conduction loss than diode. As a result, the concept of synchronous rectification (SR) came to reduce the conduction loss and maximize the conversion efficiency on the secondary side [3]. In SR, rectifying diodes are replaced by synchronous MOSFETs. Corresponding topology for the current doubler circuit is shown in Figure 9.10.

The synchronous MOSFETs operate in the third quadrant. The body diode of the MOSFET conducts prior to the turn on of the switch. In other words, conduction loss of the body diode is

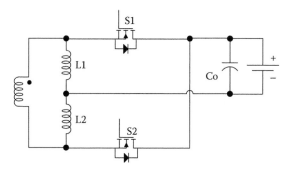

FIGURE 9.10    Synchronous rectifying current doubler.

generated just before the synchronous MOSFET turns on. However, it can be turned on in ZVS, which results in negligible switching loss at turn-on. At turn-off, the MOSFET stops conducting prior to the body diode, which means that the synchronous rectifier still has the reverse recovery losses from its body diode [10].

If the voltage stress across the semiconductor is relatively high, MOSFETs with high voltage rating need to be used. High-voltage MOSFETs have larger on-state resistance, $R_{ds}$, which might reduce the system efficiency. In this case, a Schottky diode-based configuration might provide a comparable efficiency in the secondary side with a lower cost as compared to SR MOSFET-based configuration.

Typically, there are two different techniques to control the SR: external-driven SR (EDSR) and self-driven SR (SDSR) [11]. As shown in Figure 9.11a, in the EDSR technique, the control signals are generated by an external controller, which guarantees the appropriate timing. By doing so, the switches can be turned on during the whole rectification period, and the efficiency can be maximized [12]. However, circuitry to generate the gate pulses and drivers to charge the gate capacitance of the MOSFETs are required [11].

Unlike the EDSR, the control signals as well as the energy to drive the SDSR switches are obtained from the secondary side of the transformer and no driver is needed [11], as shown in Figure 9.11b. As a result, a simple, low-cost rectification control can be implemented. However, there are mainly two drawbacks for SDSR. The first one is the voltage with which the MOSFETs are driven is variable, and it depends on the input voltage. Second, not too many topologies are suitable for SDSR. The most suitable topologies for using SDSR are the ones that drive the transformer asymmetrically with no dead time: flyback and half bridge with complementary control, and so on [9]. The concept of a half-bridge converter with SDSR control and its main waveforms are shown in Figures 9.12 and 9.13a, respectively [9]. For topologies with symmetrically driven transformers, as the full-bridge and push–pull converters, the synchronous rectifiers are not activated during the dead time of the transformer. The main waveforms are shown in Figure 9.13b. It is clear that during the dead time of the transformer, the body diode of the MOSFET, which usually creates very large forward voltage drop in the circuit, has to conduct. This fact causes a noticeable decrease in efficiency.

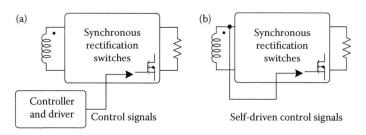

FIGURE 9.11    (a) EDSR and (b) SDSR.

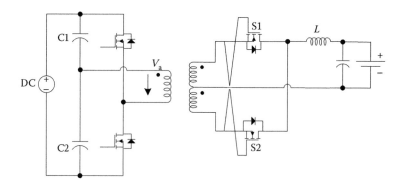

**FIGURE 9.12** Half-bridge converter with SDSR. (From A. Fernandez et al. *IEEE Transactions on Industry Applications*, vol. 41, no. 5, pp. 1307–1315, Sep. 2005.)

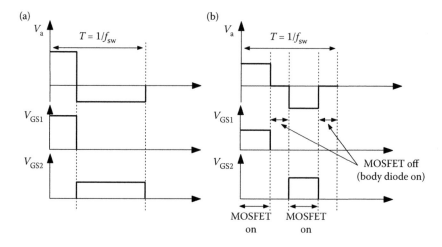

**FIGURE 9.13** Transformer voltage and SDSR gate drive signal waveform of (a) asymmetrical-driven waveform and (b) symmetrical-driven waveform.

Therefore, it is important to extend the conduction period of the SDSR MOSFETs over the period when the voltage across the transformer is null. The basic idea to improve the system efficiency with SDSR under symmetric transformer waveform is shown in Figure 9.14.

One possible implementation method to generate these extended gate driver signals is to apply an additional winding and an additional voltage source $V_A$ to force the synchronous rectifiers to be on

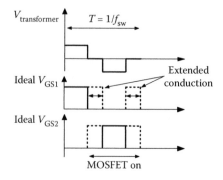

**FIGURE 9.14** Ideal SDSR gate drive signal voltage for symmetrical transformer voltage waveform.

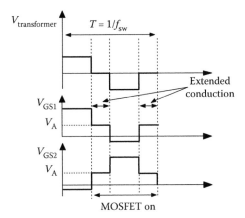

FIGURE 9.15 An implementation method of SDSR gate drive signal voltage for symmetrical transformer voltage waveform. (From A. Fernandez et al. *IEEE Transactions on Industry Applications*, vol. 41, no. 5, pp. 1307–1315, Sep. 2005.)

during the dead time, as the waveform is presented in Figure 9.15 [9,11]. In this case, the converter should be designed carefully. This method requires a well-regulated additional voltage source.

## PROBLEMS

9.1 The flyback converter in Figure 9.3 has $V_o = 12$ V and the turn ratio $N_1/N_2$ is 15:2. If duty cycle is 0.2, determine the input voltage and the voltage stress on the switch.

9.2 The forward converter in Figure 9.4 has $V_{in} = 200$ V and $V_o = 12$ V. (1) If the selected MOSFET can only handle 300 V voltage stress, determine the feasible duty cycle range. (2) If the desired duty cycle range is up to 0.5, what is the minimum voltage rating of the switch?

9.3 The push-pull converter topology is shown in Figure 9.16. Analyze the steady-state operating conditions of the converter, if the input voltage is 400 V and the output voltage is 12 V, and each switch operates at 50% duty cycle. Define the transformer ratio and the switch Sw1 and Sw2 voltage stress.

9.4 A full-bridge converter (Figure 9.6a) with current doubler (Figure 9.8b) is selected as the topology for the APM. If each switch operates at 50% duty cycle and the phase shift angle between S1 and S4 is 60°, sketch waveforms for the two inductors and output current.

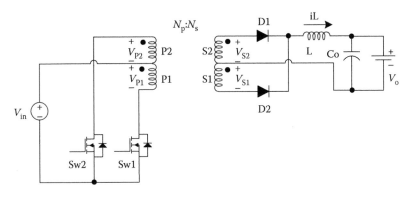

FIGURE 9.16 Push-pull converter.

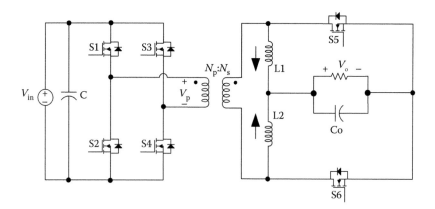

FIGURE 9.17   Full bridge with SR current doubler.

9.5 The current flow through a semiconductor is 100 A. A selected SBD obtains a voltage drop of 0.6 V. If synchronous rectification is preferred to be used, what is the $R_{ds}$ requirement of the desired synchronous rectifier MOSFETs to achieve better efficiency (only consider the conduction loss)?

9.6 A full bridge with SR current doubler is selected as the topology for the APM, as shown in Figure 9.17. If the input voltage is 400 V and the output voltage is 12 V, the primary four switches operate at 50% duty cycle, and the transformer ratio $N_p/N_s$ is set as 10:1. (1) Perform the steady-state analysis and obtain the phase shift angle. (2) Create the six MOSFETs control scheme to achieve the highest efficiency.

9.7 If the APM is built as in Figure 9.17, there are now two more MOSFETs available. Can these two MOSFETs be added into the APM to achieve higher efficiency without changing the control scheme (assume these two MOSFETs' rating fit anywhere)? If yes, draw the circuits; if not, explain the reason.

## REFERENCES

1. A. Emadi, S. S. Williamson, and A. Khaligh, Power electronics intensive solutions for advanced electric, hybrid electric, and fuel cell vehicular power systems, *IEEE Transactions on Power Electronics*, vol. 21, no. 3, pp. 567–577, May 2006.
2. Texas Instruments, Hybrid and Electric Vehicle Solutions Guide, 2013. [Online]. Available: http://www.ti.com/lit/ml/szza058c/szza058c.pdf.
3. A. Gorgerino, A. Guerra, D. Kinzer, and J. Marcinkowski, Comparison of high voltage switches in automotive DC-DC converter, in *Proceedings of Power Conversion Conference*, Nagoya, Japan, Apr. 2007, pp. 360–367.
4. D. Hart. *Power Electronics*. New York: McGraw-Hill, 2011.
5. U. Badstuebner, J. Biela, D. Christen, and J. Kolar, Optimization of a 5-kW telecom phase-shift DC–DC converter with magnetically integrated current doubler, *IEEE Transactions on Industrial Electronics*, vol. 58, no. 10, pp. 4736–4745, Oct. 2011.
6. A. Kawahashi, A new-generation hybrid electric vehicle and its supporting power semiconductor devices, in *Proceedings of 16th International Symposium Power Semiconductor Devices and ICs*, Kitakyushu, Japan, May 2004, pp. 23–29.
7. P. Alou, J. Oliver, O, Garcia, R. Prieto, and J. Cobos, Comparison of current doubler rectifier and center tapped rectifier for low voltage applications, in *Proceedings of Applied Power Electronics Conference and Exposition*, Dallas, TX, Mar. 2006.
8. Y. Panov and M. Jovanovic, Design and performance evaluation of low-voltage/high-current dc/dc on-board modules, *IEEE Transactions on Power Electronics*, vol. 16, no. 1, pp. 26–33, Jan. 2001.

9. A. Fernandez, J. Sebastian, M. Hernando, P. Villegas, and J. Garcia, New self-driven synchronous recti-fication system for converters with a symmetrically driven transformer, *IEEE Transactions on Industry Applications*, vol. 41, no. 5, pp. 1307–1315, Sep. 2005.
10. P. Xu, Y. Ren, M. Ye, and F. Lee, A family of novel interleaved DC/DC converters for low-voltage high-current voltage regulator module applications, in *Proceedings of IEEE Power Electronics Specialists Conference*, Vancouver, BC, Jun. 2001, pp. 1507–1511.
11. A. Fernandez, D. Lamar, M. Rodriguez, M. Hernando, and J. Arias, Self-driven synchronous rectifica-tion system with input voltage tracking for converters with a symmetrically driven transformer, *IEEE Transactions on Industrial Electronics*, vol. 56, no. 5, pp. 1440–1445, May 2009.
12. M. Rodriguez, D. Lamar, M. Azpeitia, R. Prieto, and J. Sebastian, A novel adaptive synchronous recti-fication system for low output voltage isolated converters, *IEEE Transactions on Industrial Electronics*, vol. 58, no. 8, pp. 3511–3520, Aug. 2011.

# 10 48-V Electrification
## Belt-Driven Starter Generator Systems

*Sanjaka G. Wirasingha, Mariam Khan, and Oliver Gross*

## CONTENTS

## 10.1  INTRODUCTION

There are more than 250 million and more than 900 million vehicles being driven daily in the United States and worldwide, respectively [1]. These vehicles continue to burn fossil fuels inefficiently at high operating costs and emissions. Environmental issues such as the depleting ozone layer and global warming have however fueled demands from the world community to reduce hydrocarbon emissions and that more energy-efficient vehicles are produced. At the current rate of consumption, there is a growing concern that the oil wells will be exhausted before transportation modes will be independent of oil. Inefficient vehicles also translate to higher lifetime energy consumption and cost. These reasons have led to a surge of innovation in the automotive industry. However, large-scale, properly tuned policies are required to substantially reduce these vehicles' carbon footprint and improve energy efficiency within an acceptable time span.

Since travel behavior is difficult to change, many analysts believe that modifying vehicle technology is the best means to offset the environmental impacts of continued increases in vehicle miles traveled (VMT) in areas where automobile use is dominant. There are many vehicle technologies currently being developed to address these issues. Improving conventional vehicles via manufacturer's systems is among the proposed solutions. This involves incorporating more efficient engines, emission filters, and so on and developing new vehicle technologies, which are either new to the market or still in prototype stage. Extensive research and development has been conducted on alternative fuel vehicles (AFVs), commercialization of natural gas vehicles, and electrification of the drive train. Depending on the degree of electrification, the combination of the internal combustion engine (ICE) with an electric motor offers a wide range of benefits from reduced fuel consumption and emission reduction to enhance performance and the supply of power-hungry hotel loads. Auto stop–start systems, low-voltage (LV) and high-voltage (HV) hybrid electric vehicles (HEVs), plug-in hybrid electric vehicles (PHEVs), and electric vehicles (EVs) are the direct results of drivetrain electrification [2–4]. The research has also focused on utilizing newer energy sources and combinations of different energy sources to improve the overall efficiency of vehicles. Since drivers have different driving needs, styles, and patterns, it has been hard to develop the ideal technology that will provide optimum performance to all.

A 48-V electrification system, the focus of this chapter, can be classified as a micro- or mini-HEV. It is essentially a combination of a high-power starter and low-power parallel hybrid having the ability to start the engine, provide electric assist, maintain regenerative braking, and serve as a generator. In some rare instances, it also drives in the EV mode. This chapter will provide a detailed overview of the importance of vehicle electrification and the position of 48-V belt starter generator (BSG) systems among many electrification topologies/drivetrains. An overview of a BSG system including functional objectives, topologies, requirements, and integration among other topics is provided followed by a detailed review of the key components of a BSG system. A high-level summary of the currently available BSG systems is also provided.

## 10.2  LV ELECTRIFICATION

### 10.2.1  NEED FOR ELECTRIFICATION

The "2025 fuel economy requirement" mandates that passenger cars and trucks in the United States deliver a fuel economy equivalent to 54.5 miles per gallon (mpg) by 2025. The new fuel economy standard impacts cars manufactured as early as 2017, requiring automakers to make incremental changes in fuel efficiencies to reach a combined average target of 34.1 mpg within the next 5 years. The eventual goal for 2025 is to approximately double the efficiency of vehicles on road today. These efficiency standards are supported by 13 major auto manufacturers that account for 90% of all vehicles sold in the United States.

It is estimated that a vehicle built under these new regulations will save the owner over $8000 in gas during its lifetime. The White House claims that for a car purchased in 2025, the net savings will be comparable to lowering the price of gasoline by approximately $1 per gallon [5]. On the basis of this prediction, U.S. drivers can expect to save over $1.7 trillion in gas costs collectively by 2025. Consumer costs aside, the regulations are projected to cut U.S. oil consumption by 12 billion barrels, or 2 million barrels a day by 2025 that constitutes approximately half of the U.S. imports from the Organization of the Petroleum Exporting Countries (OPEC). Moreover, the enforcement of this new fuel economy standard will create an emission reduction of 6 billion metric tons by 2025, which will lead to a growth of domestic jobs in the auto industry.

To achieve this goal, automakers are exploring a multitude of solutions including weight reduction, smaller engines, optimized auxiliary loads, and electrification of power trains. Electrified power-train systems employ electrical propulsion to offset the fuel consumption in conventional power trains by fully or partially replacing the ICE with an electric motor drive. The high-energy efficiency of the electric traction makes it a highly attracted solution for the design of fuel-efficient vehicles. A benefit of electrification is the ability to conserve energy through regenerative braking for added improvement in the fuel economy. This is accomplished by operating an electric machine as a generator to convert the inertial energy of the vehicle during braking into electrical energy and storing it in the battery to be reused for propulsion by the same machine or another traction motor integrated to the system. A BSG system will also allow the vehicle to turn the engine off during idle and other nonpropulsion events, and events with inefficient engine-operating points further improving fuel economy and reducing emissions.

### 10.2.2  DEGREE OF HYBRIDIZATION

On the basis of the extent of electrification, also termed as the degree of hybridization, power trains are classified into several classes. Stop/start systems offer the most basic level of electric function in which the vehicle is solely propelled by the ICE. However, it utilizes the conventional 12-V-based powernet to shut down the thermal engine when the vehicle is at a stop, while maintaining some level of accessory load function. The engine is restarted once the driver is ready to drive. Depending on the power-train design, this can reduce $CO_2$ emissions by 2%–5% compared to conventional vehicles. A bigger battery is sometimes integrated with the 12-V stop/start system to enable a higher capacity for storing regenerative energy that can result in a further 3%–5% improvement in fuel economy. Micro hybrids also maintain stop/start feature without any electric propulsion, but add a limited amount of energy recapture during brake and coast opportunities.

The next class of electrified vehicles is the mild hybrids which, in addition to the stop/start feature and regenerative braking capability, incorporate a limited use of electric power for propulsion assist. The electric motor/generator in mild hybrids is rated anywhere between 5 and 20 kW and requires the integration of a higher voltage drive system, typically connected parallel to the 12-V powernet. This permits a significantly higher energy recapture and the use of propulsive power in power-train operation. In LV mild hybrids, the voltage system remains below 60 V DC that is defined as the demarcation point for DC HV by the United Nations regulation number 100 on battery electric vehicles (BEVs) (UN LR100). This document also specified that the system voltage must be maintained below 30 V to be classified as an LV system. However, during the 53rd session of the inland transport committee of the "Economic Commission for Europe" the requirement was amended to state that a system can still be classified as an LV system if the AC voltage is not accessible. This is a key driver for an integrated motor and power electronics component.

HV mild hybrids utilize voltages higher than 60 V DC to support a greater degree of regenerative energy recapture and propulsive power. Owing to the low-power rating of the electric traction system, it cannot drive the vehicle on its own. It mainly operates in parallel with the engine to assist in propulsion during high-power demands.

In full hybrid systems, the electric motor and ICE are capable of functioning together or independently to propel the vehicle based on the drive requirements. The electric-only drive mode is made possible with a traction motor that can be rated as high as 80 kW. To cater to such high-electric-power requirements, full hybrids are equipped with a larger battery pack compared to mild hybrids. However, the electric propulsion in full hybrids is limited by the amount of energy that can effectively be recaptured in braking or generated during the normal operation of the engine.

PHEVs and extended range electric vehicles (EREVs) improve upon the full hybrid function with the feature to plug in to external electrical energy sources. This considerably increases the proportion of electric drive utilization over the use of thermal propulsive power. In the architecture of a BEV, the ICE is completely replaced with a fully electric drivetrain, thereby eliminating the dependence on combustion engine and delivering zero emissions at the tailpipe. Figure 10.1 illustrates the electric power ratings and the proportion of electric and thermal power consumption for the progressive stages of electrification in power trains utilizing both the charge-sustaining (CS) functions, and the plug-in, charge-depleting (CD) functions.

The relative decrease in fuel consumption with each level of hybridization is shown in Table 10.1. The extent of improvement in fuel economy within each class also varies depending on a number of vehicle characteristics, such as mass, rolling friction and accessory load, as well as the powertrain architecture and control strategy. With a broad range of factors affecting the fuel economy, it becomes essential to devise a classification method for hybrid vehicles over a standard baseline. One such method to classify HEVs is called the hybridization factor (HF). The HF for a parallel hybrid topology is represented in Equation 10.1 where $P_{EM}$ and $P_{ICE}$ are the maximum traction power delivered by the electric motor and the engine, respectively [6].

$$HF = \frac{P_{EM}}{P_{EM} + P_{ICE}} \tag{10.1}$$

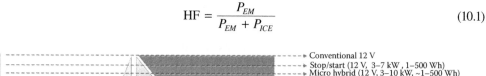

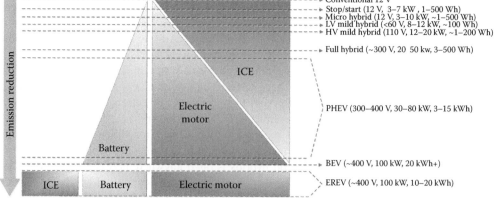

FIGURE 10.1  Classification of electrified vehicles.

TABLE 10.1

**Fuel Economy for Increasing Level of Vehicle Electrification**

| Electrified Vehicle Technology | % Fuel Consumption Reduction |
|---|---|
| 12 V stop/start | 2–5 |
| 12 V micro hybrid | 3–10 |
| LV mild hybrid/BSG | 8–15 |
| HV mild hybrid | 10–16 |
| Full hybrid | 20–50 |
| Plug-in hybrid | 40–80 |
| BEV | 100 |

HF varies from a value of 0 for the conventional vehicle to 1 for a fully electric vehicle. In hybrids with plug-in feature, connection to the grid becomes an important aspect to be taken into account when determining its classification. This is why the plug-in hybrid electric factor (Pihef) expressed in Equation 10.2 has been proposed to classify PHEVs and EREVs [7].

$$\text{Pihef} = \frac{E_{\text{grid}}}{E_{\text{grid}} + E_{\text{fuel}}} \tag{10.2}$$

where $E_{\text{grid}}$ is the average energy supplied by the grid and $E_{\text{fuel}}$ is the energy extracted from fuel combustion. A Pihef equal to 0 implies that no energy is being supplied from the grid for propulsion and any value higher than zero suggests that at least some portion of the propulsion is being powered by the grid.

Fuel consumption in a vehicle of a given mass can be reduced through a commensurate degree of hybridization signified by a higher HF of Pihef. However, hybridization effectively adds a second power train onto the existing ICE-based power train that translates into an additional component cost. Therefore, an increase in the degree of hybridization, while increasingly improving the fuel economy, also raises the proportionate cost of the hybridized power train. Vehicle manufacturers have devised methods to determine the incremental cost by which the goals for reduction in fuel consumption can be economically accomplished. This is often referred to as the best value curve. Figure 10.2 illustrates a best value curve for a given vehicle and the associated degrees of electrification.

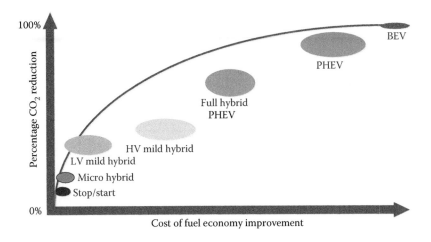

FIGURE 10.2    Best value curve for increasing the degree of hybridization.

### 10.2.3   LV versus HV Electrification

Increasing the degree of hybridization yields an improved fuel economy by maximizing the utilization of the electric traction during the vehicle drive cycle. This is achieved by increasing the power of the electric traction system such that it can assist and even fully take over the vehicle propulsion. Consequently, a higher voltage system needs to be incorporated in the vehicle that can meet the larger electric power requirements. HV hybrids, due to their higher electric propulsion power and a more extended electric drive function offer significantly reduced fuel consumption. An increase in system voltage permits a more effective support of power demands, without a proportionate increase in operating electrical current. Figure 10.3 illustrates the effect of voltage on the current required to be supported by a stop–start system. Curves at different specific power levels for a start–stop system are provided. These curves demonstrate one key advantage of using a 100+ bus voltage for a start–stop system.

The electric power required to implement electrification topologies such as HEV, PHEV, and EV range from 50 kW to almost 200 kW depending on vehicle application. This power supports vehicle functions such as electric drive, electric assist, and regenerative capability improving fuel economy of the vehicle. Voltage levels in these vehicle power trains are maintained at the 200–400+ V range to be able to satisfy the high-power demands while maintaining manageable currents. Continuous current requirements at the high voltages are still higher than in a LV system. The selection of subcomponents and interface connections rated for these high voltages and currents add a huge burden on the vehicle cost. Moreover, safety becomes a prime concern in HV systems as manufacturers are mandated to comply with federal and regional safety standards. HV is defined as a DC voltage greater than 60 V and require special wiring that have unique insulation and visual requirements. For example, they are orange in color. Multiple series-connected HV battery modules require a casing with superior isolation and relays to ensure disconnection in case of fault. A high-voltage interlock loop (HVIL) system that serially connects all the HV devices, monitors the HV bus, and reports the status to the battery controller is required. In the event of a vehicle collision or system malfunction, the controller will use this information to enforce an immediate system disconnection by cutting-off switches and relays. In this event, the system is required to reduce the system voltage to under 60 V in 5 s. Galvanic isolation becomes necessary to insulate the high- and LV electrical subsystems and vehicle ground plane/chassis. While these fault detection and protection systems are critical in preventing inadvertent access to HV energy, or failure of subsystem isolation, they add to the relatively higher cost of components in a high-power electrical system. Moreover, thermal management becomes more complex for HV hybrid systems. A summary of the impact of HV on a hybrid system is illustrated in Table 10.2. It is evident that while a full HEV, PHEV, and EV can provide the most

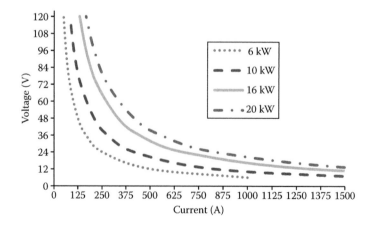

FIGURE 10.3   Effect of increasing voltage on current requirements of a stop/start system.

**TABLE 10.2**

**Comparison between Requirements for LV and HV Hybrid Systems**

|  |  | LV Mild Hybrid | HV Mild Hybrid | Full Hybrid |
|---|---|---|---|---|
| Voltage (V) |  | <60 | ~110 | 200–450 |
| Power (W) |  | 8–12 | 12–20 | 20–100 |
| HV interlock loop system |  | None | Required | Required |
| HV wiring harness |  | None | Required | Required |
| Power electronics cooling |  | Air or liquid | Liquid | Liquid |
| Galvanic isolations |  | None | Required | Required |
| Battery | Battery modules | Single | Multiple | Multiple |
|  | Battery management system | Central | Central or distributed | Central or distributed |
|  | Thermal management | Passive air | Passive or forced air | Forced air or liquid cooled |
|  | FMVSS305[a] Compliance | None | Required | Required |

[a]  FMVS305: Electric-powered vehicles, electrolyte spillage, and electrical shock protection standard that applies to vehicles with a weight of 10 k lbs or less and a nominal voltage higher than 48 V.

fuel economy improvement allowing a fleet to achieve the mandated targets, each of these vehicles will have a significant impact on vehicle cost.

Although lower than more electrified vehicles, LV hybrid systems demonstrate a substantial improvement in fuel economy by up to 15% at a significantly lower cost. They are also easy to integrate into the compact class segment of vehicles and offer a cost-effective approach that can be applied to a large percentage of a vehicle fleet. This will give a larger consumer base access to fuel-efficient cars. An LV mild hybrid therefore offers an excellent balance between reduction in fuel consumption and the system complexity or cost. The primary focus of this chapter will be the design aspects of LV mild hybrids.

### 10.2.4  12-V versus 48-V LV Electrification

The design of LV mild hybrids typically consists of an electric machine that serves as a starter/generator. There are several approaches of incorporating the starter/generator to the power train. The machine can be used to replace the flywheel and integrated directly onto the crank shaft between the engine and the clutch or at the accessory side. This topology is called the integrated starter generator (ISG) and provides good torque smoothing. An alternate approach is to integrate the starter/generator to the power train through a mechanical link such as accessory belt, chain, or gear. In a BSG where the electric machine is connected to the engine through a belt, the starter/generator occupies roughly the same space as the alternator that it replaces. Therefore, a BSG can be integrated as a compact package without any substantial changes to the engine-based power train. The electric machine of the BSG generates the torque needed to crank the engine, acts as a generator to charge the batteries during braking and normal engine operation, and provides a limited amount of electric assist during high acceleration demands.

While a BSG can be connected onto the 12-V powernet, there are a number of drivers that necessitate a higher voltage system. Increasingly, stringent emission standards along with tax and bonus incentives push for a higher fuel economy that cannot be achieved with the limited regenerative capability and stop/start feature in the 12-V vehicles. A higher degree of electric function and consequently higher power is necessary to meet these fuel economy standards. Therefore, an alternate approach to implementing LV electrifications is the design of a 48-V BSG system. While the electrical power rating in a 12-V system is limited to a few kW, a 48-V BSG can provide up to 10 kW continuous and 15 kW peak power and possibly even higher. The availability of higher power increases the capacity for storing regenerative energy and enhances the capability of torque assist, resulting in better performance and fuel economy. Torque assist capability allows the downsizing of combustion engine and the superior stop/start feature in 48-V systems takes a shorter time for starting the ICE. 48-V systems also offer the potential for downsizing auxiliary loads such as seat heaters, electric power steering (EPS) and fan blowers, and enable the use of electric-powered air compressors that cannot operate at 12 V. They allow a limited all-electric drive in the low-speed range that a 12-V system cannot deliver. On the other hand, an advantage of a 12-V BSG system is that it can be implemented in a vehicle with very little change to the conventional architecture, which remains largely the same apart from the battery and larger cables. For a 48-V BSG system, additional electrical components are required including inverter, larger cables, and a battery of higher capacity. A breakdown of the system features in a 12-V and 48-V BSG and their corresponding improvement in fuel economy is provided in Table 10.3.

48-V systems offer a clear advantage in terms of fuel economy; however, they are accompanied by a penalty of higher cost and packaging requirements driven by the number of components and power requirements. On the other hand, higher voltages reduce the current requirements and permit the selection of cheaper components rated for lower currents. The 48-V BSG, inverter, DC/DC converter, and the 48-V electric A/C compressors are the major contributors to the system cost. However, the design and integration of 48-V components into the vehicle remains within a reasonable range of manufacturing cost and the $/kWh and $/L ratios for battery packaging also remain affordable, especially when compared to HEV, PHEV, and EV drivetrains.

It can be concluded from this comparative assessment that 48-V BSG systems are particularly advantageous without the added complexity and higher cost of HV systems. Although the cost impact of increasing the system voltage from 12 to 48 V is sizeable, it is matched by a worthwhile reduction in fuel consumption. Figure 10.4 illustrates the architecture of a typical conventional, 12-V start–stop and 48-V BSG system. The latter is powered by a 48-V battery through a DC/AC inverter. Another DC/DC converter is required to step down the 48-V supply for the auxiliary loads connected on the 12-V powernet.

48-V systems, therefore, offer a sensible option of electrification at low cost for a large segment of vehicle classes. Taking these advantages into consideration, most vehicle companies are looking to implement 48-V BSG systems in their upcoming products. Tier 1 automotive supplier has also

**TABLE 10.3**

**Features and Fuel Economy of 12-V and 48-V BSG System**

|  | **12-V Stop/Start** | **48-V Stop/Start** |
|---|---|---|
| Power | ~3 kW | ~10 kW |
| Stop/start | Yes | Yes, comparatively faster and smoother |
| Regeneration | Yes, limited by peak power limits of components and max charge current limits of 12-V battery | Yes, higher regenerative energy capture, resulting in higher fuel economy improvement |
| Assist | Yes, limited by peak power limits of components and max discharge current limits of 12-V battery | Yes, higher peak power and longer duration at peak increasing fuel economy improvement |
| Generation | Conventional alternator | Using BSG components potentially more efficient |
| Components | Architecture largely unchanged; larger battery and cables required | Larger motor and cables |
|  |  | Addition of an inverter, DC/DC converter, 48-V battery, and control unit |
| Weight and packaging | Limited impact | Significant impact to vehicle weight class and packaging complexity |
| Cost | Component and integration cost | Estimated 2–3 times the cost of a 12-V BSG system |
| Fuel economy | Up to −10%, 14 g $CO_2$ | Up to −19%, 27 g $CO_2$ |

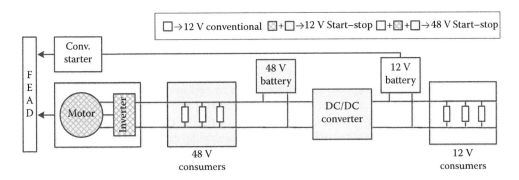

**FIGURE 10.4** Topology of a typical 48-V BSG system.

recognized this trend and the potential impact of start–stop systems on vehicle fleets with some predicting that such systems will be installed in as many as 70% of all new cars in western Europe by 2017 [8]. To keep up with these market projections, the suppliers have started developing component technologies and turn-key solutions. Details on these technologies and market trends toward a 48-V BSG will be discussed later in the chapter.

Initial guidelines for the design of LV hybrid systems have already been developed. The German Automobile Manufacturers Alliance (VDA) has established a set of performance guidelines for LV systems under 60 V, optimized for operation around 48 V. The guidelines have been documented in a specification titled LV148, and are summarized in Figure 10.5. LV148 can be considered as an addition to the LV124 document, for example, electric and electronic components in passenger cars up to 3.5; general requirements, test conditions, and tests. The intent of the guideline is to generate a system that can take maximum opportunity from the heightened operating voltage while safely preventing violation of the 60-V limit.

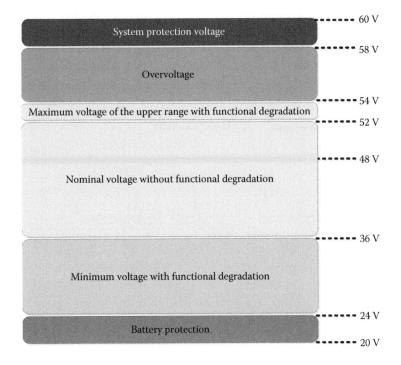

**FIGURE 10.5** Guidelines for LV electrification in vehicles.

## 10.3   BSG SYSTEM OVERVIEW

There are a number of performance requirements that determine the design of a BSG system. The system should deliver significant improvements to the vehicle fuel economy on city drive cycle such as the New European Drive Cycle (NEDC) and the Federal Test Procedure drive cycle (FTP) with stop–start functionality and additional improvements during coasting and torque assist. The BSG integration should provide an affordable alternative to HV electrification. The design should be scalable to multiple engine technology and size and it should provide a solution that is independent of transmission.

The design of the BSG should also improve customer driving experience and acceleration transients. A BSG system will also allow for engine downsizing by supporting peak torque and power demands. In addition, the system must minimize noise for comfort start and stop, improve shift and launch quality, torque assist, stall protection, and enable the introduction of 48-V auxiliary loads such as EPS, electric HVAC, active body control, and so on.

### 10.3.1   FUNCTIONAL OVERVIEW OF A BSG SYSTEM

A BSG system has four primary functional objectives:

- Support auto-stop and start
- Support regenerative braking
- Provide electric assist during high torque loads
- Generate power to support auxiliary loads

In addition to the above, a BSG system has the secondary objectives that will further improve fuel economy, performance, customer comfort, and reduce cost and packaging constraints of the vehicle. They include but are not limited to eliminating the need for an alternator, potentially eliminating the need for a starter, fuel cutoff during coasting and deceleration, smoothing engine torque, and last but not the least, enabling electric-only drive.

The stop/start feature, regenerative braking, and torque assist by the BSG system are the most significant functional objectives of the BSG system and are discussed individually in this section. Alternator function is also discussed as it has a significant impact on vehicle performance. Figure 10.6 demonstrates these different features using vehicle speed and not power requirements for a segment of the Environmental Protection Agency (EPA) city drive cycle. High torque is required to crank the

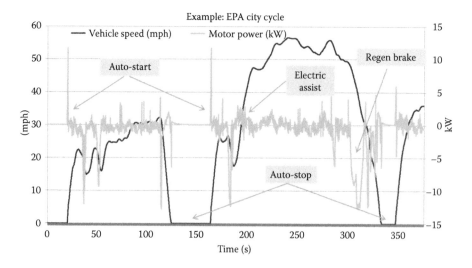

FIGURE 10.6   Power and torque profile of BSG electric motor/generator during the vehicle drive cycle.

engine to turn on when the driver removes and releases the brake pedal or presses the accelerator. During the normal operation of the vehicle, the electric machine acts as a generator to store energy in the battery. When the driver presses the brake pedal and decelerates, signified by the region with negative motor torque and power, the BSG controller shuts down the engine until the vehicle comes to a gradual stop while capturing the regenerative energy. The engine remains shutoff during idling while the auxiliary loads are powered by the 48-V battery with the DC/DC converter stepping it down to support 12-V loads.

### 10.3.1.1  Auto Stop/Start

In conventional vehicles, a dedicated starter, typically mounted to the engine, turns the engine on when initiated by the "key-on" function. In a typical BSG application, the conventional starter will continue to crank the engine during key starts while the BSG system will crank the engine during all auto-starts.

All vehicles are required to operate at ambient temperatures ranging from −40°C to 125°C. However, 48-V battery systems are in general not designed to operate at temperatures as low as −40°C. A BSG system is therefore not available across the complete temperature range, resulting in a conventional starter being maintained and used in a 48-V BSG system. Battery suppliers are working on new cell and packaging technologies that will allow the removal of the conventional starter in future designs.

A stop/start event occurs when the engine is not required to provide propulsion torque such as idle, coasting, and sometimes deceleration. A typical BSG system will primarily focus on turning the engine off during idle only because controls and calibration requirements to turn off during the other events are degrees more complex. Figure 10.7 below illustrates an example of a cranking profile of the engine during an auto-start from idle. The BSG motor provides maximum torque for as long as the crank shaft speed is increasing. This duration at max torque will vary with engine technology and is a key design criterion for the motor and the power inverter module.

In this event, the engine shuts off to conserve fuel and starts again when initiated by the driver or the vehicle system. The engine restart is either initiated by the driver or by the system. Driver-initiated auto-start is determined by the release of the brake pedal or the position of the accelerator. System-initiated auto-start is based on vehicle-operating conditions such as the engine coolant temperature, transmission oil temperature, 48-V battery state of charge (SOC), brake vacuum pressure, cabin comfort demands, and occupant detections.

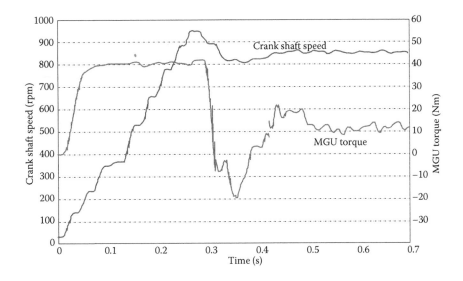

FIGURE 10.7   Typical cranking profile.

To maximize customer satisfaction and driver experience, the auto stop and start function will be calibrated to minimize change of mind restart initiation delay, reduce engine vibration in shutdown through engine breaking and torque management, enhance restart time and quality, and have the option of direct start.

The start profile, represented by the speed and torque over time, varies based on the torque requested by the driver and can be characterized as smooth or aggressive. Smooth auto-start is transparent and seamless to the driver and is initiated after engine torque potential is reduced. The engine speed versus time profile for the smooth auto-start is shown in Figure 10.8.

Aggressive auto-start ignites the engine as fast as possible to provide propulsion torque as shown in Figure 10.8. The combustion in aggressive auto-start begins after one crank shaft revolution.

On the basis of the driver's auto-start requirements, the BSG controller interpolates between the two different start types and ensures the most transparent auto-starts while maintaining a fast start option. It commands the electric motor to apply torque to spin up the engine and compensates for compression torque until combustion starts.

### 10.3.1.2  Assist/Boost

The BSG system can provide additional power to the driveline during heavy acceleration and grade driving. This is achieved by the BSG system providing supporting torque in addition to the torque by the ICE to enhance the vehicle performance during starting, acceleration, and high-grade driving. The power and duration of electric assist is limited by the battery capacity, the power rating of the components, the torque-speed map of the motor, and thermal limitations of the system.

During high acceleration demands indicated by pedal request from the driver, the boost feature supplements the ICE-based vehicle propulsion by adding electric motor torque over the engine maximum torque to increase the power-train torque. The assist feature compels the electric motor to provide torque when the engine needs to enrich the air/fuel mixture. Since the electric motor can contribute to the vehicle propulsion by providing supporting torque, it allows the driver's requirements to be met without the need to gear down. Thus, the electric assist capability of the BSG systems allows additional fuel economy by enabling the engine to operate more efficiently by optimizing the gear shift schedule. Moreover, it can potentially enhance the drive performance during acceleration events.

In smaller vehicles, provided that the BSG system has been sized accordingly, the BSG system is capable of providing enough propulsion torque to drive the vehicle in electric-only mode. This

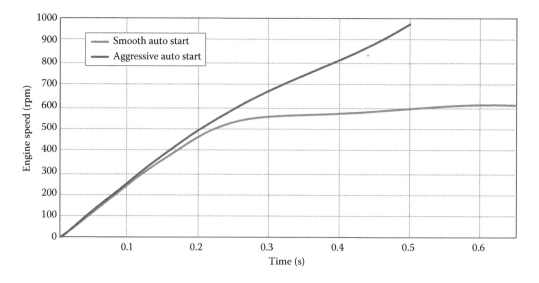

FIGURE 10.8   Engine speed versus time profile for smooth and aggressive auto-start.

allows the vehicle to further increase fuel economy and reduce emissions. However, as the vehicle platform gets larger, this is not possible as the current required will be too high for a 48-V system without having significant fuel economy loss, cost, and packaging implications.

### 10.3.1.3 Regenerative Braking

When braking in a conventional vehicle, the friction converts much of the kinetic energy into heat that is released into the air. Electrified vehicles are designed to capture this kinetic energy during deceleration and store it in the battery pack to be used for propulsion during acceleration. According to the simulated data, the overall economy of the vehicle is improved by 5%–8% in NEDC and 8%–12% in the FTP-75 drive cycles. However, adding features such as down speeding and start–stop coasting will also add to this improvement. There are various methods of capturing regenerative braking energy. Two such methods are the fully blended and the overlay-regenerating braking systems. A fully blended approach while more challenging to develop, improves fuel economy and range and helps preserve a more natural braking feel. An overlay approach is the most cost-effective solution.

Every braking event constitutes both regenerative and friction braking. During the initial pedal travel, braking is purely regenerative, that is, almost the entire braking energy is captured as electrical energy. Friction braking, where braking energy is dissipated as heat, starts after some pedal travel. To ensure safety and smooth drivability, regeneration is restricted by the deceleration requirements. Moreover, the regenerative energy storage is limited by the battery capacity and peak power limits of the power inverter module and motor.

Regenerative braking has a sizeable impact on fuel savings in electrified vehicles. Moreover, fuel cutoff during coasting is also a beneficial feature for fuel economy. Various combinations of the above-discussed features can be considered to optimize cost versus fuel economy and carbon emissions.

### 10.3.1.4 Generation

48-V systems are being designed and specified so as to support all the 12-V loads of the vehicle. The conventional alternator can then be removed positively impacting integration, cost, and packaging efforts. The motor in the generating mode will be required to provide a continuous generating power of an estimated 2 kW across the operation speed range. Note that the power requirement will vary based on the vehicle and the available auxiliary loads of that vehicle.

A typical torque versus speed map and power versus speed map of a BSG motor is presented in Figure 10.9. The different operation ranges described above have been identified. It is clear that the BSG requirements must be specified so as to maximize the advantages from each function.

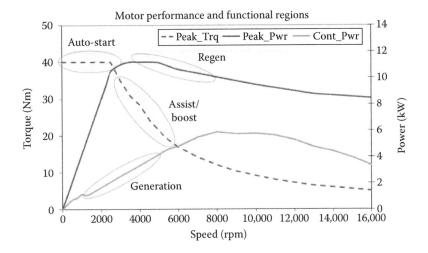

FIGURE 10.9    T–S and P–S map of a typical BSG motor.

## 10.3.2  48 V Electrification Topologies

A 48-V electrification system can be integrated with a conventional drivetrain using different topologies as illustrated in Figure 10.10. The packaging and installation of the electric motor and how it connects to the engine will be the key difference between each topology. Four key topologies are discussed in this section along with unique requirements and functional capabilities. The impact on fuel economy, implementation cost, and ease of integration is also addressed.

### 10.3.2.1  P1 Topology

The electric motor in this topology is directly integrated to the engine. Therefore, the engine will always start with the electric motor unless driven by conditions such as low ambient temperatures where a conventional starter is required. Assist/boost, idle charging, coasting, and regenerative braking features are available with this topology. However, regenerative braking is limited due to engine drag and coasting is limited only to an e-clutch system. The electric-only drive feature is not possible with this configuration. While integration complexity is low with this topology, the fuel economy improvement is small and system cost is high.

### 10.3.2.2  P2 Topology

This topology has several advantages. It offers improved regenerative capacity, idle charging, electric-only drive, coasting, and torque assist/boost feature. Stop/start is initiated by the electric motor and the electric-only drive is determined by the battery capacity and power of the electric motor. In a typical P2 topology, the motor is integrated with the engine via the front-end accessory drive (FEAD) system and can in most instances be mounted to the engine in place of the conventional alternator. Therefore, integration complexity is low. It also provides a relatively high fuel economy improvement.

### 10.3.2.3  P3 Topology

In a P3 topology, the electric motor is coupled with the drive shaft after the differential. This can be achieved either via mechanical direct couple or via a belt. This configuration allows electric-only drive, regenerative braking, and electric assist/boost features. The clutch is always electrically actuated and coasting is possible in manual transmission. In case of automated manual transmission (AMT), coasting is possible if control logic is adapted to select electrical actuation of the clutch. Integrating a P3 topology to an existing drive train will be more complex.

### 10.3.2.4  P4 Topology

A P4 topology is commonly referred to as a parallel through the road topology where the electric motor is coupled with the second set of wheels, that is, rear wheels in a front wheel drive vehicle. While this architecture does not permit idle charging or a stop/start initiated by the motor, it features fuel cutoff during coasting, high regenerative capability, torque assist/boost, and electric-only

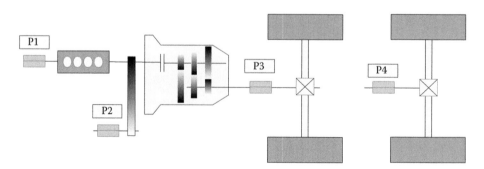

FIGURE 10.10   Possible topologies for BSG integration.

drive. Owing to the integration of the electric motor to the rear wheels, the vehicle becomes an all-wheel drive during electric assist mode that is an added advantage.

## 10.4 BSG REQUIREMENTS AND IMPLEMENTATION

### 10.4.1 BSG PERFORMANCE REQUIREMENTS

The BSG system has to be designed to meet a number of vehicle performance, cost, and timing requirements. The most significant performance requirements are listed in Table 10.4. The most primary function of the BSG system is to provide fast auto-start feature during both warm and cold cranking. The selected electric motor/generator fixed on the FEAD to replace the alternator and potentially a starter requires a power rating between 8 and 15 kW to meet the power requirements for auto-start, regenerative braking, electric assist, as well as a small portion of electric-only drive.

The key benefit of the stop/start system is the improvement in fuel economy by minimizing the fuel consumption during vehicle idle periods. This feature will provide automakers with fuel economy credits by achieving a stop/start at each vehicle idle period of the EPA FTP drive cycle with the exception of idle periods soon after a key start. This section discusses the component and system requirements, control strategy, and implementation overview to maximize this benefit. Other opportunities to further improve the fuel economy of the vehicle, reduce emissions, and ease vehicle integration complexities will also be discussed. There are multiple enabling conditions in a vehicle that define stop/start function. These conditions are unique to the vehicle and the automotive supplier. Some high-level enabling conditions include:

For auto-stop

- Brake and gear position is valid
- Ambient and coolant (liquid or air) temperatures are within the range for all components
- Battery SOC is above the threshold
- Engine and vehicle speed is below the threshold
- No inhibit
- Onboard diagnostic (OBD) conditions are met

For auto-start

- Brake and gear position is valid
- Completed key start
- OBD conditions are met

The approach to stop/start operation will vary based on multiple vehicle features, initially being the transmission system.

TABLE 10.4

**Performance Requirements from a 48-V BSG System**

| Requirement | Value |
| --- | --- |
| Operating voltage | 48 V |
| Peak power | 8–15 kW |
| Number of auto-starts | 350 k–450 K depending on location and drive cycle |
| Engine start time | 300–500 ms |
| Number of key starts | 30,000 |
| Engine start time—cold start at −25°C | 1 s—since key start only, longer duration is accepted |
| Service life | 15 years/100,000 miles |

### 10.4.1.1 Automatic Transmission

The system will generally provide automatic engine stops and zero vehicle speeds when the brake pedal is applied and all vehicle enablers have been met. These enablers will be unique to the auto-maker and the vehicle. Releasing the brake pedal will restart the engine. The engine may also restart during the brake condition if the 48-V battery SOC is too low, if it is required by the auxiliary loads, or if the component requirements for a start are at their threshold.

### 10.4.1.2 Manual Transmission

Unlike in an automatic transmission, the system will implement an engine stop at low vehicle speeds, provided that the transmission is in "Neutral" and the clutch is not engaged. In case the vehicle is geared and the clutch is engaged, the brake pedal must be applied prior to an engine stop so as to prevent vehicle roll. Engine start will be based on both clutch and brake position requirements unique to both neutral and gear conditions. Similar to the above, other vehicle conditions will also initiate engine start, provided the vehicle safety is met.

Vehicle-level objectives of a start–stop system will be generated by the respective vehicle groups. They will in general include requirements for

- Vehicle speed at shutdown
- Engine restart time—pedal driven
- Engine restart time—change of mind
- Start/stop vibration
- Noise—inside the vehicle

System-level objectives of a start–stop system will be generated by the power-train team for each unique application. They will in general include requirements for

- Engine restart time
- Cranking time
- Time to engine starting speed (RPM)
- Motor ramp rate
- Power-train jerk
- Power-train vibration
- Power-train-radiated noise

Component requirements

- Max torque capability
- Continuous torque capability
- Generating capability
- Slew rate

### 10.4.2 DESIGN CHANGES FOR A BSG SYSTEM

A BSG system will add components to the base vehicle, such as the inverter, motor, and a 48-V battery pack. However, it must be noted that implementing a 48-V BSG system will also modify, and respecify requirements of some fundamental components. In some cases, these components can be optimized as a result of the functional capabilities of the BSG to further improve the fuel economy of the vehicle or removed reducing the delta cost increase of the system.

Some of these key component areas that will be impacted by the addition of the BSG system are

- Alternator: The conventional alternator will potentially be removed.
- Accessory drive: Pulleys (alternator-decoupled pulley), tensioners, idlers, belt, and other propulsion system components.
- Controller hardware: Processing capability of controllers and input/output capability.
- Underhood environment: Modifications to the underhood packaging to create space for the motor and inverter.
- Revised wiring: Additional of both 48-V and 12-V wiring and rerouting existing wire harnesses.
- Redesigned underhood coolant system: On the basis of the coolant strategy for the added motor and inverter, either the existing engine coolant loop must be modified or a new coolant system must be added.
- Potential addition of 48-V electric air compressor (EAC).
- Potential addition of 48-V EPS: The current 12-V power- steering systems are one of the largest loads on the auxiliary system. Designing to meet edge-to-edge steering (~50+ A) has driven the DC/DC to be much larger than required for normal operation. By switching to 48 V, the current required can be reduced and a DC/DC converter will not be required to support it.
- Engine mounts: Since the motor and most likely the inverter is directly mounted to the engine, the mounting point location and mounts themselves must be reevaluated for structural and noise, vibration, and harmonics (NVH) requirements.
- Engine housing: Modifications to the engine housing as additional mounting points maybe required.
- Exhaust system: On the basis of the packaging feasibility of the component and coolant strategy, the exhaust may have to be redesigned.

The belt drive of the BSG system should be capable of transmitting the high torque based on the vehicle demands. As opposed to conventional vehicles, BSG requires a bidirectional drive-tensioning system to cater for the negative torque during regenerative braking. The belt must also be wider and made of material that supports high load and tension.

Components in a typical electrified vehicle are not integrated and implemented in aggressive environments. However, with BSG systems, electrification components are prone to the more extreme environments. The BSG design therefore needs to sustain these harsh environmental factors and based on where the components are mounted, these environmental requirements tend to vary. The two locations for component packaging in the vehicle are underhood or trunk/cabin. Irrespective of their location, all components of the vehicle must meet the NVH requirements, particularly those that directly impact propulsion. The motor and the power electronics in the case of integrated components are directly mounted to the engine. The vibration of these components over lift and the shock at each start will be extreme. It is essential that any device under specification does not emit any unwanted, undesirable, whining, disturbing, or annoying noise that a customer can hear during normal vehicle operations over the entire vehicle design life. The importance of meeting NVH requirements for BSG components, specifically the motor, is extremely important. The motor/generator is directly coupled with the engine during starts and any NVH effect will be experienced by the driver. It needs to be ensured that vibrations caused by an electric motor should not be more disturbing under any operating conditions than a pure combustion engine operation. These operating conditions cover the entire vehicle speed range and the entire ambient temperature range, including but not limited to, the motoring mode, the regeneration mode, acceleration/deceleration, slow start-up or wide open throttle, tip-in or tip-out, and so on. Note that tip-in and tip-out refers to engaging (or disengaging) the engine by stepping in (or out) of the pedal.

Another environmental factor that needs to be taken into consideration is the high temperature under the hood. The proximity of the components to the engine and the exhaust results in high ambient temperatures above the common 105°C and will be required to qualify testing at 125°C or 150°C. This high-temperature environment and the considerably high-power ratings in BSG components warrant a suitable cooling system. The components can be either forced air or liquid cooled depending on the design specifications. Typically, air cooling is considered suitable for components rated under 10 kW and liquid cooling for higher power ratings.

### 10.4.3 Design Challenges and Implementation

There are several challenges in designing the BSG system in a cost-effective manner without adding complexity to the process. To ensure an acceptable power density, the packaging of the components needs to be compact. The motor/generator and the power inverter module must fit underhood and the power pack unit should fit in the cabin/trunk and has to be designed accordingly. A BSG system will add weight to the base car at times, resulting in the vehicle moving up on weight classifications. Component and interface weight must be factored into the assessment of such a system. Key weight factors include the battery pack, motor, inverter, mounting brackets, and wiring. The breakdown of extra weight due to additional components or component redesign is given in Table 10.5. Minimizing this added weight is one of the crucial concerns in BSG design. Moreover, it is highly desirable to develop a global system that can be reused and integrated across different platforms and into various classes of vehicles to make this technology available to a wider consumer base.

As with any new technology, it is important to validate the design prior to implementation. This is conducted by simulation, functional, performance and reliability tests, and studies among others. The usage cycle is one of the most important requirements to validate the design. The usage cycle must comprise of the load of the system (by component) over the expected life of the vehicle. It should also account for environmental profiles and performance degradation of specific functions where applicable. On average, automotive life is defined as the total number of years and miles. However, for a start–stop system, the number of starts over this period, the duration of the start, and the time between starts and the environmental conditions at each event is most important. Assuming a 10-year, 200-km life cycle in Europe, the total number of starts can be calculated as shown in Table 10.6.

OBD requirement applies to all vehicles sold in North America and is a regulatory requirement that must be considered when implementing a 48-V BSG system. The stop/start is a power-train feature that will impact the overall emissions of the vehicle and must thus be compliant.

Tailpipe emission certification will be conducted with both start–stop active and disabled to capture worst-case values. If the emission delta between the cases is within a provided margin, OBD compliance requirement may be waived. However, a properly designed BSG will not (should not) fall within this range. The BSG will therefore require an indicator on the dashboard to indicate that

---

### TABLE 10.5
### BSG System-Driven Weight Analysis

| Component /Integration | Weight (kg) |
|---|---|
| Motor, brackets, and cooling | ~+8 |
| Inverter, brackets and cooling | ~+2 |
| ESS, brackets, and cooling | ~+10 |
| DC/DC converter, brackets and cooling | ~+5 |
| Engine adaptation | ~+5 |
| Interfaces (wiring, coolant lines, etc) | ~+8 |
| Removal of alternator | ~–8 |
| Total | ~30 |

**TABLE 10.6**

**Number of Auto-Starts for a BSG System for a 10-Year Life cycle**

| Cycle/Description | Mileage (km/Year) | Starts per kilometer (Starts/km) | Total No. of Starts—Lifetime (K) |
|---|---|---|---|
| Key starts | — | — | 36 |
| Urban | 7916 | 4 | 316 |
| Extra urban | 6951 | 0.2 | 14 |
| Highway | 2896 | 0.05 | 1.5 |
| Hill | 1545 | 0.2 | 3 |
| Total auto-starts | | | 335 |

the stop/start feature has been disabled for any reason. A typical system will utilize the "Malfunction Indicator Lamp" (MIL) for this purpose. If the BSG system is unable to perform as designed with the functions becoming inappropriately, unintentionally, or accidently nonfunctional, the MIL will serve as the indicator to the consumer. The reasons for this behavior include BSG subcomponents failure, BSG subcomponent fault codes, BSG enablers, inhibit codes, and so on.

## 10.5  KEY BSG SUBSYSTEM COMPONENTS

The key components of a BSG system, shown in Figure 10.11, consist of the 48-V energy-storage system, an electric motor, the power inverter module/controller, a 48-V/12-V DC/DC converter, and the FEAD module. Detailed technical information of the components will be provided in other chapters of this book. This section will focus on any unique requirements, designs, and functions of these components as they pertain to a 48-V system.

### 10.5.1  ENERGY-STORAGE SYSTEM

While a number of energy-storage technologies such as ultracapacitors and flywheels are under investigation, batteries are most typically used in energy-storage systems (ESS) for automotive

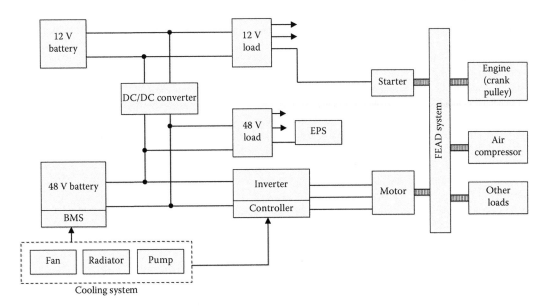

**FIGURE 10.11**   Design overview of the BSG subsystem components.

applications. The selection and sizing of a battery for the BSG system is of central importance since its parameters directly impact the vehicle performance and its capacity for electric function. The key factor in battery selection is the power rating. The battery is required to meet the peak power requirements for regenerative braking, cold cranking, and assist functions. It should provide both high-power and high-energy density. The SOC characteristics of the battery determine the battery management strategy that controls the use of available electric power. Depending on the choice of technology, the reliability and life of the battery may be compromised if operated at low SOC. The SOC, in such cases, needs to be maintained at higher levels, limiting the use of electric function and thereby affecting the overall fuel economy delivered by the vehicle. Table 10.7 provides an overview of the ESS characteristics and features for a 48-V BSG system. The specifications are compliant with the LV148 systems guidelines, taking into account typical voltage losses over power cables during operation, and a nominal suite of accessory load functions.

The ESS will normally comprise the following system components:

- *Electrochemical cells.* These may be assembled into one or more discrete modules.
- *48-V power connections.* They provide the power interface to the vehicle. A 48-V positive terminal is mandatory for the battery, while the negative terminal may be integrated into the ESS chassis and bonded to the vehicle chassis ground.
- *LV (signal) connector.* It provides an interface to communicate with the vehicles central controller.
- *Cell temperature, voltage and current, and sensors.* These perform real-time monitoring of current, cell voltage, and temperature.
- *Cabin air temperature sensors in/out.* These sensors are required if the ESS is air cooled and the design of the cooling system requires monitoring of air temperature.
- *Thermal management system.*

---

TABLE 10.7

**Features of the 48-V ESS for a 48-V BSG System**

| Parameter | Definitions | Units | Minimum |
|---|---|---|---|
| Voltage max during normal operation | Maximum operating voltage during 10-s peak charge, under normal operating temperatures | V | 52 |
| Voltage min during normal operation | Minimum operating voltage during 1-s peak discharge, under normal operating temperatures | V | 38 |
| Voltage max during extended operation | Maximum operating voltage during 10-s peak charge, under extended operating temperatures | V | 54 |
| Voltage min during extended operation | Minimum operating voltage during 1-s peak discharge, under extended operating temperatures | V | 24 |
| Discharge power, 1 s | 10-s discharge power required in target operating temperature | kW | 10 |
| Discharge power, 10 s | 1-s discharge power required in target operating temperature | kW | 8 |
| Charge power, 10 s | 10-s charge power required in target operating temperature | kW | 9 |
| Charge power, 1 s | 1-s charge power required in target operating temperature | kW | 10 |
| Max. current (1-s pulse) | Maximum current that system will see for 1 s, within normal voltage range | A | 350 |
| RMS current (>10 min) | Maximum continuous current under hybrid operation | A | 80 |
| Available energy (peak discharge) | Minimum energy required to meet the 1-s discharge and 10-s charge requirements, plus accessory functions and the voltage stay in normal operating range | Wh | 150 |
| Cell-operating temp. range (°C) | System maximum cell temperature data −30° to +50° (°C) | °C | −30/55 |
| Survival temperature range | Temperature range over which the ESS will not experience any permanent damage (nonoperational) | °C | −40/66 |

- *Cell-balancing circuits.* These circuits may be required, depending on the battery chemistry chosen, and can be possibly integrated into the battery control module.
- *Precharge contactor and resistor.* In some cases, this circuitry is an integrated part of the electronic components of the power train and is therefore not required within the ESS.
- *Fuse(s).* These are often made serviceable to disconnect the 48-V powernet during service and maintenance.
- *Battery pack control module (BPCM).* The control module contains the hardware controller including the drivers for the contactors. It performs current, cell voltage and temperature measurement, evaluates the battery SOC, state of function (SOF), and state of health (SOH), generates the corresponding control logic, and supervises the diagnostics, error management, and communication.
- *Housing.*

As discussed earlier in the chapter, HV connectors, safety relays, and HVIL can be avoided within the 48-V systems. Liquid cooling is not necessarily required for 48-V batteries and housing does not require HV galvanic isolation.

Three major battery technologies have been successfully incorporated into automotive hybrid systems: lead acid (PbA), nickel metal hydride (NiMH), and lithium ion (Li ion). Both PbA and Li-ion have seen the development of a significant number of variants in technology, including several oriented toward higher power applications. Figure 10.12 is a Ragone chart that illustrates the performance of these battery types in terms of specific power and energy.

Li-ion batteries for HEVs (HEV Li-ion) have low storage capacity under <10 Ah when compared to high-energy Li-ion cells. However, they offer high-power discharge up to 2000 W/kg, making it desirable for rapid, shallow cycling. Moreover, they are capable of delivering power at low SOC without significant degradation to their reliability and life. PbA batteries on the other hand require the SOC to be maintained at a high level. They suffer from sulfation during prolonged low SOC operation, and deliver relatively poor charge acceptance. Lead carbon (PbC) battery technology

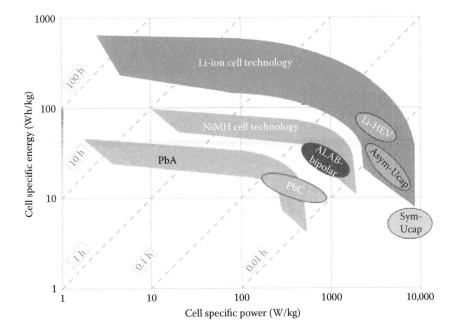

**FIGURE 10.12** Ragone chart to demonstrate battery performance in terms of specific power and specific energy.

seeks to resolve the issues associated with PbA batteries. This is achieved in PbC batteries that are combined with activated carbons, thus merging the ultracapacitive charge characteristics with the PbA battery characteristics. However, PbC batteries offer low energy and power density, when compared to PbA. Another technology is the bipolar advanced lead acid (ALAB). This advanced bipolar battery architecture has been applied to PbA construction, leading to a noticeable reduction in cell resistance, and also allowing a near doubling of battery energy density. This architecture, combined with improvements in electrode design for partial SOC operation could potentially provide a cost-effective alternative to advanced batteries. There have been noticeable technological challenges associated with ALAB batteries, particularly with battery sealing, which have delayed its introduction into large-scale commercial use.

Another type of energy-storage technology is the ultracapacitors. Currently, the available electrolytic double-layer capacitors provide significant power density, and low-energy density. Given the fact that most LV hybrid applications require very small levels of energy for operation, the ultracapacitor is a potentially good fit. Progressively decreasing cost and a wide operating temperature range make such ultracapacitor technologies such as symmetric ultracapacitor (symUcap) increasingly attractive to automotive applications. The low-energy density, however, poses challenges for sustained accessory drive support. Asymmetric ultracapacitor (AsymUCap) technology combines the electrostatic energy-storage mechanism of a symUcap with the electrochemical energy-storage mechanism of a battery. Such systems can have a doubling of energy density over symUcap, while retaining most of the ultracapacitor's advantages. This is a new technology, being applied to many battery chemistries, most notably being PbA (such as lead–carbon), and li-ion. This technology could form the long-term ideal power and energy-storage solution for LV hybrid vehicle architectures.

Table 10.8 illustrates the relative merits and shortcomings for several energy-storage technologies currently available for consideration with LV hybrid power trains. Li-ion batteries, as seen from the table, demonstrate the most suitable characteristics for HEV applications among the available battery technologies. They have very-high-energy density and sufficiently high-power density.

Figures 10.13 through 10.15 demonstrate the power, current, and voltage profiles of a Li-ion battery system in a 48-V power train under a typical drive cycle. Lithium iron phosphate (LFP) with a carbonaceous anode technology is assumed for the simulation. As seen from the graph, the peak power of the ESS is only utilized during a start event, an event that occurs intermittently while the

**TABLE 10.8**

**Comparison between Battery Technologies for Automotive Applications**

| Name | Lead–Acid | Lead–Carbon | Nickel–Metal Hydride | Nickel–Zinc | Lithium-Ion | Ultracapacitor |
|---|---|---|---|---|---|---|
| | AGM | Pb–C | NiMH | NiZn | Li-ion | Ucap |
| Maturity | Mature (multiple OEMs) | Advanced development (under study by OEMs) | Mature (HEV) | Advanced development | Mature (HEV+) | Advanced development |
| Power density (kW/l, kW/kg) | | | 0 | | | ++ |
| Energy Density (Wh/l, Wh/kg) | –– | – | 0 | | ++ | –– |
| Durability/Life (MWh, years) | –– | 0 | | | ++ | ++ |
| Temperature sensitivity | ++ | | 0 | 0 | – | |
| Cost ($/kW) | ++ | ++ | | ++ | – | – |
| Cost ($/Wh) | ++ | | – | 0 | – | –– |

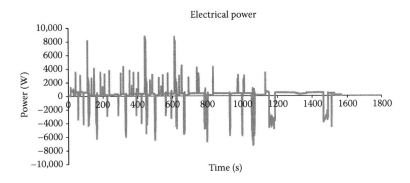

**FIGURE 10.13**   Power profile of a Li-ion battery system in a 48-V power train under a typical drive cycle.

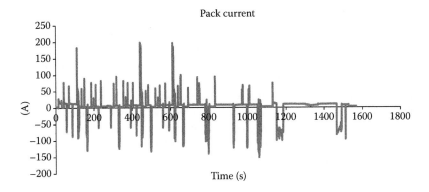

**FIGURE 10.14**   Current profile of a Li-ion battery system in a 48-V power train under a typical drive cycle.

average power is less than half of its peak value. Therefore, it is important to select a cell technology that has high-power density.

Figure 10.16 demonstrates the energy supplied by the battery while the SOC is maintained above 75%. This is an evidence that the ESS of a 48-V BSG system is not required to provide significant amounts of energy over an average drive cycle. The limited SOC discharge and average power demands reinforce the need for power-dense, energy, and size-optimized system.

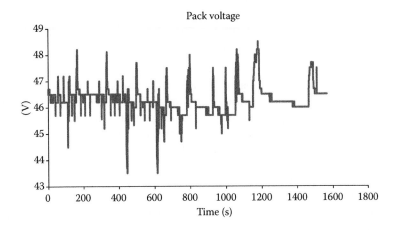

**FIGURE 10.15**   Voltage profile of a Li-ion battery system in a 48-V power train under a typical drive cycle.

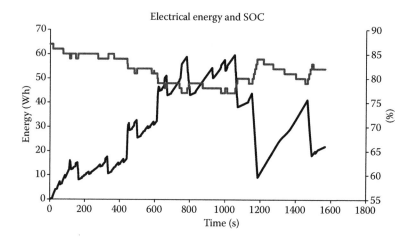

**FIGURE 10.16**   Electrical energy supplied by the battery and its SOC.

Longer-term ESS will strive to replace the current 12-V starter onboard existing power trains, and expand the support for vehicle electrification. The former goal requires a significant improvement in battery low-temperature performance. The latter will drive batteries capable of higher sustained currents and wider usable SOC ranges. This objective allows vehicles to utilize expanded heating and air conditioning with the engine off, autonomous driving functions, and high-speed coasting capability. The United States Advanced Battery Consortium (USABC) has developed and published a set of objectives for such a battery [9], shown in Table 10.9.

### 10.5.2   Motor

The primary function of the motor of the BSG system is the capability to provide a smooth and seamless auto-start with a fast response. The requirement for maximum torque and motor speed at maximum torque is dependent on the torque required at the crank shaft to start the engine and the pulley ratio of the FEAD system.

For a start function, the torque and speed of the motor are defined in Equations 10.3 and 10.4, respectively.

$$\text{Torque}_{\text{Motor}} = \frac{\text{Torque}_{\text{Crank shaft}}}{\text{Pulley ratio}} \tag{10.3}$$

$$\text{Speed}_{\text{Motor}} = \text{Speed}_{\text{Engine}} \times \text{Pulley ratio} \tag{10.4}$$

The torque slew rate of the motor is based on the starting response time specified by the vehicle. It is important to note that while an electric motor can go from zero to max torque comparatively faster, the slew rate will be limited by the supporting components of the FEAD system.

Most start–stop/BSG systems enable auto-starts only above 0°C ambient temperatures. A key reason for this is the limited power discharge capability of ESS at cold temperatures. Suppliers and original equipment managers (OEMs) are looking for cost-effective approaches to expand the start capability to the full temperature range of the vehicle, that is, −40°C to 125°C ambient temperature range. This will allow for additional fuel economy improvement. It will also allow for package and cost optimization by removing the need for a conventional starter. The torque needed for cold cranking varies anywhere between 1.5 and 1.8 times the torque required for auto-start under nominal temperatures [10]. The motor, along with the start feature, is also expected to perform the

TABLE 10.9

**USABC Requirements of ESS for 48-V HEVs at EOL**

| Characteristics | Units | Target |
|---|---|---|
| Peak pulse discharge power (10 s) | kW | 9 |
| Peak pulse discharge power (1 s) | kW | 11 |
| Peak region pulse power (5 s) | kW | 11 |
| Available energy for cycling[a] | Wh | 105 |
| Minimum round-trip energy efficiency | % | 95 |
| Cold cranking power at −30°C (three 4.5-s pulses, 10-s rests between pulses at min SOC) | kW | 6 kW for 0.5 s followed by 4 kW for 4 s |
| Accessory load (2.5-min duration)[a] | kW | 5 |
| CS 48-V HEV cycle life[b] | Cycles/MWh | 75,000/21 |
| Calendar life, 30°C | Year | 15 |
| Maximum system weight | kg | ≤8 |
| Maximum system volume | Liter | ≤8 |
| Maximum operating voltage | Vdc | 52 |
| Minimum operating voltage | Vdc | 38 |
| Minimum voltage during cold crank | Vdc | 26 |
| Maximum self-discharge | Wh/day | 1 |
| Unassisted operating temp range (power available to allow 5-s charge and 1-s discharge pulse) at min. and max. operating SOC and voltage | °C | −30 to +52 |
| 30°–52°C | kW | 11 |
| 0°C | kW | 5.5 |
| −10°C | kW | 3.3 |
| −20°C | kW | 1.7 |
| −30°C | kW | 1.1 |
| Survival temperature range | °C | −46 to +66 |
| Max system production price at 250 k units/year | $ | $275 |

[a] Total usable energy will include cycling energy and accessory load energy. The usable energy will be 313 Wh.
[b] Each individual cycle profile includes six (6) start–stop events, for a total of 450-k events over the duration of the test.

function of an alternator by generating power required for auxiliary loads, support regenerative breaking, and provide sufficient peak power to assist the vehicle propulsion during high torque demands such as acceleration. Low torque ripple, high efficiency, reduced noise, and wide speed range are desirable characteristics for the BSG motor.

In addition to meeting the power and torque requirements, volume efficiency is a design requirement for a BSG motor. The total efficiency of the electrical system including motor, inverter, and battery should be >75% during the stop/start mode, 85% during assist/boost function, and higher than 90% during generation. In most cases, the motor needs to fit within the footprint of the alternator. The machine also needs to be lightweight since it is directly mounted to the engine block.

As explained in the operational principle of a BSG system, the speed of the motor is directly linked to the engine speed through the pulley ratio of the system. The electric machine must therefore be capable of operating up to the corresponding engine speed at fuel cutoff. Most engines have fuel cutoff in or around 6000–7000 rpm and have pulley ratios proposed between 2 and 3. This implies that the typical BSG motors must be capable of 16,000 rpm. A 20% margin is incorporated for overshoots, that a motor speed of around 20,000 rpm is required for the BSG application. Speed feedbacks of most motors in electrified applications are based off resolver technology. However, with BSG systems, lower cost encoders, hall effect sensors and in some instances advanced sensor less strategies are being evaluated.

TABLE 10.10

**Comparison between Motor Technologies for BSG Applications**

| | Permanent Magnet | Wound Rotor | Claw Pole | Switched Reluctance | Induction Copper Cage |
|---|---|---|---|---|---|
| Size and weight | + | 0 | 0 | − | − |
| Efficiency | + | 0 | 0 | − | − |
| Losses at no load | − | + | + | + | + |
| Starting current | + | + | + | 0 | − |
| Back EMF | | | | | |
| Thermal | + | 0 | + | + | 0 |
| Manufacturing maturity | − | 0 | + | − | 0 |
| Cost | − | 0 | + | + | 0 |
| Torque ripple | ++ | + | + | − | + |
| Acoustic noise | ++ | + | + | − | + |
| PM material | Yes | No | Yes/no | No | No |
| Torque density | + | 0 | 0 | − | 0 |
| Fault tolerant | 0 | 0 | 0 | + | 0 |

There are a number of motor technologies that are suited to vehicle electrification, each with their own set of advantages and challenges. Design considerations for each of these motor and its performance analysis under varying drive conditions have been presented in detail in the previous chapter. In this section, the evaluation of electric motors is focused on identifying the best motor technology for BSG applications.

There are a number of factors that qualify a motor to be suitable for BSG design. Motor selection requires a compromise between cost, performance, efficiency, packaging, maturity, and simplicity in design. A comparative assessment of various motor technologies based on these criteria is presented in Table 10.10.

It can be seen that claw pole machine represents the best trade-off for BSG application. It offers compact size and good efficiency with proven high-volume manufacturing record. Most leading automotive alternator suppliers such as Valeo, Mitsubishi Electric, and Denso among a multitude of others develop millions of claw pole machines annually making them a mature, cost-effective, and reliable solution. The power factor can be adjusted close to 1 through rotor excitation that cannot be achieved in other motors. The power factor is particularly low in the case of the PM machine for smaller loads. Claw pole machine exhibits reduced losses for both within the machine and in the inverter at low loads. The starting current profile is acceptable and the machine demonstrates good torque versus current profiles if high-excitation currents are used.

Modified conventional starters are being considered for BSG applications as they come with lower investment and risk. These machines are modified by strengthening brushes and slip-ring system, an optional addition of permanent magnets (PMs), modification of electromagnetic and thermal design, and the concept of galvanically separated mass according to 48-V power net requirements. However, one key concern is the life of the rings as a start–stop system will have a much more aggressive duty cycle.

### 10.5.3 POWER INVERTER MODULE

The primary function of the power inverter module is to control the electric motor. It is tasked with receiving commands from the vehicle and engine control unit (ECU), assessing the status of system components, and providing the required phase current to provide the requested torque at the shaft.

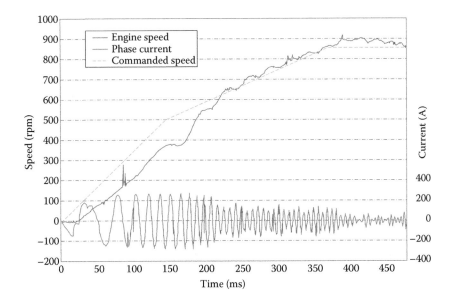

FIGURE 10.17    Engine speed and inverter phase current profile during a start.

The inverter is also responsible for ensuring that the BSG system complies with federal torque security requirements.

Another key functional requirement of the power inverter module is to ensure that the voltage at any accessible interface is lower than 60 V. In the event of a PM motor, the back electromotive force (EMF) can be >60 V at higher speeds. Both software and hardware features are specified to protect the consumer in the event.

Figures 10.17 through 10.20 illustrate the engine speed and inverter phase current profiles of a BSG system during (a) start, (b) stop, (c) change of mind, and (d) engine idle scenarios. The phase current profile behavior is a direct representation of the power profile of the inverter. The figures

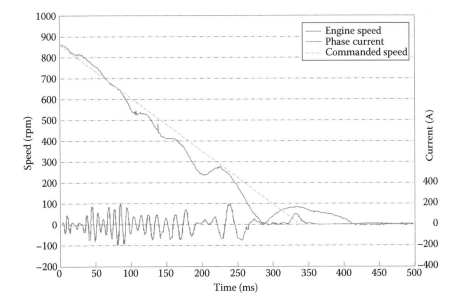

FIGURE 10.18    Engine speed and inverter phase current profile during a stop.

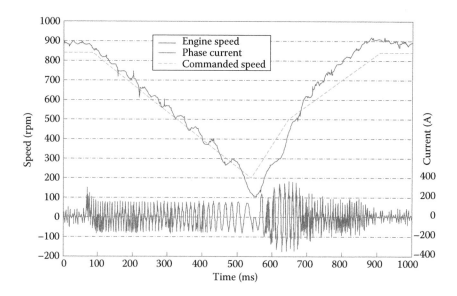

FIGURE 10.19   Engine speed and inverter phase current profile during a change-of-mind event.

illustrate the high inverter current during motoring mode of the electric machine for engine start and assist, and during regenerative braking when the inverter transfers generated power from the machine to the battery. The speed of the motor during each of these events will be a multiplication of the engine speed and pulley ratio.

The components in the power inverter module for the 48-V motor and the architecture of the motor controller in a BSG system are shown in Figure 10.21. The power inverter module consists of (1) the power module and driver circuit, (2) a DC link capacitor, (3) filter, (4) controller and LV components, (5) sensors, (6) interphases, and (7) thermal components. Figure 10.21 provides a block diagram schematic of the component, including key subcomponents and both internal and external interfaces.

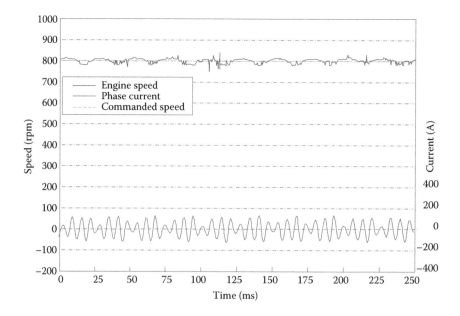

FIGURE 10.20   Engine speed and inverter phase current profile during an engine idle event.

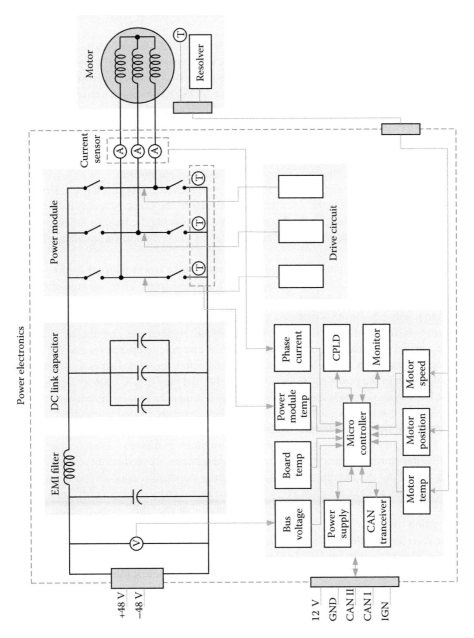

FIGURE 10.21 Power electronic architecture and control unit of the BSG system.

There are two leading switching technologies used in electrification systems, for example, insulated gate bipolar transistor (IGBT) and metal–oxide–semiconductor field-effect transistor (MOSFET). IGBTs are commonly used in power inverter modules found in electrified vehicles. A key reason is that PHEVs, HEVs, and EVs generally operate at voltages in the range of ~300–400 V. IGBTs are best suited at this voltage range and also provide a cost-effective, thermally feasible solution.

IGBTs are used in high-power applications >5 kW that require low-duty cycle, low frequency under 20 kHz, and expect small variations in load. IGBTs can operate at a high junction temperature >100°C.

MOSFETS, on the other hand, are best suited for low-power, LV designs under 250 V that operate at high switching frequency with long-duty cycles and expect large variation in load. At low voltages, to meet the power demands of the BSG motor, the switching devices need to have high current ratings. However, current-handling capability of MOSFETs is comparatively lower than in an IGBT that results in multiple Die solutions. The MOSFET is a voltage-controlled device with a positive temperature coefficient, stopping thermal runaway. The on-state resistance has no theoretical limit; hence, the on-state losses can be far lower. The MOSFET also has a body-drain diode, which is particularly useful in dealing with limited free-wheeling currents. Much like the electric motor, volume efficiency is an important factor in the design of the inverter.

The DC link capacitor is a critical component of the power inverter module. It is used to decouple the effects of the inductance from the 48-V ESS and wiring to the power stage. It provides a low-impedance path for the ripple currents. It also plays a role in reducing the leakage inductance. There are two key capacitor technologies that are currently used in electrification systems, electrolytic capacitors, and film capacitors.

Electrolytic capacitors are commonly used in the automotive world with all leading suppliers having a large range of automotive-rated components to choose from. They are cost-effective and robust but also have high equivalent series resistance (ESR), low ripple current capability, low ambient temperature limits, and limited lifetime.

BSG systems are required to perform at −40°C ambient temperatures and also at 125°C in the event the component is mounted underhood. The low ESR and cold temperatures and the high thermal dissipation and thermally robust film capacitors are therefore ideal candidates for such a system.

The HV switching in the power modules of the inverter and the DC/DC converter create voltage ripples on the DC bus that degrade the battery life and performance. To minimize these voltage ripples, a DC link capacitor is connected at the DC bus. The capacitance value is selected high enough to ensure the voltage ripples are limited within the allowable limits. However, the capacitor size must also be kept small to meet the power-density requirements and allow a compact packaging.

Additionally, the stray inductance in the inverter connections coupled with the high-frequency current in the switching devices generates high-frequency conducted noise at the input side. For the vehicle subsystems to pass the mandatory electromagnetic compatibility (EMC) tests, electromagnetic interference (EMI) filters are required to ensure that the EMC noise remains below the allowable emission limits.

The power inverter module in a 48-V application is generally mounted underhood. The proximity to the motor maximizes efficiency and reduces integration cost due to the short three-phase cable and allows for faster control. However, the component will now need to withstand environment conditions including vibration, in multiple gravitational accelerations depending on mounting location and ambient temperatures that are significantly more aggressive than power electronic components are generally exposed to. The ambient temperatures are on average above 100°C due to the engine that serves as a heat source and during operation in aggressive conditions such as Death Valley or hot locations such as Vegas, the underhood ambient temperatures can peak close to 125°C.

Some of the key steps that can be taken to minimize the temperature include adding heat shields, optimizing subcomponent placement, and mounting the inverter on the input side of the exhaust if possible. While these steps will reduce the temperature around the component, an effective coolant system is critical to ensure proper vehicle performance and component robustness.

There are two key approaches to cooling the inverter, air cooled and liquid cooled. The common approach in present-day electrified vehicles is liquid cooled using a unique coolant loop. This allows the system to operate at lower temperatures in the range of 60°–75°C. This enables the designer to use lower-rated subcomponents, minimize losses, and allow for aggressive duty cycles. However, in the BSG systems, this will result in additional components. Therefore, the BSG system architecture is forced to look at more aggressive, cost-efficient cooling methods. Two alternative options are air cooling and liquid cooling using engine coolant. The mounting location of the inverter is exponentially more important in an air-cooled system. It needs to be mounted in an area where airflow is available and more importantly, the air temperature is relatively lower. This is usually true on the input side of the exhaust. Engineers must also design the inverter so as to have the subcomponents more sensitive to high temperature further away from heat sources and closer to the housing that will serve as a coolant plate. Improved air cooling can also be achieved through integrated fans. The key heat sources in the inverter are the power semiconductor devices, drivers, capacitor, and windings. It is important to design the cooling system, either air or liquid to optimize the heat transfer from these components to the coolant.

An integrated design where the inverter is mounted onto the electric motor can offer several advantages. An integrated motor/inverter unit gives a compact design that can be easily used across all platforms with minimal integration complexity and helps avoid any conflict during packaging and handling. It reduces the volume under the hood, resulting in an improved overall power density of the system. The integrated design costs less than the sum of the cost of individual components, thus offering a cost-effective solution. It eliminates the expense of brackets required for mechanical integration and HV wiring that forms a sizeable portion of the system cost, typically $100 per phase. In the integrated motor/inverter for the BSG, these wiring costs can be eliminated and efficiency can be improved by avoiding losses due to parasitics in the wiring. Moreover, close proximity of the controller, power modules, and the motor to each other and consequent mitigation of wiring stray inductance reduces EMI noise. An integrated solution also allows for the system to be classified as an LV system as the three-phase connection where one can see above 30-V AC is mechanically isolated. The inverter can either be radially or axially connected to the motor. The challenge with the integrated structure is the printed circuit board (PCB) design and mounting that becomes complex due to higher susceptibility of the inverter to vibrations and exposure to high temperature.

### 10.5.4 DC/DC Converter

The function of the DC/DC converter is to convert the 48-V level into 12 V to power the auxiliary loads and charge the conventional 12-V battery. In some applications, it is required to perform bidirectional operation and it must always offer high efficiency as it directly impacts the fuel economy of a vehicle. The converter design requires a precharge circuit for the DC link capacitor and it is desirable to develop a converter design that is scalable to different output power levels. The main components of the converter include the PCB with power modules, and another with the control circuitry. The converter also has a vehicle communication interface. The thermal management of the converter is either based on air or liquid cooling depending on the requirements, design specifications, and mounting location.

The main functions implemented in the DC/DC converter include over-voltage and over-current protection, voltage and current measurement at 12- and 48-V level, temperature management (multiple sensors), reverse battery protection at 12-V level, precharge function of 48-V DC link capacitor, controller area network communication, commonly referred to as CAN, and diagnostic and error management. Table 10.11 presents the high-level specifications for a typical DC/DC converter for a BSG-integrated vehicle system.

Thus, the operation of a DC/DC converter in a 48-V system is similar to that of an HV electrification system. From a design perspective, the difference is limited to the power stage where LV subcomponents and MOSFETs are utilized.

TABLE 10.11

**High-Level Specifications for DC/DC Converter**

| Requirement | Value |
|---|---|
| Input voltage range | 36–54 V |
| Controllable output range | 11–15 V |
| Output power at 12.5–14 V | 1.4–2.2 kW (parallel power output) |
| Output current | 120–175 A |
| Efficiency | Dependent on current |
| Dimensions | Program specific |
| Weight/volume | <3.0 kg/<3 L |
| Diagnostics | Yes |

### 10.5.5 FRONT-END ACCESSORY DRIVE

FEAD, such as the one shown in Figure 10.22, is integral in the implementation of a BSG system. Its primary function is to transfer energy from the motor to the engine crank shaft; however, it also supports loads that are not directly related to the functions of a BSG system such as the AC compressor and power steering. The FEAD constitutes the mechanical connection between the engine and the starter/generator via a belt that passes over tension and idler pulleys. To ensure the necessary tension on the belt to transmit the required torque, the belt is passed through a tensioning unit also called the tensioner. This section will discuss these components that are directly related to the BSG functionality, namely the pulley, tensioner, and belt.

#### 10.5.5.1 Pulley

Calculating the pulley ratio is a balance of multiple requirements and finding the optimized value. The larger the pulley ratio, the smaller the required output torque of the motor needs to be. However, a larger pulley ratio results in increasing the base speed requirement of the electric machine. The *motor pulley* is directly coupled to the shaft of the motor, the pulley dimension is driven by the

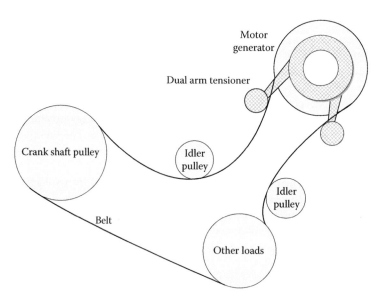

FIGURE 10.22 FEAD.

pulley ratio of the system, the available surface area, and the bearing loads. The *crank shaft pulley* as discussed in the previous sections of the chapter, the torque required at the crank shaft varies by engine technology and power. Owing to packaging requirements, the max dimension of the pulley will vary. An idler pulley is used in an FEAD system to regulate how the belt runs between the other pulleys by maximizing energy transfer.

### 10.5.5.2 Tensioner

The primary goal of a tensioner is to provide sufficient tension on the belt of the FEAD system when torque from the motor is required and prevents the pulley from becoming loose. The oscillation of the tensioner around the motor axis reduces speed irregularities at the motor and its rotational movement alternates the slack side and tight side. There are multiple tensioner solutions including single arm, dual, and e-tensioners.

### 10.5.5.3 Belt

Belt is the most commonly used component for power transmission that typically demonstrates efficiency around 97%. In BSG drive systems, the belt has to sustain higher load than a conventional vehicle. Multirib belt for applications with belt-driven start–stop function is suitable for boost and recuperation function. The belt needs to be designed such that it can handle the high torque for engine stop/start and the transmission to high-power loads. Belt drive systems in advanced designs for conventional vehicles can withstand up to 250,000 engine starts, each requiring a torque of 100 Nm [11]. The belt drive in a BSG system is required to exhibit comparable life span. The surface of the belt needs to be tough and resistant to wear, offering a high lifetime. Moreover, it should be capable of operation without degradation in a temperature range from −40°C to +140°C.

With the replacement of the conventional alternator with a BSG system, the width of the belt and the pulleys needs to redesigned to improve the mechanical load-handling capability. As opposed to the starter/alternator system that only provides positive torque, the tensioner in the BSG-integrated FEAD also has to be redesigned to ensure sufficient tension for bidirectional flow of power and thus cater for regenerative torque.

### 10.5.6 ENGINE CONTROL UNIT

The control unit for the engine requires several modifications to adapt to the BSG system. It needs to be integrated with the hybrid supervisory controller and requires brake control software for regenerative braking. It is also linked to the cabin HVAC software and the customer displays and gauges. The ECU also has additional connection to the vehicle wiring, power distribution, and CAN communication.

## 10.6 BENCHMARKING

LV mild hybrids have been under investigation for nearly two decades. In the early 2000s, a consortium of automakers and component suppliers investigated 42-V nominal voltage architecture in light-duty vehicles. Although a number of vehicles were developed by various auto manufacturers, only a few among them reached production. The most successful attempt in the first generation of LV hybrids was made by General Motors and their Belt Alternator Starter (BAS) system. This was employed into several vehicles between 2006 and 2009.

### 10.6.1 GENERAL MOTORS

General Motors developed several 36-V BAS systems including the Saturn Aura, Saturn Vue, and Chevy Malibu. The system employed a motor–generator unit, which also acted as the engine starter. The nominal system voltage was 36 V, and was therefore lower than the original 42-V target. The vehicles demonstrated an estimated improvement of 5+ mpg in the city and highway fuel economy and a 20% improvement in the EPA-combined cycles. The BSG system consists of a 36-V NiMH battery pack with 10-kW charge and discharge power mounted behind the rear passenger seat, a PM electric motor with 60+ Nm peak torque, and 3-kW continuous generator capability packaged underhood. The liquid-cooled power electronics components are also mounted underhood in close proximity to the motor.

### 10.6.2 PSA PEUGEOT CITROEN

PSA Peugeot Citroën is currently developing a 48-V mild hybrid solution compatible with both gasoline and diesel fuel engines to be launched in vehicles by 2017. This hybrid design developed in collaboration with suppliers such as Valeo, Bosch, and Continental is expected to reduce $CO_2$ emissions by 15 g/km and improve fuel economy by 10%–15% over conventional vehicles. A 10-kW electric motor and a 48-V Li-ion battery pack is integrated with the power-train. This allows for boost/assist functionality and electric drive capability under 20 mph. The power-train architecture is compatible with both manual and automatic gearboxes of B, C, and D segment vehicles, thus providing a low-cost, fuel-efficient alternative to a wide customer base [12].

The air-cooled 48-V battery is packaged in the trunk of the vehicle with the electric motor and power inverter module underhood. The DC/DC converter is packaged in the trunk alongside the battery. Therefore, you now have both a 48-V and 12-V cables running the length of the vehicle. While this could lead to some additional losses, it eases packaging concerns of the component.

### 10.6.3 THE 48-V LC SUPER HYBRID

The Advanced Lead-Acid Battery Consortium (ALABC) and Controlled Power Technologies (CPT) have unveiled a 48-V LC Super Hybrid [13]. The hybrid power train is mounted to a 1.4-L Volkswagen Passat with a target of 120 g/km $CO_2$ emissions while achieving 0–100 km/h acceleration within 9 s.

The power train designed by AVL features 1 kWh lead–carbon batteries with a battery management system developed by a UK-based company Provector. The BSG has a drive belt tensioner system by Mubea and a switched reluctance motor/generator developed by CPT. Valeo has acquired this motor/generator technology for mass production.

This mild hybrid technology provides assist/boost during auto-start and acceleration, captures a significant share of regenerative energy, optimizes fuel consumption during idling, and cruising modes with the help of electric assist. The 48-V system aims to improve the fuel economy by 4%–8% over the 12-V LC Super Hybrid, which already achieves 42 mpg and carbon emissions of 130 g/km. The 12-V Super Hybrid has already been tested thoroughly and the performance data for the 48-V system are under evaluation. These vehicles are expected to be on road by 2015.

### 10.6.4 THE GREEN HYBRID

BYD Auto Co. Ltd. has introduced its Green Hybrid initiative that targets improvement in vehicle fuel economy >20% by 2014 [14]. As part of this venture, BYD has become the first OEM to implement high-efficiency 48-V power trains that feature regenerative braking, auto-stop/start system, low rolling resistance, and improved aerodynamics all of which contribute toward achieving a fuel economy improvement of up to 20%.

The vehicle design integrates an LV, high torque, and double-winding motor onto the power train that provides battery-powered assist/boost during acceleration. The electric power converted from regenerative braking is stored in the iron–phosphate batteries developed by BYD. This new battery technology has been developed by BYD and guarantees battery lifetime comparable to the expected life of the vehicle. This power-train design has demonstrated an improvement of 7 mpg in fuel economy.

### 10.6.5 48-V Eco Drive

Continental has developed its new 48-V Eco Drive system that offers fuel economy improvement by 13% and will be available in the market by 2016 [15]. The system architecture consists of the belt system, an electric motor and inverter integrated into a single housing, a bidirectional DC/DC converter, and a Li-ion battery, all of which can be incorporated into any vehicle conveniently without having to redesign the engine or transmission configuration. The motor/generator is an induction machine that offers 14-kW peak and 4.2-kW continuous powers, can be directly mounted onto the transmission and allows flexibility in designing the motor based on available space and power requirements. The 3-kW DC/DC converter is passively air cooled and has a range of 6–16 V buck and 24–54 V boost output. The dimensions of the 460-Wh Li-ion battery are quite comparable with the conventional 12-V PbA batteries. The electric motor/generator of the 48-V vehicle provides fast auto-start even during cold cranking. Regenerative braking, electric boost capability, and auto- stop/start feature during constant speed cruising and coasting provide considerable fuel savings at low cost with the flexibility of installation in a wide range of vehicle segments.

### 10.6.6 Hybrid4All

Valeo has launched a low-cost hybrid system intended for mass production. It consists of a compact 48-V BSG that utilizes a 15-kW motor/generator. The system can be integrated to any conventional gasoline or diesel-fueled engine at low cost. This LV hybrid design features regenerative braking, idle stop/start, and boost/assist functions. When installed in Peugeot 207 1.6 L THP, the electric functions of the BSG system allow a reduction in $CO_2$ emissions by 15%. Volume production of vehicles integrated with this technology is expected in 2017.

### 10.6.7 Bosch

Like other major electric drive component suppliers, Bosch too has stop/start systems in its electrification road map, both in 12- and 48-V architectures. The 48-V systems referred to as the boost recuperation system (BRS) offers regenerative braking and electric assist features in addition to stop/start capability, giving a 9% improvement in fuel economy on a FTP75 drive cycle [16]. This is achieved with the support of a 10-kW electric motor/generator [17]. However, the size of the Li-ion battery is kept small and can be fully recharged within five braking events.

### 10.6.8 48-V Town and Country Hybrid Power Train

An advanced 48-V power train is being developed for volume manufacture and assembly as a joint initiative undertaken by the Newcastle University and Infineon Technologies in collaboration with Libralato Ltd., Tata Steel, and UK Advanced Manufacturing Supply Chain Initiative (AMSCI) Proving Factory. The 48-V Town and Country Hybrid Power Train (TC48) project aims at designing a low-cost 48-V system by integrating a 5-kWh Li-ion battery pack and a novel 33-kW SRM technology that does not require rare-earth metals with a 50-kW gasoline engine. The rotary engine offers the fuel efficiency of a diesel-fueled engine but carries double the power-to-weight ratio. The entire power train and vehicle is controlled with two ECUs and the hybrid components fit within the standard engine ways, resulting in a compact design. With a larger battery capacity and electric motor with higher power rating, this design delivers a more extended range for all-electric driving and higher electric power assist compared to other BSG systems.

This initiative aims at tackling cost, range, recharging infrastructure, and performance challenges in the market acceptance of hybrid vehicles. The target of the project is to electrify a compact car such that it is capable of 15-miles all-electric range and emits 52 g/km $CO_2$ only for a marginal cost of $2769 that can break even through fuel savings in 2 years [18].

With the ever-increasing demand for reduction in carbon emissions and the market constraints on manufacturing cost and customer affordability, BSG systems provide a highly suitable solution. The simplicity of the system design and LV operation range makes it cost-effective and easy to integrate with most class segments of cars manufactured by many of the auto makers, making them highly attractive and available to a broad base of consumers. At the same time, the stop/start feature augmented with regenerative braking, power assist capability, and limited electric-only drive in BSG systems provides a considerable improvement in fuel economy. This is why most of the prominent auto manufacturers and suppliers are developing BSG systems for their upcoming vehicles that will be available in the market in the near future.

## QUESTIONS

10.1     Identify the fuel economy requirement for 2025 and the impact of this improvement on the daily oil consumption in the United States.

10.2     Per the UN/ECE 100, what is the definition of LV system?

10.3     If you were tasked with improving the fuel economy of a small vehicle by ~5%, identify the electrification topology.

10.4     Provide three reasons for selecting an LV versus an HV electrification system.

10.5     List the four most important functional objectives of a BSG system and identify how they drive BSG requirements.

10.6     List four vehicle enablers for an auto-stop.

10.7     Describe the advantages of a 48-V EPS system.

10.8     List the key subcomponents of a 48-V electrification system.

10.9     What is the specific disadvantage to operating a battery only at a high SOC, and how does a 48-V hybrid system utilize lower SOC batteries?

10.10    Using the requirements listed in Table 10.9, determine the apparent 1-s DC resistance required by the battery, to meet the discharge performance requirements, at room temperature.

10.11    Determine the room-temperature specific energy (Wh/kg) and specific power (kW/kg) for the USABC 48-V HEV battery, based on peak power requirements, and assume the battery usable energy is 50% of the total battery energy.

10.12    An engine requires 150 Nm at 750 rpm at the crank shaft. If the pulley ratio of the FEAD system is 2.5, what are the torque and speed requirements of the motor?

## REFERENCES

1. Automobiles and trucks overview: Automobile trends, Plunkett Research, Ltd, 2009. [Online]. Available: from http://www.plunkettresearch.com.
2. A. Emadi, M. Ehsani, and J. M. Miller, *Vehicular Electric Power Systems: Land, Sea, Air, and Space Vehicles*. New York: Marcel Dekker, 2003.
3. S. S. Williamson, S. M. Lukic, and A. Emadi, Comprehensive drive train efficiency analysis of hybrid electric and fuel cell vehicles based on motor-controller efficiency modeling, *IEEE Transaction on Power Electronics*, 21, 730–740, May 2006.
4. P. Moon, A. Burnham, and M. Wang, Vehicle-cycle energy and emission effects of conventional and advanced vehicles, in *Proceedings of the SAE World Congress*, Detroit, MI, April 2006.
5. A. George, (August 28, 2012), *Obama: 54.5 Miles per Gallon by 2025, By Alexander George.* [Online]. Available: http://www.wired.com/autopia/2012/08/2025-mpg-regulation/.
6. C. Holder and J. Gover, Optimizing the hybridization factor for a parallel hybrid electric small car, in *Proceedings of the IEEE Vehicle Power and Propulsion Conference (VPPC)*, Windsor, UK, September 2006, pp.1–5.
7. S. G. Wirasingha and A. Emadi, Pihef: Plug-in hybrid electric factor, *IEEE Transaction on Vehicular Technology*, 60, 1279–1284, March 2011.
8. Bosch sees future requiring multiple powertrain technologies; the larger the vehicle, the more the electrification. [Online]. Available: http://www.greencarcongress.com/2013/06/bosch-20130618.html.

9. USCAR. (January, 3 2014). *Press Release* [Online]. Available: http://www.uscar.org/guest/news/735/Press-Release-USABC-ISSUES-RFPI-FOR-DEVELOPMENT-OF-ADVANCED-HIGH--PERFORMANCE-BATTERIES-FOR-48 V-HEV-APPLICATIONS.

10. S. Baldizzone, Performance and fuel economy analysis of a mild hybrid vehicle equipped with belt starter generator, MASc. thesis, Department of Mechanical Engineering, University of Windsor, Windsor, ON, 2012.

11. M. Arnold and M. El-Mahmoud, Belt-driven starter–generator concept for a 4-cylinder gasoline engine, *AutoTechnology*, 3, 64–67, May 2003.

12. *PSA Peugeot Citroën developing 48 V mild hybrid solution for 2017* [Online]. Available: http://www.greencarcongress.com/2013/01/psa-20130123.html.

13. *ALABC and CPT to introduce 48 V LC Super Hybrid Demonstrator at Vienna Motor Symposium* [Online]. Available: http://www.greencarcongress.com/2013/04/lcsh-20130424.html.

14. *BYD Launches Green Hybrid Initiative across 2014 Vehicle Line-Up.* [Online]. Available: http://evworld.com/news.cfm?newsid=30135.

15. J.-L. Mate, 48 V Eco-Hybrid Systems, in *Proceedings of the European Conference on Nanoelectronics and Embedded Systems for Electric Mobility*, Toulouse, France, September 2013.

16. H. Yilmaz. (September 28, 2012). Bosch Powertrain Technologies. [Online]. Available: http://www1.eere.energy.gov/vehiclesandfuels/pdfs/deer_2012/wednesday/presentations/deer12_yilmaz.pdf.

17. B. Chabot. (August 19, 2013). *Electric Drive—Bosch sees a future trending toward electrified drivetrains.* [Online]. Available: http://www.motor.com/newsletters/20130819/WebFiles/ID2_ElectricDrive.html.

18. D. Aris. (July 1, 2013). *48 V Town and Country Hybrid Powertrain (TC48).* [Online]. Available: http://contest.techbriefs.com/2013/entries/transportation-and-automotive/3940.

# 11 Fundamentals of Hybrid Electric Powertrains

*Mengyang Zhang, Piranavan Suntharalingam, Yinye Yang, and Weisheng Jiang*

## CONTENTS

## 11.1 INTRODUCTION

This chapter provides a comprehensive review on hybrid powertrains and their characteristics. The chapter begins with the history of hybrid powertrains and describes the state-of-the-art technologies in this field. Fundamental components, subsystems, and their limitations are discussed in the second section of this chapter. Regenerative braking concept, blended regenerative braking, and its essential requirements are described in the third section. Finally, very detailed hybrid powertrain control concepts are addressed at the end of this chapter.

## 11.2 INTRODUCTION TO HYBRID ELECTRIC VEHICLES AND HYBRID ELECTRIC POWERTRAINS

Since Prius, the very first mass-production hybrid electric vehicle (HEV), went on sale in Japan in December 1997, HEVs have been under extensive development by many automakers for better fuel efficiency, emissions reduction, and performance improvement in global markets. Though HEV is still in its early adoption phase, and is heavily concentrated in the United States and Japan, Toyota has clearly led the competition so far and has reached very impressive milestones in HEV sales. In 2012, Toyota sold 1.2 million hybrid vehicles globally. As of March 31, 2013, cumulative global sales of its hybrid vehicles topped the 5 million unit mark (Tokyo, April 17 [Reuters]). There has been a strong trend of HEV proliferation, in both hybrid technologies and across vehicle segments. In 2013, 43 hybrid models (2013MY) were offered in the US market by 13 automakers; the models include plug-in hybrids, full hybrids, and mild hybrids with front wheel drive (FWD), rear wheel drive (RWD), and four wheel drive (4WD) configurations, as shown in Table 11.1.

However, the very concept of HEV can be traced back more than a century ago. In 1900, Ferdinand Porsche developed the first gasoline–electric hybrid automobile called Mixte, a series hybrid 4WD vehicle (from Wikipedia, the free encyclopedia). The emergence of HEV in the late 1990s was motivated by the pursuit of cleaner and more efficient vehicles, and was enabled by the advancement in component technologies such as NiMH batteries and compact and efficient electric machines, and advancement in system controls that delivered good drive quality, reliability, and fuel efficiency much desired by customers. Currently, the majority of HEVs employ two distinctive hybrid electric powertrain configurations, that is, power-split hybrid and parallel hybrid. The details of hybrid powertrain configurations will be discussed in later sections; it is suffice to know that a power-split hybrid has a unique electrically variable transmission (EVT) that couples an internal combustion engine (ICE) engine and two electric machines through a planetary gear set, as if the mechanical power from the ICE is split into two paths: one is a direct mechanical path to the output and the other path involves mechanical–electric–mechanical energy conversions through two electric machines before reaching the output. For this reason, the power-split hybrid is also called a series–parallel hybrid. Prius is a power-split hybrid with balanced city and highway fuel efficiency. A parallel hybrid is conceptually simpler and involves only one electric machine installed in a conventional transmission such that the engine torque and the motor torque are additive (or in parallel). Typically, a disconnect clutch is needed to disconnect the engine from the transmission if engine-off is desired. VW Jetta HEV and Hyundai Sonata HEV are examples of parallel hybrids; they usually have good highway fuel efficiencies.

Fuel economy of a passenger vehicle, either in miles per gallon or in liters per 100 km, is derived from the total consumption of a standard fuel over a standard drive cycle on a chassis dynamometer under specified test conditions, known as fuel economy certification tests. In the United States, Urban Dynamometer Driving Schedules (UDDS or FTP74), shown in Figure 11.1a, is used for the city fuel economy test, and EPA highway cycle, shown in Figure 11.1b, is used for highway fuel economy test. For the European fuel economy certification test, New European Driving Cycle (NEDC), shown in Figure 11.1c, is used to measure both city and highway fuel economy. Figure 11.1d shows the Japanese drive test cycle JC08 that will be the new standard for Japan effective from 2015. Fuel economy labels for a typical midsize sedan is about 20 mpg for city and 30 mpg for highway. A typical metric for vehicle performance is the time needed for accelerating a vehicle from still to 60 mph or 100 kph, known as 0–60 time. While 10-s 0–60 time is quite common for many vehicles in the US market, a typical sporty car can achieve 5–8 s, and many performance cars can achieve below 5 s. There is a strong correlation between fuel economy and performance; averagely, the higher the performance, the poorer the fuel economy.

HEVs have achieved significant improvement in fuel economy over conventional vehicles powered by internal combustion engines alone for the same performance; the magnitude of

TABLE 11.1

## List of the U.S. Hybrid Models in 2013MY

| MY | OEM | Division | Vehicle |
|---|---|---|---|
| 2013 | Audi | AUDI | Q5 Hybrid |
| 2013 | BMW | BMW | ActiveHybrid 3 |
| 2013 | BMW | BMW | ActiveHybrid 5 |
| 2013 | BMW | BMW | ActiveHybrid 7L |
| 2013 | Ford Motor Company | Ford | FUSION HYBRID FWD |
| 2013 | Ford Motor Company | Ford | C-MAX Hybrid FWD |
| 2013 | Ford Motor Company | Ford | C-MAX PHEV FWD |
| 2013 | Ford Motor Company | Lincoln | MKZ HYBRID FWD |
| 2013 | Ford Motor Company | Ford | FUSION PHEV FWD |
| 2013 | General Motors | Buick | LACROSSE |
| 2013 | General Motors | Buick | REGAL |
| 2013 | General Motors | Chevrolet | C15 SILVERADO 2WD HYBRID |
| 2013 | General Motors | Chevrolet | TAHOE 2WD HYBRID |
| 2013 | General Motors | Chevrolet | MALIBU |
| 2013 | General Motors | GMC | K1500 YUKON 4WD HYBRID |
| 2013 | General Motors | CADILLAC | ESCALADE 4WD HYBRID |
| 2013 | General Motors | CHEVROLET | VOLT |
| 2013 | Honda | ACURA | ILX HYBRID |
| 2013 | Honda | Honda | CR-Z |
| 2013 | Honda | Honda | CIVIC HYBRID |
| 2013 | Honda | Honda | INSIGHT |
| 2013 | Hyundai | HYUNDAI MOTOR COMPANY | SONATA HYBRID |
| 2013 | Kia | KIA MOTORS CORPORATION | OPTIMA HYBRID |
| 2013 | Mercedes-Benz | Mercedes-Benz | E 400 Hybrid |
| 2013 | Mercedes-Benz | Mercedes-Benz | S400 HYBRID |
| 2013 | Nissan | INFINITI | INFINITI M35H |
| 2013 | Porsche | Porsche | Panamera S Hybrid |
| 2013 | Porsche | Porsche | Cayenne S Hybrid |
| 2013 | Toyota | LEXUS | ES 300 h |
| 2013 | Toyota | LEXUS | GS 450 h |
| 2013 | Toyota | LEXUS | LS 600 h L |
| 2013 | Toyota | LEXUS | RX 450 h |
| 2013 | Toyota | LEXUS | CT 200 h |
| 2013 | Toyota | TOYOTA | PRIUS |
| 2013 | Toyota | TOYOTA | PRIUS Plug-in Hybrid |
| 2013 | Toyota | TOYOTA | HIGHLANDER HYBRID 4WD |
| 2013 | Toyota | TOYOTA | PRIUS c |
| 2013 | Toyota | TOYOTA | AVALON HYBRID |
| 2013 | Toyota | TOYOTA | CAMRY HYBRID LE |
| 2013 | Toyota | TOYOTA | PRIUS v |
| 2013 | Volkswagen | Volkswagen | Touareg Hybrid |
| 2013 | Volkswagen | Volkswagen | Jetta Hybrid |
| 2013 | Fisker | Fisker Automotive, Inc | Fisker Karma |

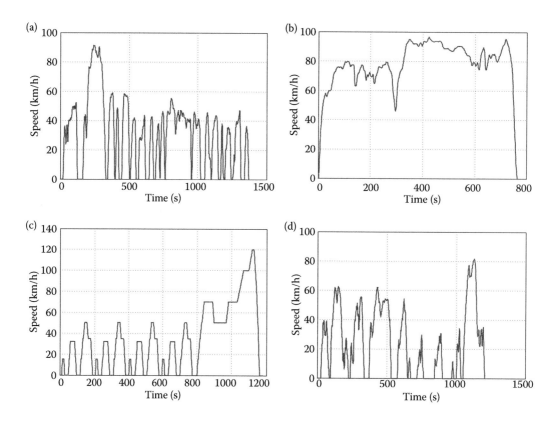

FIGURE 11.1    Various driving cycles. (a) UDDC, (b) HWFET, (c) NEDC, (d) JC08.

the improvements depends on hybrid electric powertrain capability and efficiency, and driving cycles. For example, on the one hand, a strong hybrid such as Prius can almost double the city fuel economy (51 mph) and achieve a modest improvement in highway fuel economy (48 mpg). On the other hand, a mild HEV achieves less fuel economy improvement due to the limited capability in energy recuperation through regenerative braking and due to the limited electric drive capability (refer to the case study in Chapter 13). For some luxury HEV models such as Lexus 600 h and Infiniti 35H, both the performance and the fuel economy are improved significantly. Hybrid electric powertrain is a game-changer technology that enables a new level of performance–fuel economy trade-offs, beyond what a conventional powertrain is capable of. In a conventional vehicle, the more powerful the engine, the less efficient the engine in low-power regions and the higher the fuel consumption in idle, and therefore the lower the fuel economy. Hybrid electric powertrains do not have the same trade-offs, as the designed-in electric capability of the HEV powertrains contributes to both performance by means of electric boost and fuel economy by means of regenerative braking, engine shut-off, and more efficient engine operations. As shown in Table 11.2, EPA fuel economy labels of a few outstanding C/D segment HEVs are truly impressive; these vehicles are not only fuel-efficient but also excellent in comfort and drivability. It is very important for HEVs that regenerative braking and blending, and engine start–stop transitions are smooth and transparent to drivers.

HEVs in the market are generally cleaner in terms of pollutant emissions such as NMHC, $NO_x$, CO, and particulate matters. Some of the HEV models have achieved super low emission standards (SULEV or Bin3), by means of much better cold-phase emission controls and more stable and optimal engine operations. Specifically, a hybrid electric powertrain can help catalyze light-off process and engine fuel controls, and can afford more stable engine operations.

**TABLE 11.2**

**EPA Fuel Efficiency for Selected HEV Models**

| HEV Models | EPA City Fuel Economy (mpg) | EPA Highway Fuel Economy (mpg) |
|---|---|---|
| 2013 Prius C | 53 | 46 |
| 2013 Prius C | 51 | 48 |
| 2013 Prius V | 44 | 40 |
| 2013 Lexus CT 200 h | 43 | 40 |
| 2013 Ford Fusion HEV | 47 | 47 |
| 2013 Honda Civic HEV | 44 | 44 |
| 2013 VW Jetta HEV | 42 | 48 |

In general, HEVs are more expensive than comparable conventional vehicles. High-voltage battery pack and battery management system, electric machines and power electronics, ancillary power module (replacing alternator functions), electric compressor, high voltage (HV) cables, regenerative braking system, and electric vacuum boost are adding cost and weight to HEV. However, fuel saving achieved in vehicle life is indeed significant; a recent Ford Fusion HEV claims 2-year payback that is very attractive to consumers. As the costs of the key components are improving, more HEV penetrations are expected.

Lying at the heart of HEVs are hybrid powertrains and energy storage systems. Before more detailed discussions on the topic, it is appropriate to define the boundary of hybrid electric powertrain within the context of automobiles. A conventional powertrain includes a gasoline or diesel engine (with accessories and a control system), a transmission (with clutches, hydraulics, and a controller), and a drivetrain (final drive, 2WD, 4WD). The primary function of the powertrain is to produce propulsive torque at driving wheels to meet driver's demand and vehicle system demands, by means of regulating engine crank torque and transmission gear selection. Clearly, the ICE is the energy converter to produce mechanical work output by converting fuel chemical energy and the fuel tank is the energy storage. Hybrid electric powertrain can be defined in the same way, spanning from energy sources to mechanical work at the driving wheels, as shown in Figure 11.2. However, hybrid electric powertrain differs from a conventional powertrain in the following aspects:

1. Hybrid electric powertrain interfaces two energy storages, and electric–mechanical energy conversion is bidirectional.

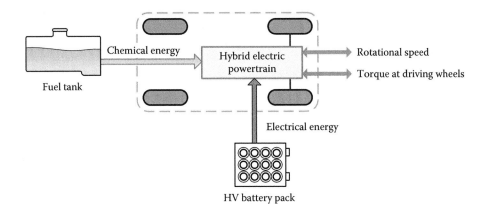

**FIGURE 11.2** Boundary of hybrid powertrain.

2. HV battery is mainly a power source.
3. Significant negative torque can be produced by electric machines (generators) at the driving wheels to decelerate the vehicle; energy recuperation relies on mechanical to electric conversion during regenerative braking.
4. The ICE can be in the off state and on–off transitions are automatic.

These four characteristics are true for all hybrid electric powertrain, regardless of specific powertrain configurations. At the vehicle level, there are four main contributors to HEV fuel economy as follows:

1. *Mechanical to electric energy conversion* when the driver desires to slow down the vehicle by braking. This is achieved by regenerative braking; a coordinated operation of the hybrid powertrain and the brake system that attempts to use less friction brake in general. The hybrid powertrain does the energy recuperation and the brake system applies the friction brake only as needed.
2. *Automatically shutting off the engine* to reduce engine energy loss. There is a significant energy loss associated with engine rotation and pumping, as reflected in engine idle fuel consumption. Shutting off engine eliminated this energy loss. Owing to the same reasons, the engine at very light loads is less efficient; if the hybrid powertrain can use its electric path to provide the propulsion, whole system efficiency is improved.
3. *Operating the engine more efficiently.* This is a hybrid powertrain control function as it can displace engine operating points by controlling engine speed, or engine load, or both. Atkinson cycle technology, as seen in many hybrid vehicles, is also a big contributor to engine efficiency improvement in HEV as it achieves brake-specific fuel consumption (BSFC) of 220 g/kWh.
4. *Reducing vehicle accessory load and road load.* Most of the hybrid vehicles in the markets have reduced the road load by using active grill shutter, more aerodynamic designs, and low rolling friction tires, and using lightweight materials. Active accessory load management also plays a part in reducing overall vehicle power consumptions.

As explained above, points 2 and 3 are hybrid powertrain functions while point 1 is mainly a powertrain function but coordinated by the regenerative braking system. Point 4 is mainly managed at the vehicle level. Before we discuss how a hybrid powertrain of a specific configuration realizes the first three functions, let me introduce three typical hybrid powertrain configurations; they are series hybrid, parallel hybrid, and power-split hybrid.

## 11.2.1 SERIES HYBRID

In a series hybrid, the propulsion is provided by a traction motor or motors; ICE and generator are decoupled from the drivetrain, as shown in Figure 11.3.

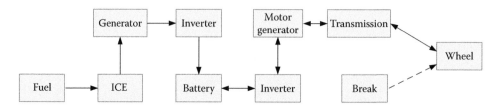

FIGURE 11.3  Series hybrid configuration.

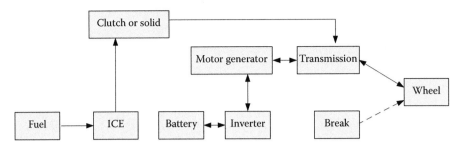

FIGURE 11.4    Parallel hybrid configuration.

## 11.2.2  PARALLEL HYBRID

In a parallel hybrid, both the electric motor and ICE can contribute to the propulsion directly. If the clutch is engaged, the engine torque and the motor torque are additive; in other words, they are paralleled, as shown in Figure 11.4.

## 11.2.3  POWER-SPLIT HYBRID

In a power-split hybrid, an engine and two electric machines are connected to a planetary gear set (PGS) through the carrier, the sun gear, and the ring gear. In Prius and Ford Fusion Hybrid, ICE is connected to the carrier as the input to the transaxle, traction motor (also known as motor B) is connected to the ring gear as the output of the transaxle, and the generator (known as motor A) is connected to the sun gear. There are fundamental properties associated with the power-split device:

1. It is an electrically continuous variable transmission (EVT) that allows the engine speed to be independent of the output speed, which is proportional to the vehicle speed.
2. The transaxle output torque is produced by the motor B torque and a fraction of the engine torque.

Recall that in a series hybrid, the engine torque does not contribute to the output torque, and that in a parallel hybrid, all engine torque is transmitted to the output. Now, it should be clear that power-split hybrid is something between series hybrid and parallel hybrid, controlled by that "fraction" determined by gear ratios as shown in Figure 11.5. It can be designed to achieve a better compromise for city and highway driving cycles than a series hybrid or a parallel hybrid.

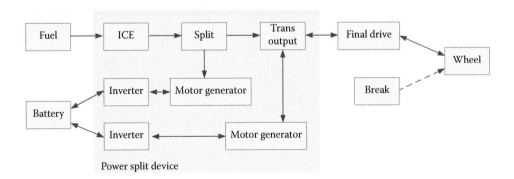

FIGURE 11.5    Power-split hybrid (series–parallel configuration). (Adapted from J. Miller, *Propulsion Systems for Hybrid Vehicles*. IEE Power and Energy Series 45, The Institute of Electrical Engineers, United Kingdom, 2003.)

Now, we discuss how these three basic hybrid configurations realize three main powertrain functions, including regenerative braking, engine start–stop, and engine operating point's optimization. To be brief and right to the points, pros and cons are organized in a tabular form in Table 11.3; detailed control functions are left out.

### 11.2.3.1 Case Study

Comparisons between a series hybrid, a parallel hybrid, and a power-split hybrid are carried out by using powertrain simulations software Autonomie, see Appendix. All the powertrains are assigned with the same vehicle masses, aerodynamic parameters, tire parameters, and the same road grades, as shown in Table 11.4. All the vehicles operate over one UDDS driving cycle and one HWFET driving cycle to simulate city driving and highway driving, respectively.

Different control strategies are applied to the three types of hybrids. In the series hybrid, load following control strategy is applied since the motor provides all the power and torque demanded from the road. Engine provides the power and runs the generator to regulate the battery state-of-charge (SOC) at a certain constant level. In the parallel hybrid, rule-based control strategy is applied to regulate the motor turn-on and turn-off time power assistance and regenerative braking. In the power-split hybrid, lower fuel consumption control strategy combined with performance propelling control is applied to achieve the best engine operating regions. For all three types of

**TABLE 11.3**

**Pros and Cons of Three Hybrid Configurations**

|  | Series Hybrid | Parallel Hybrid | Power-Split Hybrid |
|---|---|---|---|
| Energy recuperation by regenerative braking | Accomplished by the traction motor. Sufficient regen capability | Accomplished by the electric motor. Limited regen capability because of motor and battery capability. Motor speed compromised by gear ratios | Accomplished by the electric motor. Limited regen capability because of battery capability |
| Engine start–stop | Can shut off ICE at any vehicle speed and load. Controlled by the generator. Decoupled start–stop controls with high quality | Usually requires a disconnect clutch. Can shut off ICE at any vehicle speed and load. Without a starter motor, engine start is challenged and may impact drivability | Can shut off ICE below certain vehicle speed because of transaxle kinematic constraints. Controlled by both electric machines. Good start–stop quality |
| Engine operating point optimization | Can optimize in both speed and torque. Can optimize with ICE technology and generator technology | Can optimize in torque (load leveling). Complicated with gear selection and shift controls | Can optimize in both speed and torque, may compromised by electric losses. Clutchless CVT for smooth operations |
| Overall strength and weakness | Good for very transient drive cycle. Compromised efficiency in highway cruise or steady state because of double energy conversion (mechanical–electric–mechanical). Higher requirement on the traction motor. Higher requirement on battery power | Good for highway and steady state. Compromised fuel economy in transient driving cycles. Lower requirement on the traction motor. Lower requirement on battery power. More cost-effective. Trade-off between efficiency and drivability | Balanced city and highway fuel economy be design. Compact and no major disruption to vehicle architecture. Reasonable requirements on motors and battery pack. Excellent drive quality. Fuel economy degradation for very aggressive driving styles |

### TABLE 11.4

### Simulation Input Parameters Comparison

| Components | Parameters | Series Hybrid | Parallel Hybrid | Split Full Hybrid | |
|---|---|---|---|---|---|
| Chassis | Mass (kg) | 1850 | 1850 | 1850 | |
| | Front weight ratio | 0.64 | 0.64 | 0.64 | |
| | Center of gravity height (m) | 0.5 | 0.5 | 0.5 | |
| Wheel | Radius (m) | 0.30 | 0.30 | 0.30 | |
| | Rolling resistance coefficient | 0.008 | 0.008 | 0.008 | |
| Engine | Base model | 1.5 L 57 kW 04Prius | 1.5 L 57 kW 04Prius | 1.5 L 57 kW 04Prius | |
| | Scaled maximum power (kW) | 80 | 80 | 80 | |
| Electric machine | Base model | 25/50 04Prius | 25/50 04Prius | 15/30 04Prius | 25/50 04Prius |
| | Scaled maximum power (kW) | 120 | 25 | 51 | 63 |
| Gearbox | Type | Fixed gear | Automatic | Planetary | |
| | Gear ratio | 1 | 3.45, 1.94, 1.29, 0.97, 0.75 | 2.6 | |
| Final drive | Final drive ratio | 4.438 | 3.63 | 3.93 | |
| Mechanical accessory | Power (W) | 10 | 10 | 10 | |
| Electrical accessory | Power (W) | 217 | 217 | 217 | |

the configurations, regenerative braking is only allowed below the chassis deceleration speed of 2 m/s$^2$, above which only conventional friction braking is used.

Table 11.5 summarizes the fuel economy and engine efficiency comparison results based on the above-described control strategies and the two driving cycles. It should be noted that different control strategies and different driving cycles will result in different fuel consumption results. For a given vehicle and an engine, the fuel consumption results depend on component characteristics such as gear ratios and electric machine loss maps, and the control strategies. Thus, these comparison simulations are intended to elaborate the potential trade-off between electric losses and engine efficiencies for three basic hybrid electric configurations. Gear ratios, subsystem loss maps, engine efficiency maps, and control strategies could be quite different from the ones in real production HEVs.

Figure 11.6 shows the comparison results of the power demanded and the power compositions in each hybrid configuration under UDDS driving cycle from time 0 to 500 s. In a series hybrid, the motor provides all the traction power demanded from the powertrain while the engine charges the battery to power the motor. The engine is regulated to operate at higher efficiency points to improve the fuel economy. In a parallel hybrid, both the engine and the motor provide the traction power;

### TABLE 11.5

### Simulation Result Comparisons

| | UDDS | | | HWFET | | |
|---|---|---|---|---|---|---|
| Performance | Series Hybrid | Parallel Hybrid | Power-Split Hybrid | Series Hybrid | Parallel Hybrid | Power-Split Hybrid |
| Distance (mile) | 7.42 | 7.44 | 7.44 | 10.25 | 10.25 | 10.25 |
| Fuel consumption rate (g/s) | 0.37 | 0.40 | 0.31 | 0.98 | 0.92 | 0.80 |
| Motor/inverter loss (Wh) | 810.4 | 103.5 | 381.5 | 837.15 | 34.5 | 438.6 |
| Engine average efficiency (%) | 33.56 | 26.25 | 32.79 | 35.27 | 27.46 | 34.48 |

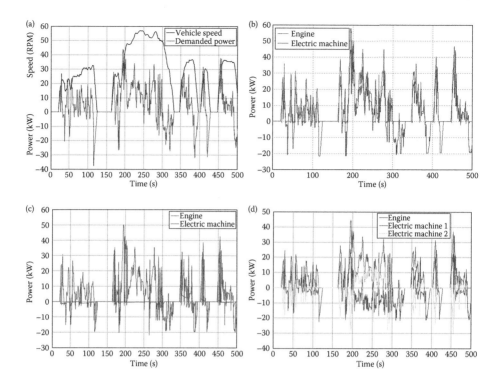

**FIGURE 11.6** Power comparisons under UDDS driving cycle. (a) Demanded power, (b) series hybrid, (c) parallel hybrid, (d) power-split hybrid.

however, the engine does not operate in the fuel optimal region since it cannot be decoupled from the final drive, as shown in Figure 11.7b. In a power-split hybrid, the traction power comes from the engine and both the motors. Both the motors can either serve as motor or generator, and the engine is regulated at its fuel optimal operating regions and thus the fuel efficiency is improved. In addition, engines in all these three types of hybrids can be shut off during idling or braking.

Figure 11.7 compares the engine operation points under UDDS driving cycle for the series, parallel, and power-split hybrid. It can be observed that the engine operates in high-efficiency regions for both the series and power-split hybrid, while the parallel hybrid cannot offer this benefit since the engine is always coupled with the output shaft.

Figure 11.8 shows the comparison results of the power demanded and the power compositions in each hybrid configuration under HWFET driving cycle. The motor in the series hybrid still provides all the traction power needed at the driving wheels and retrieves regenerative braking energy. In the parallel hybrid, the engine provides the majority of traction power while the motor assists the power performance occasionally and retrieves the regenerative braking energy. In the power-split hybrid, the engine is still regulated by the two motors, with one mainly serving as a motor and the other serving as a generator. The power-split hybrid has slightly lower fuel efficiency on highways due to higher losses on the electric energy path. Shown in Figure 11.8, when the vehicle decelerates at 280 s and 730 s, the engine shuts down and the regenerative braking comes into play.

### 11.2.3.2 Constant-Speed Cruise Simulation

Table 11.6 summarizes the fuel economy and engine efficiency comparison results for constant-speed cruise simulation for the three types of hybrid configurations. The vehicle speed starts from 0 to 20 m/s and reaches 28.89 m/s (45 miles/h and 65 miles/h) within the first 150 s and then keeps

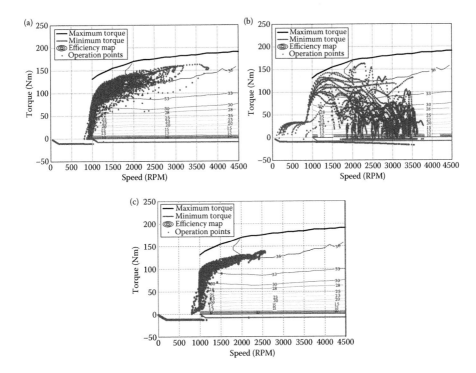

FIGURE 11.7    Engine operation points under UDDS driving cycle comparisons. (a) Series hybrid, (b) parallel hybrid, (c) power split hybrid.

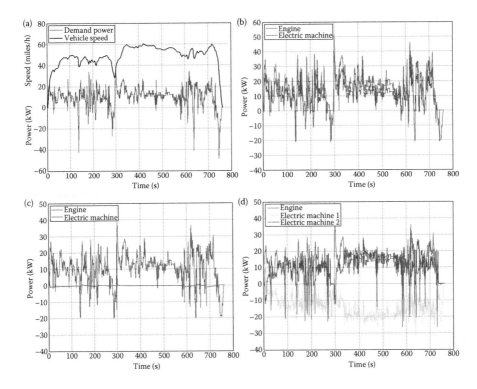

FIGURE 11.8    Power comparisons under HWFET driving cycle. (a) Demanded power, (b) series hybrid, (c) parallel hybrid, (d) power-split hybrid.

TABLE 11.6

**Steady-State Simulation Result Comparisons**

| Performance | Series Hybrid | | Parallel Hybrid | | Power-Split Hybrid | | | |
|---|---|---|---|---|---|---|---|---|
| Vehicle speed (miles/h) | 45 | 65 | 45 | 65 | 45 | | 65 | |
| Engine speed (RPM) | 1092 | 1750 | 2233 | 2493 | 1000 | | 1683 | |
| Engine torque (Nm) | 96 | 123 | 33.5 | 65.4 | 94 | | 114 | |
| Engine power (kW) | 11.0 | 22.5 | 7.8 | 17.1 | 9.9 | | 20.2 | |
| Engine average efficiency (%) | 34.26 | 35.9 | 24.04 | 31.37 | 33.4 | | 35.5 | |
| Fuel consumption rate (g/s) | 0.71 | 1.39 | 0.74 | 1.20 | 0.62 | | 1.21 | |
| Electric machine speed (RPM) | 4502 | 6503 | 2233 | 2493 | 2492 | −2880 | 3600 | −3298 |
| Electric machine torque (Nm) | 16 | 25 | −1 | −0.9 | −39 | −26.2 | −38 | −32 |
| Electric machine power (kW) | 7.5 | 17.0 | −0.2 | −0.2 | −10.3 | 7.9 | −14.3 | 11.0 |
| Electric machine power loss (kW) | 2.89 | 4.65 | 0.04 | 0.04 | 0.94 | 1.24 | 1.35 | 1.74 |

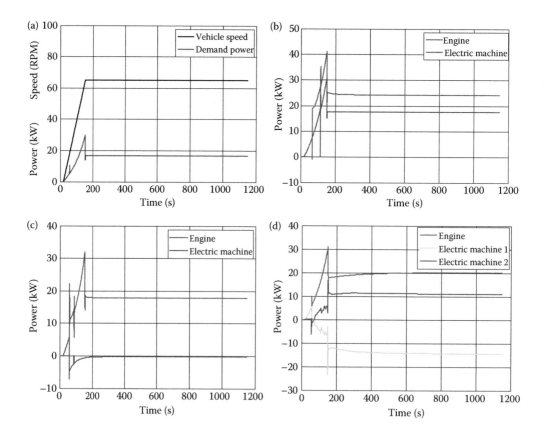

FIGURE 11.9  Power comparisons under steady state (65 mph). (a) Demand power and speed, (b) series hybrid, (c) parallel hybrid, (d) power split hybrid.

at the speed thereafter. Figure 11.9 shows the comparison results of the power demanded and the power from each component.

It can be observed that in series hybrid, the motor provided all the power required from the road load while the engine operates to charge the battery. In the parallel hybrid, the engine provides all the power at constant cruising speed. In the power-split hybrid, the two electric machines regulate the engine speed and power in its fuel optimum regions by using one serving as a generator and the other serving as a motor. Both high engine efficiency and high vehicle fuel economy were achieved in the power-split hybrid configuration.

Comparing the engine power, the series HEV requires the highest (22.5 kW) due to the power losses incurred in two power conversions; lowest engine power (17.1 kW) is required for the parallel HEV because of much lower power loss; and engine power required for the power-split hybrid is in the middle (20.2 kW). Second, compare the engine efficiency, which depends on engine operating points for a given power output and engine technology. The series HEV and the power-split HEV have a relatively higher engine efficiency compared to the parallel HEV, because in parallel HEV, the engine is always coupled with the transmission output shaft and thus the operating points are not regulated in the fuel optimal regions.

In sum, all three types of HEV configurations have their unique advantages in certain applications. The series hybrids offer the highest engine efficiency due to the exclusively decoupling of engine from the transmission output drive shaft. They also have higher fuel economy on city drives than highways because of more frequent implementation of regenerative braking. However, series HEVs have significant losses in the power conversions from the mechanical power to electric power and back to mechanical power again. The parallel hybrids have relatively smaller electric systems on par with conventional ICEs. The fuel economy is higher on highways because engine operates with a higher efficiency and the electric machine assists with the power performance. The electric machine also functions with certain degrees of regenerative braking to retrieve the kinetic energy back to the battery. However, the fuel economy and the engine efficiency of parallel hybrids on city drives are compensated due to the low efficiency of the engine. The power-split hybrids combine the benefits of both the series hybrids and parallel hybrids to offer a balanced solution. Satisfactory engine operating efficiency and vehicle fuel economy are achieved on both city drives and highway drives. Both engine speed and torque can be decoupled from the transmission output shaft. Less power are lost in the power conversions due to the integration of mechanical power and electric power in power-split devices. However, the power-split hybrids have slightly lower fuel economy on highways than in cities because of less regenerative braking opportunities and more losses in electric machine and inverter at higher speeds.

## 11.3  INTRODUCTION TO HYBRID POWERTRAIN COMPONENTS

### 11.3.1  INTERNAL COMBUSTION ENGINES FOR HEV

For a typical HEV, the energy capacity of the HV battery pack is very limited, about 1–2 kWh. Only 30% of it is used as an energy buffer, equivalent to a range of two miles in the electric mode. Therefore, over a practical driving distance, the internal combustion engine provides all the mechanical work to be consumed by the vehicle; the average engine efficiency plays a significant role in HEV fuel economy.

For a given ICE, the efficiency is not a constant. It depends on the engine operating point in terms of engine speed and engine torque, and it also depends on engine operating conditions that may affect the engine friction, the spark timing, valve timings, and fuel air ratio. Figure 11.10 shows the general characteristics of engine operations across speed–torque domains. Under nominal conditions, the engine minimum fuel consumption for a brake power output varies with the engine output power, and the engine efficiency peaks in the medium power range; usually, this power level is

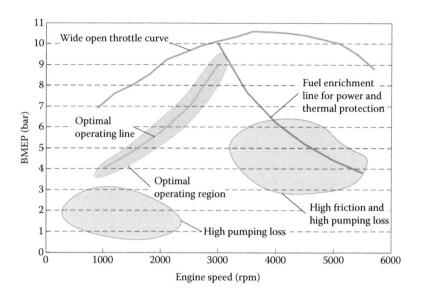

FIGURE 11.10   Optimal engine operating region.

higher than what is desired for fuel economy. When the engine output power (brake power) is too low or too high, the efficiency is lower. When an engine idles with zero mechanical power output, the efficiency is zero, the very reason that an HEV will shut off the engine and use electric mode. The essence of this is shown in Figure 11.11.

Vehicle acceleration performance such as 0–60 time is heavily influenced by the peak power of the powertrain. For most of the parallel hybrids and power-split hybrids, engine peak power is still the first design factor for performance. In a midsize hybrid, the peak battery discharge power is

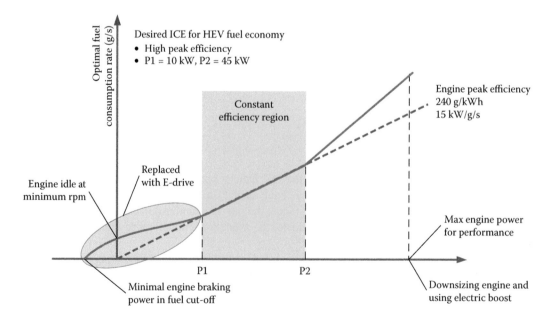

FIGURE 11.11   ICE characteristics important to HEV efficiency.

likely in the range of 20–35 kW, while the peak engine power is probably rated at 80–120 kW. For series hybrids, the peak powertrain power can be as high as the sum of peak battery power and the electric power generated by the internal combustion engine.

The ICE peak power, peak efficiency, and the efficiencies in the low-power region are all important for designing an HEV to meet the fuel economy and performance objectives. ICE improvements for HEV applications have been concentrated in these areas. "Atkinsonized" engines have achieved remarkable successes in increasing the peak efficiency (below 220 g/kWh BSFC) and in moving the high-efficiency region toward low-power operations at the expense of peak engine power. These engines have high compression ratios (12–13), late intake valve closing and late exhaust opening, and other features. It is worth noting that engine optimization for HEV applications is a complex systems engineering and the achievements are impressive.

Combustion engines emit hydrocarbons (HC), $NO_x$, carbon monoxide (CO), and particulates out of tailpipes. HC can also be produced from evaporative processes. HEVs must meet tailpipe emission standards and evaporative emission standards. Many recent hybrid vehicles have met California SULEV (super low emission vehicle) standard, shown in Table 11.7. These SULEV HEV are indeed very clean, resulting from precise engine and emission controls by hybrid powertrains. There are a few unique aspects of emission controls in an HEV. In a power-split HEV, because of EVT and available battery power, catalyst light-off process can be much better controlled through spark, air, and fuel controls in the cold phase, where the majority of tailpipe emission is produced because of poor conversion efficiency of the cold catalyst. Second, engine in HEV operates in much less transient than in a conventional vehicle; the combustion is better controlled, with less pollutants present in engine exhaust gases. Since HEV starts and stops the engine quite frequently, there are certain challenges in these transient emission controls; nevertheless, through proper fuel enrichment and closed-loop fuel controls along with spark controls, very satisfactory results have been achieved. Maintaining catalyst substrate temperature and precise oxygen level are also critically important; they are achieved with good designs and controls based on real-time catalyst state estimations.

Engine torque control performance is very important to hybrid powertrain controls as it is a torque-based control. Generally speaking, HEV controls require more stringent engine torque controls than a typical conventional vehicle. Engine torque accuracy can affect battery power control accuracy, which can be crucial under extreme conditions. It can also affect drivability under certain conditions. Engine controls by its own right is a very rich and challenging field, because the success requires coordination of many device controls. A framework for modern engine controls is shown in Figure 11.12.

In summary, engine sizing, design, engine technologies, and engine control must be considered along with hybrid powertrain functions and controls for achieving vehicle-level objectives in fuel economy, emission, performance, and drivability. It is the interactions between the powertrain level and subsystem level that make hybrid powertrains very interesting and yet challenging, particularly in the real world with a wide range of customer preferences and vehicle operating conditions.

TABLE 11.7

**California LEVII, FTP, PC and LDV < 8500 lb, g/mile, 50 K (120 K)**

|  | NMOG | CO | NOX | PM | HCHO |
|---|---|---|---|---|---|
| LEV | .075 (0.090) | 3.4 (4.2) | .05 (.07) | (0.01) | .015 (.018) |
| ULEV | .04 (0.055) | 1.7 (2.1) | .05 (.07) | (0.01) | .008 (.011) |
| SULEV | (0.01) | (1.0) | (0.02) | (0.01) | (0.004) |

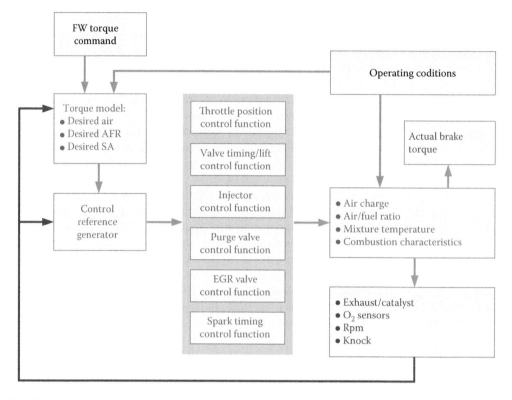

FIGURE 11.12   Basic gasoline engine torque control framework.

### 11.3.2   ELECTRIC MACHINES FOR HYBRID ELECTRIC POWERTRAINS

Interior permanent magnet synchronous machines (IPMSM or IPM) are dominant in modern hybrid electric powertrains, because of their high power density and torque density, high efficiency, and superior torque control performance. 2010 Prius transmission houses two IPM machines (mentioned in the previous sections as motor A and B); the designs and the benchmarking performance have been investigated by many researchers. Table 11.8 shows some published specifications and performances.

---

TABLE 11.8

**2010 Prius Motor Specifications**

| **Motor (B)** | **Generator (A)** |
|---|---|
| Interior PMSM | Interior PMSM |
| Distributed winding | Concentrated winding |
| 60 kW peak at 650 V | 42 kW peak at 650 V |
| 207 Nm peak | Stator weight: 8.6 kg |
| Maximum speed: 13,500 rpm | Rotor weight: 4 kg |
| Peak efficiency: 96% | High specific power: 3.3 kW/kg |
| Stator weight: 16 kg | |
| Rotor weight: 6.7 kg | |
| High specific power: 2.6 kW/kg | |
| High specific torque: 9.1 Nm/kg | |
| Volume: 15 L | |

---

Electric machines play two critical roles in a hybrid electric powertrain and controls: electric mechanical energy conversion and system controls. Typical hybrid powertrain controls are torque based; the intended propulsion torque and desired battery power are achieved through desired motor torque command and engine torque command to be realized by motor torque controls and engine torque controls. Here electric motors and the engine are used as torque actuators to control engine speed, HV battery power and voltage limit control, regenerative braking process, and engine start–stop controls. Some of the high bandwidth control functions such as active driveline damping controls rely on fast responses of motor torque controls, therefore motor torque control performances are very important. Optimal hybrid powertrain operations, and thus the efficiency, are heavily influenced by the efficiencies of electric machines in the hybrid electric powertrain.

There are five aspects of electric motor operations that require more considerations for hybrid powertrains. They are discussed briefly as follows:

1. *Wide speed range.* In a power-split hybrid powertrain and series hybrid powertrain, the speed of the traction motor (Motor B) is proportional to the vehicle speed. The maximum motor B speed of 2010 Prius is 13,500 rpm, and the corresponding vehicle speed is about 110 mph. Torque control performance near zero speed largely determines vehicle hill-hold performance and drive quality when the vehicle creeps. Motor torque smoothness and proper thermal derating performance are particularly important. On the other hand, motor torque control performance and motor efficiency near maximum speed affect vehicle's high-speed drive quality and acceleration performance, and thus deserve more attention than typical industrial applications where motor speed range may be much narrower.

2. *Wide torque range.* Motor torque demands by the hybrid powertrain operations are quite dynamic and both motoring and regenerating performance (and efficiency) are important for powertrain controls and HEV drive quality and efficiency. During aggressive accelerations such as 0–60 mph acceleration, the traction motor likely operates near the maximum torque envelope for 10 s or so; stable motor controls and high motor efficiency must be maintained. During aggressive deceleration, the traction motor instead operates near the minimum torque envelope; therefore, the regenerating performance (and efficiency) directly impacts HEV regenerative braking performance and efficiency. In most HEVs, engine start–stop transitions are very transparent and seamless to drivers, and are achieved by precise controls relying on motor controls. Motor A (or the generator) provides initial positive torque with position dependence to spin up the engine smoothly and provide negative torque when the engine achieves stable operations. This transition is usually completed within 500 ms, and the motor A torque control plays a major role in the process.

3. *Dynamic DC bus voltage.* As HV battery power varies dynamically to meet the vehicle operations, the battery voltage changes dynamically, primarily due to the battery internal resistance that may depend on cell temperatures and cell SOC. As the internal resistance increases significantly at low temperature and at very low or very high SOC, much larger bus voltage variation can be expected. This imposes certain challenges in motor controls and powertrain controls, because motor maximum power, efficiency, and even motor control methods may depend on the DC bus voltage. In some of the popular HEVs such as Prius and Ford Fusion, the DC bus voltage is actively controlled at the output of a boost converter to improve powertrain performance and efficiency. In 2010 Prius, the DC bus voltage is boosted from 200 V (HV battery voltage) to as high as 650 V during heavy acceleration and deceleration, and engine starts, as shown in Figure 11.13. Ford Fusion HEV boosts the DC voltage to 400 V during accelerations.

4. *Precise motor torque controls.* Performances and efficiencies of hybrid electric powertrains depend on the accuracy and bandwidth of motor torque controls. Let us first examine how motor torque accuracy will impact battery power controls. In a power-split hybrid

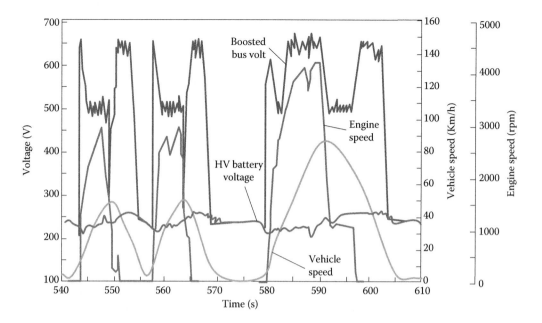

FIGURE 11.13  Dynamic bus voltage boost of 2010 Prius.

powertrain, both the motor power and the generator power contribute to the HV battery power that must be controlled, as described by Equation 11.1:

$$P_{batt} = \omega_A * T_A + \omega_B * T_B + P_{lossA}(\omega_A, T_A, V) + P_{lossB}(\omega_B, T_B, V) \tag{11.1}$$

If we wish to control $P_{batt}$ within 5 kW in transients of 500 ms, when both $\omega_A$ and $\omega_B$ are at about 5000 rpm, then the transient torque errors of motor A and motor B should be within 5 Nm.

A rule of thumb is that the motor torque error should be less than 5% or 5 Nm. Dynamic responses in terms of rise time, settling time, and overshoot are equally important because motor torques are the controls for many system-level functions such as engine start–stop controls, engine speed controls, driveline damping controls, and battery power and voltage controls. As two motors are subjected to a wide range of operating conditions, satisfactory motor torque controls require more and more sophisticated controls and extensive calibrations, very challenging in their own rights.

It is often uncounted in real applications that noise, vibration, and harshness (NVH) issues under certain operating conditions require specific treatments in motor controls to address phase current waveforms, torque ripples, and even pulse width modulation (PWM) techniques. Smooth and humming-free powertrain operations are highly desired. Safety requirement of hybrid powertrains also impose requirements on motor torque monitoring and diagnostics, generally known as motor torque security and functional safety practices.

5. *Real-time motor information for hybrid controls.* Hybrid electric powertrain controls need real-time motor information such as speed, torque, DC voltage, maximum and minimum torque envelopes, temperatures, and mode indicators; correct information per specifications is essential for the powertrain controller to make correct decisions about the operations and even shutdowns for safety reasons. System power-up and power-down sequence typically requires particular information and execution orders from all critical components,

including motors. Quality hybrid powertrains are possible only through proper system design and precise information exchange in real time; correct information of electric motor states and capabilities are essential.

### 11.3.3  2010 TOYOTA PRIUS TRANSAXLE

2010 Prius transaxle has a very compact design. It includes two electric machines (the generator is also known as motor A and MG1; the traction motor is also known as motor B or MG2): the first being the PGS for power-split and the second being the PGS for connecting the traction motor to the ring of the first PGS, a torsional damper, oil pump, and final drive gear set, as shown in Figure 11.14a [5]. Speed and static torque relationships for this transaxle are given as follows:

$$\omega_e = 0.2778\omega_A + 0.7222\omega_r \tag{11.2}$$

$$\omega_B = 2.636\omega_r \tag{11.3}$$

$$T_r = 0.7222T_e + 2.636T_B \tag{11.4}$$

$$T_A = -0.2778T_e \tag{11.5}$$

$$\omega_r = 3.267\omega_{fd} \tag{11.6}$$

$$T_{fd} = 3.267T_r \tag{11.7}$$

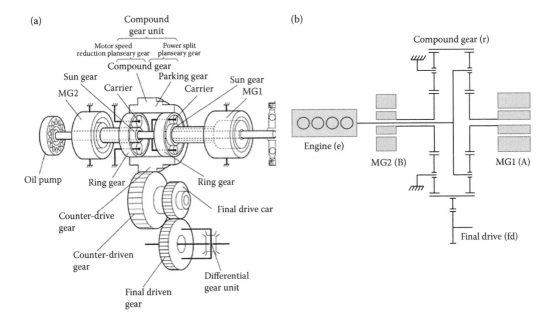

**FIGURE 11.14**  2010 Toyota Prius transaxle. (a) Comprehensive unit. (b) Simplified diagram. (From https://techinfo.toyota.com/t3Portal/document/ncf/NM14C0U/xhtml/RM0000042WY003 (accessed: 17 February 2010).)

It is evident that 72% of engine torque contributes to the output torque at the ring. It is also interesting to work out the mechanical powers of motor A and B as follows:

$$P_A = -P_e + 0.7222\omega_r T_e \tag{11.8}$$

$$P_B = P_r - 0.7222\omega_r T_e \tag{11.9}$$

Or they can be rearranged to reveal the power-split, the mechanical path, and the electric path involved in this particular hybrid transmission.

$$P_e = -P_A + 0.7222\omega_r T_e \tag{11.10}$$

$$P_r = P_B + 0.7222\omega_r T_e \tag{11.11}$$

It is clear that the engine power is split into two parts, motor A power and the power term $0.7222$ $\omega_r T_e$ that is the direct mechanical power transferred from the engine to the output (mechanical path). At the output (the ring), this mechanical power and motor B power are combined to form the output mechanical power. The electric path includes generator–battery–motor, and it plays a dominant role when vehicle power demand is high (either positive or negative). On the one hand, for steady-state driving at highway speeds, the mechanical path dominates; therefore, it is fairly efficient for highway fuel economy. On the other hand, for transient driving at low speeds such as city driving, the electric path dominates, and its efficiency becomes important. Prius Transaxle design has achieved a good trade-off between city cycle fuel economy and highway cycle fuel economy.

Cooling and lubrication are extremely important to this compact transaxle; they are thoroughly designed to minimize the churning losses and yet provide sufficient cooling and lubrication to all parts, including motors, gears, and bearings. Oil feed and return for cooling and lubrication are facilitated by various orifices and passages. In the hybrid mode, a mechanical pump driven by the engine supplies the oil for lubing and cooling. In the EV mode, the final drive moves the oil from the sump into an oil catch tank; cooling and lubing are facilitated through a different set of orifices and passages, as illustrated in Figure 11.15 [5]. It is worth mentioning that the engine coolant can be used to warm the cold transmission oil quickly for reducing the viscous loss of the oil, and thus improving fuel economy.

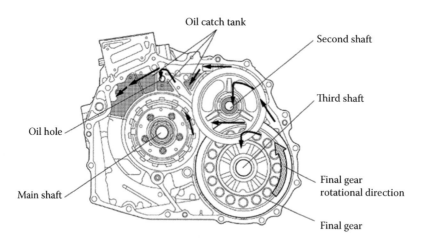

**FIGURE 11.15** Cooling and lubrication in EV mode. (From https://techinfo.toyota.com/t3Portal/document/ncf/NM14C0U/xhtml/RM0000042WY003 (accessed: 17 February 2010).)

### 11.3.4 SUMMARY OF HYBRID ELECTRIC POWERTRAINS

We have briefly discussed hybrid powertrain configurations and key components. It is important to recognize that the hybrid electric powertrain is just one of the automotive powertrain technologies that span over multiple energy sources and energy carriers and the degree of electrifications, as shown in Figure 11.16. Although hybrid electric powertrains focus on improving powertrain efficiency, the underlying technologies are relevant to many other powertrains, and advancement in hybrid powertrains, in system controls and component technologies, will benefit the development of more electrified powertrains.

As the fuel economy benefits continue to be the primary driver for HEVs, it is proper to dedicate efforts to discuss the physics of HEV fuel efficiency at a system level without having to resort to drive cycle simulations. Figure 11.17 provides a framework for analyzing powertrain efficiencies. The power relationship is given by Equation 11.12:

$$P_{supply} - P_{loss} - P_{acc} = v(A + Bv + Cv^2) + mv\dot{v} \qquad (11.12)$$

where A, B, and C are the vehicle road load coefficients that describe static and rolling frictions and aerodynamic drags. $P_{supply}$ is the total power supplied by the powertrain, $P_{acc}$ is the accessory power needed for vehicle operations and is usually provided by a 12-V system, and $P_{loss}$ is the total powertrain and vehicle system losses that include powertrain mechanical losses, electric–mechanical conversion losses, HV battery loss, and brake power loss due to friction brakes. $v$ and $\dot{v}$ are the velocity and acceleration/deceleration rate of the vehicle. By integrating the power conservation equation over a driving cycle (with zero initial and terminal vehicle speeds), we obtain the energy conservation Equation 11.13.

$$E_{supply} - E_{loss} - E_{acc} = \int_{0}^{t_f} v(A + Bv + Cv^2)dt \qquad (11.13)$$

where $E$ is the total mechanical energy produced by the engine; it can be linked to fuel consumption through average engine efficiency.

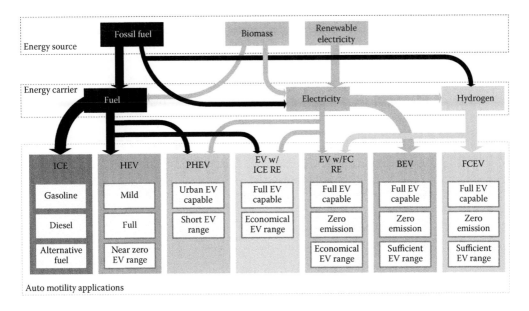

FIGURE 11.16    Automotive powertrains for tomorrow.

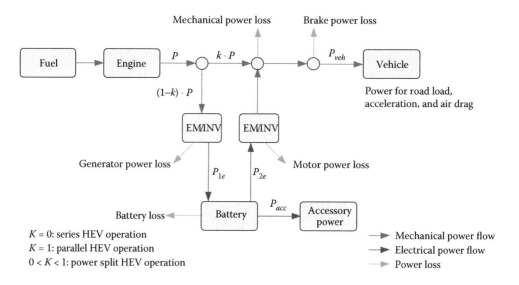

**FIGURE 11.17**   Simplified HEV power flow model.

Now we can define the powertrain efficiency as

$$
\text{Powertrain efficiency} = \frac{\left( E_{acc} + \displaystyle\int_0^{t_f} v(A + Bv + Cv^2)dt \right)}{E_{supply}} = \frac{E_{used}}{E_{supply}} \tag{11.14}
$$

where $E_{used}$ is the total amount of energy consumed in road load and accessory load over a driving cycle. Figures 11.18 and 11.19 show the powertrain efficiency of a few recent HEV models for city cycle and highway cycle, respectively. Power-split HEV powertrain achieves the powertrain efficiency of near 20% for the EPA city cycle. By the same measure, parallel hybrid powertrains

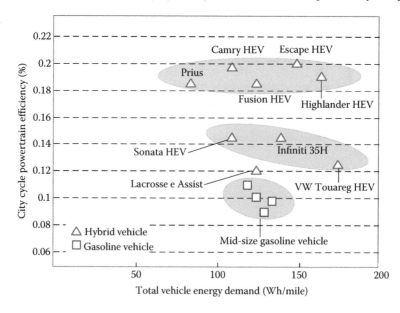

**FIGURE 11.18**   Hybrid powertrain efficiency in EPA city cycle.

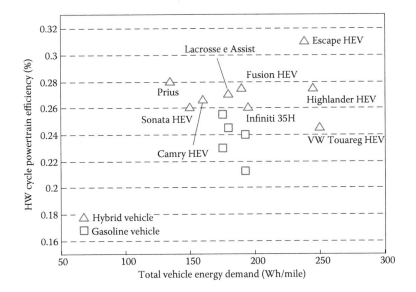

FIGURE 11.19 Hybrid powertrain efficiency in EPA highway cycle.

achieve the efficiency of about 15%, and that for conventional gasoline vehicles hovers around 10%. Highway cycle efficiency is interesting; on an average, hybrid powertrain efficiency is 15% higher than that of gasoline powertrains, but they clearly overlap. Ford Escape HEV stands out with a powertrain efficiency of 31%.

Hybrid powertrain efficiency is reduced for more transient and aggressive driving cycles, as shown in Figure 11.20. Both Prius and Ford Fusion HEV show similar efficiency degradation caused by driving aggressiveness in terms of the ratio of vehicle kinetic energy over vehicle road load energy demand. Consumer Report City Cycle (CR city) is the most aggressive test cycle; the powertrain efficiency for CR city is the lowest (about 10%), almost a half of the EPA city efficiency. The efficiency degradation is mostly due to limited regenerative capability and more losses are associated with high

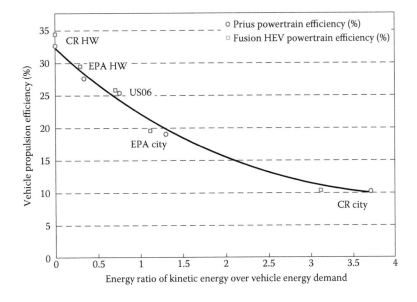

FIGURE 11.20 Hybrid powertrain efficiency degradation.

power during accelerations. It can be expected that a more powerful and efficient electric path in the hybrid powertrain improves the powertrain efficiency for aggressive driving cycles.

It is an interesting exercise to estimate the total powertrain energy loss and the average engine efficiency from measured Camry HEV fuel economy tests. Assume that the total loss is related to the total energy throughput by the following equation:

$$L = (E_p + abs(E_n))(1 - \eta_{tm} + K(1 - \eta_e)) \tag{11.15}$$

where $E_p$ is the supplied energy of the vehicle to overcome the vehicle road load $\left(\int_{t_s}^{t_f} v(A + Bv + Cv^2)dt\right)$ when the vehicle is in acceleration or constant-speed driving mode, that is, $\dot{v} \geq 0$.

$E_n$ is the supplied energy of the vehicle to overcome the vehicle road load $\left(\int_{t_s}^{t_f} v(A + Bv + Cv^2)dt\right)$ when the vehicle is in deceleration mode, that is, $\dot{v} < 0$.

$\eta_{tm}$ is the efficiency assigned to the mechanical path and $\eta_e$ is the efficiency assigned to the electric path. $K$ accounts for the fact that the electric path energy throughput is less than the total energy throughput. Now the total mechanical energy provided by the engine is obtained as

$$E_{ice\_mech} = E_{acc} + \int_{t_s}^{t_f} v(A + Bv + Cv^2)\, dt + L \tag{11.16}$$

And the total fuel consumption over the cycle is given by $E_{ICE}mech/\eta_{ICE}/33{,}700$ ($E_{ICE}mech$ is in Wh and 33,700 is the gasoline energy density in Wh/gallon, and $\eta_{ICE}$ is the average engine efficiency). Figure 11.21 presents 2012 Camry HEV city and highway fuel economy test results as measured unadjusted mpg. "bag3" and "bag4" are the first portion of UDDS, namely, hot 505, and the second portion of UDDS, namely, hot transient that lasts 867 s. 10% SOC difference (final SOC—initial SOC) was observed for bag3, and −10% SOC difference was observed for bag4; their effects on engine mechanical energy output are corrected. It should be kept in mind that this example is only for showing how the loss and engine efficiency may be estimated from measured fuel consumption and known vehicle road load coefficients. The estimated average engine efficiency over the cycle is as low as 30%, and there are significant losses associated with electric path involving mechanical to electric and electric to mechanical conversions. Again, this example is only for illustrating a reasonable approach for estimating the engine efficiency and powertrain losses; a detailed model and simulation should be used to obtain more insights.

Hybrid efficiency may be improved through improving average ICE efficiency and reducing system losses. As shown in the example, the average engine efficiency is still significantly below the engine peak efficiency of 38%, typical for some Atkinson engines used in recent hybrid vehicles. The electric loss has significant impact on the powertrain efficiency. More efficient electric path not only improves the efficiency in EV mode but also pushes the engine operating points closer to the peak efficiency areas. Thus, the benefits from electric improvements are magnified. There are certainly opportunities in system level, that is, powertrain architecture and system optimization, and powertrain operating strategy optimization. In general, hybrid powertrain architecture, component efficiency, and control strategy are highly coupled, with conflicting objectives and trade-offs; the optimization is computational intensive and calibration intensive.

## 11.4   REGENERATIVE BRAKING SYSTEMS

Regenerative braking is one of the key features of the electrified powertrain systems, which enables the vehicle to increase its fuel economy and driving range while reducing the carbon footprint to the

| 2012 Camry HEV | | | 1.6 | kWh | | | | | | | |
| | Total demand Wh/mile | Pos energy Wh/mile | Neg energy Wh/mile | SOC delta % | Propulsion energy flow (Pos + Neg) Wh/mile | mech eff | elec eff | k | Energy output from ICE (Wh/mile) | Eff ICE | Estimated Unadj mpg | Measured Unadj mpg (EPA) |
| City | 111.6 | 198 | 98 | 0 | 296 | 0.95 | 0.8 | 0.76 | 171 | 0.3 | 59 | 58.9 |
| HW | 161.7 | 176.7 | 22 | 0 | 198.7 | 0.95 | 0.8 | 0.76 | 202 | 0.34 | 56.8 | 56 |
| bag 3 | 132.8 | 211 | 90 | 10 | 301 | 0.95 | 0.8 | 0.76 | 238 | 0.32 | 45.3 | 45.4 |
| bag 4 | 91.8 | 186 | 105 | −10 | 291 | 0.95 | 0.8 | 0.76 | 106 | 0.3 | 95.4 | 96.1 |
| | ↑ Road resistance accessory | | | | ↑ Vehicle mass cycle aggressiveness | | ↑ Electric losses | ↑ | | ↑ Roughly estimated average ICE efficiency | | |

FIGURE 11.21  2012 Camry HEV city and highway fuel economy.

environment [1,3,4]. Recovery of the stored kinetic energy of the vehicle during braking is the fundamental concept behind regenerative braking. During the braking phase of the vehicle, the electric machine acts as a generator to convert the kinetic energy of the vehicle to electrical energy and store it in the onboard storage device (typically advanced batteries such as NiMH and Li-ion batteries, supercapacitor, and flywheel system), and this energy can be used for the consequent acceleration or driving of the vehicle. Depending on the storage availability and the power/torque-handling capability of the electric machine, the regenerative braking efficiency of the electrified powertrain system would vary. Moreover, some braking power requirements (during a heavy braking event) may not be handled only by the onboard electric propulsion system. Therefore, friction braking system also needs to assist the electric braking system to achieve such braking objective. When an HEV is still or at very low speed (a few mph), friction brakes are much preferred to hold the vehicle and the hybrid powertrain outputs near zero torque. When the brake pedal is released, the powertrain torque will increase to achieve consistent vehicle creeping behaviors. The process of utilizing dual braking system for braking is called brake blending, and precise brake blending is important for braking performance and for fuel economy as well.

A full blended regenerative braking system is used in many strong hybrid vehicles and electric vehicles. It is a fairly sophisticated system that interprets driver's brake pedal input and achieves vehicle deceleration consistent to the pedal input with safety and efficiency by proper actuations of the friction brake system (foundation brake) and the hybrid powertrain regenerative braking functions. The basic working principle is illustrated in Figures 11.22 and 11.23. The blending factor is defined as the ratio of friction brake torque over the total brake force demanded by the driver; it is a complex function of brake pedal input, vehicle speed, safety events, and mechanization specifics. As shown in the figures, the desired friction brake force is determined from the total brake force demanded by the driver and the blending factor subjected to the powertrain regenerative brake capacity that may vary dynamically, and the regenerative brake torque command to the powertrain is the difference between the total brake torque and the friction brake torque. A typical regenerative braking system can accommodate 0.25–0.3 g deceleration; for brake-by-wire systems, the desired friction brake force can be decoupled from the brake pedal and can be more precisely controlled, therefore more efficient energy recuperation can be expected.

The brake pedal force feedback is important, and is usually realized with a particular force emulator before the pedal travel reaches a limit where a portion of true friction brake force is coupled. Current regenerative braking systems, whether fully blended or more economical overlay regenerative braking, are still challenged to meet the stringent braking performance and yet still offer

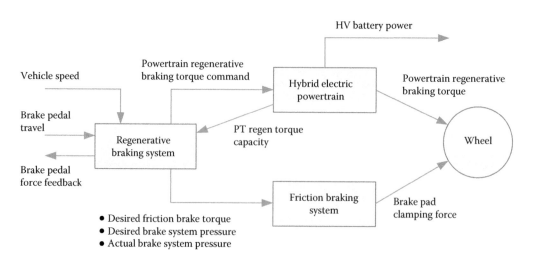

FIGURE 11.22   Working principle of a regenerative braking system.

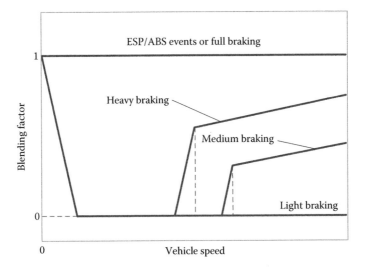

FIGURE 11.23   Blending strategy of a regenerative braking system.

high efficiency of kinetic energy recuperation. A few new mechanizations for regenerative braking have been proposed that gear toward providing higher deceleration level (0.4 g) and more favorable blending with by-wire technologies.

### 11.4.1   Primary Reasons for Dual Braking System in Electrified Vehicles

In theory, all of the kinetic energy stored in the vehicle can be harvested by the help of regenerative braking system. However, in reality, it may not be feasible due to the following reasons.

*Fluctuation in braking power demand and torque/power limitation of the electric propulsion system:* In general, braking events can be categorized into three types: mild, medium, and heavy braking, and their typical velocity–time relationship is depicted in Figure 11.24. During a mild braking event, the braking power requirement of the vehicle will be much lower and this can be supplied by the electric propulsion system alone, that is, pure regenerative braking. Braking power requirement of a medium braking event is much higher than mild braking. Therefore, in medium braking condition, regenerative braking capability of the vehicle is mainly dependent on the percentage share of the electric propulsion system in the powertrain. For example, to accomplish a

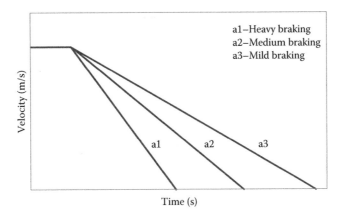

FIGURE 11.24   Possible braking requirements.

medium braking event in a mild HEV, the combination of friction and regenerative braking systems must be used (due to the higher braking power requirement). A typical passenger car may require a few hundred kilowatts of braking power in a heavy braking event and which may not be able to be handled by the electric propulsion system alone (typical power rating of the electric propulsion system in passenger HEV is around 50–60 kW).

Therefore, to handle various braking requirements of the vehicle, the involvement of dual braking system is essential.

*Operational characteristic of the electric machine and its influence:* Torque capability of the electric machine in generating mode is highly dependent on the rotational speed of the machine, and this capability may decrease when the machine runs at low speed. In addition to that, both the electric machine and the inverter have poor efficiency (due to the thermal stress of the power modules) at low-speed braking conditions. Therefore, friction braking systems must be employed at low-speed braking events.

*Limitations of the energy storage system:* The recovered kinetic energy of the vehicle will be transferred to the energy storage device. In some braking occasions, the energy storage system will reach its maximum capacity and it may not accept further charge from the electric machine. In such situations, the involvement of friction braking system is essential to achieve the braking objective. Not only does the energy storage device limit the involvement of regenerative braking system, but the power/torque-handling capability of the electric machine and the power electronic system also act as primary limiting factors. Although it is possible to design an electrified powertrain system to handle all kinds of braking requirements in theory, it cannot be accomplished without the weight, volume, and, most importantly, overall cost increment of the vehicle. Therefore, to achieve adequate regenerative braking performance within the realistic limitations, dual braking system is essential.

*Weight distribution and braking dynamics:* The center of gravity of the vehicle and its location related to the front and rear wheels are very important in braking performance. To understand this, let us walk through the fundamentals behind the braking dynamics of vehicles. Figure 11.25 depicts a two-axle braking dynamics of a vehicle, where

$h$ is the height between the center of gravity (CG) and the ground
$l$ is the distance between the CG and rear axle
$L$ is the distance between the front and rear axle
$a$ is the deceleration of the vehicle (m/s²)
$V$ is the velocity of the vehicle
$\theta$ is the road inclination angle

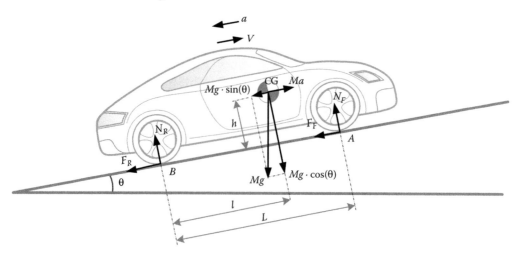

FIGURE 11.25   Two-axle braking dynamic of the vehicle.

$N_R$ is the normal force acting on the rear wheel
$N_F$ is the normal force acting on the front wheel
$F_R$ is the braking force acting on the rear wheel due to ground friction
$F_F$ is the braking force acting on the front wheel due to ground friction

The applicable braking force on a given terrain in a given wheel is proportional to the normal force acting on that wheel. Therefore, by knowing the normal force, the braking force distribution can be obtained. By taking moment from points A and B, the action of normal forces on the rear and front wheels $N_R$ and $N_F$ can be derived as

$$N_R = \frac{Mg\sin(\theta)h + Mg\cos(\theta)(L - l) - Mah}{L} \tag{11.17}$$

$$N_F = \frac{Mg\cos(\theta)l + Mah - Mg\sin(\theta)h}{L} \tag{11.18}$$

From Equations 11.17 and 11.18, maximum braking forces in the front and rear wheels can be derived as

$$F_R = \mu N_R \tag{11.19}$$

$$F_F = \mu N_F \tag{11.20}$$

Here, $\mu$ is the friction coefficient of the road.

From Equations 11.17 through 11.20, one can clearly understand how the coordinates of the center of gravity and deceleration rate of the vehicle influence the braking force distribution between the front and rear wheels. The variation of normal force distribution with respect to the deceleration rate is known as dynamic force transfer (an interested reader is advised to read Reference 2 for more information about ground vehicle braking dynamics). Now, it is straightforward to understand how a single-axle drive vehicle's regenerative performance is affected by the dynamic force transfer. Moreover, this information can be effectively utilized to design the electrified drivetrain system in the vehicle to maximize the kinetic energy recovery. It also suggests that the energy recovery efficiency can be significantly improved by integrating electric drive systems in both axles. However, this increases the overall cost of the vehicle.

Apart from the regenerative braking capability, the theory derived in this section provides more insights of the ground vehicles braking dynamics and a valid reason why a dual braking system is essential in electrified vehicles. Also, it suggests where the CG of the vehicle must be placed in order to increase the regenerative braking performance.

*Antilock braking system:* Antilock braking system will be on action on a slippery terrain braking condition. During this occasion, only the friction braking system will be activated with the assistance of antilock braking system. When the antilock braking system is in action, the regenerative braking system will be terminated due to the braking control complexity.

Therefore, the primary reasons for the dual braking systems in HEVs can be summarized as

1. Limited power/torque-handling capability of the electric propulsion system
2. Insufficient torque-generating capability of the electric machine at low speed (due to the low back-EMF)
3. Storage limitations of the energy storage system in the vehicle
4. Dynamic force transfer of the vehicle
5. Involvement of antilock braking system on slippery terrain braking condition

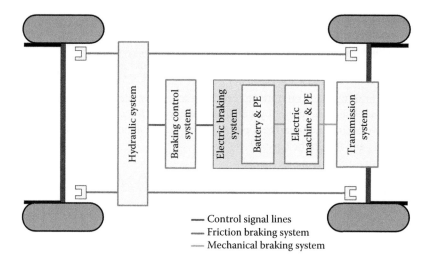

FIGURE 11.26    Braking system in a typical single-axle drive hybrid electric/electric car.

## 11.4.2    BRAKING SYSTEM IN ELECTRIFIED VEHICLES

A dual braking system in an electrified vehicle is depicted in Figure 11.26. Here, the drive wheels have the dual braking capability while driven wheels are integrated with the mechanical friction brakes. Friction brakes are operated by the hydraulic system. The braking system of the electrified vehicle is controlled by the braking control system, which prioritizes the braking systems depending on the expected deceleration requirement of the vehicle. The primary objectives of the braking control system are (1) to maximize the involvement of regenerative braking system on any given braking condition and (2) to ensure that the expected braking torque is provided to achieve the expected deceleration requirement in a safe manner. Also, achieving braking feel is one of the key design challenges in braking control system.

Figure 11.27 shows the braking power flow diagram of a FWD vehicle. Here, braking power is shared between the front and rear wheels. Since the front axle is integrated with the regenerative

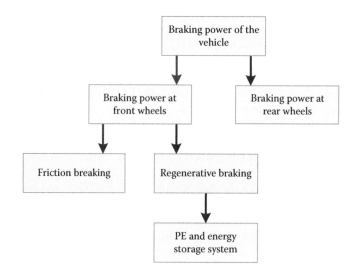

FIGURE 11.27    Regenerative braking and friction braking power flow in a front wheel drive hybrid electric/electric car.

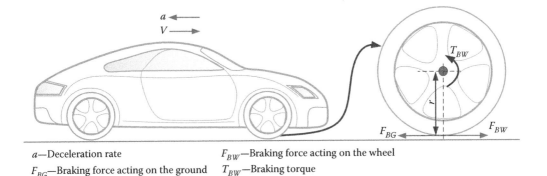

a—Deceleration rate  $F_{BW}$—Braking force acting on the wheel

$F_{BG}$—Braking force acting on the ground  $T_{BW}$—Braking torque

FIGURE 11.28   Force/torque acting on the braking wheel.

braking system, part of the front axle braking is achieved by a regenerative braking system as well as a friction braking system. A friction braking system is employed for rear axle braking.

Figure 11.28 shows the simplified braking force diagram of a wheel. According to Newton's second law of motion, a constant braking force $F_{BW}$ should act on the wheel (which is equivalent to the force exerted by the wheel on the ground $F_{BG}$) for the vehicle to achieve a constant deceleration rate $a$. To generate a constant braking force, the torque generated by the braking system $T_{BW}$ should be constant and equivalent to $rF_{BG}$. Regardless of this, the torque generated by the electric propulsion system is highly nonlinear and varies with the operating speed of the electric machine. Therefore, to achieve a constant deceleration rate, the dual braking system must be blended together in an optimized way, which maximizes the utilization of electric braking while supplying the excess braking power by the friction braking system.

### 11.4.3   ENERGY STORAGE FOR REGENERATIVE BRAKING SYSTEM

Flywheel system and supercapacitor banks are commonly used energy storage devices in regenerative braking systems [3–4]. Owing to the poor power-handling capability, electric battery is not a good candidate for this application. Flywheel system stores the recovered energy of the vehicle in the form of kinetic energy while supercapacitors store it in the form of electric energy. As depicted in Figure 11.29a and b, flywheel system can be used in two different ways for energy storage. Figure 11.29a shows the pure mechanical regenerative braking system. Here, the transmission system connected with the drive axle is directly coupled with the mechanical flywheel. Hence, the speed difference between the flywheel and the drive axle must be managed by the transmission system. Therefore, the transmission system design is one of the primary challenges in this particular design and it is still in the early stage in its development. Owing to the less number of energy conversions in this design, it achieves higher energy recovery efficiency and it is considered as one of the primary advantages of this design.

Figure 11.29b shows another design that includes a flywheel system as the regenerative braking energy storage. Contrary to pure mechanical arrangement, this design uses an electric machine to charge/recharge the flywheel system. The recovered electric power from the electrified propulsion system will be transferred to the electric machine via suitable power regulations during the braking phase of the vehicle. The primary advantage of this design is that it can be easily incorporated in any existing electrified vehicles. However, the higher number of energy conversions in this design reduces its overall efficiency.

A supercapacitor-based regenerative energy storage design is depicted in Figure 11.29c. Here, the electrified flywheel system has been replaced by the supercapacitor bank. This design can also be conveniently integrated in any electrified vehicle. It also suffers from a higher number of energy conversions.

In terms of specific power and power density, battery systems are inferior to supercapacitor and flywheel systems, and therefore batteries are not a good candidate for this application. However,

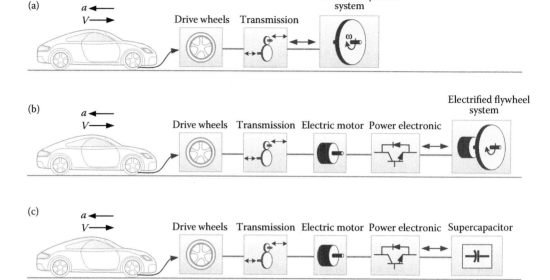

**FIGURE 11.29** Regenerative braking system: (a) pure mechanical flywheel system as regenerative energy storage, (b) electrified flywheel system as regenerative energy storage, (c) supercapacitor-based regenerative energy storage.

battery systems are also used for this application with the appropriate design of power and energy management systems.

### 11.4.4 SUMMARY

Regenerative braking is one of the essential features of the electrified powertrain system, which enables the vehicle to achieve higher fuel economy and to reduce emission. The power/torque-handling capability of the electric propulsion system and the braking condition of the vehicle are primary factors determining the energy recovery efficiency. Owing to several reasons, dual braking system (regenerative braking and friction braking systems) must be incorporated in electrified vehicles. Some research publications claim that kinetic energy recovery of the electric propulsion system increases the driving range by 25%–30%. Toyota Prius HEV achieves a 20% increase in the driving range with the help of kinetic energy recovery. Apart from energy harnessing, the electric braking system also provides an opportunity to minimize the utilization of the friction braking system and therefore extends its life.

## 11.5  INTRODUCTION TO HYBRID POWERTRAIN CONTROLS

Hybrid powertrain control is usually a torque-based control; the control objective is to meet driver's demand as efficiently as possible by coordinating three torque actuators and clutch controls, subject to all system constraints and component constraints considering safety, drivability, emission, and component protection. Figure 11.30 depicts a generic HEV control architecture; the main functions include

1. Drive demand generator. It generates the desired powertrain propulsion torque and response type based on driver's pedal inputs, mode selection, and the needs from other vehicle systems such as cruise system and vehicle stability control system.

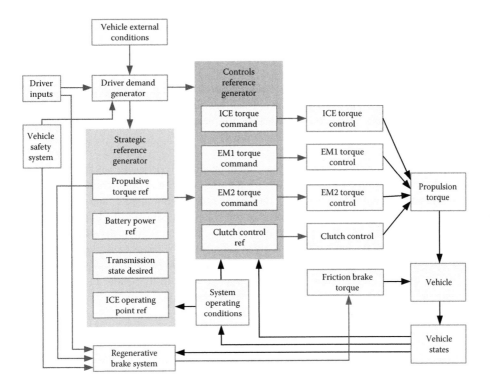

FIGURE 11.30    Generic HEV controls.

2. Strategic reference generator. It generates strategic control references such as engine on–off command, desired engine speed and operating mode, desired battery power, and propulsion torque target based on efficiency and other considerations.

3. Controls reference generator. It generates control references such as motor torque command, engine torque command, and clutch control command (pressure or torque) that satisfy the strategic control targets and fulfill tactical control needs such as engine start–stop controls. The control references are the control targets to be achieved by subsystem controls such as motor torque controls by a motor controller.

4. Regenerative braking system control. It generates the total brake torque based on the brake pedal input and vehicle conditions and determines the regenerative brake torque command to be executed by the powertrain controls and the friction brake torque command to be executed by the brake system.

Vehicle speed, vehicle operating conditions, and system operating conditions such as motor temperatures are the necessary inputs to the functions mentioned above, and they will be discussed in more detail. In general, the partition between vehicle level and hybrid powertrain controls is not unique; for an example, some hybrid powertrain controls may include driver demand interpretation function mentioned above as function 1, while others may not. Nevertheless, functions 2 and 3 are clearly within hybrid powertrain controls, and therefore they are the focus of the following discussions.

For a given powertrain propulsion torque command (can be positive or negative), there are only two groups of questions to be answered by the hybrid powertrain strategic controls (strategic reference generator) in real time:

1. Which mode is desired, engine-off mode or engine-on mode? And by what criteria?
2. What engine operating points (speed–torque) by which engine operating mode are desired in the engine-on mode?

For a desired strategic solution, the hybrid powertrain tactical controls (controls reference generator) will determine torque commands to achieve the desired strategic solution. In general, a torque command may have three components:

1. A component needed for achieving the powertrain propulsion torque demand
2. A component needed for various control functions such as regulating engine speeds
3. A component that is designed to compensate torque errors from motor torque controls and engine torque controls, or errors in plant models

At this stage, specific hybrid configurations and plant characteristics are involved in the controls. In a P2 parallel configuration (here P2 stands for parallel two-clutch powertrain technology implemented in Hyundai Sonata hybrid vehicles), the engine is coupled to the P2 motor through a disconnect clutch, so engine start controls may involve the clutch fill, slip, and lock during engine start. When the clutch is disengaged, the engine is decoupled from the rest of the drivetrain. In contrast, in a power-split powertrain, the engine speed is always controlled by the system, even in the engine-off mode; the engine speed control is still active. By now you should be quite familiar with power-split hybrid powertrain. We will continue to use the Prius powertrain to explore the basic concepts of hybrid controls. The same concepts should still hold for other hybrid configurations.

### 11.5.1 ENGINE OFF/ON DECISIONS

The starting point is the intended acceleration—vehicle speed envelope for EV mode under nominal conditions, as shown in Figure 11.31. The equivalent wheel torque–wheel speed envelope for the EV mode may be useful for control purposes, and it is shown in Figure 11.32. If the wheel torque demand is within the envelope, the engine-off mode is desired. The envelope in Figure 11.31 is characterized by the maximum EV mode acceleration (say 0.15 g), the maximum power demand (say 10 kW), the minimum acceleration (or maximum deceleration, say −0.25 g), and the minimum power demand (say −25 kW). The nominal conditions mean that battery power capability, SOC considerations, and motor capabilities are not the limiting factors. Figure 11.32 is based on Figure 11.31 and 2010 Prius vehicle mass (1530 kg), road load coefficients (A = 18.5 lbs, B = 0.02235 lbs/mph, C = 0.01811 lbs/mph²), and tire rolling radius of 0.3 m. Readers can easily verify that the envelope is

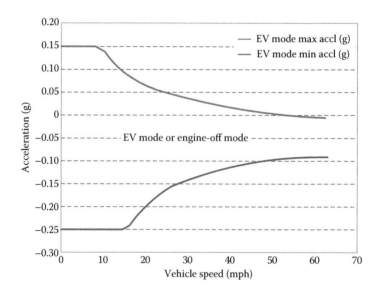

FIGURE 11.31   Nominal EV mode envelope.

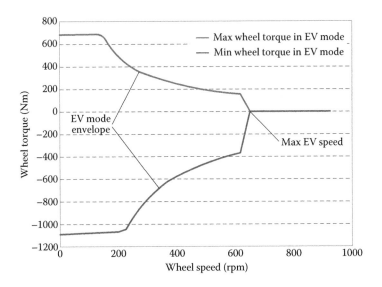

FIGURE 11.32   Nominal EV mode wheel torque–wheel speed envelope.

well within motor B capabilities (207 N m, 60 kW, motor B to ground ratio of 8.61) and within HV battery discharge and charge power capabilities. For many midsize hybrids, the EV power threshold is likely between 5 and 15 kW, and it depends on engine efficiency details, motor and inverter efficiencies, and driving cycle considerations.

Now, we can bring more considerations to modify the nominal EV envelope. First, this envelope may contract or expand, depending on the factors deviating from the nominal levels. As the battery discharge power capability is reduced, or SOC is approaching the lower SOC limit, it may be desired to reduce the EV power threshold above which engine-on mode is preferred. Second, certain hysteresis must be introduced to the envelope to prevent excessive engine-off and engine-on mode changes. Stability can also be introduced in time domain, for an example, by a denounce feature. Third, certain considerations can override the mode selections that might be efficiency centric. If the engine or catalyst is cold, it may be desired to complete engine–catalyst warm-up processes before enabling the engine-off mode. If the engine heat is needed for cabin heating, of course, the engine-on mode should be the solution. Fourth, drivability consideration may push for selecting engine-on solution at the power level that is lower than the nominal one. One example of that is the quick accelerator pedal movement toward high demand, where engine-on should be selected immediately. In general, starting engine is much more challenging and more power demanding than shutting off the engine; therefore, it is important to gate the engine-off decision with restart quality considerations.

Now, it is quite clear that engine-off and engine-on mode selection is not just driven by the efficiency considerations, but many factors are involved. Nevertheless, algorithms based on quantitative evaluation and logics can be devised to achieve consistent and near optimal results after significant amount of simulations, calibrations, and tests, as evidenced by many efficient and fun-to-drive hybrid models in the markets.

### 11.5.2   Engine Operating Points Optimization

This is the second question to be answered by the strategic controller. Under nominal engine operating conditions, the fuel consumption rate as a function of engine speed and engine torque is known, and is designated as $\dot{m}(Ne, Te)$. We have briefly discussed this previously in Section 11.2.1. Now let us try to construct a process for solving it. For a given wheel torque demand $Tw$ and a given wheel speed $Nw$, and a desired battery power $P$ (assuming we know it for now), there will be a pair or multiple pairs of engine speed and torque (*Ne,Te*) that can achieve the desired powertrain output and the desired battery powertrain with minimal fuel consumption rate. Again, we will use the power-split hybrid powertrain to elaborate the optimization process.

The 2D optimization problem can be defined as

$$\min_{N_e, T_e} \dot{m}(N_e, T_e) \tag{11.21}$$

Subjecting to

$$P = NaTa + LossA(Na,Ta) + NbTb + LossB(Na,Ta)$$

$$Tw = Tw(Te,Tb)$$

$$Na = Na(Ne,Nw)$$

$$Ta = -0.2778 \, Te$$

$$Nb = 8.61 \, Nw$$

$$Tb = Tb(Te,Tw)$$

where *LossA(Na,Ta)* is the total loss of motor A loss and inverter A loss associated with the operating point (*Na,Ta*), and *LossB(Nb,Tb)* is the total loss of motor B loss and inverter B loss associated with the operating point (*Nb,Tb*). And all speeds and torques are within the known maximum and minimum limits.

This optimization problem can be particularly challenging since the maximum and minimum torque for motors A and B, and the engine is a function of respective speed, and the battery power is related to the operating points of motors A and B in a complex way. Although online implementations are possible, robustness of the optimal solutions must be examined in great detail. Quite often, offline simulations can reveal sufficient insights about the optimal solutions and produce reasonable approximations to be used for simplifying the online optimization problem such that the robustness and the computational efficiency may be improved.

One way for removing the battery power constraint is to minimize a new objective function such as

$$f(Ne,Te) = \dot{m}(Ne,Te) + \lambda P \tag{11.22}$$

where $\lambda$ is a constant to be calibrated through offline simulations and analysis. Based on offline studies and known characteristics of the engine fuel consumption map and the motor loss maps, it is possible to construct an algorithm to select a few relatively concentrated operating points to evaluate *f(Ne,Te)*, and then construct a model to predict where *f(Ne,Te)* reaches the minimal.

It is challenging enough to just find the optimal solutions for fuel minimization alone. In real applications, there are other factors to be considered and more trade-offs to be made. Some engine operation specifics must be taken into consideration in determining the desired engine operating points and possible engine operating modes. The specifics may include emission control needs, purge control needs, and engine fast and slow torque response types. Some examples are mentioned as follows:

- Emission control after cold start prefers particular engine speed/torque/air fuel ratio with retarded spark, mostly 1200–1400 rpm
- Purge control prefers higher vacuum (lower torque) and higher speed
- NVH considerations

- Engine fast torque path management via spark and fuel cutoff
- Rate of engine torque/speed changes based on engine control characteristics

A hybrid operating strategy is about best utilizing key components to achieve the system targets in efficiency, emission, drivability, onboard diagnostics (OBD), safety, and component protection. Most of the operating strategies are implemented in the strategic controller (strategic reference generator). The following is a summary of topics and trade-offs that are often addressed in real applications for the purpose of inviting more readings and explorations:

1. When to use EV mode (or engine-off mode)
2. Engine operating modes and operating points in engine-on mode
3. Trade-off regenerative braking efficiency and safety and braking quality
4. Performance mode
5. Trade-off between efficiency and responsiveness
6. Strategy for emission controls and OBD operations
7. Trade-off between efficiency and NVH
8. Component protection strategies
9. System-degraded performance and shutdown strategies
10. Interaction with vehicle systems such as electronic stability program (ESP)
11. Operating strategy for extreme conditions such as ambient temperature and grade
12. Interaction with drivers such as human–machine interface (HMI) and feedback systems

The objective of the tactical controller (controls reference generator) is to generate the motor torque commands and the engine torque command that together will achieve the desired strategic solutions and the system control targets through some specific control functions. The functions may include

1. Engine start–stop control
2. Engine speed control
3. Clutch control and shift execution
4. Battery power control
5. Regenerative braking control
6. Driveline damping control
7. Battery SOC control
8. HV contact control
9. Thermal system control
10. Transmission pump control
11. Accessory power control

It is not this chapter's intention to discuss these control functions in great detail. Readers are strongly encouraged to build up more understanding about these functions. Instead, engine start control and regenerative braking control are chosen for more elaborations because of their apparent importance and technical challenges.

### 11.5.3 ENGINE START CONTROL

The objective of the engine start control is to complete the transition as quickly as possible without creating noticeable disturbances to the driver and to the driveline. This transition usually includes four phases, that is, engine break-away, engine spinning up, engine firing, and engine stabilization in speed and torque. The transition controls are accomplished by the coordinated actions of motor A, motor B, and the engine. The transition from engine-off to engine-on is likely triggered by driver's rising demand; the slow transition just could not provide the responses expected by the driver. There are six dynamic events occurring rapidly during the transition and must be controlled properly:

1. High engine break-away torque and it changes fast
2. Engine compression pulse is significant and must be compensated
3. Spinning the engine fast to reduce torsional damper resonance
4. Engine first firing pulse may be high even with spark retardation
5. Engine torque varies fast as spark is advancing and the manifold is pumping down
6. Battery voltage and power vary fast

The first five dynamics must be modeled and calibrated precisely and be included in the feed-forward term. The speed control loop has difference gains as the transition advances through each of the phases. Engine decompression techniques, small manifold, proper dual mass torsional damper design, and engine crank position estimation all aid the controls. Prius and Ford Fusion HEV and other similar models have demonstrated seamless engine start. The quality is remarkable and it is consistent for every start.

### 11.5.4 REGENERATIVE BRAKING CONTROL

Regenerative braking is a major efficiency enabler, and it should be used as much as possible. Meanwhile, HEV drivers continue to have high expectation about linear brake feel, as if it was a conventional brake system. Regenerative braking is still challenged in delivering seamless brake feel; coordinated actuation of two drastically different brake systems is not trivial and requires accurate information exchange and precise execution of friction brakes and powertrain regenerative braking. Within powertrain controls, the regenerative braking function accomplishes three tasks:

1. Estimating regenerative braking capacity for all modes, EV mode, engine-on mode, and the transitions
2. Dynamically shaping the capacity to remove the oscillatory component
3. Executing the regenerative braking torque as commanded by regenerative braking system

Blending is still a big challenge, as it is prone to multiple error sources. Not only do both braking force controls have errors, but also the timing can be out of sync. Therefore, the resulting errors in total braking force can be multiple times higher than just friction brake alone. It can be expected that the controls using modeled responses of both brake actuation processes will improve the blending process.

Part of the complexity in hybrid powertrain controls is caused by more subsystems and increased number of system operating conditions that must be included in the controls. The powertrain controls must be robust and effective under all external conditions and system conditions as typified by the following list:

1. Battery temperature and cell temperature(s)
2. Battery SOC
3. Battery voltage and cell voltage(s)
4. Cabin air temperature
5. Engine coolant temperature
6. Engine oil temperature
7. Exhaust temperature
8. Catalyst temperature
9. Motor temperature(s)

10. Inverter temperature(s)
11. Converter temperature
12. Transmission fluid temperature

The powertrain controls' design, development, calibration, and tests must cover all conditions to ensure safe, reliable, efficient, and fun-to-drive HEVs.

In many hybrid vehicles, the hybrid powertrain controller is networked with many control modules such as engine control module, motor control modules, battery control module, and transmission control through vehicle controller area networks (CAN) networks. Control architecture must be carefully designed to ensure that the relevant information is updated in real time. Hybrid powertrain control system is a software-intensive embedded system with millions of lines of code.

In conclusion, hybrid electric powertrains encompass hybrid powertrain architectural design, controls engineering, embedded systems engineering, and embedded software engineering. It involves modeling and simulation, hardware development, controls development, software development, vehicle integration, calibration, and verification and validation tests. HEV has been providing R&D opportunities in key components and systems, as they continue to advance. HEVs will be more efficient, more affordable, and more fun to drive. Hybrid powertrains are technically challenging and fun to work with for automotive engineers.

## APPENDIX: AUTONOMIE SIMULATION CONFIGURATIONS AND INPUT PARAMETERS

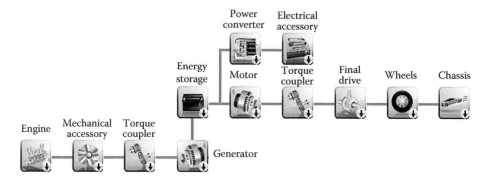

FIGURE 11.A1    Series HEV simulation block diagram.

### TABLE 11.A1

### Series HEV Model: Series Engine Midsize Fixed Gear HEV 2wd Default

| System | Initialization | Scaling | Value |
| --- | --- | --- | --- |
| Driver | drv_ctrl_normal_1000_05.m | | |
| Environment | env_plant_common.m | | |
| Chassis | chas_plant_990_225_03_ midsize.m | | |
| Electrical accessory | accelec_plant_200.m | | |
| Energy storage | ess_plant_li_6_75_saft.m | | |
| Engine | eng_plant_si_1497_57_ US_04Prius.m | eng_plant_s_pwr_lin.m | 80 kW |
| Final drive | fd_plant_444_accord.m | | |

*continued*

**TABLE 11.A1    (continued)**

**Series HEV Model: Series Engine Midsize Fixed Gear HEV 2wd Default**

| System | Initialization | Scaling | Value |
|---|---|---|---|
| Generator | gen_plant_pm_45_75_UQM_ PowerPhase75_SR218N.m | gen_plant_pwr_s.m | 75 kW |
| Mechanical accessory | accmech_plant_0.m | | |
| Motor | mot_plant_pm_25_50_prius.m | mot_plant_pwr_scale.m | 120 kW |
| Power converter | pc_plant_095_150.m | | |
| Torque converter 1 | tc_plant_1.m | | |
| Torque converter 2 | tc_plant_16.m | | |
| Wheel | whl_plant_0317_P195_65_R15.m | | |
| Brake controller | vpc_brake_ser_eng_p234_init.m | | |
| Propel controller | vpc_prop_ser_eng_load_ following_no_tx_init.m | | |

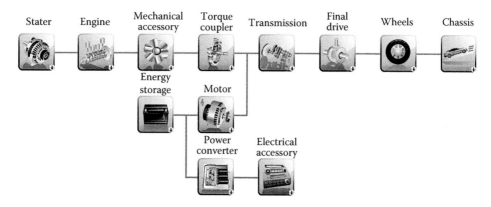

FIGURE 11.A2    Parallel HEV simulation block diagram.

**TABLE 11.A2**

**Parallel HEV Model: Parallel Engine Midsize AMT HEV 2wd Default**

| System | Initialization | Scaling | Value |
|---|---|---|---|
| Driver | drv_ctrl_normal_1000_05.m | | |
| Environment | env_plant_common.m | | |
| Chassis | chas_plant_990_225_03_midsize.m | | |
| Electrical accessory | accelec_plant_200.m | | |
| Energy storage | ess_plant_li_6_75_saft.m | | |
| Engine | eng_plant_si_1497_57_US_04Prius.m | eng_plant_s_pwr_lin.m | 80 kW |
| Final drive | fd_plant_363_cavalier.m | | |
| Gearbox | gb_plant_5_dm_345_19_129_09_075.m | | |
| Starter | str_plant_10_10.m | | |
| Mechanical accessory | accmech_plant_0.m | | |
| Motor | mot_plant_pm_25_50_prius.m | mot_plant_pwr_scale.m | 25 kW |
| Power converter | pc_plant_095_12.m | | |
| Wheel | whl_plant_0317_P195_65_R15.m | | |
| Brake controller | vpc_brake_par_pretx_1mot_init.m | | |
| Propel controller | vpc_prop_par_direct_pwr_1mot_pretx_init.m | | |

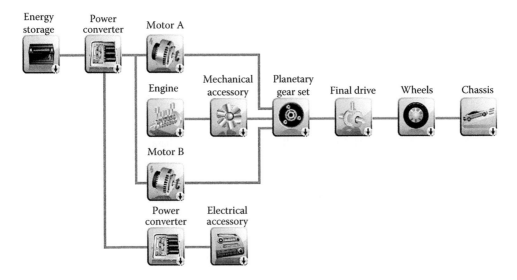

FIGURE 11.A3    Power-split HEV simulation block diagram.

TABLE 11.A3

**Power-Split HEV Model: Split Midsize Single Mode HEV 2wd Default**

| System | Initialization | Scaling | Value |
|---|---|---|---|
| Driver | drv_ctrl_normal_1000_05.m | | |
| Environment | env_plant_common.m | | |
| Chassis | chas_plant_990_225_03_ midsize.m | | |
| Electrical accessory | accelec_plant_200.m | | |
| Energy storage | ess_plant_li_6_75_saft.m | | |
| Engine | eng_plant_si_1497_57_ US_04Prius.m | eng_plant_s_pwr_lin.m | 80 kW |
| Final drive | fd_plant_393_prius.m | | |
| Gearbox | gb_plant_planetary_30_78.m | | |
| Mechanical accessory | accmech_plant_0.m | | |
| Motor | mot_plant_pm_25_50.m | mot_plant_pwr_scale.m | 63 kW |
| Motor 2 | mot_plant_pm_15_30.m | mot_plant_pwr_scale.m | 51 kW |
| Power converter | pc_plant_095_12.m | | |
| Power converter 2 | pc_plant_boost_095.m | | |
| Wheel | whl_plant_0317_P195_65_R15.m | | |
| Brake controller | vpc_brake_split_best_eng_init.m | | |
| Propel controller | Split Prius MY04 | | |

**QUESTIONS**

11.1    Classify the vehicles listed in Table 11.1 based on mild hybrids, full hybrids, and plug-in hybrids. Compare their powertrain configurations, ranges, and electric systems.

11.2    Compare mild hybrids with full hybrids. Based on the contents of this chapter and the case study in Chapter 13, specify the differences in configurations, components sizing, and power requirements.

11.3   Which mechanical–hybrid powertrain concepts give a competitive trade-off between fuel saving, cost, and control complexity?

11.4   What are the key parameters determining the hybrid powertrain topology selection? Perform a case study to investigate which powertrain topology is highly desirable for fuel economy enhancement in UDDS city driving cycle.

## REFERENCES

1. K. Henry, J. A. Anderson, M. Duoba, and R. Larsen, Engine start characteristics of two hybrid electric vehicles HEVs, Honda Insight and Toyota Prius, in *SAE Future Transportation Technology Conference and Exposition*, no. 2492, Aug 2001.
2. J. Wong, *Theory of Ground Vehicles*. John Wiley and Sons, Canada, 1993.
3. J. Miller, *Propulsion Systems for Hybrid Vehicles*. IEE Power and Energy Series 45, The Institute of Electrical Engineers, United Kingdom, 2003.
4. S. McCluer and J. Christin, Comparing data centre batteries, flywheels, and ultracapacitors, Transportation Research Part D: Transport and Environment, Technical Report, 2008.
5. https://techinfo.toyota.com/t3Portal/document/ncf/NM14C0U/xhtml/RM0000042WY003 (accessed: 17 February 2010).

# 12 Hybrid Electric Vehicles

*Piranavan Suntharalingam, Yinye Yang, and Weisheng Jiang*

## CONTENTS

## 12.1 INTRODUCTION

This chapter provides an overview of hybrid electric vehicles and their characteristics. Initially, the difference between hybrid electric and conventional vehicles is distinguished. This includes how hybrid electric vehicles achieve improved fuel economy compared to conventional vehicles. In Section 12.2, the influence of driving cycles on fuel economy improvement is analyzed. A worked example is also provided with the relevant ground works to emphasize the correlation between driving cycle and fuel economy. In Section 12.3, the terrain information is also taken into consideration with the driving cycle, and variation in propulsion requirements is highlighted. Major hybrid electric vehicle technologies are addressed in Section 12.4. Section 12.5 classifies road vehicles based on their powertrain system. The main challenges in the successful realization of hybrid electric vehicles are highlighted in Section 12.6 and finally, the key areas for research and development of this technology are provided.

## 12.2  HYBRID ELECTRIC VEHICLES

Hybrid electric vehicles (HEVs) incorporate more than one propulsive power source to energize the vehicle. They have the ability to optimize the propulsive energy of the vehicle by effectively utilizing dual power sources and achieve higher mileage and reduce greenhouse gas emissions [1–2]. HEVs are integrated with an internal combustion engine (IC engine) and electric machines, which are energized by onboard electric battery, ultracapacitor, or a combination of both. The involvement of different propulsive components in HEVs and conventional vehicles is depicted in Figure 12.1. The dual-mode operating capability of the electric machine (function as motor as well as generator) is one of the primary advantages of the HEVs in terms of recovering energy during the braking phase of the vehicle. Also, the electric propulsion system provides higher acceleration performance at low speed, which cannot be achieved in conventional vehicles due to the various mechanical constraints of the IC engine. Owing to the increased number of propulsive components and complexity, the power management and control system in the HEVs plays a vital role in maximizing its overall energy efficiency and achieving higher operating performance, while significantly reducing emissions. The key features and the primary difference between hybrid and conventional vehicles are listed in Table 12.1.

### 12.2.1  Fuel Economy Improvement in HEVs

In plenty of occasions, it has been emphasized that HEVs achieve higher fuel economy and release less greenhouse gas emissions into the environment [3–5]. However, how this is achieved is not clearly explained. Hence, this section focuses on how HEVs take advantages over conventional vehicles in real driving environments and under which circumstances it may or may not show differences in fuel economy.

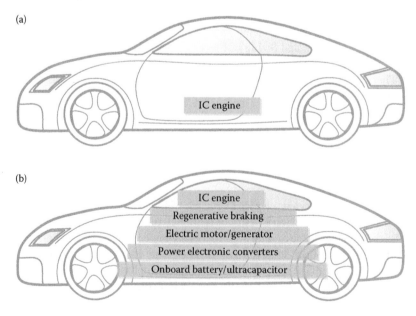

**FIGURE 12.1**  Primary components of hybrid electric vehicle and conventional vehicle. (a) Conventional vehicle and (b) hybrid electric vehicle.

## TABLE 12.1

## Comparison between Conventional and Hybrid Electric Vehicles

| Features/Requirements | Conventional Vehicles | Hybrid Electric Vehicles |
|---|---|---|
| Powertrain | IC engine | IC engine + EM |
| Energy source | Gasoline | Gasoline + electricity |
| Energy container | Fuel tank | Fuel tank + battery |
| Transmission system | Mechanically and hydraulically actuated transmission system | Electrically, mechanically, and hydraulically actuated transmission system |
| Propulsive redundancy | One propulsive system (no redundancy) | More than one propulsive system (more redundancy) |
| Power density of the powertrain | Higher power-to-weight ratio | Comparatively low power-to-weight ratio |
| Emissions | Higher emissions during operation | Lower emissions during operation |
| Operational energy efficiency | Low energy efficiency during stop-and-go (city) driving conditions | Higher energy efficiency during stop-and-go (city) driving conditions |
| Technology | Matured technology | Growing technology |
| Cost | Inexpensive | Expensive |
| Maintenance | Relatively easier | Complex due the involvement of multiple systems |
| Space | More spacious | Relatively less spacious |
| Energy recovery system | No energy-recovering capability during braking | Energy-recovering capability during braking |
| Low-level control | Relatively easier | Relatively complex |

The involvement of electric propulsion system in HEVs provides significant advantages to increase the fuel efficiency, which cannot be otherwise achieved. There are four different ways the electric propulsion system effectively acts to improve the energy efficiency of the vehicle:

- Higher operating efficiency of electric machines over their entire speed torque envelop provides a remarkable advantage on fuel efficiency improvement in HEVs. Generally, the efficiency of IC engines is less than half of what electric machine has. Besides, IC engines are designed to operate under certain speed–torque operating conditions to achieve optimum efficiency and its efficiency varies considerably throughout the entire speed–torque envelop of the engine. Regardless of this issue, the real-world driving demands fluctuating power and torque and they have to be somehow supplied in an efficient manner to drive the wheels. It is accomplished in conventional vehicles by incorporating a sophisticated transmission system, which employs mechanical advantage in the gear system to match the expected outcome while maintaining higher operating efficiency of the IC engine. This very same objective is easily achieved in HEVs by effectively exploiting the advantages of the electric motors with the involvement of a less complicated transmission system. However, to maximize the full potential of the electromechanical system of HEVs, the involvement of advanced control systems is mandatory.
- Kinetic energy recovery is another key attribute of HEVs, which is not available in conventional vehicles. The kinetic energy of the vehicle is generally dissipated as heat in the brake disks during the braking phase of the conventional vehicle. However, it is not the case in HEVs where the dual-mode operating capability of the electric machine (motor and generator) provides a great opportunity to recover this waste energy and store it in the battery. This free energy can be effectively used to fuel the consequent acceleration demand of the vehicle. It could be significant during stop-and-go city driving conditions of the vehicle compared to steady and smooth highway driving conditions.
- A remarkable amount of energy is wasted during the idling mode operation of the vehicle, which happens a lot during peak driving hours (in both city and highway driving

environments). Conventional vehicles are not designed to stop the engine during idling and it simply lets the engine to idle. Hybrid vehicles effectively exploit this opportunity by simply shutting off the engine during the idling mode of the vehicle and uses pure electric propulsion system to maneuver the vehicle in a slow-moving traffic. This particular feature significantly increases fuel efficiency while reducing greenhouse gas emissions in the environment.

• Some HEV designs incorporate bigger rechargeable onboard electric batteries to fuel electric motors. These types of HEVs can achieve considerable electric-only range without the involvement of the IC engine. This concept is widely known as plug-in HEVs, and the onboard battery of the vehicle is recharged by the grid energy during night and it is used during day time driving. Therefore, this kind of vehicle does not consume any gasoline to achieve significant driving range, which results in zero emission release to the environment. It should be noted that this feature is not common in all varieties of HEVs; they will be discussed in great detail in Chapter 14.

## 12.3  DRIVING CYCLES

Driving cycles and terrain conditions are critical factors, which influence energy saving and emission reduction to a large extent.

Driving cycle is defined as the velocity variation of the vehicle with respect to the time under different driving conditions. For example, you could have noticed that highway driving differs significantly compared to city driving. In city driving, there are a lot of traffic signals and speed limits, and depending upon the traffic conditions, the driver has to go through frequent stop-and-go-type driving. On the other hand, this is not the case in highway driving conditions. In highway driving, the vehicle does not accelerate too often and most of the time it cruises with a constant velocity (except for overtaking and lane changing). In general, common driving cycles can be categorized into four different types, as shown in Figure 12.2. It includes three types of highway driving cycles such as a smooth highway driving, highway driving cycles with frequent lane changing and overtaking, and highway driving cycles with unexpected traffic idling, and frequent stop-and-go-type city driving cycles. It can be noted from Figure 12.2a that in smooth highway driving, the vehicle accelerates at the beginning to reach an expected speed limit and it continuously cruises with a constant velocity for a significant period and then decelerates to exit the highway. Figure 12.2b shows another type of highway driving cycle, where the driver demands the vehicle to go through frequent overtaking and lane-changing activities. As depicted in Figure 12.2c, in the third type of highway driving cycle, the vehicle goes through unexpected traffic idling. Although it has been divided into three different categories, a combination of all three could also be possible in real highway driving conditions. Figure 12.2d shows frequent stop-and-go-type city driving conditions.

There are many studies that suggest that driving pattern variations are not only decided by external factors such as traffic and other conditions, but also by the driver's age, weather condition, type of vehicle they drive, and so on. Therefore, it is very hard to generalize a single driving cycle, which represents all possible real-world driving scenarios. In this section, we have discussed a few common driving cycles and how they are different from each other. The next section describes the connection between the driving cycle fuel economy enhancement and emission reduction of the vehicle.

### 12.3.1  Effect of Driving Cycles on Fuel Economy Enhancement and Emission Reduction

To understand the effect of driving cycle on fuel economy enhancement and emission reduction, the understanding of vehicle dynamics is very important. Therefore, this section provides a small introduction to simplified vehicle dynamics, which provides the necessary ground work.

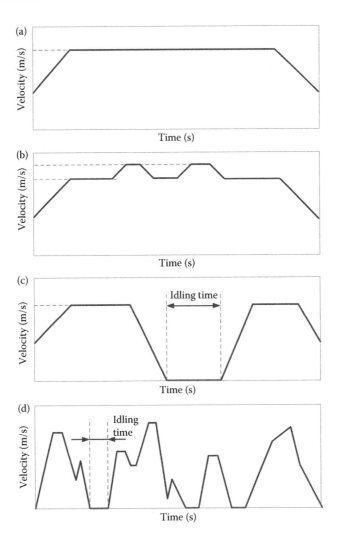

FIGURE 12.2 Different types of real-world driving cycles: (a) Smooth highway driving; (b) highway driving with frequent overtaking and lane-changing activities; (c) highway driving with the unexpected traffic idling; (d) frequent stop-and-go-type city driving.

Figure 12.3 shows a simplified vehicle dynamics, where the vehicle is accelerating with an acceleration rate of $a$ from a velocity $V$ in a road, which makes an angle $B$ to the horizontal. In this driving condition, the propulsive force required to achieve this objective can be given by

$$F_p - F_{r\_f} - F_{r\_r} - F_{aero} - Mg\sin(B) = Ma \tag{12.1}$$

Here

$F_p$: force exerted on the drive wheels by the propulsive system of the vehicle (N)
$F_{r\_f}$, $F_{r\_r}$: the ground resistance on the front and rear wheels of the vehicle (N)
$F_{aero}$: aerodynamic drag force (N)
$a$: acceleration of the vehicle (m/s²)
$M$: mass of the vehicle (kg)
$V$: velocity of the vehicle (m/s)

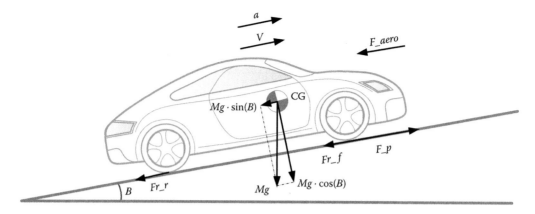

FIGURE 12.3   Simplified force diagram of the vehicle.

$B$: inclination angle of the road
$CG$: center of gravity of the vehicle

From Equation 12.1, the propulsive power requirement of the vehicle can be derived as

$$P_p = (F_{r\_f} + F_{r\_r} + F_{aero} + Mg\sin(B) + Ma)V \qquad (12.2)$$

Here, $P_p$ is propulsive power of the vehicle (W).

The force due to ground resistance can be given by Equation 12.3. This is a function of the mass of the vehicle, road inclination angle, and the adhesive coefficient $\upsilon$ of the road. In general, the adhesive coefficient of the road is very less (between 0.2 and 0.4); therefore, the force due to ground resistance will not go beyond a kilowatt for a typical passenger car under any driving circumstance.

$$F_{r\_f} + F_{r\_r} = \upsilon Mg\cos(B) \qquad (12.3)$$

To make it simpler, let us neglect the ground resistance and assume that the vehicle is traveling on a flat road. Therefore, the power requirement of the vehicle can be further simplified as Equation 12.4.

$$P_p = (F_{aero} + Ma)V \qquad (12.4)$$

The power required to overcome the aerodynamic resistance can be written as

$$P_{aero} = \frac{1}{2}\rho AC_D V^3 \qquad (12.5)$$

Here, the cross-sectional area of the vehicle is given by $A$, air density is given by $\rho$, and the aerodynamic drag coefficient of the vehicle is given by $C_D$. As one can note from Equation 12.5, aerodynamic power holds a linear relationship with the third order of the vehicle velocity. Therefore, the power required to overcome the aerodynamic resistance is significantly high at high-speed conditions and is negligible at low speeds.

The acceleration power of the vehicle can be given by Equation 12.6.

$$P_{acc} = MaV \qquad (12.6)$$

As one can note here, the power required to accelerate the vehicle is a function of the vehicle mass $M$, acceleration rate of the vehicle $a$, and the velocity of the vehicle $V$. Since the mass of the vehicle will remain constant over a given driving condition, the vehicle's velocity $V$ and the expected acceleration rate $a$ at that velocity will determine the acceleration power requirement of the vehicle.

Integrating the propulsive power requirement with respect to time will provide the energy requirement of the vehicle. Therefore, the energy required to achieve a particular driving requirement can be given by Equation 12.7.

$$E_{propulsion} = \int_{t_0}^{t_1} \left( MaV + \frac{1}{2}\rho AC_D V^3 \right) dt \tag{12.7}$$

It should be noted here that vehicle acceleration $a$ and velocity $V$ are time-dependent variables.

Applying these equations to a particular driving cycle, one can calculate the power and energy requirements of the propulsive system to fulfill different phases of the driving cycle. To find out the variation in power and energy requirement, let us investigate a case study.

### 12.3.1.1 Case Study 1

Table 12.2 depicts some basic parameters of a typical passenger car. Assume that, initially, the vehicle is accelerating on a flat road from standstill. As a performance requirement, the vehicle has to reach a velocity of 60 mph within 6 s (which is roughly 26 m/s). After it reaches the velocity of 60 mph, it just cruises for another 6 s and then decelerates for another 6 s to come back to standstill. Let us calculate the maximum propulsive power requirement of the vehicle in each driving condition and also calculate the relevant energy requirements.

The driving cycle of this particular driving scenario can be developed as shown in Figure 12.4a. Here, Phase 1 represents the acceleration, Phase 2 represents the constant velocity cruising, and Phase 3 represents the deceleration of the vehicle.

Let us apply Equations 12.4 and 12.7 to each phase to calculate the propulsive power and energy requirements.

Since the rate of change of velocity is defined as acceleration, the acceleration of the vehicle can be calculated as 4.33 m/s².

*Phase 1*

Applying Equation 12.5, the maximum power required to overcome the aerodynamic resistance can be calculated as 8.5 kW. Similarly, applying Equation 12.6, the maximum power required to overcome the acceleration requirement can be calculated as 180 kW. It clearly shows that the power required to achieve the acceleration requirement of the vehicle is much higher than the power required to overcome the aerodynamic resistance. Therefore, the powertrain should be able to supply more than 188.5 kW at the drive axle in order to achieve this propulsive requirement. As depicted in Figure 12.4b, the vehicle has traveled only a distance of 78 m in Phase 1.

*Phase 2*

Let us apply the same equations to calculate the propulsive requirement of the vehicle in Phase 2. Since there is no acceleration in this phase, it requires only 8.5 kW power to overcome the

---

TABLE 12.2

**Basic Parameters of a Typical Passenger Car**

| Parameter | Value |
|---|---|
| Weight of the car including driver | 1600 kg |
| Aerodynamic drag coefficient | 0.5 |
| Air density | 1.29 kg/m³ |
| Cross-sectional area of the vehicle | 1.5 m² |

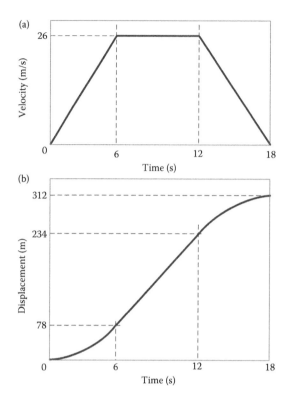

**FIGURE 12.4** (a) Driving cycle of the vehicle. (b) Displacement–time diagram.

aerodynamic resistance, which is significantly lower compared to Phase 1. In this phase, the vehicle has traveled a maximum distance of 156 m. When we compare Phases 1 and 2, it can be clearly seen that the acceleration phase requires a high amount of power to travel a small distance while the cruising phase requires a very little amount of power to travel a longer distance.

*Phase 3*

In Phase 3, since the vehicle is decelerating, the propulsion system need not supply any power. Instead, the brake system should be activated in order to dissipate the stored kinetic energy of the vehicle. By applying Equation 12.4, the maximum power dissipated in the brake system can be calculated as 171.5 kW. It also dissipates a higher amount of power while achieving less distance.

As one can note from Figure 12.5, the energy requirement to achieve the acceleration target and the dissipated energy during braking is significantly higher than the energy requirement for the cruising phase of the vehicle. It clearly shows how a vehicle can be energy thirsty when we consider a driving cycle with frequent acceleration and deceleration phases.

It should be noted here that many different realistic issues are not considered in this study such as

- Operating efficiency of the powertrain system
- Ground resistance of the vehicle
- Road inclination angle (up/down hill driving conditions)
- Ground wind effect on the aerodynamic resistance
- Energy dissipated in the suspension

Therefore, when we consider all these issues, the net energy requirement to achieve propulsive requirement will be even higher than the values depicted in Figure 12.5. However, this simplified study gives more insights on how driving cycle could be crucial to determine the propulsive power

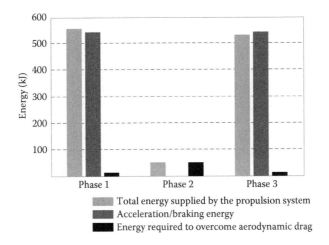

FIGURE 12.5    Energy requirement of the propulsion system to fulfill different phases of the driving cycle.

requirements of the vehicle. This study also suggests that the driving behavior can contribute a lot on fuel saving and emission reduction.

### 12.3.2    HOW HYBRID VEHICLES CAN BENEFIT FROM CONVENTIONAL VEHICLES TO ACHIEVE FUEL ECONOMY

Let us revisit Figure 12.2. With the help of case study 1, it can be very clearly seen how a smooth highway driving cycle can achieve higher fuel economy and emission reduction in any vehicle. Since there is no frequent acceleration and braking events taking place in this driving cycle, both hybrid powertrains and conventional powertrains will provide similar outcomes.

The second category of the highway driving cycle shown in Figure 12.2b incorporates a few lane-changing and overtaking activities. One can understand how this particular driving cycle can consume a significant amount of fuel to achieve the speed ripple requirement of the vehicle. These types of driving cycle demand very high propulsive power from the powertrain. The kinetic energy recovery system in the HEVs can benefit in this driving cycle. However, this kind of driving pattern is not advisable to increase fuel economy regardless of which kind of vehicle someone drives.

On the one hand, when we consider the third category shown in Figure 12.2c, HEVs can benefit a lot compared to conventional vehicles. As this driving cycle goes through a few acceleration and braking events, HEVs can recover a certain amount of kinetic energy of the vehicle to enhance fuel economy. Also, simply shutting off the engine during traffic idling will significantly increase fuel economy and emission in the highway. Although it seems a very simple rule to implement even in a conventional vehicle, drivers do not prefer to do this on their own. On the other hand, the incorporated control system in the hybrid powertrain system very carefully exploits every single opportunity to improve the fuel economy and emission reduction of the vehicle.

In the stop-and-go-type city driving cycles, HEVs show significant advantages over conventional vehicles. Owing to traffic conditions, the nature of the driving cycles in urban areas cannot be changed and the behavioral change of the driver will not improve fuel economy. Frequent acceleration, deceleration, and idling characteristic of the driving cycle provide great opportunity for HEVs to increase fuel economy compared to conventional vehicles. Proper power blending between the IC engine and electric machines enables HEVs to increase the propulsive energy efficiency during the acceleration phase of the vehicle. Similarly, HEVs employ regenerative braking to recover the braking energy of the vehicle, and they simply shut off the powertrain during traffic idling. However, these three opportunities are not available in conventional vehicles. Therefore, HEVs show significant fuel saving in city driving cycles compared to highway driving cycles.

## 12.4 DRIVING CYCLES AND ROAD CONDITIONS ON FUEL ECONOMY

So far, we have assumed flat road driving condition to perform our analysis. However, it is not a valid assumption, since real-world road conditions are not flat as we prefer. In practice, vehicles should be able to drive uphill, downhill, and in flat terrain environments. It implies that vehicles should be able to achieve expected driving cycles in different terrain environments. Therefore, realistic driving cycles should incorporate the geographical information of the road and the velocity variation of the vehicle over the entire duration of the driving.

Figure 12.6 shows all of the driving cycles discussed in Section 12.2, with all possible geographical information. Now, the propulsive power requirement to accomplish the same driving cycles in

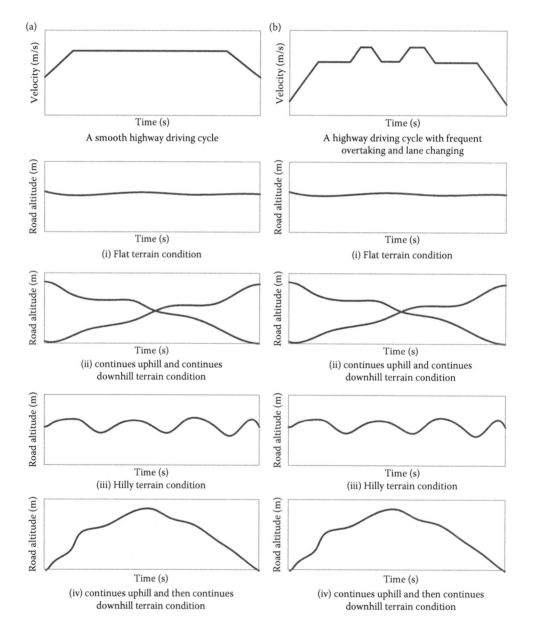

**FIGURE 12.6** Driving cycle with terrain information: (a) Smooth highway driving in different road conditions and (b) highway driving with frequent overtaking and lane-changing activities.

different road condition will vary. However, it should be verified how significant it is on the fuel efficiency of the vehicle. Therefore, let us perform another case study to investigate the effect of driving terrain condition on fuel economy.

### 12.4.1 CASE STUDY 2

Let us consider the driving cycle and vehicle parameters investigated in case study 1 with the terrain information shown in Figure 12.7. For simplicity, only two different terrain conditions are considered here (uphill and downhill). Here, both roads make a 4° angle (which is equivalent to $4\pi/180$ radians) to the horizontal line. Calculate the peak propulsive power and energy requirement of the vehicle to achieve this driving cycle in both road conditions. For simplicity, the ground resistance is neglected in this analysis.

Since there is an inclination, the gravitational acceleration cannot be neglected in this study. Therefore, to calculate the propulsive power, Equation 12.2 has to be modified as Equation 12.8.

$$P_p = (F_{aero} + Mg\sin(B) + Ma)V \tag{12.8}$$

The energy requirement of the vehicle can be calculated by simply integrating Equation 12.8 with respect to time.

$$E_{propulsion} = \int_{t_0}^{t_1}\left( MaV + Mg\sin(B)V + \frac{1}{2}\rho AC_DV^3 \right)dt \tag{12.9}$$

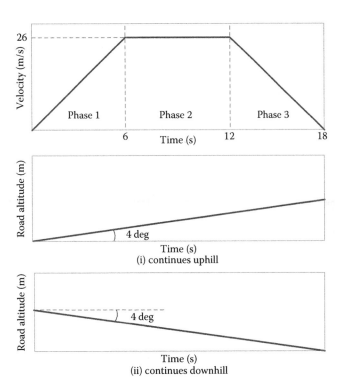

FIGURE 12.7 Driving cycle with the terrain information: (i) uphill road condition, (ii) downhill road condition.

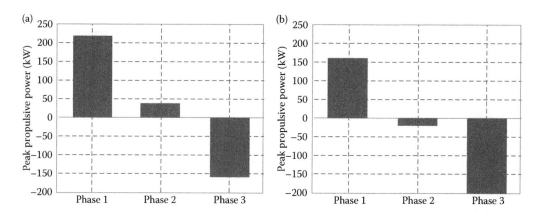

**FIGURE 12.8**   Peak propulsive/regenerative power requirement of the vehicle on (a) uphill driving, (b) downhill driving.

Figure 12.8 shows the peak propulsive power variation of the vehicle in different phases of driving cycles on different road conditions. It can be noted here that even a 4° inclination, either in the negative or positive direction, could significantly change the propulsive power requirement of the vehicle. When we consider the acceleration phase of the vehicle, the uphill road requires 220 kW power while the downhill road requires 160 kW propulsive powers. Also, the potential for energy recovery is considerably high in downhill driving than in uphill driving. Downhill driving can even recover a certain amount of regenerative braking power during constant-speed cursing. Similar trend can be observed in propulsive/regenerative energy variation of the vehicle under different road conditions, as shown in Figure 12.9.

This study clearly shows how the terrain condition can influence the propulsive power/energy requirement of the vehicle and how an HEV can take advantage of different terrain conditions to enhance the fuel economy of the vehicle.

As we incorporate more and more realistic issues in our study, the power and energy requirement of the vehicle increases significantly. From this case study, one can clearly see how driving cycles, terrain conditions, ground wind, and so on can be critical for fuel economy improvement and emission reduction of the vehicle. Also, one can see how HEVs can take advantage over conventional vehicles in different driving environments.

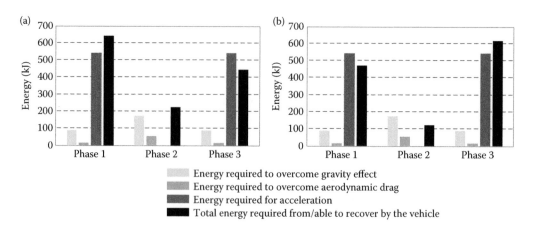

**FIGURE 12.9**   Different energy components and the propulsive/regenerative energy of the vehicle in (a) uphill driving condition, (b) downhill driving condition.

It is obvious that when the driving cycle incorporates more frequent acceleration, deceleration, and idling characteristic, and it has to be achieved in different terrain environments as depicted in Figure 2.6, the energy saving of the HEVs will be significant compared to conventional vehicles.

## 12.5 HEV TECHNOLOGIES

With the development of hybrid technology, varieties of HEV configurations have been created. These hybrid configurations have been widely applied on a range of vehicle types, including passenger cars, sports utility vehicles, trucks, and transit buses. Despite these different configurations and different platforms, there are several ways to classify HEVs into separate groups. One of the generally accepted categorizing methods is to evaluate the ratio of electric systems power to the overall systems power in an HEV. Depending on how the systems are integrated and how much the electric power portion is compared with the overall power, HEVs can be categorized into three groups: micro hybrids, mild hybrids, and full hybrids.

### 12.5.1 MICRO HYBRIDS

Micro hybrids employ a modest electric portion in the power systems. The typical power rating for a micro hybrid sedan is between 3 and 5 kW. Micro hybrids normally refer to hybrid vehicles with the start–stop or idle–stop systems, which automatically shut down the engines when vehicles are coasting, braking, or stopped according to certain road conditions, and restart the engines when the speed is regained. The added electric power system can also be used to help supply power to driving accessories such as power steering and air conditioning. Some micro hybrids are also capable of certain levels of regenerative braking. Although micro hybrids are one of the simplest hybrids among the various hybrid configurations, they can provide up to 10% of fuel economy benefits,[1] especially in urban driving situations, where frequent stop-and-go is inevitable.

Micro hybrids incorporate only a small portion of the electric system, normally a small motor. This results in a relatively simple structural change and cheaper re-engineering costs while significantly increasing fuel economy and cutting air emissions. Therefore, many auto manufacturers applied this technology during the initial transit from conventional petroleum-powered vehicles to HEVs.[2]

BMW's micro hybrids incorporated the efficient dynamics technologies aimed at reducing fuel consumption and air emissions. Both start-and-stop and regenerative braking functions are available in its micro hybrids. Volkswagen also equipped its micro hybrid fleets with similar features under the name of Blue Motion Technologies. FIAT introduced the PUR-O2 in a range of its micro hybrid models, and Mercedes developed the micro hybrid drive (MHD) onto its Smart hybrid, which is reported to increase fuel economy by nearly 8%.

### 12.5.2 MILD HYBRIDS

Mild hybrids have a higher level of electric power rating, typically ranging from 7 to 15 kW for a sedan. Consequently, a higher level of fuel economy gain can be achieved, saving up to 20% in fuel compared with conventional combustion vehicles. Propulsion systems in mild hybrids normally consist of electrical motor–generators between the engine crank shafts and the transmission input shafts. The added motor–generators provide the vehicle with the start–stop function, regenerative braking function, and additional electric power to drive the accessories. Some of the mild hybrids can also provide a modest level of power assistance to the engine.

Similar to micro hybrids, mild hybrids are relatively cost-effective because they require minimal vehicle platform reconstructions and typically maintain the fundamental manufacturing process. The high diameter-to-length ratio of the electric machine results in a high motor inertia such that the original flywheel of the engine can be replaced by the electric machine. Moreover, the electric machine can also function to start the engine and charge the battery; thus, the added costs of the electric machine and its supporting power electronics are offset by the removal of the starter motor and the alternator from the vehicle. As the manufacturing lines are largely retained and no significant changes are required, total costs essentially remain unchanged.

Mild hybrids have been developed by many auto manufacturers. Honda developed the integrated motor assist (IMA) system in 1999 and applied it onto the Honda Insight Hybrid, which was capable of stop-and-start, regenerative braking, and power assisting up to 30% of the engine power. It scored high fuel economy as well as low air emissions, and in 2000 was ranked the most efficient gasoline-fueled vehicle certified by the United States Environmental Protection Agency (EPA). The IMA system was also applied onto Honda Civic hybrid and CR-Z.

The General Motors (GM) Belt Alternator Starter (BAS) system can also be grouped in the mild hybrid category. Similarly, it took advantage of stop-and-start and regenerative braking technologies to improve fuel and driving performance. The 2007 model Chevrolet Silverado Hybrid pickup truck could achieve an overall fuel savings of 12% compared with its nonhybrid version.

Besides Japanese and American automakers, European automakers also came up with mild hybrids. Mercedes equipped its flagship S-Class with the Blue Motion Technologies as mild hybrids. BMW also released its Active Hybrid into their 7-Series mild hybrids.

## 12.5.3 FULL HYBRIDS

Full hybrids have the highest electric portion compared with micro hybrids and mild hybrids. The power rating for a full hybrid sedan is 30 kW or higher. Full hybrids are defined as those gasoline–electric vehicles that can run on either engine-only mode, battery-only mode, or a combination of the two. In addition to the functions that micro hybrids and mild hybrids are capable of, full hybrids can also operate on an all-electric range where only electric motors are used to propel the vehicle and supply all the internal power loads. However, owing to the limited size of the electric machine and the battery pack, full HEVs normally have a relatively short all-electric range with limited power output. Typically, full hybrids can achieve more than 40% of fuel economy gains in city drives and have more electric power assistance to increase driving performance.

Compared with micro and mild hybrids, full hybrids employ the largest electric power portions in the HEV powertrain systems. A larger battery pack is required to achieve the desired electric drive level. Meanwhile, since the motor is directly coupled with the output drive shaft in the electric-only mode, a robust motor with sufficient speed and enough torque is demanded. Full hybrids also have relatively more complicated configurations. Most full hybrids integrate the electric power path with the mechanical power path by means of power split devices such as planetary gear sets. Power split devices serve to divide the power from the onboard power plants, that is, the engine and batteries, and redistribute the power flow between the electric path and the mechanical path to achieve optimal fuel efficiency and driving performance. Power split devices and other added mechanical components all add to the complexity of full hybrid systems. Therefore, though the full hybrids achieve significantly higher fuel economy and better performance, the manufacturing costs also increase as larger battery packs, more powerful electric machines, and more complicated configurations are implemented.

Full hybrids have attained the highest acceptance compared with the other two as full hybrids fulfill the demanded purpose of reducing fuel consumption and air emissions. Up to the second quarter of 2013, Toyota had achieved phenomenal success with its full hybrid models, the Toyota Prius family, which had sold more than 3 million units throughout the world. Toyota Camry Hybrid and Honda Civic Hybrids have also gained quite considerable popularities. GM, Daimler Chrysler, and BMW also released several models based on their two-mode hybrid transmission system, which is a complex full hybrid system capable of both high efficiency and high performance. Many governments around the world all released either targets or regulations to adjust the automobile industry into more hybridized forms, and large incentives were provided to compensate for the initial high costs of the full hybrid technologies. Table 12.3 summarizes the three categories of HEVs.

**TABLE 12.3**

**Three Categories of HEVs and Their Features**

| | Electric Rating | Fuel Saving | Costs Increase | Added Features | Examples |
|---|---|---|---|---|---|
| Micro hybrids | 3–5 kW | ~10% | Low | Start-and-stop<br>Moderate regenerative braking<br>Accessories powering | Mercedes Smart<br>BMW Efficient Dynamics<br>Volkswagen Blue Motion |
| Mild hybrids | 7–15 kW | ~20% | Moderate | Start-and-stop<br>Regenerative braking<br>Accessories powering<br>Moderate electric assistance | Honda Insight<br>Chevrolet Silverado Hybrid<br>Mercedes S-Class Hybrid<br>BMW 7-Series Hybrid |
| Full hybrids | > 30 kW | ~40% | High | Start-and-stop<br>Regenerative braking<br>Accessories powering<br>Electric assistance<br>Electric-only drive | Toyota Prius<br>Ford Escape Hybrid<br>Chevrolet Tahoe Hybrid<br>BMW X6 Active Hybrid |

## 12.6 CLASSIFICATION OF ROAD VEHICLES BASED ON THEIR POWERTRAIN SYSTEM

There are five different types of HEVs available in the market and they can be categorized as mild hybrid, parallel hybrid, plug-in hybrid, series hybrid, and series–parallel hybrid electric vehicles. Moreover, plug-in hybrid electric vehicle is another type of HEV that incorporates a bigger onboard battery for the electric power supply and it can be charged from the electric grid. The available vehicular propulsion technologies can be categorized based on the percentage share of electric and mechanical propulsive systems in the powertrain. Figure 12.10 shows the general trend on how the conventional and electrical propulsion systems contribute to different vehicle architecture.

As shown in Figure 12.10, the name of the vehicle reflects which percentage of the propulsive power is contributed by the IC engine and electric machine. For example, mild hybrid vehicle is mainly powered by the IC engine, while the plug-in hybrid electric vehicle is mainly powered by the electric machines.

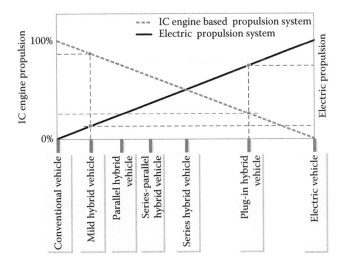

**FIGURE 12.10** Different vehicular architectures and their electric and conventional propulsion systems.

### 12.6.1  Mild Hybrid Electric Vehicles

Mild hybrid electric vehicle incorporates a smaller electric propulsion system and generally it is integrated in the IC engine as depicted in Figure 12.11a. The flywheel system of the IC engine is designed as an electric machine; therefore, the IC engine can be started by electric propulsion and the electric machine can also power assist the IC engine when the power demand increases. Also, it can recover a certain amount of kinetic energy during the braking phase of the vehicle. However, from Figure 12.10, one can easily understand that the fuel economy of the vehicle cannot be significantly improved in different driving cycles since mild hybrid electric vehicles incorporate a small percentage of the electric propulsion system.

### 12.6.2  Parallel Hybrid Electric Vehicles

Significant portion of the propulsive power in parallel hybrid electric vehicle is supplied by the electric machine. Typical electromechanical integration of the parallel hybrid electric vehicle is depicted in Figure 12.11b and it has two independent propulsive power flow paths such as electrical and mechanical paths. Therefore, depending upon the propulsive power requirement of the vehicle, electric only, mechanical only, or a combination of both can be chosen. Since parallel hybrid electric vehicles incorporate bigger electric machines, they have better regenerative braking capability compared to the mild hybrids, and therefore they show better fuel economy for city driving cycle and certain highway driving conditions. As depicted in Figure 12.12, there are four different operating modes that can be obtained in parallel hybrid electric vehicle.

a. IC engine supplies the propulsive torque/power demand.
b. Electric machine is the only propulsive medium.

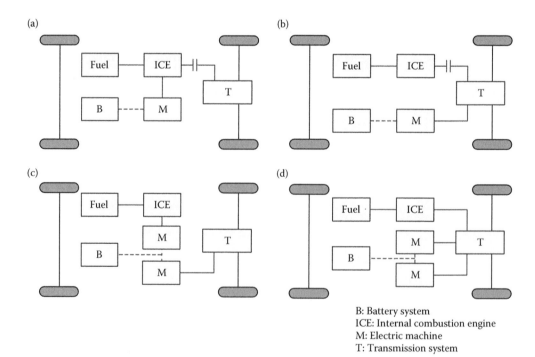

B: Battery system
ICE: Internal combustion engine
M: Electric machine
T: Transmission system

**FIGURE 12.11**  Vehicular architectures and their electromechanical integration methods: (a) mild hybrid electric vehicle, (b) parallel hybrid electric vehicle, (c) series hybrid electric vehicle, and (d) series–parallel hybrid electric vehicle.

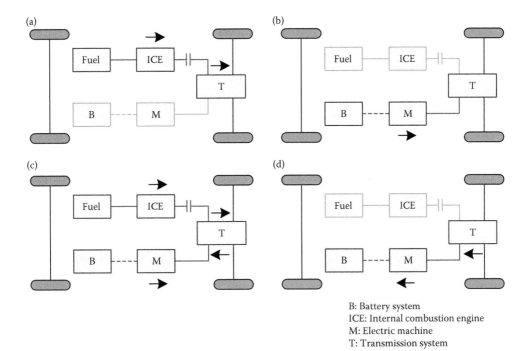

B: Battery system
ICE: Internal combustion engine
M: Electric machine
T: Transmission system

FIGURE 12.12 Operating modes of parallel hybrid electric vehicles: (a) IC engine supplies the propulsive torque/power demand; (b) electric machine is the only propulsive medium; (c) IC engine and electric machines supply the propulsive torque/power demand; (d) regenerative braking mode of the vehicle.

   c. IC engine and electric machines supply the propulsive torque/power demand.
   d. Regenerative braking mode of the vehicle.

### 12.6.3 SERIES HYBRID ELECTRIC VEHICLES

Series hybrid electric vehicles are powered by electric machines. As depicted in Figure 12.11c, series hybrid vehicles incorporate two electric machines and an IC engine in the propulsion system. One electric machine is directly attached with the IC engine and most of the time it functions as an electric generator (during starting of the IC engine, it acts as an electric motor). Another electric machine is integrated with the transmission system to deliver the propulsive power requirement of the vehicle. Since series hybrid electric vehicles incorporate a bigger electric machine in the propulsion system, their energy recovery capability is significantly higher than other types of HEVs. Series hybrid electric vehicles achieve six different operating modes as depicted in Figure 12.13. They can be described as

   a. Electric machine is the only propulsive medium.
   b. IC engine is supplying the propulsive torque/power demand.
   c. IC engine and electric machines are supplying the propulsive torque/power demand.
   d. Regenerative braking mode of the vehicle.
   e. IC engine supplies the propulsive torque/power demand of the vehicle. The excess power of the IC engine is converted into electricity and stored in the battery system.
   f. IC engine charges the battery system (also, battery can receive power from regenerative braking).

Since electric machines are capable of handling a wide range of speed–torque requirements of the vehicle, series hybrid electric vehicle architecture provides benefit for heavy-duty vehicles in

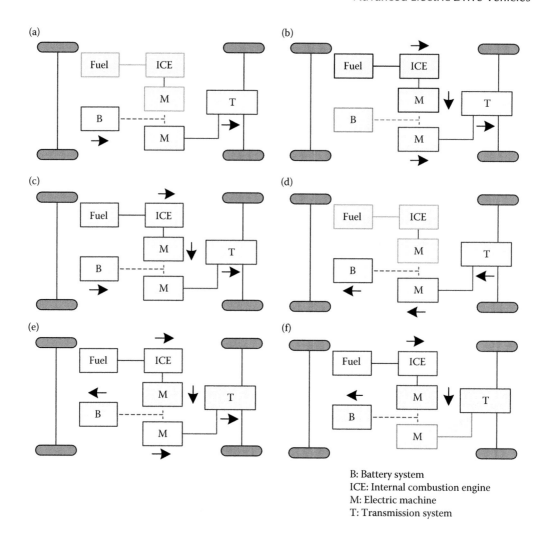

B: Battery system
ICE: Internal combustion engine
M: Electric machine
T: Transmission system

**FIGURE 12.13**  Operating modes of series hybrid electric vehicles: (a) Electric machine is the only propulsive medium; (b) IC engine is supplying the propulsive torque/power demand; (c) IC engine and electric machines are supplying the propulsive torque/power demand; (d) regenerative braking mode of the vehicle; (e) IC engine supplies the propulsive torque/power demand of the vehicle; (f) IC engine charges the battery system (also, battery can receive power from regenerative braking).

terms of simplifying the transmission system design. Since this architecture incorporates bigger propulsive components (e.g., to provide X kW power to the drive wheels, they incorporate 3X kW power sources), this architecture is not appropriate for light-duty vehicles. In heavy-duty vehicles, the space constraint is not very significant; however, space is limited in light-duty vehicles and it has to be effectively managed.

### 12.6.4  SERIES–PARALLEL HYBRID ELECTRIC VEHICLES

The most popular hybrid electric vehicle in the market (Toyota Prius) is a series–parallel hybrid electric vehicle type, which incorporates two electric machines and an internal combustion engine for propulsion purposes, as shown in Figure 12.11d. Contrary to the series hybrid electric vehicles, series–parallel hybrid electric vehicles use low-power electric machines and both of them can act as motor and generator. Here, two electric machines and an IC engine are connected to the drive

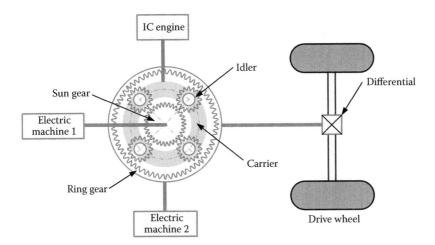

FIGURE 12.14  Planetary gear system and the connection diagram of propulsive systems in series–parallel hybrid electric vehicles. (Adapted from A. Emadi, M. Ehsani, and J. M. Miller, *Vehicular Electric Power Systems: Land, Sea, Air, and Space Vehicles*. New York: Marcel Dekker, December 2003.)

axle via a planetary gear system, as shown in Figure 12.14. The ingenious arrangement of this gear system allows the propulsive system to have much more flexibility while achieving different propulsive requirements. It also achieves many different power flow patterns by choosing different combinations of power sources. As depicted in Figure 12.15, series–parallel hybrid electric vehicles also achieve similar operating modes as series hybrid electric vehicles.

However, the control complexity of the vehicle increases as the number of power flow increases; therefore, many research works are focusing on how to effectively utilize different power sources to achieve various power flow patterns with higher energy efficiency.

### 12.6.5  PLUG-IN HYBRID ELECTRIC VEHICLES

Plug-in hybrid electric vehicles are generally either parallel hybrid electric vehicles or series–parallel hybrid electric vehicles. However, the primary difference between a parallel hybrid electric vehicle and a plug-in hybrid electric vehicle is shown in Figure 12.16. Here, a typical parallel hybrid electric vehicle incorporates a smaller electric propulsion system and a bigger IC engine-based propulsion system. On the other hand, the plug-in hybrid electric vehicle incorporates a bigger electric propulsion system and a smaller IC engine-based propulsion system. Also, the plug-in hybrid electric vehicle incorporates an external charging adopter and it can be used to charge the battery from the grid. Owing to the higher percentage of the electric propulsion system, plug-in HEVs have higher regenerative braking capability compared to traditional HEVs. Plug-in HEVs are highly desirable for city driving cycles.

## 12.7  MAIN CHALLENGES IN HYBRID ELECTRIC VEHICLE DESIGN AND REALIZATION

Although HEVs provide superior energy efficiency compared to conventional vehicles, their design and development process is not very straightforward. This is mainly due to variations in driving cycles and involvement of many different propulsive components. As we have discussed in the driving cycle sections, different driving cycles demand different propulsive power and energy requirements and it is difficult to incorporate all aspects of the driving cycles in a single powertrain design. For example, an HEV with higher percentage of electric propulsion system will achieve better fuel economy and emissions reduction in a city driving cycle. However, in terms of driving range, the very same vehicle will show a poor performance in a smooth flat terrain highway driving cycle.

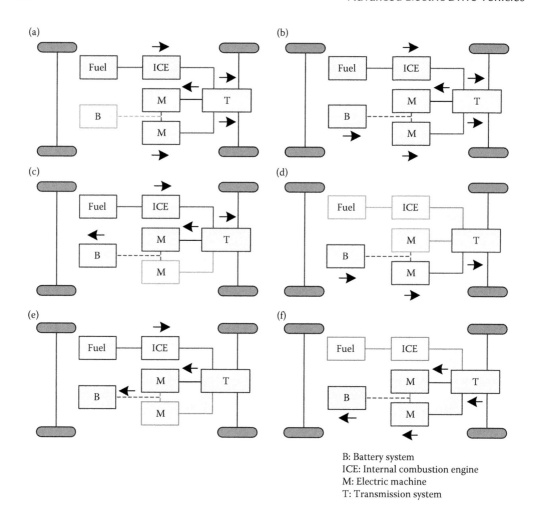

B: Battery system
ICE: Internal combustion engine
M: Electric machine
T: Transmission system

**FIGURE 12.15** Operating modes of series–parallel hybrid electric vehicles: (a) IC engine is the only propulsive medium. Here part of the power produced by the IC engine is converter via motor generator to the traction wheels. (b) IC engine and electric batteries are supplying the propulsive torque/power demand. (c) IC engine is supplying the propulsive torque/power demand while the battery is charging. (d) Batteries are supplying the propulsive demand. (e) IC engine charges the battery while the vehicle is standstill. (f) Regenerative braking of the vehicle.

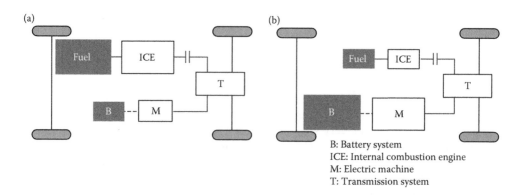

B: Battery system
ICE: Internal combustion engine
M: Electric machine
T: Transmission system

**FIGURE 12.16** Difference between parallel and plug-in parallel hybrid electric vehicles: (a) parallel hybrid electric vehicle, (b) plug-in hybrid electric vehicle.

Therefore, it is very difficult to develop a unique powertrain system, which can provide all of the benefits in different driving environments. This is one of the factors deciding which hybrid vehicular topology is more suitable for which driving cycle.

Let us consider an HEV topology for an investigation. As it has been highlighted in Equation 12.1, the power requirement of the vehicle will increase with the increment of six different factors such as acceleration requirement, vehicle mass, ground resistance, aerodynamic characteristic of the vehicle, uphill driving performance, and the expected maximum velocity of the vehicle. Also, the energy requirement of the vehicle will increase with the increment of power requirement and the driving duration of the vehicle, as shown in Equation 12.9. On the other hand, to supply higher propulsive power requirement of the vehicle, the capacity of the power/energy sources of the vehicle should be increased. For example, to achieve higher acceleration performance, the vehicle should be integrated with a high-power electric machine and an IC engine. In two different ways, this will contribute to the weight increment of the vehicle:

1. Power output of the propulsive component is proportional to the weight increment of the powertrain system. The weight increment is due to the direct contribution of a particular propulsive component.
2. Second, high-power propulsive component requires bigger thermal management system, power electronic converters, transmission system, and so on. This also increases the overall weight of the powertrain.

Therefore, an increment in one propulsive component will have a chain effect on other associated components and it will significantly influence the weight increment of the vehicle. However, this weight increment of the powertrain system will increase the total weight of the vehicle and it demands more power to achieve the expected performance.

This simple weight analysis shows how things are interconnected with each other and how a slight change in one component influences the change in the entire system. Similarly, when we consider the electric-only driving range, the on-board electric battery should be sized appropriately to handle the power/energy requirement of the electric machines and the driving range of the vehicle. Owing to the poor energy density of electric batteries compared to gasoline, to achieve certain electric-only mileage, the battery size needs to be increased significantly. The increment in battery size will increase the overall weight of the vehicle and also limit the available passenger space. Therefore, as depicted in Figure 12.17, the optimization process should be carried out in a comprehensive manner with the consideration of vehicle's performance and its influence on hybrid powertrain design and development.

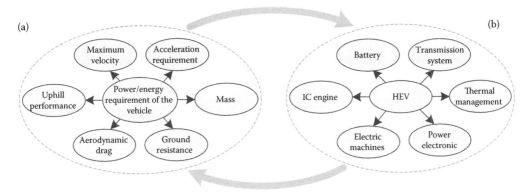

**FIGURE 12.17**  Power and energy requirement of the vehicle and their effect on HEV design. (a) Key design parameters to define the power and energy requirements of the vehicle. (b) Power and energy supplying components of the powertrain.

Apart from the correlation between weight, volume, and performance requirements of the vehicle, there are many other challenges that must be overcome. Production and operational cost of the vehicle is one of the major issues that influence the technology market penetration. No matter how good the technology is, if it is expensive, or if the operational cost is high, then consumers are reluctant for the technology. Therefore, HEVs should be available in an affordable price and it should also possess very less operational cost to get the attention of customers.

## 12.8  AREAS FOR RESEARCH AND DEVELOPMENT

### 12.8.1  Design, Selection, and Sizing of Electric Machines

Selection and sizing of the electric machine is one of the critical design processes in HEV development. There are mainly three electric machine technologies available, such as induction machines, permanent magnet machines, and switch reluctant machines, and they have different operational fundamentals. Permanent magnet machines include surface permanent magnet machines (SPMs), surface insert permanent magnet machines (SIPMs), and interior permanent magnet machines (IPMs). Owing to the extensive operating range, high power/torque density, and their higher operating efficiency, permanent magnet machines are the preferred candidates for many hybrid electric cars. However, there is no fixed design category identified yet. This is partly due to copyright reasons of certain designs and the increasing price of permanent magnet materials.

Permanent magnets show different magnetic properties in different operating temperatures and this has a significant effect on the machine's performance. Therefore, many research investigations are carried out by many experts in this field. Machine sizing and selection is not only limited by the propulsive power/torque requirements of the vehicle but also depends heavily on the design process of power converters and electrical system to gain synergetic advantages. Since this technology is relatively new and very important for HEVs, this area has a great potential for further research and development activities.

### 12.8.2  Energy Storage System

Electrical energy storage system is very critical for HEV performance. Depending on the expected range and required driving cycle, a particular energy storage system will be chosen for a particular vehicle. In general, higher-power and energy-dense storage systems are preferred for HEV applications. As depicted in Figure 12.11, with the increase in electric propulsive component, energy storage requirement also increases. Owing to the safety requirement, nickel metal hybrid battery technology is chosen by Toyota in their first generation of Prius. On the other hand, higher power and energy density of lithium-ion technology appeared to be more attractive and it is becoming the preferred candidate for many automakers. Extensive research activities are performed to ensure the safe operation of the battery system in different operating conditions. Also, hybrid energy storage is another promising area of research, which combines the high-power-dense supercapacitors and high-energy-dense electric batteries together to provide various peak power and energy requirements of the vehicle.

### 12.8.3  Thermal Management System

Various propulsive systems in HEVs require different temperature controls for thermal management system design. For example, the coolant temperature requirement for the IC engine is not similar to the coolant temperature requirement for electric machines. In reality, some components are air cooled; some are water cooled, while others are oil cooled. Therefore, the thermal management system needs to be developed to meet all these requirements.

Since HEVs use different propulsive components in real-time driving, some of the components may not be utilized for certain driving requirements. For example, consider a smooth highway driving condition; here, the involvement of electric propulsion system is less required. In this situation, the thermal management system related to the electric propulsion system should be shut down while the thermal management system related to IC engine should function. This particular attribute of the vehicle demands an adaptive thermal management system design to handle various driving requirements and this is one of the challenging research areas in HEV technologies advancement.

Second, the cabin thermal management is another important issue in HEVs. In conventional vehicles, the heat generated by the IC engine is used to heat the cabin area of the vehicle during winter and an air-conditioning unit is attached to the IC engine to control the cabin temperature in summer. Since the IC engine is continuously running in the conventional car, it is very convenient to control the cabin temperature under any weather condition. However, it is not the case in some of the HEVs, especially for plug-in hybrid electric vehicles. Let us consider a stop-and-go city driving cycle of an HEV. Here, primarily the vehicle will use the electric propulsion system. Therefore, cabin heating/cooling energy should be supplied by electric heaters and electrically operated AC units. This increases the complexity and reduces the energy efficiency of the vehicle.

## 12.9    CASE STUDY 3

Various HEV configurations result in diverse efficiency gains and emissions reductions because of the different combinations of the power usage ratios. Two types of HEV powertrains are compared here as a case study by using the powertrain simulation software Autonomie. The first HEV powertrain is a mild hybrid composed of an integrated starter generator (ISG) paralleled with the IC engine before the transmission. The second HEV powertrain is a full hybrid with the series–parallel configuration by using power split devices. Figure 12.18 illustrates the two HEV powertrain configurations.

In order to compare the performance of the two HEV powertrains, common parameters were assigned in the blocks of chassis, wheels, electrical accessory, and mechanical accessory with the same values. Engine powers and motor powers varied, as shown in Table 12.4.

Two separate driving cycles are applied to the two HEV powertrains, respectively. One is to access the vehicle powertrain local driving performance composed of three U.S. Urban Dynamometer Driving Schedule (UDDS) cycles and the other is for highway driving performance evaluation composed of three Highway Fuel Economy Driving Schedule (HWFET). Table 12.5 summarizes the fuel economy comparison results and the emission results throughout the drive cycles.

Figure 12.19 presents the power compositions of engines and motors going through three UDDS driving cycles for both the parallel ISG mild hybrid and the split full hybrid. It can be observed that the engine in the parallel ISG mild hybrid provided the majority power while the motor provided power assistance and regenerative braking around 10 kW. The loads were further shared by the two motors in the split full hybrid. The engine was operating more in its fuel-efficient regions and the two rotors assisted the performance with higher power output. Besides, more regenerative braking power was retrieved in the split full hybrid to further improve the fuel economy.

Figure 12.20 presents the power compositions of engines and motors going through three HWFET driving cycles for both the parallel ISG mild hybrid and the split full hybrid. The engine almost exclusively provided all the power in the parallel ISG mild hybrid under the highway driving scenario. The motor functioned to retrieve kinetic energy by applying regenerative braking during vehicle deceleration. In the configuration of the split full hybrid, both the engine and Motor B provided the traction power while Motor A largely operated as a generator to charge the battery. The

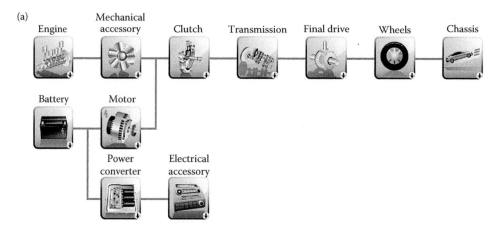

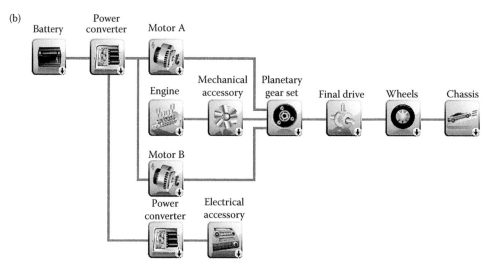

FIGURE 12.18   Two HEV powertrain configuration comparisons. (a) Parallel ISG mild hybrid configuration and (b) Split full hybrid configuration.

TABLE 12.4

## Simulation Input Parameters Comparison

| Components | Parameters | Parallel ISG Mild Hybrid | Split Full Hybrid | |
|---|---|---|---|---|
| Chassis | Mass (kg) | 1600 | 1600 | |
| | Front weight ratio | 0.64 | 0.64 | |
| | Center of gravity height (m) | 0.5 | 0.5 | |
| Wheel | Radius (m) | 0.29 | 0.29 | |
| | Rolling resistance coefficient | 0.007 | 0.007 | |
| Engine | Maximum power (kW) | 115 | 57 | |
| | Maximum torque (Nm) | 220 | 123 | |
| Motor | Maximum power (kW) | 10 | 50 | 15 |
| | Maximum torque (Nm) | 140 | 200 | 77 |
| Final drive | Final drive ratio | 3.63 | 4.113 | |
| Mechanical accessory | Power (W) | 10 | 10 | |
| Electrical accessory | Power (W) | 217 | 217 | |

TABLE 12.5

## Simulation Result Comparisons

| | UDDS | | HWFET | |
| | Parallel ISG | | Parallel ISG | |
| Performance | Mild Hybrid | Split Full Hybrid | Mild Hybrid | Split Full Hybrid |
|---|---|---|---|---|
| Distance (mile) | 22.33 | 22.33 | 30.76 | 30.76 |
| Fuel consumption (gallon) | 0.64 | 0.32 | 0.68 | 0.48 |
| Fuel economy (mile/gallon) | 35.23 | 69.83 | 45.06 | 63.04 |
| Engine average efficiency (%) | 23.95 | 34.25 | 26.98 | 35.03 |
| Regenerative braking recovered percentage (%) | 59.75 | 72.22 | 68.55 | 74.83 |

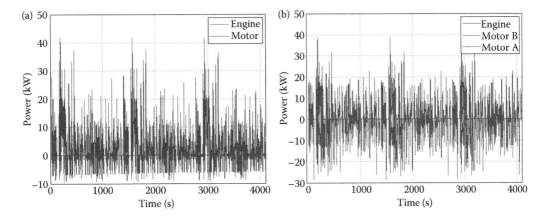

FIGURE 12.19 Engine and motor output power under UDDS driving cycles. (a) Parallel ISG mild hybrid and (b) Split full hybrid.

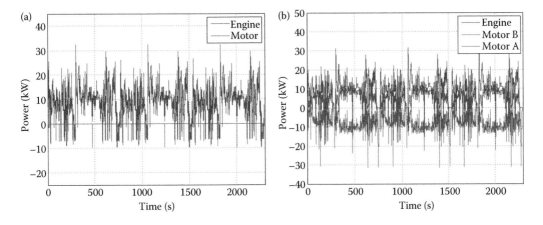

FIGURE 12.20 Engine and motor output power under HWFET driving cycles. (a) Parallel ISG mild hybrid and (b) Split full hybrid.

motors could also switch their operation roles so that Motor A functioned to retrieve regenerative braking energy while motor B operated in the motoring mode.

## QUESTIONS

12.1 Consider the NEDC driving cycle and calculate the theoretical energy requirement of the vehicle shown in Table 12.2. To perform this study, initially assume that the test vehicle is a parallel hybrid electric vehicle with 30% electrical and 70% IC engine-based propulsion system and is driven in a flat road (the peak power of the propulsion system is 200 kW). Following that, consider that the test vehicle is a series hybrid electric vehicle and perform the same study. From the obtained results, compare all three results to identify the suitable vehicle technology for this driving condition.

12.2 Table 12.6 shows the time–velocity information of a sample driving cycle. Consider a conventional vehicle with the vehicle parameters shown in Table 12.2 and calculate the peak power and energy requirement of the vehicle to achieve this driving cycle. To

TABLE 12.6

**Time versus Velocity Information of the Driving Cycle**

| Time (s) | Velocity (m/s) |
| --- | --- |
| 0 | 0 |
| 7 | 26 |
| 17 | 26 |
| 25 | 0 |
| 35 | 0 |
| 44 | 30 |
| 59 | 30 |
| 63 | 23 |
| 83 | 23 |
| 89 | 0 |
| 99 | 0 |
| 105 | 22 |
| 115 | 22 |
| 122 | 0 |
| 137 | 0 |
| 143 | 20 |
| 153 | 20 |
| 160 | 0 |
| 170 | 0 |
| 179 | 27 |
| 194 | 30 |
| 198 | 23 |
| 218 | 23 |
| 224 | 0 |
| 234 | 0 |
| 240 | 22 |
| 250 | 20 |
| 257 | 0 |
| 272 | 0 |
| 278 | 20 |
| 288 | 20 |
| 295 | 0 |

perform the analyses, assume a flat road driving condition. Also, assume that the idling power of the vehicle is equivalent to 15% of the peak power.

a. From this analysis, calculate the acceleration energy requirement, energy dissipated during braking, energy wasted during idling mode of the vehicle, and the total energy requirement of the vehicle to achieve this driving cycle.

b. Consider a different road condition, where the first 50% of the driving distance is a continuous uphill road with the inclination angle of 3° to ground and the following 50% of the driving distance is a continuous downhill with the same inclination angle. Based on this information, calculate the acceleration energy requirement, energy dissipated during braking, energy wasted during idling mode of the vehicle, and total energy requirement of the vehicle to achieve this driving cycle.

c. From the results obtained in the above two studies, determine the theoretical electric propulsion system requirement for a hybrid electric vehicle to recover the braking energy of the vehicle in both road conditions.

## REFERENCES

1. J. M. Miller, Hybrid Electric Vehicle Propulsion System Architectures of the e-CVT Type, *IEEE Transactions on Power Electronics*, 21(3), 756–767, May 2006.
2. F. Orecchini and A. Santiangeli, Automakers' powertrain options for hybrid and electric vehicles, In *Electric and Hybrid Vehicles: Power Sources, Models, Sustainability, Infrastructure and the Market*, Elsevier, United Kingdom, ISBN: 978-0-444-53565-8, 2010.
3. C. C. Chan, The State of the Art of Electric, Hybrid, and Fuel Cell Vehicles, *Proceeding of the IEEE*, 95(4), 704, 718, April 2007.
4. A. E. Fuhs, *Hybrid Vehicles and the Future of Personal Transportation*. Boca Raton, FL: CRC, 2009.
5. A. Emadi, M. Ehsani, and J. M. Miller, *Vehicular Electric Power Systems: Land, Sea, Air, and Space Vehicles*. New York: Marcel Dekker, December 2003.
6. M. Ehsani, Y. Gao, and A. Emadi, *Modern Electric, Hybrid Electric, and Fuel Cell Vehicles: Fundamentals, Theory, and Design*, 2nd Edition, Boca Raton, FL: CRC Press, ISBN: 9781420053982, 2009.

# 13 Fundamentals of Chargers

*Fariborz Musavi*

## 13.1  INTRODUCTION

As the demand for energy drastically increased in the twentieth century, fossil fuels became the main source of energy due to convenience and cost. Over the years, however, the price of oil and problems caused by pollution have increased considerably, putting pressure on governments and industries to invest on other solutions to replace fossil fuels. Consequently, interest in other means of transportation, such as plug-in hybrid electric vehicles (PHEVs) and electric vehicles (EVs), has increased again.

EV technology has existed since the early 1900s. However, the high cost and low energy density of available energy storage systems, primarily batteries, along with the very low cost of oil, had limited the interest in EV and PHEV. Recent innovations in lithium-ion batteries, higher price of gas, and air pollution associated with fossil fuels have significantly impacted the alternative transportation industry.

As a result, PHEVs and EVs are an emerging trend in automotive circles, and consumer's interest is growing rapidly. With the development of PHEVs and EVs, battery chargers for automotive applications are also becoming an essential part of transportation electrification. In addition, the improvement of overall charger efficiency and cost are critical for the emergence and acceptance of these vehicular technologies; as the charger efficiency increases, the charge time and utility cost decreases.

In this chapter, several conventional PHEV charger front-end AC–DC converter topologies and isolated DC–DC topologies are reviewed. Considerations to improve the efficiency and performance, which is critical to minimize the charger size, charging time, and the amount and cost of electricity drawn from the utility, are discussed. A detailed practical example along with an analytical model for these topologies is developed, along with various questions and homework assignments at the end of the chapter.

## 13.2  CHARGER CLASSIFICATION AND STANDARDS

A PHEV is a hybrid vehicle with a storage system that can be recharged by connecting a plug to an external electric power source through an AC or DC charging system. The AC charging system is commonly an on-board charger mounted inside the vehicle and is connected to the grid. The DC charging system is commonly an off-board charger mounted at fixed locations, supplying required regulated DC power directly to the batteries inside the vehicle.

### 13.2.1  AC CHARGING SYSTEMS

The charging AC outlet inevitably needs an on-board AC–DC charger with a power factor correction (PFC). Table 13.1 illustrates charge method electrical ratings according to SAE EV AC charging power levels.

These chargers are classified by the level of power they can provide to the battery pack [1]:

- Level 1: Common household circuit, rated up to 120 V AC and up to 16 A. These chargers use the standard three-prong household connection, and they are usually considered portable equipment.
- Level 2: Permanently wired electric vehicle supply equipment (EVSE) used especially for electric vehicle charging; rated up to 240 V AC, up to 60 A, and up to 14.4 kW.

**TABLE 13.1**

**Charge Method Electrical Ratings—SAE EV AC Charging Power Levels**

| Charge Method | Nominal Supply Voltage | Maximum Current | Branch Circuit Breaker Rating | Output Power Level |
|---|---|---|---|---|
| AC level 1 | 120 V AC, 1-phase | 12 A | 15 A | 1080 W |
| | 120 V AC, 1-phase | 16 A | 20 A | 1440 W |
| AC level 2 | 208–240 V AC, 1-phase | 16 A | 20 A | 3300 W |
| | 208–240 V AC, 1-phase | 32 A | 40 A | 6600 W |
| | 208–240 V AC, 1-phase | ≤80 A | Per NEC 635 | ≤14.4 kW |

- Level 3: Permanently wired EVSE used especially for electric vehicle charging; rated greater than 14.4 kW. Fast chargers are rated as level 3, but not all level 3 chargers are fast chargers. This designation depends on the size of the battery pack to be charged and how much time is required to charge the battery pack. A charger can be considered a fast charger if it can charge an average electric vehicle battery pack in 30 min or less.

In summary

- AC chargers are commonly on-board the vehicle
  - AC is supplied to the vehicle
  - Charger supplies DC to the battery
  - Must be automotive-grade components
- Considerations for reliability, thermal cycling, vibration, lifetime/warranty, and so on
- High cost to produce and low profit margins for suppliers
- AC levels 1 and 2 are the dominant technologies in production today

### 13.2.2 DC CHARGING SYSTEMS

The DC charging systems are mounted at fixed locations, like the garage or dedicated charging stations. Built with dedicated wiring, these chargers can handle much more power and can charge the batteries more quickly. However, as the output of these chargers is DC, each battery system requires the output to be changed for that car. Modern charging stations have a system for identifying the voltage of the battery pack and adjusting accordingly. Table 13.2 illustrates charge method electrical ratings according to SAE EV DC charging power levels.

These chargers are classified by the level of power they can provide to the battery pack [1]:

- Level 1: Permanently wired EVSE includes the charger; rated 200–450 V DC, up to 80 A, and up to 36kW

**TABLE 13.2**

**Charge Method Electrical Ratings—SAE EV DC Charging Power Levels**

| Charge Method | Supplied DC Voltage Range | Maximum Current | Power Level |
|---|---|---|---|
| DC level 1 | 200–450 V DC | ≤80 A DC | ≤36 kW |
| DC level 2 | 200–450 V DC | ≤200 A DC | ≤90 kW |
| DC level 3 | 200–600 V DC | ≤400 A DC | ≤240 kW |

- Level 2: Permanently wired EVSE includes the charger; rated 200–450 V DC, up to 200 A, and up to 90 kW
- Level 3: Permanently wired EVSE includes the charger; rated 200–600 V DC, up to 400 A, and up to 240 kW

In summary

- DC chargers are off-board (not in the vehicle)
    - AC supplied to a charging box
    - Charger supplies DC to the vehicle
- Consumer-grade components
    - Considerations for reliability, thermal cycling, vibration, and so on not as demanding
    - Lower cost to produce and potentially increased profit margins
- DC level 3 Tesla superchargers limited availability
- EVSE includes an off-board charger

## 13.3  CHARGER REQUIREMENTS

Several considerations and regulatory standards must be met. The charger must comply with the following standards for safety:

- UL 2202: EV Charging System Equipment
- IEC 60950: Safety of Information Technology Equipment
- IEC 61851-21: Electric Vehicle Conductive Charging System—Part 21: Electric Vehicle Requirements for Conductive Connection to an AC–DC Supply
- IEC 61000: Electromagnetic compatibility (EMC)
- ECE R100: Protection against Electric Shock
- ISO 6469-3: Electric Road Vehicles—Safety Specifications—Part 3: Protection of Persons against Electric Hazards
- ISO 26262: Road Vehicles—Functional Safety
- SAE J2929: Electric and Hybrid Vehicle Propulsion Battery System Safety Standard
- FCC Part 15 Class B: The Federal Code of Regulation (CFR) FCC Part 15 for EMC Emission Measurement Services for Information Technology Equipment

In addition, it may be affected by high temperatures, vibration, dust, and other parameters, which comprise the operating environment. Therefore, the charger must meet the following operating environment:

- Engine compartment capable
- IP6K9K, IP6K7 protection class
- −40°C to 105°C ambient air temperature
- −40°C to 70°C liquid coolant temperature

The input and output requirements for a level 2, 3.3 kW charger are also given below.
Input:

- Input voltage range: 85–265 VAC
- Input frequency range: 45–70 Hz
- Input current: 16 ARMS max
- Power factor: ≥0.98

Output:

- Output voltage range: 170–440 V DC
- Output power: 3.3 kW max
- Output current: 12 A DC max
- High efficiency: >94%

## 13.4  TOPOLOGY SELECTION FOR LEVEL 1 AND 2 AC CHARGERS

The front-end AC–DC converter is a key component of the charger system. A variety of circuit topologies and control methods have been developed for the PFC application [2,3]. The single-phase active PFC techniques can be divided into two categories: the single-stage approach and the two-stage approach. The single-stage approach is suitable for low-power applications. In addition, owing to large low-frequency ripple in the output current, only lead acid batteries are chargeable. Furthermore, galvanic isolation is required in on-board battery chargers in order to meet the double fault protection for the safety of the users of PHEV. Therefore, the two-stage approach is the proper candidate for PHEV battery chargers, where the power rating is relatively high, and lithium-ion batteries are used as the main energy storage system. The front-end PFC section is then followed by a DC–DC section to complete the charger system.

Figure 13.1 illustrates a simplified block diagram of a universal input two-stage battery charger used for PHEVs and EVs.

The PFC stage rectifies the input AC voltage and transfers it into a regulated intermediate DC link bus. At the same time, PFC function is achieved. The following DC–DC stage then converts the DC bus voltage into a regulated output DC voltage for charging batteries, which is required to meet the regulation and transient requirements.

### 13.4.1  FRONT-END AC–DC CONVERTER TOPOLOGIES

As a key component of a charger system, the front-end AC–DC converter must achieve high efficiency and high power density. Additionally, to meet the efficiency and power factor requirements and regulatory standards for the AC supply mains, PFC is essential.

As the adoption rate of these vehicles increases, the stress on the utility grid is projected to increase significantly at times of peak demand. Therefore, efficient and high power factor charging is critical in order to minimize the utility load stress, and reduce the charging time. In addition, a high power factor is needed to limit the input current harmonics drawn by these chargers and to meet regulatory standards, such as IEC 1000-3-2 [4].

According to the requirements of input current harmonics and output voltage regulation, a front-end converter is normally implemented by a PFC stage. Conventionally, most of the power conversion equipment employs either a diode rectifier or a thyristor rectifier with a bulk capacitor to

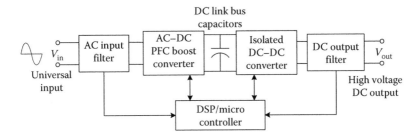

FIGURE 13.1  Simplified system block diagram of a universal on-board two-stage battery charger.

convert AC voltage to DC voltage before processing it. Such rectifiers produce input current with rich harmonic content, which pollute the power system and the utility lines. Power quality is becoming a major concern for many electrical users.

The simplest form of PFC is passive (passive PFC). A passive PFC uses a filter at the AC input to correct poor power factor. The passive PFC circuitry uses only passive components—an inductor and some capacitors. Although pleasantly simple and robust, a passive PFC rarely achieves low total harmonic distortion (THD). Furthermore, because the circuit operates at the low line power frequency of 50 or 60 Hz, the passive elements are normally bulky and heavy. Figure 13.2 shows input voltage and current for a passive PFC and the harmonic spectrum of input current.

The input power factor (PF) is defined as the ratio of the real power over apparent power as

$$\text{Power factor (PF)} = \frac{\text{Real power (W)}}{\text{Apparent power (VA)}} \tag{13.1}$$

Assuming an ideal sinusoidal input voltage source, the power factor can be expressed as the product of two factors, the distortion factor and the displacement factor, given as

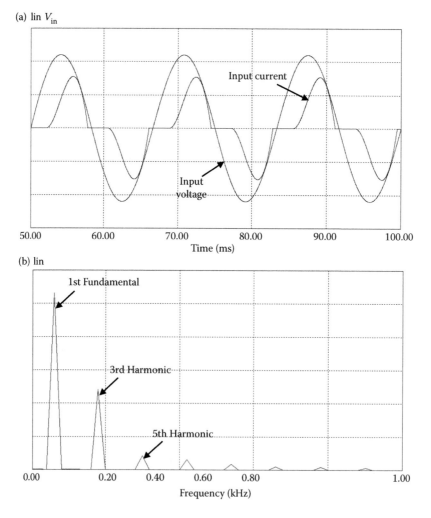

**FIGURE 13.2** Passive power factor correction AC main voltage and current waveforms. (a) Input voltage and input current. (b) Harmonic spectrum of input current.

$$PF = K_\mathrm{d}K_\theta \qquad (13.2)$$

The distortion factor, $K_\mathrm{d}$, is the ratio of the fundamental root mean square (RMS) current to the total RMS current.

a. Input voltage and input current
b. Harmonic spectrum of input current

The displacement factor, $K_\theta$, is the cosine of the displacement angle between the fundamental input current and the input voltage fundamental RMS current.

$$K_\mathrm{d} = \frac{I_{1\mathrm{rms}}}{I_\mathrm{rms}} \qquad (13.3)$$

$$K_\theta = \cos\theta_1 \qquad (13.4)$$

where $I_{1\mathrm{rms}}$ is the fundamental component of the line current, $I_\mathrm{rms}$ is the total line current, and $\theta_1$ is the phase shift of the current fundamental relative to the sinusoidal line voltage.

The distortion factor is close to unity, even for waveforms with noticeable distortion; therefore, it is not a very convenient measure of distortion for practical use. The distortion factor is uniquely related to another figure of merit: the THD.

$$\mathrm{THD} = \sqrt{\frac{I_\mathrm{rms}^2 - I_{1\mathrm{rms}}^2}{I_{1\mathrm{rms}}^2}} \qquad (13.5)$$

$$K_d = \sqrt{\frac{1}{1 + \mathrm{THD}^2}} \qquad (13.6)$$

$K_d$ is regulated by IEC 1000-3-2 for lower power levels and by IEEE Std 519-1992 [5] for higher power levels, where $K_\theta$ is regulated by utility companies.

Significant reduction of current harmonics in single-phase circuits can only be achieved by using rectifiers based on pulse width modulated (PWM) switching converters. These converters can be designed to emulate a resistive load and, therefore, produce very little distortion of the current. By using PWM or other modulation techniques, these converters draw a nearly sinusoidal current from the AC line in phase with the line voltage. As a result, the rectifier operates with very low current harmonic distortion and very high, practically unity power factor. This technique is commonly known as PFC. As a result of this research, the existing PFC technology based on the boost converter topology with average-current-mode control was significantly improved. The proposed improvements allowed an extended range of operating conditions and additional functionality. The following section illustrates several common PFC topologies suitable for PHEV charger applications.

### 13.4.1.1 Conventional Boost PFC Converter

The conventional boost topology is the most popular topology for PFC applications. It uses a dedicated diode bridge to rectify the AC input voltage to DC, which is then followed by the boost section, as shown in Figure 13.3.

In this topology, the output capacitor ripple current is very high and is the difference between diode current and the DC output current. Furthermore, as the power level increases, the diode bridge losses significantly degrade the efficiency, so dealing with the heat dissipation in a limited area

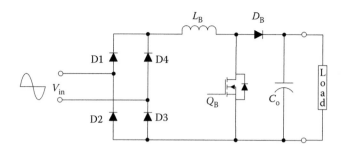

FIGURE 13.3   Conventional PFC boost converter.

becomes problematic. The inductor volume also becomes a problematic design issue at high power. Another challenge is the power rating limitation for current sense resistors at high power.

### 13.4.1.2   Interleaved Boost PFC Converter

The interleaved boost converter, illustrated in Figure 13.4, consists of two boost converters in parallel operating at 180° out of phase [6–8].

The input current is the sum of the two input inductor currents. Because the inductors' ripple currents are out of phase, they tend to cancel each other and reduce the input ripple current caused by the boost switching action. The interleaved boost converter has the advantage of paralleled semiconductors. Furthermore, by switching 180° out of phase, it doubles the effective switching frequency and introduces smaller input current ripple, so the input EMI filter is relatively small [9,10]. With ripple cancellation at the output, it also reduces stress on output capacitors.

### 13.4.1.3   Bridgeless Boost PFC Converter

The bridgeless boost topology, illustrated in Figure 13.5, is the second topology considered for this application. The gates of the powertrain switches are tied together, so the gating signals are identical, as is illustrated in Figure 13.6. It avoids the need for the rectifier input bridge, yet maintains

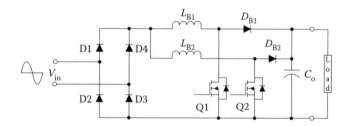

FIGURE 13.4   Interleaved PFC boost converter.

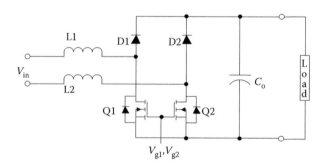

FIGURE 13.5   Bridgeless PFC boost topology.

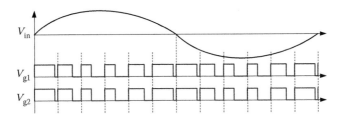

FIGURE 13.6 Gating scheme for the bridgeless PFC boost topology illustrating the identical gating signals for both MOSFETs.

the classic boost topology [11–14]. It is an attractive solution for applications >1 kW, where power density and efficiency are important. The bridgeless boost converter, also known as dual-boost PFC converter, solves the problem of heat management in the input rectifier diode bridge, but it introduces increased electromagnetic interference (EMI) [15–17]. This is because the amplitude of the noise source applied to the stray capacitor from high-voltage DC bus and power ground is a lot higher in bridgeless PFC; as a result, the common mode (CM) noise generated by bridgeless PFC is much higher than conventional boost PFC topology. Another disadvantage of this topology is the floating input line with respect to the PFC stage ground, which makes it impossible to sense the input voltage without a low-frequency transformer or an optical coupler.

### 13.4.1.4 Dual-Boost PFC Converter

The dual-boost converter, illustrated in Figure 13.7, is an alternative adaptation of the bridgeless boost topology [18]. In this topology, the MOSFET gates are decoupled, enabling one of the switches to remain on and operate as a synchronous MOSFET for half-line cycle. Figure 13.8 illustrates the gating scheme for a dual-boost PFC topology. The dual-boost topology reduces gate loss, and at light loads, conduction loss can be reduced until the voltage drop across the MOSFET channel $R_{DS(ON)}$ becomes equal to the voltage drop across the MOSFET body diode, at which point any additional current conducts through the body diode. The light load efficiency improvement comes at the expense of the cost of an additional driver and increased controller complexity.

### 13.4.1.5 Semi-Bridgeless Boost PFC Converter

The semi-bridgeless configuration, shown in Figure 13.9, includes the conventional bridgeless topology with two additional slow diodes, Da and Db, that connect the input to the PFC ground [19]. The slow diodes were added to address EMI-related issues [15,16]. The current does not always return through these diodes, so their associated conduction losses are low. This occurs since the inductors exhibit low impedance at the line frequency, so a large portion of the current flows through the MOSFET intrinsic body diodes. The semi-bridgeless configuration also resolves the floating input line problem with respect to the PFC stage ground. The topology change enables input voltage sensing using a string of simple voltage dividers.

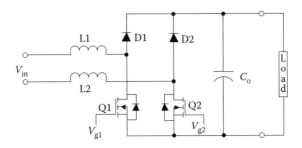

FIGURE 13.7 Dual-boost PFC topology.

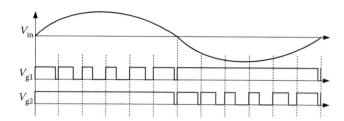

FIGURE 13.8   Gating scheme for the dual-boost PFC topology illustrating half-line cycle synchronous rectification.

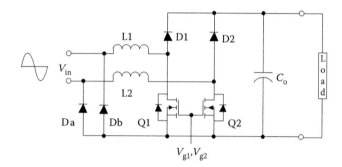

FIGURE 13.9   Semi-bridgeless PFC boost topology.

### 13.4.1.6   Bridgeless Interleaved Boost PFC Converter

The bridgeless interleaved topology, shown in Figure 13.10, was proposed as a solution to operate at power levels above 3.5 kW. In comparison to the interleaved boost PFC, it introduces two MOSFETs and also replaces four slow diodes with two fast diodes. The gating signals are 180° out of phase, similar to the interleaved boost. A detailed converter description and steady-state operation analysis are given in Reference 20. This converter topology shows a high input power factor, high efficiency over the entire load range, and low input current harmonics.

Since the proposed topology shows high input power factor, high efficiency over the entire load range, and low input current harmonics, it is a potential option for single-phase PFC in high-power level 2 battery charging applications.

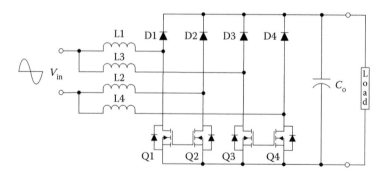

FIGURE 13.10   Bridgeless interleaved PFC boost converter. (From F. Musavi; W. Eberle; W.G. Dunford, *IEEE Transactions on Industry Applications*, 47, July/August 2011.)

## 13.4.2 ISOLATED DC–DC CONVERTER TOPOLOGIES

Many high-efficiency full-bridge DC–DC converter solutions have been proposed that are potential candidates for the isolated DC–DC converter in a PHEV charger. The DC–DC converter requirements for battery chargers are

- Galvanic isolation (regulatory requirement)
- Suitable for high power (>1 kW)
- High efficiency (>95%)
- Soft-switching (ZVS and zero current switching (ZCS)) (>100 kHz operation)
- Low EMI
- Low output voltage/current ripple (avoid battery heating)
- Small size
- Cost effective

### 13.4.2.1 Zero Voltage Switching Full-Bridge Phase-Shifted Converter

The phase-shifted zero voltage switching (ZVS) PWM DC-to-DC full-bridge converter, illustrated in Figure 13.11, was presented in References 21–23. ZVS for the switches is realized by using the leakage inductance of the transformer in addition to an external inductor and the output capacitance of the switch.

There are several issues with this topology. Although various improvements have been suggested for this converter, these solutions increase the component count and suffer from one or more disadvantages, including a limited ZVS range, high-voltage ringing on the secondary-side rectifier diodes, or duty cycle loss. The wide ZVS range of operation is discussed in References 24 and 25, and the high-voltage ringing on the secondary-side rectifier diodes is addressed in References 26–29. Duty cycle loss is reviewed in Reference 30. A new complementary gating scheme for the full-bridge DC-to-DC PWM converter is presented in Reference 31. This gating scheme requires an additional zero voltage transition (ZVT) circuit to achieve ZVS for all the switches for a wide variation in the load current. This topology is more suitable for low output voltage, high output current applications.

### 13.4.2.2 Zero Voltage Switching Full-Bridge Trailing-Edge PWM Converter

The full-bridge ZVS converter with trailing-edge PWM converter presented in Figure 13.12 behaves similar to a traditional hard-switched topology, but rather than simultaneously driving the diagonal bridge switches, the lower switches (Q3 and Q4) are driven at a fixed 50% duty cycle, and the upper switches (Q1 and Q2) are PWM on the trailing edge [32].

A clamp network consisting of $D_C$, $R_C$, and $C_C$ is needed across the output rectifier to clamp the voltage ringing due to diode junction capacitance with the leakage inductance of the transformer. This DC–DC converter also suffers from duty cycle loss. Duty cycle loss occurs for converters

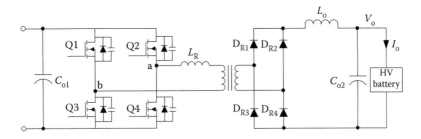

FIGURE 13.11   ZVS F.B. phase-shifted DC–DC converter topology.

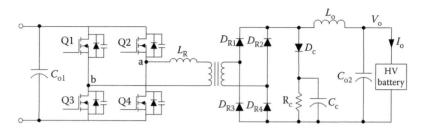

**FIGURE 13.12**  Improved ZVS F.B. trailing-edge DC–DC converter topology.

requiring inductive output filters when the output rectifiers commutate, enabling all of the diodes to conduct, which effectively shorts the secondary winding. This causes a decrease in the output voltage; thus, a higher transformer turn ratio is needed, which increases the primary peak current. This topology is also more suitable for low output voltage, high output current applications.

### 13.4.2.3  Zero Voltage Switching Full Bridge with Capacitive Output Filter Converter

The ZVS full-bridge converter topology with capacitive output filter is illustrated in Figure 13.13. Current-fed topologies with capacitive output filter inherently minimize diode rectifier ringing since the transformer leakage inductance is effectively placed in series with the supply-side inductor [33]. In addition, high efficiency can be achieved with ZVS; in particular, the trailing-edge PWM full-bridge gating scheme proposed in Reference 34 is an attractive solution to achieve ZVS.

The converter primary-side circuit consists of a traditional full-bridge inverter, but rather than driving the diagonal bridge switches simultaneously, the lower switches (Q3 and Q4) are driven at a fixed 50% duty cycle and the upper switches (Q1 and Q2) are pulse width modulated on the trailing edge. Although the proposed converter can operate in either discontinuous conduction mode (DCM), boundary conduction mode (BCM), or continuous conduction mode (CCM), only the DCM and BCM modes are desirable.

### 13.4.2.4  Interleaved Zero Voltage Switching Full Bridge with Capacitive Output Filter Converter, Operating in BCM

The current-fed topologies suffer from high ripple current stress at the output filter capacitors. The interleaved ZVS full bridge with capacitive output filter converter, illustrated in Figure 13.14, reduces the input and output filtering requirements and also reduces the reverse recovery losses in the secondary rectifier diodes.

An interleaved, multicell configuration that uses two cells (each rated at 1.65 kW) in parallel (at both input and output) with each cell being phase-shifted by 180° (=360°/2) is adopted for this high-power application [35]. Owing to interleaving, each cell shares equal power and the thermal losses are distributed uniformly among the cells and also the input/output ripple is four times the switching frequency.

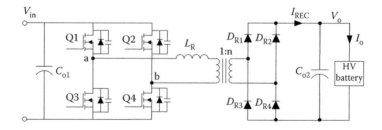

**FIGURE 13.13**  ZVS F.B. trailing-edge DC–DC converter topology with capacitive output filter.

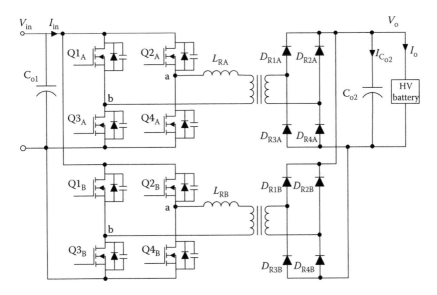

**FIGURE 13.14**   Interleaved ZVS F.B. trailing-edge DC–DC converter with capacitive output filter.

### 13.4.2.5   Interleaved Zero Voltage Switching Full Bridge with Voltage Doubler, Operating in BCM

An interleaved ZVS full bridge with voltage doubler, illustrated in Figure 13.15, is proposed to further reduce the ripple current and voltage stress on the output filter capacitors, as well as component cost reduction.

Owing to interleaving, each cell shares equal power and the thermal losses are distributed uniformly among the cells and also the input ripple is four times the switching frequency. Moreover, the output voltage doubler rectifier significantly reduces the number of secondary diodes and the voltage rating of the diodes is equal to the maximum output voltage [36].

Although the proposed converter can operate in either DCM, BCM, or CCM, only the DCM and BCM modes are desirable. Operation in CCM results in the lowest RMS currents and ZVS

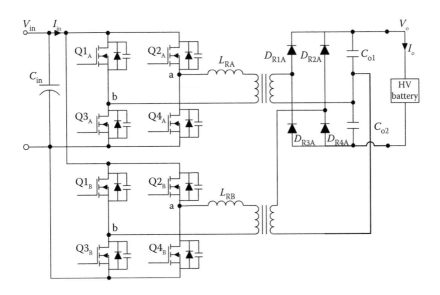

**FIGURE 13.15**   Interleaved ZVS F.B. trailing-edge DC–DC converter with voltage doubler.

can be achieved for all switches, but the high *di/dt* results in large reverse recovery losses in the secondary-side rectifier diodes and high-voltage ringing. Moreover, to operate this converter in CCM, it requires a larger resonant inductor, which also increases the transformer turns ratio and thus increases stress on the primary-side switches. Thus, this converter should be designed to operate in DCM, or BCM.

### 13.4.2.6  Full-Bridge LLC Resonant DC–DC Converter

Figure 13.16 illustrates a full-bridge LLC resonant converter. The LLC resonant converter is widely used in the telecom industry for its high efficiency at the resonant frequency and its ability to regulate the output voltage during the holdup time, where the output voltage is constant and the input voltage might drop significantly. However, the wide output voltage range requirements for a battery charger are drastically different and challenging compared to telecom applications, which operate in a narrow output voltage range.

Figure 13.17 illustrates a family of typical DC gain characteristics for an LLC converter as a function of normalized switching frequency for seven different load conditions varying from no-load to short circuit. Resonance occurs at unity gain, where the resonant capacitors and series resonant inductor are tuned.

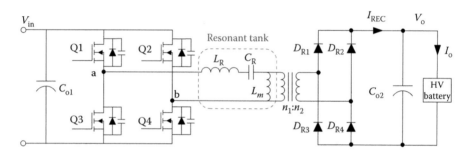

**FIGURE 13.16**  Full-bridge LLC resonant DC–DC converter topology.

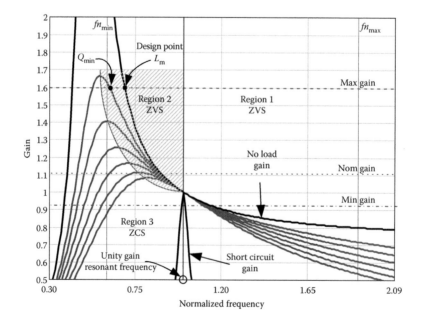

**FIGURE 13.17**  Typical DC gain characteristics of a LLC converter using FHA.

A detailed design procedure and resonant tank selection for LLC resonant converters in battery charging application is given in Reference 37.

## 13.5 TOPOLOGY SELECTION FOR LEVEL 3 CHARGERS

Level 3 chargers are mostly permanently wired EVSE used especially for electric vehicle charging and rated greater than 14.4 kW. These chargers are mainly off-board connected to a three-phase supply. Fast chargers are rated as level 3, but not all level 3 chargers are fast chargers. A charger can be considered a fast charger if it can charge an average electric vehicle battery pack in 30 min or less. This charger is also considered DC charger, since the interface between the charger and vehicle is through DC connector.

Figure 13.18 illustrates a typical block diagram for EV, PHEV charger level 3 using digital power controllers, communication devices, high-performance drivers, and interface devices [38].

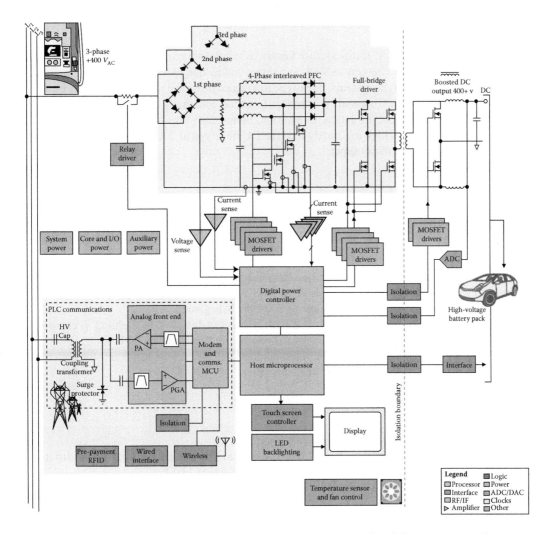

**FIGURE 13.18** Typical block diagram for EV, PHEV charger level 3 using digital power controllers, communication devices, high-performance drivers, and interface devices. (From http://www.ti.com/solution/ev_hev_charger_level_3, Texas Instrument.)

As it can be noted, the supplied voltage to the charger is three-phase 400 V AC, so it requires a three-phase PFC converter for the mains interfaces, followed by a high-power high-voltage DC–DC converter(s).

### 13.5.1 FRONT-END AC–DC CONVERTER TOPOLOGIES

These EV chargers, supplied from three-phase AC lines, typically require a peak power ranging from 10 to 150 kW in order to inject direct current into the battery sets at variable voltage levels according to the vehicle (50–600 V).

The three-phase unity power factor converter options for the mains interface of high-power level 3 chargers are [39,40]:

a. Three single-phase PFC converters connected in Y or Δ configuration
b. Buck-type three-phase PFC rectifiers
c. Boost-type three-phase PFC rectifiers

A simple, reliable solution would be connecting three single-phase PFC boost converters in either Y or Δ configuration.

Buck-type three-phase PFC rectifiers, also known as current source rectifiers (CSRs), are appropriate for these high-power chargers as well, as a direct connection to the DC bus could be used.

Compared to the boost-type systems, buck-type topologies provide a wider output voltage control range, while maintaining PFC capability at the input and can potentially enable direct startup, while allowing for dynamic current limitation. In addition, three-phase boost-type rectifiers generate an output voltage, which is too high to directly feed the DC bus (typical 700–800 V), requiring a step-down DC–DC converter at their output.

### 13.5.2 ISOLATED DC–DC CONVERTER TOPOLOGIES

The second DC–DC stage of these high-power chargers is essentially the modular building blocks consisting of solutions given in Section 13.4.2.

## 13.6 PRACTICAL EXAMPLE

In this section, a practical design example for a level 2, 3.3 kW universal input two-stage PHEV battery charger, with an output voltage range of 200–450 V is given.

Figure 13.19 illustrates a 3.3 kW on-board battery charger designed by Delta-Q Technologies Corporation. The charger requirements and specifications are illustrated in Table 13.3.

A step-by-step design consideration and methodology will be reviewed in detail for both front-end PFC boost converter and isolated DC–DC converter.

### 13.6.1 FRONT-END PFC BOOST CONVERTER DESIGN

#### 13.6.1.1 Topology Selection

The first step for any converter design is topology selection. The topologies of choice for front-end PFC boost converters in level 2 chargers are discussed in Section 13.4.1.

The selected topology in this example is a two-channel interleaved boost converter, as illustrated in Figure 13.4.

The switching frequencies for the PFC and the DC-to-DC converter stages are selected to be 70 and 200 kHz, respectively. To achieve high efficiency (e.g., >97%) for the hard-switched interleaved PFC and limit the fundamental switching frequency ripple to below 150 kHz in order to meet the EMI requirements, a switching frequency of 70 kHz was selected.

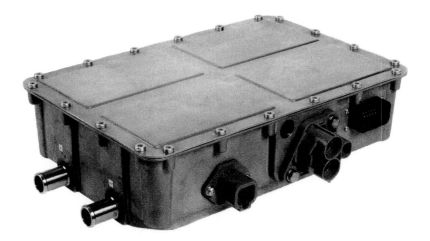

FIGURE 13.19 A 3.3 kW on-board PHEV battery charger designed by Delta-Q Technologies Corp.

In order to design the interleaved PFC converter, it can be treated as two conventional boost PFC converters with each operating at half of the load power rating. With this approach, all equations for the inductor, switch, and diode in the conventional PFC remain valid, since the stresses are unchanged with the only exception being the reduced ripple current through the output capacitors.

### 13.6.1.2 PFC Boost Converter Inductor Design

The minimum required boost inductor value in each phase for low line is given by Equation 13.7, where the minimum duty cycle at low line is defined by Equation 13.8 and $\Delta I_{\text{L--LL}}$ is the desired inductor current ripple at low line.

$$L_{\text{B}} = \frac{\sqrt{2} V_{\text{in\_min}} D_{\text{min\_LL}}}{f_{\text{s}} \Delta I_{\text{L--LL}}} \approx 400 \, \mu\text{H} \tag{13.7}$$

TABLE 13.3

**Level 2, 3.3 kW Charger Requirements and Specifications**

| Parameters | Value (Units) |
|---|---|
| Input AC voltage | 85–265 (V) |
| Maximum input AC current | 16 (A) |
| Power factor @ F.L. and 240 V input | 99 (%) |
| AC input frequency | 47–70 (Hz) |
| THD at full load and 240 V input | <5 (%) |
| Overall charger efficiency | Up to 94 (%) |
| Output DC voltage range | 200–450 (V) |
| Maximum output DC current | 11 (A) |
| Maximum output power | 3.3 (kW) |
| Output voltage ripple | <2 (Vp-p) |
| Cooling | Liquid |
| Dimensions | 273 × 200 × 100 (mm) |
| Mass/volume | 6.2 (kg)/5.46 (L) |
| Operating temperature | −40°C to +105°C ambient |
| Coolant temperature | −40°C to +70°C |

$$D_{\min\_LL} = 1 - \frac{\sqrt{2}V_{in\_min}}{V_{PFC\_bus}} \sin\left(\frac{\pi}{2}\right) \tag{13.8}$$

### 13.6.1.3  PFC Bus Capacitor Selection

The PFC bus capacitor is determined by Equation 13.9, where the maximum holdup time required for the PFC bus is given by Equation 13.10 and $\Delta V_{PFC\text{-}bus}$ is the intended low-frequency ripple across the PFC bus capacitor:

$$C_{PFC} = \frac{2P_o T_{Hold\_Up}}{V_{PFC\_bus}^2 - (V_{PFC\_bus} - \Delta V_{PFC\_bus})^2} \tag{13.9}$$

$$T_{Hold\_Up} = \frac{1}{4}\frac{1}{2f_{Line}} \tag{13.10}$$

In addition to the capacitor value, the capacitor ripple current handling must be considered as well. The high-frequency ripple current in the PFC capacitor is given by Equation 13.11, where $I_o$ is given by Equation 13.12 and $\eta_{PFC}$ is the efficiency of the PFC stage.

$$I_{C\_HF} = I_o\sqrt{\frac{16V_{PFC\_bus}}{6\pi\sqrt{2}V_{in\_min}} - \eta_{PFC}^2} \tag{13.11}$$

$$I_o = \frac{P_o}{V_{PFC\_bus}} \tag{13.12}$$

The low-frequency ripple current in the PFC capacitor is given by Equation 13.13.

$$I_{C\_LF} = \frac{I_o}{2} \tag{13.13}$$

### 13.6.2  ISOLATED DC–DC CONVERTER DESIGN

#### 13.6.2.1  Topology Selection

The topologies of choice for front-end PFC boost converters in level 2 chargers are discussed in Section 13.4.2. The selected topology in this example is a full-bridge ZVS converter with trailing-edge PWM converter, as illustrated in Figure 13.12.

The full-bridge DC–DC converter was designed to operate at a PFC bus voltage, $V_{PFC}$, of 400 $V$ and an output voltage, $V_o$, of 400 V at full load. Initially, a peak-to-peak output ripple current, $\Delta I_o$, of 1 A was assumed. Including dead-time and duty cycle loss, an effective duty cycle of 0.75 was assumed.

#### 13.6.2.2  Transformer Design

Following the assumptions made, the transformer turns ratio is determined to be 0.75 using Equation 13.14.

$$n_t = \frac{D_{eff}V_{in}}{V_o} \tag{13.14}$$

A custom planar-type ferrite transformer was designed using turns ratio of 12(Np):16(Ns).

### 13.6.2.3   Output Filter Inductor Design

An output filter inductor value of 400 µH was selected using Equation 13.14.

$$L_o = \frac{(V_{in}/n_t - V_o)D_{eff}}{\Delta I_d 2 f_s} \tag{13.15}$$

### 13.6.2.4   Resonant Inductor Design

The calculated resonant inductor is given in Equation 13.16. A 6-µH resonant inductor was selected, which is smaller as compared to the value calculated using Equation 13.16. A toroidal (iron powder core) inductor was used to obtain 4 µH and an additional 2 µH was obtained using the transformer leakage inductance.

$$L_R = \frac{n_t V_{in}(1 - D_{eff})}{4 \Delta I_d f_s} = 8 \, \mu H \tag{13.16}$$

## 13.7   WIRELESS CHARGERS

### 13.7.1   Introduction

On-board chargers are burdened by the need for a cable-and-plug charger, galvanic isolation of the on-board electronics, the size and weight of the charger, and safety and issues with operating in rain and snow. Wireless power transfer (WPT) is an approach that provides a means to address these problems and offers the consumers a seamless and convenient alternative to charging conductively. In addition, it provides an inherent electrical isolation and reduces on-board charging cost, weight, and volume [41].

A typical closed-loop inductive WPT charging system is illustrated in Figure 13.20 [42,43]. The basic principle of inductive WPT charging is that the two halves of the inductive coupling interface consist of the primary and secondary of a two-part transformer. The charger converts the low-frequency AC utility power to high-frequency AC power in the power conversion stage. The secondary side wirelessly receives high-frequency AC from the charger, which is converted to DC by a rectifier, which then supplies the battery pack.

A two-part transformer behaves like mutual-inductively coupled or magnetically coupled inductors configured such that a change in current flow through one winding induces a voltage across the ends of the other winding through electromagnetic induction, as shown in Figure 13.21.

The inductive coupling between two conductors is given by Equations 13.7 and 13.8.

$$v_1 = L_1 \frac{di_1}{dt} + M \frac{di_2}{dt} \tag{13.17}$$

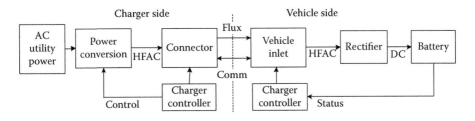

**FIGURE 13.20**   Typical closed-loop WPT charging systems.

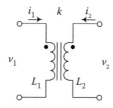

FIGURE 13.21 Coupled inductor circuit symbol.

$$v_2 = L_2 \frac{di_2}{dt} + M \frac{di_1}{dt} \tag{13.18}$$

In Equations 13.7 and 13.8, $M$ denotes the mutual inductance, as given by Equation 13.9, where $k$ is the coupling coefficient of the windings, or the quality of the magnetic circuit.

$$M = k\sqrt{L_1 L_2} \tag{13.19}$$

For a current $I_1$ in $L_1$, the open circuit voltage induced in $L_2$ is given by Equation 13.10.

$$V_{OC} = \omega M I_1 \tag{13.20}$$

With a short circuit on the right-hand side, the current is given by Equation 13.11.

$$I_{SC} = \frac{V_{OC}}{\omega L_1} = I_1 \frac{M}{L_2} \tag{13.21}$$

When the system is tuned at the operating frequency with a capacitor, the available power is $V_{OC}$ $I_{SC}$ multiplied by the circuit tuning resonant factor $Q$, and is given by Equation 13.12, where $Q$ is given by Equation 13.13 [44].

$$P = \omega \frac{M^2}{L_2} I_1^2 Q = \omega L_1 I_1 I_1 \frac{M^2}{L_1 L_2} Q = V_1 I_1 k^2 Q \tag{13.22}$$

$$Q = \frac{\omega L}{R_L} \tag{13.23}$$

In Equation 13.12, the first two terms are the input voltage and current, the third term is the magnetic coupling factor, and the final term is the secondary circuit $Q$. The power that an inductive WPT system can produce is therefore dependent on the input voltage–ampere product ($VA$) to the primary pad, the quality of the magnetic circuit ($k$), and the quality of the secondary electric circuit ($Q$).

### 13.7.2 Inductive Charging

In the 1990s, electric vehicles used inductive charging. An inductive charger uses mutual inductance to transfer electrical energy from the source to the vehicle. This works much the way a transformer works. In this system, an insulated paddle containing an electrically energized primary coil is brought close to a secondary coil within the vehicle. The magnetic field of the primary coil then induces a charge in the secondary coil.

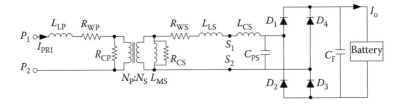

**FIGURE 13.22** Inductive interface (paddle) equivalent circuit.

The charging paddle (the primary coil) of the Magne Charge inductively coupled charger was sealed in epoxy as was the secondary. The paddle inserted into the center of the secondary coil permitted charging of the EV1 without any contacts or connectors at either 6.6 or 50 kW. It should be noted that this system is connectorless, but not wireless.

The equivalent circuit parameters at the charge coupling interface for an IPT charger are shown in Figure 13.22.

### 13.7.3 RESONANT INDUCTIVE CHARGING

Resonant inductive power transfer (RIPT) is the most popular current WPT technology [45,46]. It was pioneered by Nikola Tesla and has recently become popular again, enabled by modern electronic components. This technique uses two or more tuned resonant tanks resonating at the same frequency [47].

A typical schematic of an RIPT system is illustrated in Figure 13.23. The receiver and transmitter contain resonant capacitors, Cp and Cs. Various resonant compensation topologies are proposed in Reference 48. As noted in Reference 47, the primary functions of the resonant circuits include

- Maximizing the transferred power
- Optimizing the transmission efficiency
- Controlling the transmitted power by frequency variation
- Creating a certain source characteristic (current or voltage source)
- Compensating variation of the magnetic coupling
- Compensating the magnetizing current in the transmitter coil to reduce generator losses
- Matching the transmitter coil impedance to the generator
- Suppressing higher harmonics from the generator

Efficient resonant magnetic power coupling can be achieved at distances up to approximately 40 cm. RIPT systems have several advantages over IPT, including increased range, reduced EMI, higher-frequency operation, resonant switching of the inverter and receiver rectification circuitry, and higher efficiency. However, the main advantage of this concept is that the operating frequency is in the kHz range, which can be supported by current state-of-the-art power electronics technologies.

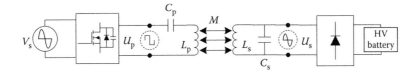

**FIGURE 13.23** Simplified typical schematic of a resonant inductive charger.

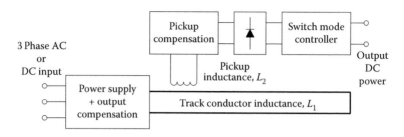

**FIGURE 13.24**  A typical on-line wireless power transfer system [49–58].

### 13.7.4  ROADWAY/ON-LINE CHARGING

The application of RIPT technology in public transit systems has been proposed in References 49–58. An on-line wireless power transfer system (OLPT) is illustrated in Figure 13.24. The concept is similar to RIPT; however, a lower resonant frequency is used and the technology has the potential for application at high power levels. Technologically, the primary coil is spread out over an area on the roadway and the power transfer happens at multiple locations within this area. Typically, the combination of the input side of the resonant converter along with the distributed primary windings is called the track and is on the road, and the secondary is called the pickup coil, which is in the vehicle. The system is supplied by a three-phase AC system, or high-voltage DC system. Considering both the short range of EVs and the associated cost of infrastructure, the feasibility of these charging systems might be unfavorable. However, one benefit is that due to frequent and convenient charging, vehicles can be built with a minimal battery capacity (about 20% compared to that of the conventional battery-powered electric vehicles), which can consequently minimize the weight and the price of the vehicle [56].

### QUESTIONS AND PROBLEMS

13.1  What are the differences between AC charger and DC charger systems?

13.2  What are the classifications of chargers based on their power level?

13.3  What are the most common topologies for front-end AC–DC converters in a level 2 charger?

13.4  What are the most common topologies for front-end DC–DC converters in a level 2 charger?

13.5  Derive a formula to show the relationship between PF and THD.

13.6  In a conventional PFC boost converter operating in CCM, consider all components to be ideal. Let $V_{in}$ be 85–265 V, $V_o = 400$ V (regulated), $P_o = 1650$ W, and $f_s = 70$ kHz. Calculate $L_{min}$, low-frequency and high-frequency ripple current in the output capacitors, and the capacitance value if the desired low-frequency voltage ripple at the output is assumed to be 10% of nominal output voltage.

13.7  Repeat problem 13.6 for a bridgeless boost converter and $P_o = 3300$ W.

13.8  In a ZVS full-bridge DC–DC converter topology with capacitive output filter, what are the modes of operation to be considered? Let $V_{in} = 400$ V, $V_o = 200–450$ V, $f_s = 200$ kHz, and $P_o = 3300$ W. Calculate $L_{R\_min}$, low-frequency and high-frequency ripple current in the output capacitors, and the capacitance value if the desired low-frequency voltage ripple at the output is assumed to be 2% of nominal output voltage (300 V).

13.9  Use the parameters given in problem 13.8 for an LLC resonant DC–DC converter. Find the operating frequency range, resonant tank component values, and low-frequency and high-frequency ripple current in the output capacitors, and the capacitance value if the desired low-frequency voltage ripple at the output is assumed to be 2% of nominal output voltage (300 V).

13.10  Using PSIM, connect the converter in problems 13.7 and 13.9 and simulate the whole system, and verify both the high-frequency and low-frequency ripple currents through the DC link capacitor. Compare your results with the values calculated in problems 13.7 and 13.9.

## REFERENCES

1. Surface Vehicle Recommended Practice J1772, SAE Electric Vehicle and Plug in Hybrid Electric Vehicle Conductive Charge Coupler. SAE International, January 2010.
2. B. Singh; B.N. Singh; A. Chandra; K. Al-Haddad; A. Pandey; D.P. Kothari, A review of single-phase improved power quality AC–DC converters, *IEEE Transactions on Industrial Electronics*, 50, 962–981, 2003.
3. C. Qiao; K.M. Smedley, A topology survey of single-stage power factor corrector with a boost type input-current-shaper, *IEEE Transactions on Power Electronics*, 16, 360–368, 2001.
4. Compliance testing to the IEC 1000-3-2 (EN 61000-3-2) and IEC 1000-3-3 (EN 61000-3-3) Standards. Agilent Technology.
5. IEEE Std 519-1992 IEEE recommended practices and requirements for harmonic control in electrical power systems, *IEEE* 1992.
6. M. O'Loughlin, An interleaved PFC preregulator for high-power converters. Topic 5, in *Texas Instrument Power Supply Design Seminar*, 2007, pp. 5–1, 5–14.
7. L. Balogh; R. Redl, Power-factor correction with interleaved boost converters in continuous-inductor-current mode, in *IEEE Applied Power Electronics Conference and Exposition*, San Diego, CA, 1993, pp. 168–174.
8. Y. Jang; M.M. Jovanovic, Interleaved boost converter with intrinsic voltage-doubler characteristic for universal-line PFC front end, *IEEE Transactions on Power Electronics*, 22, 1394–1401, 2007.
9. P. Kong; S. Wang; F.C. Lee; C. Wang, Common-mode EMI study and reduction technique for the interleaved multichannel PFC converter, *IEEE Transactions on Power Electronics*, 23, 2576–2584, 2008.
10. C. Wang; M. Xu; F.C. Lee; B. Lu, EMI study for the interleaved multi-channel PFC, in *IEEE Power Electronics Specialists Conference, PESC*, Orlando, FL, 2007, pp. 1336–1342.
11. B. Lu; R. Brown; M. Soldano, Bridgeless PFC implementation using one cycle control technique, in *IEEE Applied Power Electronics Conference and Exposition*, Austin, TX. vol. 2, 2005, pp. 812–817.
12. U. Moriconi, A Bridgeless PFC Configuration Based on L4981 PFC Controller. STMicroelectronics Application Note AN1606, 2002.
13. Y. Jang; M.M. Jovanovic, A bridgeless PFC boost rectifier with optimized magnetic utilization, *IEEE Transactions on Power Electronics*, 24, 85–93, 2009.
14. L. Huber; J. Yungtaek; M.M. Jovanovic, Performance evaluation of bridgeless PFC boost rectifiers, *IEEE Transactions on Power Electronics*, 23, 1381–1390, 2008.
15. F.C. Pengju Kong; Shuo Wang; Lee, Common mode EMI noise suppression for bridgeless PFC converters, *IEEE Transactions on Power Electronics*, 23, 291–297, January 2008.
16. T. Baur; M. Reddig; M. Schlenk, Line-conducted EMI-behaviour of a high efficient PFC-stage without input rectification, Infineon Technology—Application Note, 2006.
17. H. Ye; Z. Yang; J. Dai; C. Yan; X. Xin; J. Ying, Common mode noise modeling and analysis of dual boost PFC circuit in *IEEE International Telecommunications Energy Conference, INTELEC*, Chicago, IL, 2004, pp. 575–582.
18. T. Qi; L. Xing; J. Sun, Dual-boost single-phase PFC input current control based on output current sensing, *IEEE Transactions on Power Electronics*, 24, 2523–2530, 2009.
19. F. Musavi; W. Eberle; W.G. Dunford, A phase-shifted gating technique with simplified current sensing for the semi-bridgeless AC–DC converter, *IEEE Transactions on Vehicular Technology*, 62, 1568–1576, 2013.
20. F. Musavi; W. Eberle; W.G. Dunford, A high-performance single-phase bridgeless interleaved PFC converter for plug-in hybrid electric vehicle battery chargers, *IEEE Transactions on Industry Applications*, 47, 1833–1843, July/August 2011.
21. L.H. Mweene; C.A. Wright; M.F. Schlecht, A 1 kW, 500 kHz front-end converter for a distributed power supply system, *IEEE Transactions on Power Electronics*, 6, 398–407, 1991.
22. D.B. Dalal, A 500 kHz multi-output converter with zero voltage switching, in *IEEE Applied Power Electronics Conference and Exposition, APEC*, 1990, pp. 265–274.
23. J.A. Sabate; V. Vlatkovic; R.B. Ridley; F.C. Lee; B.H. Cho, Design considerations for high-voltage high-power full-bridge zero-voltage-switched PWM converter in *IEEE Applied Power Electronics Conference and Exposition, APEC*, 1990, pp. 275–284.

24. K. Gwan-Bon; M. Gun-Woo; Y. Myung-Joong, Analysis and design of phase shift full bridge converter with series-connected two transformers, *IEEE Transactions on Power Electronics*, 19, 411–419, 2004.
25. M. Borage; S. Tiwari; S. Bhardwaj; S. Kotaiah, A full-bridge DC-DC converter with zero-voltage-switching over the entire conversion range, *IEEE Transactions on Power Electronics*, 23, 1743–1750, 2008.
26. W. Xinke; Z. Junming; X. Xiaogao; Q. Zhaoming, Analysis and optimal design considerations for an improved full bridge ZVS DC-DC converter with high efficiency, *IEEE Transactions on Power Electronics*, 21, 1225–1234, 2006.
27. W. Xinke; X. Xiaogao; Z. Junming; R. Zhao; Q. Zhaoming, Soft switched full bridge DC-DC converter with reduced circulating loss and filter requirement, *IEEE Transactions on Power Electronics*, 22, 1949–1955, 2007.
28. C. Wu; R. Xinbo; Z. Rongrong, A novel zero-voltage-switching PWM full bridge converter, *IEEE Transactions on Power Electronics*, 23, 793–801, 2008.
29. C. Wu; R. Xinbo; C. Qianghong; G. Junji, Zero-voltage-switching PWM full-bridge converter employing auxiliary transformer to reset the clamping diode current, *IEEE Transactions on Power Electronics*, 25, 1149–1162, 2010.
30. J. Yungtaek; M.M. Jovanovic, A new PWM ZVS full-bridge converter, *IEEE Transactions on Power Electronics,* 22, 987–994, 2007.
31. A.K.S. Bhat; L. Fei, A new gating scheme controlled soft-switching DC-to-DC bridge converter, in *Power Electronics and Drive Systems, 2003. The Fifth International Conference on PEDS 2003*, Singapore, vol. 1, 2003, pp. 8–15.
32. D.S. Gautam; F. Musavi; M. Edington; W. Eberle; W.G. Dunford, An automotive onboard 3.3-kW battery charger for PHEV application, *IEEE Transactions on Vehicular Technology*, 61, 3466–3474, 2012.
33. I.D. Jitaru, A 3 kW soft switching DC-DC converter, in *IEEE Applied Power Electronics Conference and Exposition, APEC*, New Orleans, LA, vol. 1, 2000, pp. 86–92.
34. D. Gautam; F. Musavi; M. Edington; W. Eberle; W.G. Dunford, A zero voltage switching full-bridge DC-DC converter with capacitive output filter for a plug-in-hybrid electric vehicle battery charger, in *Applied Power Electronics Conference and Exposition (APEC), 2012 Twenty-Seventh Annual IEEE*, Orlando, FL, 2012, pp. 1381–1386.
35. D. Gautam; F. Musavi; M. Edington; W. Eberle; W.G. Dunford, An interleaved ZVS full-bridge DC-DC converter with capacitive output filter for a PHEV charger, in *Energy Conversion Congress and Exposition (ECCE), 2012 IEEE*, Raleigh, NC, pp. 2827–2832.
36. D. Gautam; F. Musavi; M. Edington; W. Eberle; W.G. Dunford, An isolated interleaved DC-DC converter with voltage doubler rectifier for PHEV battery charger, in *Applied Power Electronics Conference and Exposition (APEC), 2013 Twenty-Eighth Annual IEEE*, Long Beach, CA, pp. 3067–3072.
37. F. Musavi; M. Craciun; D.S. Gautam; W. Eberle; W.G. Dunford, An LLC resonant DC-DC converter for wide output voltage range battery charging applications, *IEEE Transactions on Power Electronics*, 28, 5437–5445, 2013.
38. http://www.ti.com/solution/ev_hev_charger_level_3, Texas Instrument.
39. J.W. Kolar; T. Friedli, The essence of three-phase PFC rectifier systems—Part I, *IEEE Transactions on Power Electronics*, 28, 176–198, 2013.
40. T. Friedli; M. Hartmann; J.W. Kolar, The essence of three-phase PFC rectifier systems—Part II, *IEEE Transactions on Power Electronics*, 29, 543–560, 2014.
41. F. Musavi; W. Eberle, An overview of wireless power transfer technologies for EV battery charging, *IET Power Electronics*, 7(1), 60–66, 2014.
42. SAE J1773, SAE Electric Vehicle Inductively Coupled Charging, 1999.
43. SAE J2954, Wireless Charging of Plug-in Vehicle and Positioning Communication, 2012.
44. http://www.qualcomm.com/media/documents/, Inductive Power Transfer systems (IPT) Fact Sheet: No. 1—Basic Concepts, J.T. Boys; G.A. Covic, Eds., 2013.
45. A. Karalis; A.B. Kurs; R. Moffatt; J.D. Joannopoulos; P.H. Fisher; M. Soljačić, Wireless energy transfer. USA: Massachusetts Institute of Technology, Patent # 7825543, 2008.
46. A. Kurs; A. Karalis; R. Moffatt; J.D. Joannopoulos; P. Fisher; M. Soljačić, Wireless power transfer via strongly coupled magnetic resonances, *International Science Journal, American Association for the Advancement of Science (AAAS)*, 317, 83–86, 2007.
47. E. Waffenschmidt, Inductive wireless power transmission, in *Technical Educational Seminar, IEEE Energy Conversion Congress & Exposition*, Phoenix, AZ, 2011, pp. 1–128.
48. O.H. Stielau; G.A. Covic, Design of loosely coupled inductive power transfer systems, in *International Conference on Power System Technology, PowerCon 2000*, Perth, Australia, vol. 1, 2000, pp. 85–90.
49. L. Farkas, High power wireless resonant energy transfer system, USA, Patent # 20080265684, 2007.

50. R.A. Pandya; A.A. Pandya, Wireless charging system for vehicles, USA, Patent # 8030888, 2008.
51. G.A. Covic; J.T. Boys, Power demand management in inductive power transfer systems. New Zealand: Auckland Uniservices Limited, Patent Application # PCT/NZ2010/000181, 2010.
52. C.Y. Huang; J.T. Boys; G.A. Covic; M. Budhia, Practical considerations for designing IPT system for EV battery charging, in *IEEE Vehicle Power and Propulsion Conference (VPPC)*, 2009, pp. 402–407.
53. H.H. Wu; J.T. Boys; G.A. Covic, An AC processing pickup for IPT systems, *IEEE Transactions on Power Electronics*, 25, 1275–1284, 2010.
54. J. Huh; S. Lee; C. Park; G.H. Cho; C.T. Rim, High performance inductive power transfer system with narrow rail width for on-line electric vehicles, in *IEEE Energy Conversion Congress and Exposition (ECCE)*, Atlanta, GA, 2010, pp. 647–651.
55. S. Lee; J. Huh; C. Park; N.S. Choi; G.H. Cho; C.T. Rim, On-line electric vehicle using inductive power transfer system, in *IEEE Energy Conversion Congress and Exposition (ECCE)*, Atlanta, GA, 2010, pp. 1598–1601.
56. S. Ahn; J. Kim, Magnetic field design for high efficient and low EMF wireless power transfer in on-line electric vehicle, in *Proceedings of the 5th European Conference on Antennas and Propagation (EUCAP)*, Rome, Italy, 2011, pp. 3979–3982.
57. C.S. Wang; O.H. Stielau; G.A. Covic, Design considerations for a contactless electric vehicle battery charger, *IEEE Transactions on Industrial Electronics*, 52, 1308–1314, 2005.
58. G.A. Covic; J.T. Boys; M.L.G. Kissin; H.G. Lu, A three-phase inductive power transfer system for roadway-powered vehicles, *IEEE Transactions on Industrial Electronics*, 54, 3370–3378, 2007.

# 14 Plug-In Hybrid Electric Vehicles

*Yinye Yang, Weisheng Jiang, and Piranavan Suntharalingam*

## CONTENTS

## 14.1 INTRODUCTION

Plug-in hybrid electric vehicle (PHEV) is another type of emerging vehicle that combines alternative fuels to displace the oil consumptions in conventional vehicles. As the name suggests, PHEVs are a special type of hybrid electric vehicles (HEVs). Similar to HEVs, PHEVs integrate the electric power path with the mechanical power path by using both conventional combustion engines (ICE) and electric machines. They can also be charged directly by plugging the wire into the wall to get power from the grid (hence the name).

The differences between PHEVs and HEVs primarily lie in battery capacity and recharging methods. PHEVs are equipped with larger battery capacities that are capable of operating on battery power alone for a considerable range, which is called all-electric driving range. Typically, this all-electric range (AER) is designed to meet the daily driving requirements of PHEV owners, especially city drivers and suburban commuters. It is estimated that in Europe, 50% of trips are less than 10 km (6.25 miles) and 80% of trips are less than 25 km (15 miles). In the United Kingdom, 97% of trips are less than 80 km (50 miles). In the United States, about 60% of vehicles are driven less than 50 km (31.25 miles) daily, and about 85% are driven less than 100 km (60miles).[1] Therefore, a PHEV with an electric range of 60 miles would meet most of the trip range requirements in Europe and America, which is denoted as PHEV-60 (or PHEV-100 km). Figure 14.1 shows the typical U.S. daily travel distance distributions.[2]

In addition, Figure 14.2 plots the single-trip distance within the day trips published by the 1995 National Household Travel Survey in the United States that combined the survey results for more than 400 thousand interviewees.[3]

It is apparent that the majority of the single-trip travel distance lies less than 10 miles per trip, which is kept well within the AER of almost all the PHEVs. In between any of the two trips, it is possible to recharge the vehicle batteries at home, at work, at the parking lots in front of the grocery stores, at the electric charging station in public, and so on.

The battery-recharging capability in PHEVs by plugging the vehicle directly into external electric power outlets makes another major difference compared to HEVs. This is also the key benefit of PHEVs since petroleum is no longer the only fuel source for the vehicle. In fact, electricity serves the larger part of the energy supply in PHEVs and thus, the energy dependence on petroleum products is greatly reduced. Typically, the electric energy comes from the electrical grids, which might be a selection from conventional coal energy, nuclear energy, or the renewable energies such as wind energy and solar energy. Depending on how energy is generated at different regions, different levels of well-to-wheel fuel economy and emission reductions can be achieved. Thus, compared with conventional ICEs that exclusively rely on petroleum fuel, PHEVs offer the option to choose from cheap and clean energy sources that generate electricity, reducing the reliance on either petroleum energy or any other single form of energy. In the United States, the number of renewable electricity generation plants[4] is substantially increasing as shown in Figure 14.3.

Generally speaking, PHEVs involve higher degrees of electric power portion and require higher performance from the electric power system, while the mechanical power system is reduced to a minimal level so that it can help sustain the electric power system. Table 14.1 compares PHEVs, different configurations of HEVs, and conventional vehicles.

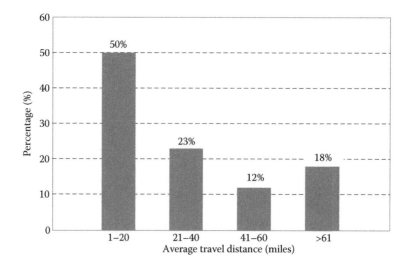

FIGURE 14.1    U.S. daily travel distances distributions. (Adapted from L. Sanna, *EPRI Journal*, Fall, 2005.)

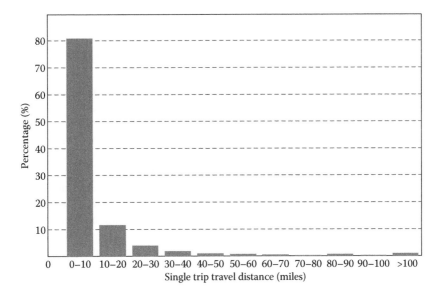

**FIGURE 14.2**    Single-trip travel distance. (Adapted from Day Trips, 1995. National Personal Transportation Survey (NPTS), Research and Innovative Technology Administration, Bureau of Transportation Statistics.)

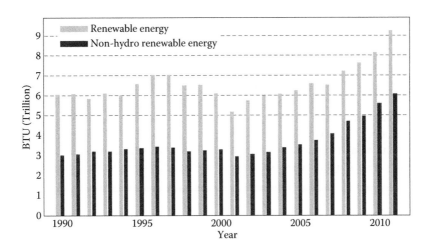

**FIGURE 14.3**    U.S. renewable electricity generation. (Adapted from U.S. Energy Information Administration, *Electric Power Annual and Electric Power Monthly* [March 2012] based on preliminary 2011 data.)

**TABLE 14.1**

## PHEVs, Different Types of HEVs, and Conventional Vehicle Comparisons

|  | Stop and Start | Regenerative Braking | Motor Assistance | Electric Driving | External Battery Charge |
|---|---|---|---|---|---|
| Conventional vehicles | Mostly no | No | No | No | No |
| Micro-HEVs | Yes | Minimum | No | No | No |
| Mild HEVs | Yes | Yes | Minimum | No | No |
| Full HEVs | Yes | Yes | Yes | Yes | No |
| PHEVs | Yes | Yes | Yes | Yes | Yes |

## 14.2 FUNCTIONS AND BENEFITS OF PHEV

PHEVs combine the function of HEVs and electric vehicles (EVs) in a large sense, running both on electricity and liquid petroleum fuels. Large battery capacities enable PHEVs to operate in all-electric mode as much as possible, thus reducing fuel consumption with the use of cheaper and cleaner electricity. However, PHEVs still have a much shorter all-electric driving range compared with pure EVs due to battery costs, which limit battery capacities. Therefore, all-electric driving mode is mostly used within urban driving or daily commuting. After discharging the battery power to a certain low level, the engine starts to charge the battery, and the PHEV switches from operating like an EV to working as an HEV instead. Both the electric power system and the mechanical power system are utilized to supply the vehicle power, and an extended driving range is achieved in the form of hybrid operation. In addition, the range anxiety associated with EV drivers that the battery might deplete while driving is much alleviated in PHEVs since the engine provides a backup source of power and extends the driving range of PHEVs as good as other conventional gas-powered vehicles.

Similar to HEVs, PHEVs have been mainly developed to deal with the three emerging issues in the vehicle transportation sector: fossil energy security, vehicle air pollution, and climate change due to greenhouse gas (GHG) emissions.

Reducing oil consumption in the transportation sector is the primary objective for which PHEVs are designed. According to the International Energy Outlook 2010[5] released by the U.S. Energy Information Administration, the world's annual oil consumption reached 495 quadrillion Btu in 2007, increasing by 36% from the 1980 level. And it is the transportation sector that accounts for the largest oil consumption and shows the largest growth in oil demand during the past few decades. Especially with the current soaring demand from developing countries, the oil consumption rate is increasing faster than ever, which can also be revealed in the substantial increase trend of the crude oil price. Figure 14.4 shows the U.S. crude oil refiner acquisition costs from 1968 to 2011.[6]

In addition, energy security is emphasized by many nations as a major priority. Take the United States as an example, the United States reliance on imported crude oil[7] has been declining since 2009 as shown in Figure 14.5, and the government is still urging for less-imported crude oil by producing more domestic crude oil and alternative energy as well as encouraging measures to increase vehicle fuel efficiency.[7] The fuel efficiency standard released by the U.S. government in 2012 in collaboration with major auto manufacturers, the United Auto Workers, consumers, and environmental groups required cars and light trucks to achieve an average of 54.5 miles per gallon by 2025,[8] saving

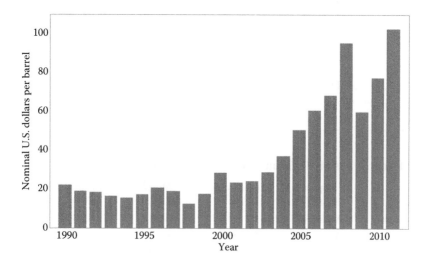

FIGURE 14.4    U.S. crude oil refiner acquisition costs. (Adapted from *Annual Energy Review* 2010, U.S. Energy Information Administration, Report number: DOE/EIA-0384, 2010.)

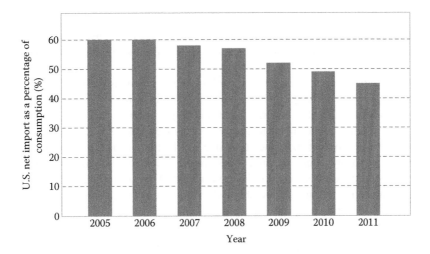

FIGURE 14.5  United States dependence on imported oil declining. (Adapted from M. Slack, Our dependence on foreign oil is declining, The White House Blog, March 2012.)

the average family an estimated $8000 at the pump and helping the United States halfway to its goal to cut imported oil by a third. European and Asian countries are also implementing similar fuel efficiency regulations to reduce oil consumption to ensure national energy security.[9,10] Therefore, vehicles such as PHEVs and EVs that provide means to achieve better fuel efficiency and less or no petroleum consumption are highly desired in the transportation sector to lessen their dependence on petroleum energy and to alleviate or even avoid the upcoming potential energy crisis.

Another important objective that PHEVs are designed to resolve is the reduction of environmental pollution from vehicle emissions. Emissions are generated as a form of vehicle exhaust after the fuel–air burning process in ICEs. They are also produced by fuel evaporation during uncompleted fuel burning or simply during fueling process. Poorly treated emissions can cause severe environmental problems and health problems such as cancers due to significant, chronic exposures. Table 14.2 lists the most commonly found pollutants in vehicle emissions.

The development of PHEVs is considered by many policy makers as one of the most promising and currently practical strategies to reduce environmental pollution from the transportation sector. Regulations have been adopted and incentives have been offered throughout the world to stimulate research and development in PHEVs. Conventionally, ICEs are highly inefficient, with an average efficiency of less than 30% due to the maximum heat–work conversion constraint, and they produce a wide range of emissions even with the assistance of after-treatment systems. By comparison, machines that use electricity as their energy source have much higher efficiency, and thus produce

---

TABLE 14.2

**Commonly Found Pollutants in Vehicles**

| | |
|---|---|
| Greenhouse and ground-level gases | Carbon dioxide ($CO_2$), carbon monoxide (CO), nitrogen oxide ($NO_x$), and sulfur dioxide ($SO_2$) |
| Air toxics | Hydrocarbons (HC) |
| Solids/liquids | Particulate matter (PM) |

*Source:* Adapted from U.S. Department of Energy, Office of Energy Efficiency and Renewable Energy, Just the Basics: Vehicle emissions, freedom CAR AND vehicle technologies program, August, 2003.

larger power output with the same input power. There are essentially no tailpipe emissions generated because the only by-product of using electric machines is the used battery, which can be recycled or reused for other applications. In general, PHEVs produce much lower tailpipe emissions than similar conventional vehicles while they produce zero tailpipe emissions during the all-electric driving range. Even when compared with the well-to-wheel emissions, PHEVs significantly reduce the emissions by a third compared with the conventional gas vehicles as shown in Figure 14.6.[12] This is because power plants that generate electricity typically have higher efficiency than ICEs; meanwhile, more and more renewable electricity has been generated such as hydropower and wind power so that the emissions at the generation side are further reduced. In addition, since many power stations are far away from cities, the emissions are away from human residential areas while conventional ICEs produce emissions in cities significantly. Thus, by taking advantage of electricity in electric machines, PHEVs can considerably reduce emissions from vehicle tailpipes.

Furthermore, there is growing public acceptance that carbon dioxide ($CO_2$) emissions are one of the primary contributors to global climate change. The burning of conventional petroleum fuels in ICEs generates $CO_2$, which contributes to a majority portion of the total U.S. GHG emissions, as illustrated in Figure 14.7.[13] In 2011, the United States generated 6.7 billion ($6.7 \times 10^9$) metric tons of equivalent $CO_2$ emissions alone, which are equivalent to the annual GHG emissions from 1.4 billion passenger vehicles, or carbon sequestered by 171 billion tree seedlings grown for 10 years.[13] Figure 14.7 illustrates the total U.S. greenhouse emissions by economic sectors in 2011, clearly showing that the transportation sector accounts for nearly a third of the total GHG emissions, making it the second largest contributor throughout all economic sectors. It has been recorded that the GHG emissions from the transportation sector have increased by about 18% since the 1990s largely due to increased demand for travel and the stagnation of fuel efficiency across the U.S. vehicle fleet.[14]

PHEVs can also help to reduce GHG emissions from vehicle tailpipes. Global efforts are strived together to regulate the $CO_2$ emissions as shown in Figure 14.8.[15] By largely displacing petroleum fuels with the use of electricity, PHEVs remarkably reduce the amount of fuel burned; thus, much less $CO_2$ is generated in the tailpipes. Moreover, the use of electricity as one of the energy carriers in PHEVs enables power selection from renewable power sources such as wind power and solar power in local districts. This clean energy helps to further reduce $CO_2$ emissions as well as air pollution during the electricity generation phase, which substantially enlarges the benefits of clean emission from PHEVs.

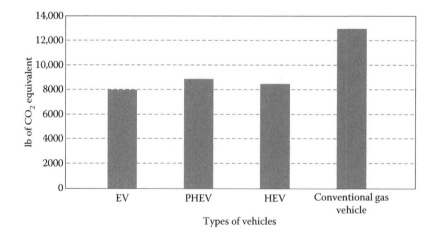

FIGURE 14.6  EV, PHEV, HEV, and conventional gas vehicle emissions comparison. (Adapted from U.S. Department of Energy, Office of Energy Efficiency and Renewable Energy, Emissions from Hybrid and Plug-In Hybrid Electric Vehicles, Alternative Fuel Data Center, Fuels and Vehicles, Electricity, Emissions.)

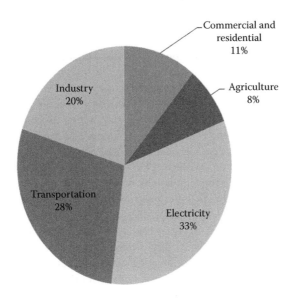

FIGURE 14.7  Total U.S. GHGs by economic sector in 2011. (Adapted from United States Environment Protection Agency, *Greenhouse Gas Equivalencies Calculator*, Updated April 25, 2013.)

In sum, PHEVs have many benefits:

1. *Petroleum consumption reduction:* The AER enables the switch from using conventional petroleum energy to electricity, which can be generated by various forms of resources. This significantly reduces the dependence on fossil fuel energy in the transportation sector, and provides a wide range of choices to charge the vehicles by generating electricity from renewable energies such as wind energy and solar energy. The benefit of the potential fuel reduction could be substantial. As a U.S. National Laboratory report found out, a 45% of

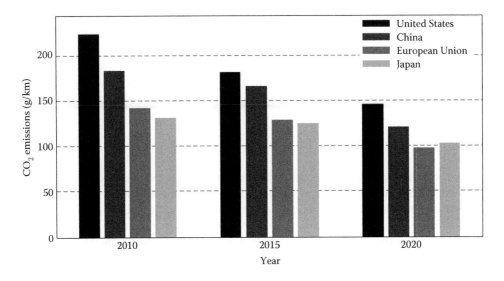

FIGURE 14.8  Global $CO_2$ regulations for passenger vehicles. (Adapted from Policy updates, European $CO_2$ Emission Performance Standards for Passenger Cars and Light Commercial Vehicles, International Council on Clean Transportation, July 12, 2012.)

fuel reduction can be achieved by replacing the conventional vehicle by a PHEV with 20 miles of electric travel range.[16]

2. *Emissions reduction:* As petroleum consumption is reduced, vehicle emissions due to the burning of the fossil fuels are remarkably reduced as a consequence. And, as discussed above, both the centralized generation of electricity and the use of renewable energy sources contribute to significant emission reductions. However, it should also be noted that since PHEVs typically work in the electric-only range that requires minimal engine operations, emissions might increase at the beginning of the engine start due to infrequent, multiple-engine cold starts. Methods and control algorithms[17] have existed to address such problems so that the overall emissions of PHEVs still remain much less than typical conventional vehicles under the same comparable size.

3. *Energy cost saving:* Besides the fuel consumption and emission reduction benefits, PHEVs also bring in the benefits of much lower energy costs. Although the exact cost saving depends on both long-term fossil fuel prices and electricity prices, it is estimated that, on an average, the fuel cost per mile of electricity is one-quarter to one-third of the cost per mile of fossil fuels for PHEV owners.[18] Meanwhile, governmental green energy incentives and certain privileges of lower auto-insurance compensate for the higher initial costs of PHEVs.

4. *Maintenance cost saving:* Since the mechanical components such as transmission and clutches are downsized and less frequently used, there are relatively fewer maintenance requirements regarding these parts, which are normally the maintenance concerns of conventional vehicles. Reducing the use of the engine also extends the engine life and reduces the frequency of oil changes. Besides, by taking advantage of regenerative braking, there is less friction wear on the mechanical brake, thus reducing the costs of frequently replacing the brake pad.

5. *Vehicle-to-grid (V2G) benefits:* PHEVs have the ability to supply the power back to the grid when they are connected to the grid; this serves to maintain a stable grid power level and to reduce power ripples. PHEVs could potentially serve as a temporary backup power source for home usage when grid power is not available.

6. *Customer benefits of home recharging:* PHEV owners enjoy the benefits of charging their vehicles in their garages or near their homes instead of looking for public-charging stations. This also allows the benefit of charging the vehicles at night when the vehicles are typically not in use and the electricity rate is the cheapest.

## 14.3  COMPONENTS OF PHEVs

PHEV power trains are composed of the electric motor, generator, battery, and engine, which are all similar to those in HEV configurations. However, different sizes and power ratings are used in PHEVs.

### 14.3.1  BATTERY

Battery serves as the major energy source in PHEVs. Owing to the increased portion of the electric power system and the desired all-electric driving range, large quantities of batteries with sufficient energy capacity and power density are required in the PHEV power train to meet the demanded all-electric driving range. It is capable of supplying all the power required to propel the vehicle throughout the entire speed range, and it should be equipped with sufficient energy capacity to sustain the desired AER. Moreover, the battery also needs to provide all the power to the accessories such as air conditioning and power steering during the all-electric driving range. Thus, higher battery performance is demanded by PHEVs.

On the other hand, the overall vehicle weight and manufacturing costs are prone to the amount of the battery, which increases significantly as the battery packs increase. Moreover, large quantities of onboard vehicle batteries also bring in safety concerns about the fire hazards or high-voltage short circuits in either normal vehicle operations or accidents. Thus, battery technology plays the most critical role in developing PHEVs with regard to performance, costs, and reliability.

Different types of batteries are used in PHEVs. Lithium-ion batteries are currently the most widely used battery in PHEVs. They provide high-energy density and high-power density so that for the same weight of the battery, they enable longer all-electric driving range and better vehicle performance. They also have low self-discharge rate that may reduce the charging frequency and perform better under low usage rate.[19] On the other hand, safety is a big issue associated with lithium-ion batteries. To operate lithium-ion batteries in a continuous stable state, well-designed battery management system (BMS) and cooling systems are required. Specific conditions such as vibration, humidity, overcharge, short circuit, extreme weather, fire, and water immersion should all be taken into account during the design and manufacturing process. These add up the cost of lithium-ion batteries and how to bring down the cost is a hot topic in both academic research and industrial manufacturing. Despite the high price currently, lithium-ion batteries dominate the PHEV battery market due to their high performance. The top three best-selling PHEVs currently are: GM's Chevy Volt, Toyota's Prius Plug-in Hybrid, and Ford's C-Max Energi, and all these use lithium-ion battery technologies. Other variations of lithium-ion battery are also developed. For instance, BYD implemented lithium iron phosphate ($LiFePO_4$) batteries into F3DM PHEV and Qin PHEV.

Nickel–metal hydride (NiMH) batteries are another type of commercialized battery that has been implemented into PHEVs. It is capable of comparably high-power density and energy density. NiMH batteries operate in a much more stable state that is abuse tolerant compared with lithium-ion batteries. It also has a much longer life cycle than the lead-acid batteries. NiMH batteries are mostly used as the energy source for the first generation of HEVs developed before 2005 such as Toyota Prius and Ford Escape Hybrid because of their lower cost. However, they are gradually replaced by lithium-ion batteries as the technologies are getting better and the cost is coming down.

There are some other types of batteries that can also be used in PHEVs. Lead-acid batteries are the oldest rechargeable battery. The technology has been developed for more than 150 years and the cost is very inexpensive compared with other types of PHEV batteries. They have been widely applied as the low-voltage batteries in the automotive industry for starting, lighting, and ignition. Electric scooters, electric bicycles, wheelchairs, golf carts, and some microhybrid vehicles can also be equipped with lead-acid batteries. In addition, lithium-air batteries are also under research and development. They are capable of extremely high-energy density compared to the conventional gasoline fuels. Toyota is collaborating with BMW on the advanced battery development including lithium-air batteries. IBM is also developing the lithium-air batteries for automotive traction applications.

## 14.3.2 Electric Machine

Electric machine is another core component in PHEVs. They serve as the primary movers in PHEVs to output speed and torque to the output shafts that are connected with vehicle wheels. Regenerative braking is also achieved by running the electric machine in generating modes so that the kinetic energy is retrieved from the electric machine into batteries. Meanwhile, because the electric machine is the only power source to propel the vehicle in all-electric driving mode, a higher power rating is required for the electric machine so that it can meet the required speed and torque. For instance, GM's Chevy Volt is capable of 35 miles of all-electric driving range in which all the propulsion power and the accessory power come from the onboard electric machine that outputs the peak power of 111 kW and peak torque of 370 Nm.

It is common in PHEVs that a second electric machine is utilized to serve as a generator and engine starter. The secondary electric machine can also operate as in the motoring mode to assist with the vehicle performance such that both the electric machines operate in the motoring mode that maximum power and torque are generated. In the operations of none all-electric driving mode, the secondary generator helps to charge the battery so that the battery state of charge (SOC) remains above the threshold level and the vehicle can operate under hybrid electric mode, thus significantly increasing the driving range of PHEVs.

Compared with conventional gasoline engines, electric machines typically have much higher efficiency that is greater than 90% in most of the speed and torque range. The lifetime of onboard electric machines is also expected to be more than 15 years, which is competitive with conventional gasoline engines and there is no need for customs to replace the electric machines within the factory warranty time. Currently, interior permanent magnet machine is the most popular choice for traction drive applications due to its high efficiency, high torque density, and high-power density. The achieved power density of electric machines in vehicle propulsion applications is 1.2 kW/kg at the current stage.[20] Research is still going on to increase the power density of electric machines to further reduce the size and increase the power. The targeted power density of electric machines for traction drive in 2020 published by the U.S. Department of Energy is 1.6 kW/kg, requiring 33% increase on electric machine power density within the next 5–7 years.[20]

### 14.3.3 Engine

Similar to HEVs, PHEVs are also equipped with onboard internal combustion engines. The engines applied differ by the configurations of PHEVs. If the engine is connected in series with the electric machines, since the electric machines serve as the primary mover to supply the majority of power, the engine only functions to support the electric machines to share the peak load or charge the battery when the vehicles operate in the hybrid electric mode to extend the PHEV range. Thus, the size and power rating of the engine can be minimized and high engine efficiency is required at constant operating regions. On the other hand, if the engine outputs power in parallel with the electric machines, the engine is responsible for a substantial portion of power demanded from the power train. Thus, the engine should still retain its power and size accordingly, based on the power ratios in PHEVs between mechanical and electric power. In some PHEVs with large battery packs, the engine may only serve as a backup when the battery is depleted so as to extend the driving range and alleviate the range anxiety of customers.

Engines in PHEVs may also apply different technologies compared with engines in conventional vehicles. Atkinson cycle is used instead of the conventional Otto cycle in some of the PHEVs to further improve the vehicle efficiency. Atkinson cycle allows the engine intake, compression, power, and exhaust strokes that all happen in one revolution of a special designed crank shaft. A greater thermal efficiency is achieved at the expense of losing power density, which is acceptable in most of the PHEVs since the engine is not the major energy source and higher efficiency is preferred. Toyota Prius Plug-in Hybrid, Ford C-Max Energi, and Honda Accord Plug-in Hybrid all use Atkinson cycle for their engine propulsion. Toyota Prius Plug-in Hybrid, for instance, achieved 38.5% thermal efficiency by using Atkinson cycle in its 1.8-L gasoline engines.[21]

In addition, since the demand for engine power is downsized in PHEVs, systems associated with mechanical power system such as exhaust systems and mechanical transmissions can also be reduced to smaller scales.

### 14.3.4 Power Electronics

Power electronics in PHEVs include inverters, DC–DC converters, chargers, and BMS, which also typically come along with battery systems. Inverters serve to transform the DC power from the batteries into AC power to propel the electric machines. It is also necessary to retrieve the regenerative energy from the electric machines back into the battery pack by using the motor drive components.

Besides, an inverter and associated controller are typically needed for the onboard air conditioners that use AC machines.[22]

Multiple DC–DC converters are used to step up and step down the voltages at different levels to suit for various applications. A boost converter is used to increase the DC bus voltage up to a high level from the voltage of the battery pack, which is desired for the electric machines so that the constant torque region is extended and higher power and higher speed can be outputted at the rated operation point. This DC–DC converter should also be capable of bidirectional power transfer so that the power retrieved from the electric machines by regenerative braking can be transferred back into the battery. Multiple DC–DC converters are also needed to adjust the battery voltage to different low-voltage levels. For instance, a DC–DC converter is used to supply the power for the 12-V accessory loads and charge the 12-V low-voltage battery, while another DC–DC converter may be used to step down the battery voltage to a higher level to operate the high- power applications such as power-steering systems and compressing pumps.

AC–DC converters are needed in battery chargers to convert the AC power from the grid into DC power to charge the battery. Power factor correction and programmable digital controllers with proper voltage–current profiles are needed for high-energy battery packs.

Proprietary BMS are used to actively monitor the battery SOC and state of health (SOH). The power and state of each individual battery cell is also regulated and balanced by the BMS system. A good thermal performance is also ensured by properly adjusting the temperature on the battery cells, as well as controlling the flow rate of intake and outtake coolant.

## 14.4 OPERATING PRINCIPLES OF PLUG-IN HYBRID VEHICLE

The operation modes of PHEVs largely depend on the battery SOC. Battery SOC is the term to describe the current state of the battery from 0% to 100%, with 0 standing for an empty battery and 100 meaning a full-charged battery. In comparison, HEVs typically remain battery SOC in a narrow range, for instance, 60%, to optimize the battery performance and ensure the required battery life. However, PHEVs typically demand greater depth of discharge (DOD) due to the higher dependence on the electricity energy source.

Because of the different operation patterns that PHEVs have from the HEVs, PHEV operations are more often classified by another set of specific operation modes[23]: charge-depleting (CD) mode, charge-sustaining (CS) mode, AER mode, and engine-maintenance mode. Figure 14.9 shows the

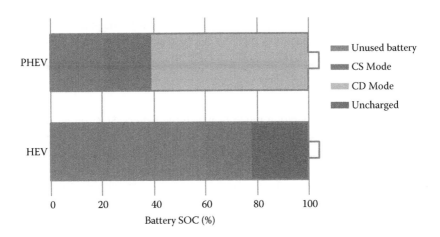

FIGURE 14.9 Battery performance comparison between HEVs and PHEVs. (From U.S. Department of Energy, Office of Energy Efficiency and Renewable Energy, Plug-In Hybrid Electric Vehicle R&D Plan, Freedom Car and Vehicle Technologies Program, June 2007.)

**TABLE 14.3**

**Battery Requirements for Different Vehicle Operations**

| | CD Only | CD and Sustaining | CS Only |
|---|---|---|---|
| Desired electrical range | 100 miles | 10–40 miles | |
| Desired cycles | 1000 deep cycles | 5000 deep cycles | 300,000 shallow |
| | | 300,000 shallow cycles at 25% SOC | cycles at 55% SOC |
| Function | Energy | Energy power assist | Power assist |

*Source:* Adapted from U.S. Department of Energy, Office of Energy Efficiency and Renewable Energy, Plug-In Hybrid Electric Vehicle R&D Plan, Freedom Car and Vehicle Technologies Program, June, 2007.

battery SOC comparison between HEVs and PHEVs. Different modes result in different requirements on batteries and also affect the performance of the vehicle. Table 14.3 relates the battery requirements with the vehicle operation modes and performance.[24]

### 14.4.1 Charge-Depleting Mode

CD mode refers to the PHEV operation mode in which the battery SOC on an average decreases while it may fluctuate along this trend. CD mode is frequently used at the first phase of PHEV operations, in which the SOC of the battery is sufficient to power the vehicle largely by electricity for a certain range. It prioritizes the use of electricity by drawing most of the power from the battery pack as long as the battery SOC stays above the preset threshold. However, if the demanded road power exceeds the battery power, the engine will also be running to assist the electric machine, thus enhancing the output tractive power.

CD mode is the primary operation mode in PHEV operations. In most city-driving and sub-urban-commuting cases, the round-trip distances are well within the PHEV battery power range. Thus, CD mode is largely utilized to take advantage of electric driving so that less fuel is used and fewer emissions are produced.

The extent of CD mode depends on the battery energy capacities and the frequencies of external battery charging. A larger battery pack with higher energy density would result in a longer CD mode range. However, this also contributes to much higher battery costs as well as vehicle weight increases. Recharging the PHEVs also helps to extend the CD range. With the installation of charging stations and the implementation of charging infrastructures at public places such as workplaces, parking lots, or in front of grocery stores, PHEVs can be readily recharged and CD ranges can be significantly increased in daily driving.

### 14.4.2 Charge-Sustaining Mode

CS mode refers to the PHEV operation mode in which the battery SOC on average maintains a certain level while it may frequently fluctuate above or below this level. CS mode utilizes both the engine and the electric machine to supply the vehicle power while keeping the SOC of the battery pack at a constant level. It is equivalent to the HEV operation mode in which the engine is mostly running within its optimal fuel efficiency range and the electric machines supply the power ripples. Engine power assistance and hybrid battery charging are realized in the CS mode to extend the driving range.

In PHEV operations, CS mode is more often used after the CD range when the battery power is discharged to a certain low threshold. Once the battery power is insufficient to power the vehicle on its own, the engine starts to supply the vehicle with petroleum combustion power. Both the engine and the electric machine operate together, coordinated under HEV operation mode. This takes advantage of the HEV operation benefits; so, high fuel efficiency is gained while the battery SOC is maintained at a certain level. Thus, the CS mode operation significantly increases the PHEV driving range compared with the CD mode without further increasing battery costs.

The combination of CD and CS mode enables energy use from two energy sources. The electricity works as the primary energy carrier to drive the vehicles in the preferred CD mode. The batteries can be recharged from external electric energy sources by plugging the vehicles into external power outlets. They can also be recharged by operating the vehicle in the CS mode, in which the engine utilizes the secondary energy carrier, the petroleum fuel, to generate power. In PHEVs, both energy sources are carried onboard the vehicles as they are stored in battery packs and fuel tanks. However, electricity is much preferred because it can be generated by a wide variety of cheaper energy sources, including coal, nuclear, natural gas, wind, hydro, and solar energy, and it greatly reduces vehicle tailpipe emissions. Thus, large packs of battery are normally required on PHEVs while relatively small fuel tanks are used.

### 14.4.3 AER Mode

As the name suggests, the AER mode uses electricity exclusively as its energy source to power the vehicles. The engine is shutoff during the AER mode while the electric machine supplies all the power by drawing energies from the battery pack. AER mode is similar to CD mode to a large extent, except that AER mode does not use the engine to assist the power output. The maximal range per charge depends on the onboard battery capacities. AER mode is often activated by manually switching under the command of the vehicle driver either to gain more fuel economy or to obey the rules in certain electric-only driving zones.

### 14.4.4 Engine-Maintenance Mode

Unlike the other operation modes, the engine-maintenance mode is not designed to propel the vehicle in PHEVs. Instead, it mainly functions to maintain the engine and prevent the fuel from being stale. This is useful for situations in which the driving range is always less than the AER and the vehicle gets recharged frequently. Thus, only AER mode is used and the engine never starts, which may cause problems for both engine components and fuel after a long time of nonuse.

### 14.5 PLUG-IN HYBRID VEHICULAR ARCHITECTURE

PHEVs combine the energy from the electric power path and the mechanical power path by utilizing two energy sources. Depending on the way these two power paths are integrated, different types of PHEV architectures are realized, which can also be categorized into series hybrids, parallel hybrids, and compound hybrids.

### 14.5.1 PHEV Series Hybrids

In the series hybrid PHEV configuration, the engine, generator, battery, and motor are connected in a series sequence, and the battery has the ability to be recharged from external power outlets, as shown in Figure 14.10. In PHEVs, battery power supplies the majority of power demands, and the motor plays the full role in propelling the vehicle. The engine combines with the generator to assist the electric motor or to charge the battery by using power from fossil fuel. Typically in series PHEV architecture, the engine has a small power rating since it is decoupled from the driving wheels and is mainly used to assist the electric motor to achieve better overall vehicle performance. The engine mostly operates in its fuel-optimal regions so that fuel efficiency significantly increases and a smaller engine is required. Consequently, the fuel tank can be reduced to a relatively small size.

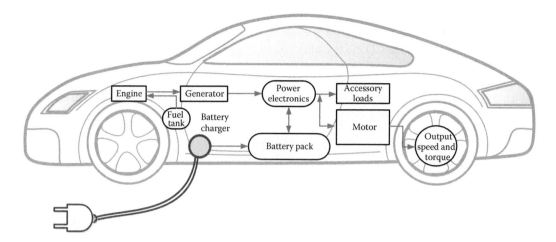

FIGURE 14.10   Configuration of series PHEV.

The relatively large battery capacity and the ability to recharge the vehicle by plugging into power outlets allow the engine to be turned off as much as possible. Thus, the series PHEV hybrids typically operate in electric-dominant mode until the battery reaches the lower SOC threshold. This helps the PHEV series hybrids have much less power conversion losses compared with the HEV series hybrids, as in HEV series hybrids, a significant portion of the engine power is lost due to the mechanical–electric–mechanical conversions. Regenerative braking can also be achieved by operating the electric machine as a generator to convert kinetic energy into electricity so that charging can be done during braking.

Figure 14.11 presents a simulation case for a typical series PHEV. The vehicle operated for more than four UDDS cycles. It can be clearly observed that the electric machine provides all the power and drives the vehicle in AER mode at first. Then the engine starts when the battery SOC decreases to a low threshold level. The vehicle then operates in the CD mode to extend the drive range while the battery SOC is regulated around the constant level.

### 14.5.2   PHEV Parallel Hybrids

PHEV parallel hybrids allow the power from both the primary and the secondary energy source to drive the vehicle. Typically, the battery would serve as the primary energy source in PHEV parallel configuration, and it can be charged by external power sources by plugging the charging cords into power outlets. The engine is used to propel the driving wheels directly, assisting the electric machine when additional power is needed, or it is used to charge the battery during hybrid operating mode. Figure 14.12 illustrates the PHEV parallel hybrid architecture.

Compared with parallel HEVs, parallel PHEVs further downsize the mechanical power system since the engine is not used to supply the majority of the power. Thus, smaller engines and smaller fuel tanks are installed on parallel PHEVs. On the other hand, parallel PHEVs significantly increase the power portion from the electric power system to prioritize electricity as the primary energy source. The electric system is required to power the vehicle on its own for a certain range without the assistance from the engine, and therefore, a much larger battery pack and a more powerful motor are required to realize a higher power rating of the electric power system compared with parallel HEVs.

In addition, since the engine is directly coupled with the output-driving wheels in the parallel PHEV configuration, the engine needs to be detached from the drive train when AER mode is demanded. This can be achieved by using clutches or torque converters so that the engine can be shutoff when the vehicle is operating in AER mode.

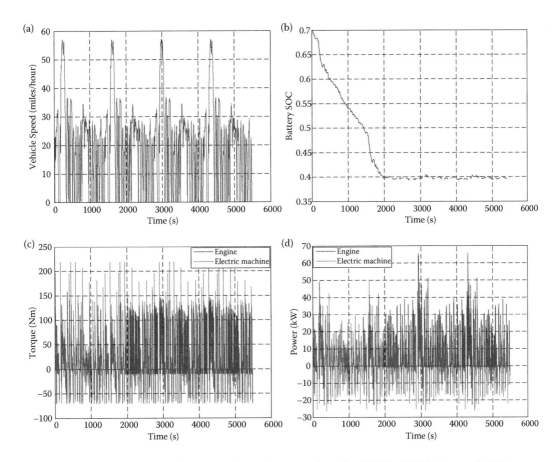

**FIGURE 14.11** Engine and electric machine performance in series PHEV. (a) Vehicle speed, (b) battery SOC, (c) engine and electric machine torque, and (d) engine and electric machine power.

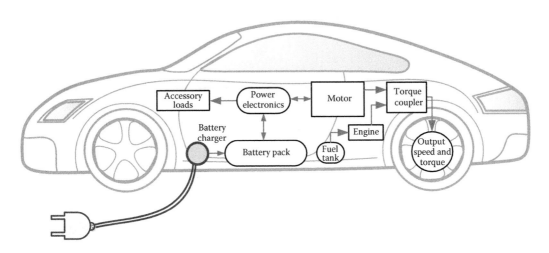

**FIGURE 14.12** Parallel hybrid architecture.

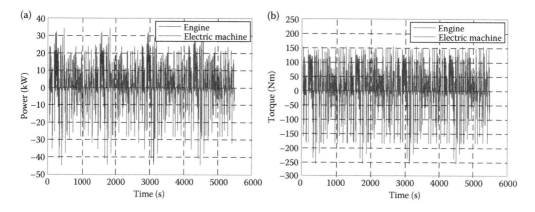

**FIGURE 14.13** Engine and electric machine performance in parallel PHEV. (a) Engine and electric machine power output in parallel PHEV; (b) engine and electric machine torque output in parallel PHEV.

Figure 14.13 presents the power and torque output from the engine and the electric machine, respectively in a typical parallel PHEV. Both the engine and the electric machine provide power and torque throughout the whole drive cycles and the vehicle is operating in the CD mode.

### 14.5.3 PHEV COMPOUND HYBRIDS

The power flow in compound PHEV does not follow a simple series pattern or a parallel pattern; instead, the mechanical and the electric power path interact with each other in a compound way, in which a planetary gear set is typically implemented to divide and combine the power. Similar to other types of PHEV hybrids, PHEV compound hybrids implement a much powerful electric power system to satisfy the performance requirements in AER mode. When driving in the compound mode where the engine starts to assist the electric machine, both the mechanical and the electric power path are integrated to supply the vehicle power together. Figure 14.14 illustrates the architecture of a compound PHEV power train.

Compound PHEVs combine the benefits of both series and parallel PHEVs when the vehicle is propelled by both the mechanical and the electric power system. Unlike in parallel PHEVs, the engine is decoupled from the output-driving shaft by taking advantage of the power-split

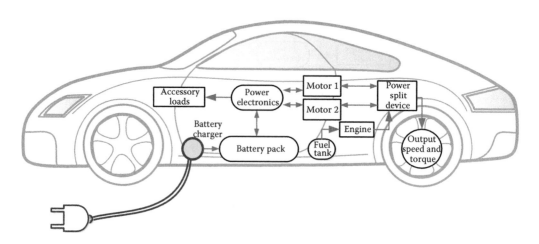

**FIGURE 14.14** Compound PHEV architecture.

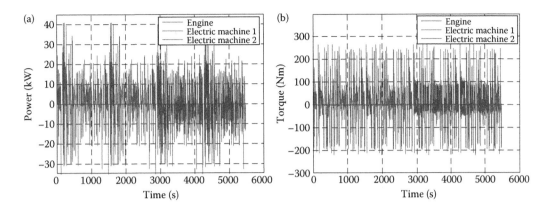

FIGURE 14.15   Engine and electric machines performance in compound PHEV. (a) Engine and electric machines power output in compound PHEV; (b) engine and electric machines torque output in compound PHEV.

devices, so that it can operate in its fuel-optimal regions. Besides, compound hybrids have less power conversion losses compared with series hybrids, as part of the engine power is directly transmitted by the mechanical path. In addition, both electric machines can operate either as motors or as generators, thus increasing the flexibility of the system control as well as vehicle-driving performance.

Figure 14.15 presents the power and torque output from the engine and the two electric machines, respectively, in a typical compound PHEV. The vehicle first operates in CD mode. Electric machine 1 provides the dominant power and torque to the output shaft while the engine only provides power occasionally when the power demand from the power train is large. The vehicle then operates in CS mode to extend the drive range. Both the engine and electric machine 1 are providing power while electric machine 2 operates in the generating mode to convert the engine power into electric power and charge the battery.

## 14.6   CONTROL STRATEGY OF PHEV

PHEV operation modes can be either manually selected by the driver or automatically controlled based on the feedback signals of various vehicle systems such as the battery SOC, power demands, road loads, and expected trip length, among others. In terms of the control strategies, two methods are typically applied in PHEVs: the AER-focused and the blended control strategy.

The AER-focused control strategy takes the greatest advantage of the electric power and runs the vehicles intensively in AER mode before the battery SOC drops below a certain threshold level, after which, the engine starts and the system operates in CS mode. The AER-focused control strategy prioritizes fuel reduction and emissions reduction in short-range trips by running exclusively on battery power. It is more suitable for city drives and short-range suburban commuting where daily round-trip distance is normally within the electric range. Since all the power in AER mode comes from the electrical systems, batteries with large energy capacity and electrical machines and power electronics with high-power density are needed to satisfy all the drive performance requirements.

The blended control strategy utilizes both the engine and the electric machines to power the vehicle. On the basis of expected travel distance, the blended control strategy picked the most appropriate fuel/electricity combination so that the battery SOC decreases smoothly in a linear trend. It operates the vehicle under CD mode with the engine running in its high efficiency region all the time until the battery SOC drops below the preset threshold level, after which, the vehicle operates in CS mode, similar to the AER-focused control strategy. The blended control strategy prioritizes

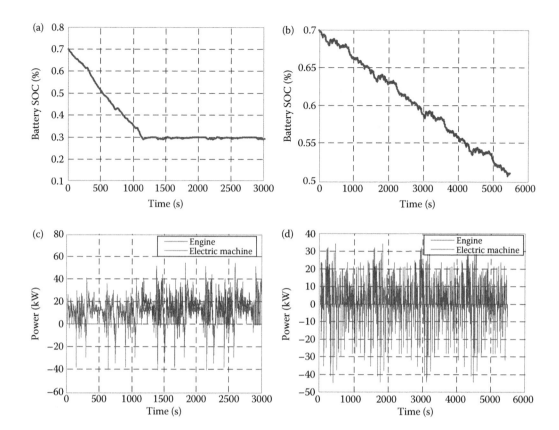

**FIGURE 14.16** Control strategy effects on battery SOC and power composition. (a) Battery SOC under AER control strategy, (b) battery SOC under blended control strategy, (c) engine and electric machine power output under AER control, and (d) engine and electric machine power output under blended control.

the range extent. It achieves an extended range in CD mode by using either the engine dominant strategy or the electric dominant strategy. In the former strategy, the engine is operating in its optimal fuel regions and the electric machine is used to subsidize the additional power demands. The latter strategy mainly utilizes electric power; the engine turns on only when the road loads exceed the electric capacity.

Figure 14.16 illustrates the battery SOC based on AER-focused control strategy and the blended control strategy, respectively. Four of the UDDS driving cycles are applied for a typical series PHEV. Power from both the engine and the electric machine is presented under each control strategy as well.

The optimum control strategy thus should rely on the trip distance that one PHEV is going to travel. If the trip distance is well within the battery AER, AER-focused control strategy should be applied so as to achieve the maximum fuel displacement. When the trip distance is greater than the AER, the blended control strategy is preferred with the engine running in its high efficiency region throughout the trip to achieve the optimum fuel efficiency.

In addition, Figure 14.17 presents the difference in engine and electric machine operation points between UDDS (Urban Dynamometer Driving Schedule) cycle and HWFET (The Highway Fuel Economy Test) cycle, which simulate the vehicle-driving behaviors on local and highway, respectively. It can be observed that the electric machine operates frequently in the low-speed regions under the local driving scenario while it operates more frequently in the high-speed regions on the highway. It can also be observed that the electric machine operates frequently in the negative torque region under local driving so as to retrieve more regenerative braking energy. For both

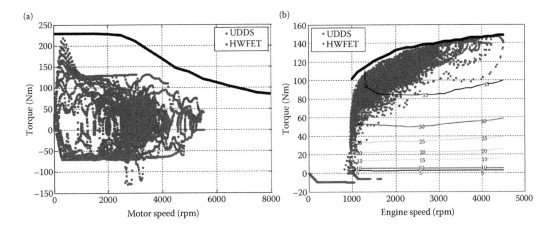

FIGURE 14.17    Operation points comparisons between drive cycles. (a) Motor operation point comparison, and (b) engine operation point comparison.

local and high way driving, the engine is largely controlled to be operated at high efficiency level. Different control strategies will result in different operation points for both the engine and the electric machine, thus affecting the fuel consumption as well as the emissions of the vehicle.

## 14.7    PHEV-RELATED TECHNOLOGIES AND CHALLENGES

Compared with HEVs, PHEVs take a further step in the transition from conventional fossil fuel combustion vehicles to electric power vehicles. They employ a large set of electric power systems, with electricity serving as the primary power source and electric machines serving as the primary propulsion drives. Engines and fuel tanks are retained onboard, but they are downsized to smaller scales and only function to assist the motors and batteries with additional power and additional drive range. All these changes bring in higher standards for the electric power system, including battery, electric motors and generators, power electronics, and their related controls. The massive component changes also require different platforms or even the reconstruction of the mechanical drive train. In addition, PHEVs change the way of refueling the vehicles, which enables drivers to charge the vehicles at home from grids instead of looking for public gas stations. However, this also brings in concerns regarding the grid's capability as well as the installation of power-charging outlets and public-charging access.

The above-discussed issues are closely related to the development of PHEVs. These issues significantly determine the performance, costs, and consumer acceptance of PHEVs in competing with conventional vehicles as well as HEVs and EVs. The following section will discuss these PHEV-related technologies and their challenges.

### 14.7.1    PHEV BATTERIES

High-performance battery is one of the most important components in realizing PHEV architectures. Vehicle performance, costs, and reliability are heavily dependent on the battery. Owing to the increased portion of the electric power system and the desired all-electric driving range, large quantities of batteries are required in PHEV power trains.

Batteries with large energy capacities are highly desired in PHEVs. In AER mode, the electric machine draws power exclusively from the battery pack, and the battery pack also needs to supply the power to all the accessories. Although the kinetic power retrieved by using the regenerative braking system helps to recharge the battery during the vehicle-driving process, it actually contributes very little to battery SOC in terms of the entire range. It is the battery energy capacity that directly determines the extent of the all-electric driving range, which in turn determines the degree of the displacement of fossil fuel and emission reductions. From the consumers' point of view, the more energy that the battery carries onboard, the fewer range anxieties the owners might have. Large battery energy capacity also provides sufficient energy for accessories such as air conditioning and radios while they are on AER mode.

In addition to battery energy capacity, battery power density is another key metric that considerably accounts for PHEV performance. Battery power density defines the maximal power that the battery can supply to the vehicle in terms of a certain weight or volume. It largely contributes to the acceleration time that the vehicles need from speed zero to the demanded driving speed with the engine shutoff. It also determines the maximal torque the vehicles can output from the electric power system when sudden acceleration is demanded at certain levels of speed. In AER mode, the battery power is designed to supply both the propulsion system and the other internal power accessories, and thus sufficient power density is highly demanded to meet vehicle driving performance without sacrificing the operations of the power accessories.

Both large energy density and high-power density are demanded in PHEVs to achieve the desired high-standard PHEV performance and the operating functions. These high requirements dramatically increase battery costs and thus the overall PHEV manufacturing costs. Compared with HEVs, these increased costs result in a significant economic hurdle for customers to overcome. Therefore, how to reduce battery costs while improving performance is a great challenge in battery technology. The increased battery requirement also leads to the increase of battery size and weight, which reduces the vehicle dynamic performance as well as the fuel economy. Furthermore, a major challenge that battery technology encounters is that the goals of large energy density and high-power density contradict each other most of the time, due to the inherent chemical trade-offs in battery technology.[25] High-power density is often achieved in advanced battery technologies by using thinner electrodes. However, high-energy density is typically realized oppositely by implementing thicker electrodes. Therefore, the development of high-performance batteries still faces the challenge of improving the power density and energy density at the same time.

Batteries in PHEVs also face another challenge regarding the discharge cycles, which are different from those in either HEV batteries or EV batteries. In HEVs, batteries typically go through shallow discharge cycles as the engine functions to operate the battery SOC at a relatively constant level in CS mode, while in EVs, batteries typically experience one deep discharge before the next recharge as a result of the larger energy capacity. However, batteries in PHEVs have to go through repeated one deep charge–discharge cycle per charge as well as a number of shallow charge–discharge cycles in both CD mode and CS mode for power assistance and regenerative braking. This brings in particularly high standards and great challenges in battery technology development so as to meet the typical 8 years or 80,000 miles automotive battery warranty standard.

The safety and reliability of the batteries are also among the top concerns for PHEVs since batteries take such heavy portions in PHEVs. High-voltage and high-current components should be carefully packaged and isolated from the chassis and accessory systems completely. Coolants are needed and should be regularly maintained to keep the batteries working in safe operating temperatures. Hazards protection methods should be taken into account during design and manufacturing process. In addition, batteries applied to the vehicles are not only required to be safe for drivers and passengers during the driving process; but they should also not cause any hazardous situations in the maintenance and repair of vehicles. For instance, it is necessary to keep all high-voltage components labeled, fused, and well insulated to reduce the risk during maintenance. On the other hand, appropriate safety gears such as safety glass, insulating boots, and insulating gloves are absolutely required when dealing with batteries. In addition, the risks surrounding batteries during accidents or collisions should be reduced to the minimal level when the PHEV configurations are designed. Trainings of emergency response should also be provided regarding the high-voltage battery and the electric system. Many codes and standards have already been established to address the technical issues relating to PHEVs. For example, in the United States, a list of SAE and NFPA codes and standards are applied to regulate PHEV vehicle safety, emergency response, and infrastructure safety.[26]

## 14.7.2  PHEV Costs

The costs of PHEVs extensively challenge the development of such vehicles, with the increased electric power system adding a significant part on the manufacturing costs. The large demands for battery energy capacity and power density require a remarkable quantity of batteries, which accounts for a substantial part of the cost increment. It is estimated that only when battery technologies advance to a further stage and can be mass produced to reduce the costs will PHEVs be competitive with conventional petroleum-powered vehicles. For example, NiMH batteries and lithium-ion batteries are currently the two most popular batteries in HEVs and PHEVs market. The production battery price by 2011 is roughly around \$700/kWh to \$900/kWh, which is many times of the United States Advanced Battery Consortium (USABC) long-term goal of \$100/kWh.[27] The large torque and speed requirements for the electric motor and generator also add considerable manufacturing costs due to the implementation of high-powered electric machines and power electronics.

PHEV reconstructions in most of the cases also significantly contribute to the total manufacturing costs. Typically, the large battery pack installation, integrated transmission power train, added electric motors and generators, and the control power electronics all together require redesigns of the vehicle chassis to accommodate these added components and their corresponding weights. Additional safety components are required to be added onboard the vehicles to prevent the hazardous situations regarding the electric power system. For instance, high-voltage shutdown mechanism is required to be added to prevent battery fire during a vehicle crash or in the event of air bag deployment.

In addition, the maintenance of the electric power system may add potential costs to the overall investment in PHEVs. Diagnosis and repair become more difficult due to the more complicated power train integrated by the mechanical power path and the electric power path. Battery replacement within the vehicle's driving life also brings in substantial maintenance costs for PHEV owners. Therefore, the reliability of the added electric system in PHEVs is critically desired to bring down the overall costs of PHEVs.

## 14.7.3  Charging of PHEVs

Since PHEVs switch the charging method from the conventional fuel station refueling to the primarily plug-in electricity recharging, charging-related issues become important in terms of PHEV development. These issues include charging strategies, charging types, and the corresponding charging infrastructures.

Charging strategies greatly influence battery SOC and the all-electric trip range based on a fixed battery energy capacity. There are several charging scenarios that can be applied onto daily PHEV driving.[28] The first scenario is that the driver charges the PHEV whenever the car is parked, which maximizes the all-electric driving range and achieves the greatest fossil fuel displacement. It can be applied on daily commute driving where recharging access can be found at work places, parking lots, grocery stores, and so on. The second scenario is to recharge the PHEV once every night, when the vehicle is typically parked at home and the price of electricity is relatively low. This enables drivers to recharge their vehicles in their garages or somewhere near their home while they stay at home or sleep. The third scenario is to recharge the vehicle during the lowest utility load demands. This recharging strategy helps balance the grid utility loads and achieves the maximal savings for PHEV owners. However, this requires control and communication between the vehicles and the grids, and the battery may end up without being fully charged. The fourth commonly used scenario is to recharge the PHEV at any time and at any day. This charging method is also called unconstrained charging. This is more like the conventional way that people fuel up their petroleum-powered cars, and it offers the largest freedom to the drivers in terms of recharging schedule. However, it generally requires a quick charging speed so that PHEV drivers would not feel uncomfortable waiting at the charging station for a relatively long time.

Different charging levels are applied to suit different charging scenarios. There are typically three charging levels that vary by charging voltage and charging currents. The first level is the home-charging level, in which PHEVs are generally charged by the home power outlets, such as those in garages. The second level supplies higher charging voltages and currents so that the charging time will be reduced. It is typically used by implementing high-power charging equipment to boost the voltage and current output. The third level is the highest charging level, which carries hundreds of volts and hundreds of amps. It significantly reduces the charging time to the level that is competitive with the conventional petroleum refill. The third charging level is typically applied in public charging stations where fast recharging is demanded and safety is prioritized.

All three levels of charging are desired to realize PHEV charging convenience, and the implementation of these charging facilities is critical to the development of PHEVs. Both the low-level and the high-level home-charging systems should be well regulated so that they would not overload the existing home power system or the local power distribution system, and they should be easily installed by customers. And large numbers of the public-charging stations or charging vehicles are needed to provide PHEVs with fast and convenient recharging choices outside their home. Thus, the development of both home-charging facilities and public-charging infrastructures considerably influences customer acceptance of PHEVs.

### 14.7.4 PHEV-Related Grid Challenges

PHEV-related grid challenges also bring in considerable concerns with regard to PHEV development. First, grid capacity will be a potential issue with the increasing use of electricity displacing conventional fossil fuels. PHEVs will bring in a large amount of utility load increase since electricity serves as the primary energy carrier to meet drivers' daily trip demands. Therefore, the grid should be able to tolerate the maximal utility loads when the worst scenario happens, that is, when all PHEVs and even EVs are plugged in at the same time.

Second, the increase in overnight utility loads caused by PHEV night recharging may result in the change of the control strategies to manage the grid balance. Currently, utility loads hit a valley at night when most people are asleep and household utilities are turned off. With the increase in PHEVs, more electricity will be consumed during the night hours; control strategies to balance utility loads will need to be adjusted accordingly.

Finally, the grid may also implement V2G technology to take advantage of the plug-in features so that the battery power of PHEVs can be turned back to the grids when the load demand is at its peak, provided the vehicle is parked. The V2G technology helps to balance the grid loads and reduces the utility expense of PHEV owners. Besides, the vehicle batteries can also be used as the backup power storage to send power to homes in case the utility is temporarily out of service. The V2G communication requires new grid technologies such as the implementation of smart meters and new power distribution and control strategies.

## 14.8 PHEV MARKET

In spite of the challenges, PHEVs still emerge as one of the most promising transitional vehicles from conventional petroleum-powered vehicles to electric-powered vehicles and combine the benefits of both. Production and sales of PHEVs gradually picked up after 2008. Table 14.4 summarizes the current PHEV production models up to September 2013 and Table 14.5 summarizes the scheduled models with market launch between 2013 and 2014.

TABLE 14.4

**PHEV Production up to September 2013**

| Models | Manufacturer | Production Since | Electric Range (km) |
|---|---|---|---|
| F3DM | BYD | 2008 | 64–97 |
| Volt | GM Chevolet | 2010 | 56 |
| Karma | Fisker | 2011 | 51 |
| Prius Plug-in Hybrid | Toyota | 2012 | 18 |
| C-Max Energi | Ford | 2012 | 34 |
| V60 Plug-in Hybrid | Volvo | 2013 | 50 |
| Accord Plug-in Hybrid | Honda | 2013 | 21 |
| Fusion Energi | Ford | 2013 | 21 |
| Panamera S E-Hybrid | Porsche | 2013 | 32 |
| Outlander P-HEV | Mitsubishi | 2013 | 60 |

With the increasing price of global crude oil as well as the increasing concerns about environmental pollution, PHEVs will be more promising in resolving these problems and will become more competitive with conventional petroleum-powered vehicles. Figure 14.18 predicts the PHEV, HEV, as well as EV sales number from 2010 to 2050 based on International Energy Agency (IEA)'s energy technology perspectives analysis that aims to achieve a 50% reduction in global $CO_2$ emissions from 2005 levels by 2050.[29] PHEVs sales number will boost exponentially from 2015 and will remain a significant portion of the light-duty passenger vehicles.

Beside the plug-in hybrid electric light-duty passenger vehicles, there are also research and developments on medium-duty PHEVs. Medium-duty vehicles are used in a broad array of fleet applications, including transit buses, school buses, and parcel delivery. These vehicles are all excellent candidates for hybrid electrification due to their transient-intensive duty cycles, operation in densely populated areas, and relatively high fuel consumption and emissions.[30,31] The home-based parking facility also facilitates overnight charging. Local governments and agencies such as Southern California Air Quality Management District and Milwaukee County in Wisconsin have been working on the conversion of PHEV shuttle buses and PHEV utility trucks.[32,33] IC bus offers diesel PHEV school buses and has already been delivered to many school districts. Parcel delivery PHEVs have also been tested and deployed to Fedex and UPS for evaluation.[31] In addition, plug-in hybrid electric motorcycle Piaggio MP3 Hybrid has been commercialized in Europe.

TABLE 14.5

**PHEV Scheduled Models with Market Launch between 2013 and 2014**

| Models | Manufacturer | Production Since | Electric Range (km) |
|---|---|---|---|
| P1 | McLaren | 2013 | 20 |
| i3 | BMW | 2013 | 130–160 |
| XL1 | Volkswagen | 2013 | 50 |
| A3 Sportback e-tron | Audi | 2013 | 50 |
| Qin | BYD | 2013 | 50 |
| 918 Spyder | Porsche | 2014 | 24 |
| ELR | Cadillac | 2014 | 64 |
| i8 | BMW | 2014 | 35 |
| S 500 Plug-in Hybrid | Mercedes-Benz | 2014 | NA |

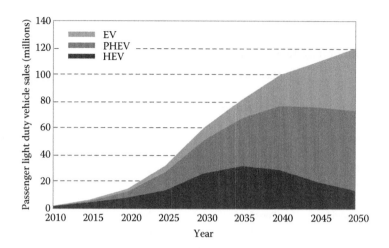

FIGURE 14.18   Light-duty HEVs, PHEVs, and EV sales prediction scenario. (Adapted from J. Axsen, A. F. Burke, and K. S. Kurani, *Electric and Hybrid Vehicles: Power Sources, Models, Sustainability, Infrastructure and the Market*, Elsevier, ISBN: 978-0-444-53565-8, 2010.)

## 14.9   CONCLUSION

PHEV is a special type of HEVs that make use of both conventional petroleum energy and electric energy as their energy sources. PHEVs have the full capability to operate on the AER for a certain demanded range and can be directly charged from off-board power grid. They provide high degrees of fuel displacement and emission reductions. The major components for PHEVs are similar to those in HEVs, but with much larger electric systems. Different configurations and control strategies exist that enable PHEVs to combine the benefits of both HEVs and EVs. In addition, PHEV-related technologies and challenges have been discussed and the current and future markets for PHEV have also been presented.

## QUESTIONS

14.1   Compare PHEV-20, PHEV-40, and PHEV-100. What would be the benefits, costs, and technology challenges in both component level and vehicular level?

14.2   Compare the range, fuel consumption, and emissions of PHEVs with the AER control strategy and the blended control strategy. Use power-train simulation software to verify the results.

14.3   Simulate the vehicle performance in CD mode. Adjust the threshold value for the engine to start. Compare the battery SOC and overall fuel consumptions.

14.4   List all the alternative fuel vehicles. Discuss the benefits and advantages of each other and compare PHEV with them.

## REFERENCES

1. Technology Roadmaps: Electric and plug-in hybrid electric vehicles (EV/PHEV), International Energy Agency, 2011.
2. L. Sanna, Driving the solution: The plug-in hybrid vehicle, *EPRI Journal*, Fall, 2005.
3. Day Trips, 1995 National Personal Transportation Survey (NPTS), Research and Innovative Technology Administration, Bureau of Transportation Statistics.

4. U.S. Energy Information Administration, *Electric Power Annual and Electric Power Monthly* (March 2012) based on preliminary 2011 data.
5. *International Energy Outlook 2010*, U.S. Energy Information Administration, Report Number: DOE/ EIA-0484(2010).
6. *Annual Energy Review 2010*, U.S. Energy Information Administration, Report Number: DOE/ EIA-0384(2010).
7. M. Slack, Our Dependence on Foreign Oil is Declining, The White House Blog, March 2012.
8. The White House Office of the Press Secretary, President Obama Announces Historic 54.5 mpg Fuel Efficiency Standard, Statements & Releases, Briefing Room, July 29, 2011.
9. Pocketbooks. Energy, transport and environment indicators. ISSN 1725-4566. 2012 edition.
10. J. German, Global vehicle fuel economy and GHG emissions regulations for light and heavy duty vehicles, *MIIT Workshop*, Beijing, China, April 14, 2011.
11. U.S. Department of Energy, Office of Energy Efficiency and Renewable Energy, Just the Basics: Vehicle Emissions, Freedom CAR and Vehicle Technologies Program, August 2003.
12. U.S. Department of Energy, Office of Energy Efficiency and Renewable Energy, Emissions from Hybrid and Plug-in Hybrid Electric Vehicles, Alternative Fuel Data Center, Fuels & Vehicles, Electricity, Emissions.
13. United States Environment Protection Agency, *Greenhouse Gas Equivalencies Calculator*, Updated April 25, 2013.
14. United States Environment Protection Agency, Sources of Greenhouse Gas Emissions, Transportation Sector Emissions, July 2013.
15. Policy updates, European $CO_2$ Emission Performance Standards for Passenger Cars and Light Commercial Vehicles, International Council on Clean Transportation, July 12, 2012.
16. A. Simpson, T. Markel, Cost-Benefit Analysis of Plug-in Hybrid Electric Vehicle Technology, *22nd International Electric Vehicle Symposium,* Yokohama, Japan. October 2006.
17. D. E. Smith, H. Lohse-Busch, and D. k. Irick, A Preliminary Investigation into the Mitigation of Plug-in Hybrid Electric Vehicle Tailpipe Emissions Through Supervisory Control Methods Part 1: Analytical Development of Energy Management Strategies, SAE International, 2010.
18. D. Sandalow, *Plug-in Electric Vehicles: What Role for Washington?* Brooking Institution Press, ISBN: 0815703058, 2009.
19. M. Winter and R. J. Brodd. What are batteries, fuel cells, and supercapacitors?, *Chemical Reviews* 2004 104 (10), pp. 4245–4270, September 2004.
20. S. Rogers, Electric Drive Status and Challenges, *EV Everywhere Grand Challenges*, U.S. Department of Energy, Office of Energy Efficiency and Renewable Energy, July 24, 2012.
21. Toyota Prius Plug-in Hybrid, *Wikipedia*. Webpage last modified on Oct. 25, 2013. Retrieved on Oct 29, 2013.
22. A. Emadi, Y.J. Lee and K. Rajashekara, Power electronics and motor drives in electric, hybrid electric, and plug-in hybrid electric vehicles, *IEEE Trans. Ind. Electron.*, 55(6), 2237–2245, 2008.
23. M. Ehsani, Y. Gao, and A. Emadi, *Modern Electric, Hybrid Electric, and Fuel Cell Vehicles: Fundamentals, Theory, and Design*, 2nd edition, CRC Press, ISBN: 1420053981, 2009.
24. U.S. Department of Energy, Office of Energy Efficiency and Renewable Energy, Plug-In Hybrid Electric Vehicle R&D Plan, Freedom Car and Vehicle Technologies Program, June 2007.
25. J. Axsen, A. F. Burke, and K. S. Kurani, Batteries for PHEVs: Comparing Goals and the State of Technology, *Electric and Hybrid Vehicles: Power Sources, Models, Sustainability, Infrastructure and the Market*, Elsevier, ISBN: 978-0-444-53565-8, 2010.
26. C. C. Grant, Summery Report, U.S. National Electric Vehicle Safety Standards Summit, November, 2010.
27. D. Howell, Vehicle Technologies Program, 2011 Annual Merit Review and Peer Evaluation Meeting, Energy Storage R&D, U.S. Department of Energy, Office of Energy Efficiency and Renewable Energy, May 9–13, 2011.
28. T. Markel, K. Smith, and A. Pesaran, Improving Petroleum Displacement Potential of PHEVs Using Enhanced Charging Scenarios, *Electric and Hybrid Vehicles: Power Sources, Models, Sustainability, Infrastructure and the Market*, Elsevier, ISBN: 978-0-444-53565-8, 2010.
29. International Energy Agency, Technology Roadmap: Electric and plug-in hybrid electric vehicles, updated June 2011.
30. R.A. Barnitt and J. Gonder, Drive Cycle Analysis, Measurement of Emissions and Fuel Consumption of a PHEV School Bus, SAE Technical Paper 2011-01-0863, 2011, doi:10.4271/2011-01-0863.

31. R.A. Barnitt, A.D. Brooker, and L. Ramroth, Model-based analysis of electric drive options for medium-duty parcel delivery vehicles, National Renewable Energy Laboratory, Golden, CO, Conference Paper NREL/CP-5400-49253, 2010.
32. J. Cox, Plug-In Hybrid Medium-Duty Truck Demonstration and Evaluation Program, South Coast Air Quality Management District, February 2012.
33. G. Bennett, 'Plugging In' to Hybrid Technology, Milwaukee County's PHEV Utility Trucks, Sustainability Summit, Milwaukee, WI, March 6, 2013.

# 15 All-Electric Vehicles and Range-Extended Electric Vehicles

*Weisheng Jiang, Yinye Yang, and Piranavan Suntharalingam*

## CONTENTS

Electric vehicles include battery electric vehicles (BEVs), fuel cell electric vehicles (FCEVs), and range-extended electric vehicles (REEVs). Electric machines are employed for traction purpose. Compared with conventional internal combustion engine vehicles (ICEVs), BEVs produce less emission, have higher efficiency, and generate less noise. However, the driving range of all-electric vehicles is limited. To overcome this problem, REEVs are introduced in the market; they employ secondary power source to either charge the battery or power the propulsion system. FCEVs use fuel cells as the primary energy source to power the vehicle, which significantly reduces carbon dioxide emissions and provides longer driving range. Solar electric vehicles and electric bicycles are also important parts of electric mobility. Hybrid electric vehicle (HEV) and plug-in hybrid electric vehicle (PHEV) technologies have been comprehensively discussed in the previous chapters. This chapter will first introduce the history of EV, its recent development, and its performances. Different powertrain electrification technologies, various traction motors, and energy storage devices will also be compared in this chapter. REEV, FCEV, solar electric vehicle, and electric bicycle are introduced and discussed briefly in the later sections.

## 15.1   EV HISTORY AND DEVELOPMENT

Electric vehicle technology has evolved for more than 100 years. Three main periods can be identified from the development of electric vehicles.

In the late 1800s, people in France, England, America, and other countries started to develop electric vehicle prototypes. In 1897, the first commercial electric vehicles entered the New York City taxi fleet. Around 1900, electricity-powered vehicles stood out and outsold other types of road cars. At the same time, Pope Manufacturing Co. became the first large-scale electric vehicle manufacturer in the United States. However, after the invention of internal combustion engine and due to the low cost of gasoline, the market share for electric vehicles started to decline in the 1920s, partly because of the mass production of ICEVs, and the reduction of gasoline price. Electric vehicles at that time were struggling to overcome the problem of limited battery capacity and short driving range, its Achilles' heel. Around 1935, electric vehicles became extinct.

The second period, from 1940 to 1990, saw a gradually improved interest in electric vehicles. Problems of exhaust emissions from conventional ICEVs drove automotive manufacturers to find alternatives. Government policies were also refined to support the development of electric vehicles. For example, in 1966, the US Congress recommended electric vehicles as a means of reducing air pollution. The rise in petroleum price due to the OPEC oil embargo and the increase in tailpipe emissions forced automakers to increase their attention toward the development of electric vehicles.

Since the 1990s, increased concerns about soaring oil prices, depleting fossil fuel reserves, and environmental issues caused by conventional vehicles have been greatly influencing the global automotive industry. Compared with conventional ICEVs, electric vehicles have several advantages such as

1. High electric machine efficiency compared to the internal combustion engine efficiency
2. Low level of environmental pollution, which improves local air quality
3. Lower noise
4. Smoother operation
5. Various electricity sources that can be obtained from renewable energies, such as hydro, nuclear, wind, and solar
6. Various onboard energy storage devices such as batteries, supercapacitors, flywheels, and hydrogen fuel cells
7. Regenerative braking to recover the kinetic energy of the vehicle

Since the 1990s, zero-emission vehicle (ZEV) requirements and rising petroleum prices rekindled people's passion in electric vehicles. To comply with California's ZEV requirements, General Motors produced and began leasing the EV1 electric car from 1996 to 1999. Tesla Motors, a California-based electric vehicle manufacturer, launched the Tesla Roadster and Model S, the first fully electric sports car, in small numbers in the United States. In 2010, the BEV Nissan LEAF was launched as well. Table 15.1 summarizes the highway-capable electric cars and light utility vehicles produced from 2008 through September 2013.[*]

---

[*] Electric car, Wikipedia, Last modified December 1, 2013. Retrieved December 2, 2013.

**TABLE 15.1**

**Highway-Capable Electric Cars and Light Utility Vehicles**

| Model | Market Launch | Global Sales | Sales Through |
|---|---|---|---|
| Nissan Leaf | December 2010 | 83,000 | September 2013 |
| Mitsubishi i-MiEV family | July 2009 | >26,000 | September 2013 |
| Tesla Model S | June 2012 | 18,200 | September 2013 |
| Renault Kangoo Z.E. | October 2011 | 11,069 | September 2013 |
| Chery QQ3 EV | March 2010 | 9512 | October 2013 |
| Renault Zoe | December 2012 | 6605 | September 2013 |
| Mitsubishi Minicab MiEV | December 2011 | 4972 | September 2013 |
| JAC J3 EV | July 1905 | 4918 | June 2013 |
| Smart electric drive | July 1905 | >4300 | September 2013 |
| Renault Fluence Z.E. | July 1905 | 3715 | September 2013 |
| BYD e6 | May 2010 | 3220 | October 2013 |
| Tesla Roadster | March 2008 | ~2500 | December 2012 |
| Bolloré Bluecar | December 2011 | 2300 | September 2013 |
| Ford Focus Electric | December 2011 | 2167 | September 2013 |

Despite all the merits mentioned above, purely electric-powered vehicles currently face significant challenges, including insufficient drive range and high costs. Battery-related challenges such as battery cost, volume and weight, short drive range per charge, and long charging time all present significant hurdles to the wide adoption of electric vehicles. To solve this problem, researchers are investigating to improve the specific energy and specific power of battery packs, invent alternative methods to reduce the cost, and also work on other electric components such as electric machines and power electronics to bring down the cost on a vehicle level.

Figure 15.1 shows the major types of vehicles, ranging from conventional vehicle to BEV. In general, vehicle powertrain electrification is becoming increasingly important to achieve improved fuel efficiency, reduce dependence on fossil fuel, and diminish carbon emissions. Currently, the REEV concept is one of the affordable electric vehicles that minimizes the risk of depleting batteries and running out of range by adding an auxiliary energy supply.

FCEV, running on hydrogen, does not emit pollutants and has a relatively long driving range. For fuel cell vehicles (FCVs), the refill time is relatively short. Compared with BEVs, ECVs can attain a much higher driving range and low carbon dioxide emissions as well, as shown in Figure 15.2. The Honda Clarity, an FCEV, is currently available for lease in the United States, Japan, and Europe, and is expected to start mass production around 2018. The price for fuel cells is high, and infrastructure, such as hydrogen filling stations, is not available in most locations. Most hydrogen produced recently comes from fossil fuels, in which process greenhouse gases, such as carbon dioxide, are released. However, Figure 15.2 shows that the hydrogen generated from natural gas, renewable biomass, and other carbon-free renewable energy sources, such as wind energy, solar energy, and nuclear energy, would reduce carbon dioxide emissions greatly.

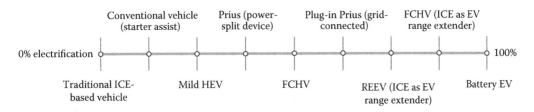

FIGURE 15.1  A wide variety of vehicles, different in terms of degrees of vehicle electrification.

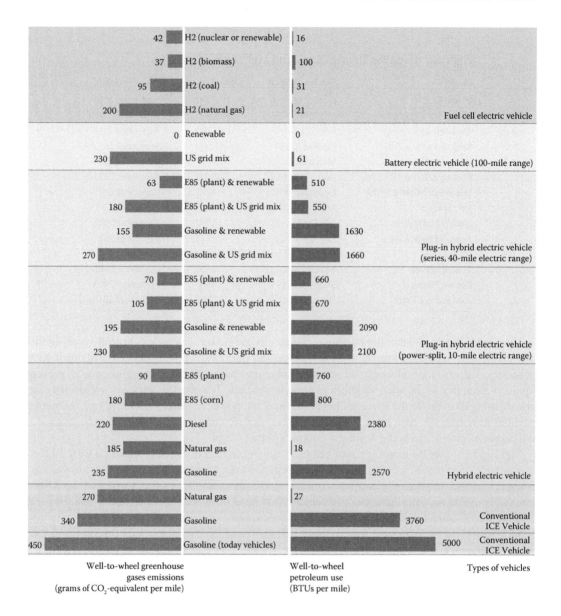

FIGURE 15.2 Well-to-wheels petroleum energy use and greenhouse gas emissions for future midsize vehicles. Renewable energy in the figure includes various types of ultralow-carbon renewable energies, such as wind energy and solar energy. British thermal unit (BTU), used in the figure, is defined as the amount of energy required to raise the temperature of one pound of water by 1°F.

Solar electric vehicles are driven by electricity generated by solar panels, which are not practical for everyday mobility, but are still a bright opportunity for research. Furthermore, solar power can be used to provide power onboard, not for traction, but for auxiliary functions, such as audio and communication, and can also be incorporated into charging stations or grids and indirectly power vehicles.

Electric bicycles, with electric machines to assist the riders or propel the bicycle fully, are also important part of electric mobility. EV, REEV, FCEV, solar-powered electric vehicle (SEV), and electric bicycle (EB) all play their roles in electric mobility, which are discussed in detail in the following sections.

## 15.2 EV CONFIGURATIONS AND MAIN COMPONENTS

### 15.2.1 CONFIGURATIONS FOR EVs

Figure 15.3 shows a front-engine, rear-wheel-drive powertrain layout for conventional vehicles. The internal combustion engine acts as the only power source, located at the front. The mechanical power from the engine is transmitted via a clutch and a short shaft to the gearbox. A propeller shaft from the gearbox delivers the power to the differential gears, which drive the wheels through two drive shafts.

The operation of a conventional electric vehicle is comparable to that of an ICEV, as shown in Figure 15.4. The major components of the electric vehicle's powertrain are an electric machine, an eclectic control unit (ECU), a battery pack, a battery management system, a power converter, an inverter, and a regenerative braking system. However, instead of being driven by an internal combustion engine, an electric vehicle is propelled by a traction motor, which is controlled by an ECU. The ECU takes signals from the driver via the accelerator pedal, the brake pedal, and so on, and feedback signals from the sensors such as vehicle's speed and acceleration to control the power requirement and power flow direction of the traction motor. In addition to that, electric vehicles incorporate a regenerative braking system, which captures most of the kinetic energy otherwise wasted when the brake pedal is applied.

EVs are much more flexible in terms of powertrain configurations, as illustrated in Figure 15.5. As explained earlier, a conventional EV drivetrain configuration mainly consists of an electric machine, a clutch, a multispeed gear transmission, a differential and driving shaft, as shown in

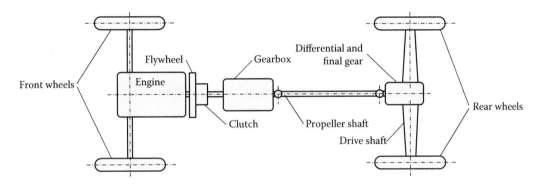

FIGURE 15.3   Front-engine rear-wheel-drive conventional vehicle powertrain.

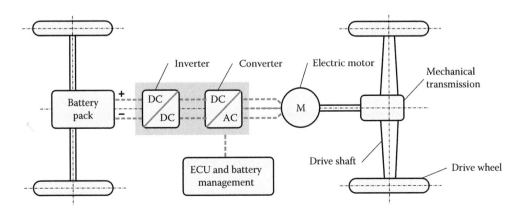

FIGURE 15.4   Conventional electric vehicle powertrain.

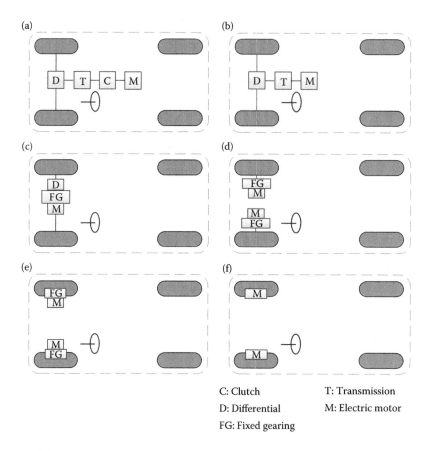

C: Clutch          T: Transmission

D: Differential     M: Electric motor

FG: Fixed gearing

**FIGURE 15.5** Selected EV configurations: (a) drivetrain with transmission and clutch, (b) fixed-gear transmission without clutch, (c) fixed gear and differential integrated into driveshaft, (d) separate electric machines with fixed-gear arrangement at drive wheels, (e) direct drive with separate electric machines and fixed gear, and (f) motor-in-wheel drive. (Adapted from C. C. Chan and K. T. Chau, *Modern Electric Vehicle Technology*, Oxford University Press, New York, 2001.)

Figure 15.5a. As seen in conventional electric vehicle powertrains, the internal combustion engine was directly replaced by an electric machine and a battery pack. Figure 15.5b shows an EV configuration with a fixed-gear transmission. In this configuration, the electric machine used as the prime mover should be capable of operating in a wide speed range. The mechanical complexity of the transmission is reduced and the drivetrain is downsized. The electric machine, the fixed-gear transmission, and the differential can be integrated in the driveshaft, which further simplified the whole drivetrain. The differential, connected to the driveshaft, enables the drive wheels to rotate at different speeds during cornering.

In Figure 15.5d, two electric machines not only operate as the prime mover for the EV but replace the role played by the differential as well. This configuration can be further simplified by integrating the fixed gear into the drive wheels, as shown in Figure 15.5e. In-wheel-motor drivetrain employs in-wheel motors, which significantly simplify the mechanical design for EVs, as shown in Figure 15.5f. In-wheel motors can be deployed in several arrangements, such as two front-wheel drive, rear-wheel drive, and four-wheel drive, as shown in Figure 15.6.

Instead of working as a drive wheel, the in-wheel motor contains the electric motor, which actually drives the vehicle. The in-wheel motors for some electric vehicles have the feature of regenerative braking as well. In this EV configuration, most conventional mechanical components, such as driveshafts, axles, transmission, and differentials in the drivetrain, are replaced by a drive-by-wire

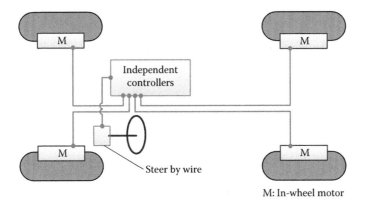

FIGURE 15.6    In-wheel motors in a four-wheel drivetrain.

system. The motor-in-wheel drivetrain can improve the vehicle's fuel economy by reducing the vehicle's weight and raising the drivetrain's efficiency, and enlarge passenger and luggage space. Four motor-in-wheel drives can increase the flexibility of steering and reduce the turning radius. However, this configuration requires the in-wheel motor to operate in a much wider range and be capable of starting and accelerating the vehicle by themselves, and the weight added to the wheels requires adjustment of the suspension system. Further, the motor-in-wheel drivetrain requires more complex control systems.

### 15.2.2   ENERGY STORAGE DEVICE FOR EVs

As discussed earlier, electric vehicles can be fueled by a wide variety of energy sources, which will reduce the dependency of transportation on fossil fuel and enhance a nation's security of energy supply. Currently, there exist several energy sources for electric vehicles, such as lead-acid batteries, nickel–cadmium batteries, lithium-ion batteries, supercapacitors, fuel cells, solar cells, and flywheels. Several energy storage devices are compared using the Ragone plot, as shown in Figure 15.7.

The specific energy shown in the Ragone plot represents the energy storage capacity for a particular device, which is related to how far the vehicle can go on a single charge, while the specific power shows the energy release rate for the device, associated with how fast the vehicle can accelerate. Supercapacitors have very high specific power, but low specific energy, while electrochemical batteries have a lower specific power and higher specific energy. Fuel cells stand out in terms of specific energy. The diagonal lines, often called "burn time," are obtained by dividing the specific energy by the specific power. For instance, if one energy storage device contains 1 Wh/kg and releases power at 1 W/kg, the stored energy would be depleted in 1 h. Thus, the burn time for this device is 1 h.

The energy supply configurations vary due to the variations of the energy source and the combinations of different energy sources. Several energy supply configurations are shown in Figure 15.8. As shown in Figure 15.8a, the battery pack operates as the sole energy source for the vehicle when it is accelerating, cruising, and climbing hills, and during regenerative braking. This requires the battery pack to have proper specific energy and specific power to satisfy these requirements.

In 2010, GE exhibited an electric bus with a dual-battery energy supply system, which combines a high-energy-density sodium battery with a high-power lithium battery, the mechanism for which is similar to the configuration shown in Figure 15.8b. The combined energy storage system satisfies the power requirements of acceleration and drive range of the vehicle.

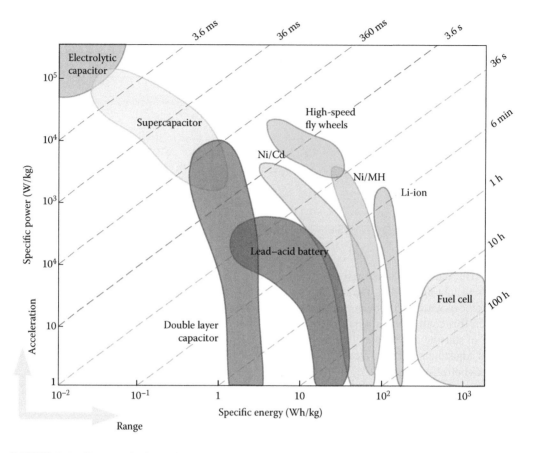

FIGURE 15.7    Ragone plot for various electrical energy storage devices.

Figure 15.8c shows that the fuel cell and battery pack can be combined and serve as the energy storage system for electric vehicles. The fuel cells combine hydrogen stored in a fuel tank onboard and oxygen extracted from air to generate electricity. The by-product of this reaction is water. The onboard battery pack can be used to store the excess energy and the regenerative braking energy. However, the key problems of this technology are hydrogen supply infrastructure development and onboard storage.

Figure 15.9 presents an alternative. The hydrogen used by a fuel cell can be generated onboard from methanol, ethanol, gasoline, or diesel by a reformer. Long-life and low-cost catalyst for the fuel conversion is crucial. The hydrogen generated needs to be purified, and CO, a poisoning gas for most fuel cells, needs to be removed from the reformate gas before being fed into the fuel cells.

The energy density of supercapacitors pales against most chemical batteries, but the super-capacitors excel regarding power density. The supercapacitors can be charged and can release power quickly. The supercapacitor can absorb the power peaks in regenerative braking and aid the battery pack when the vehicle is in need of high power. A combination of supercapacitor and battery pack would downsize the battery pack, improve the powertrain's specific energy, reduce the peak power load on the battery pack, and enhance the life of the battery pack. As shown in Figure 15.10, an additional DC–DC converter is used to interface between the battery pack and the supercapacitor.

The flywheel, operating at very high speeds in a vacuum environment, is another energy buffer that can be used for the electric vehicle's powertrain. Flywheels have high specific power and can receive high energy, which can be employed to recover kinetic energy when the electric vehicle is

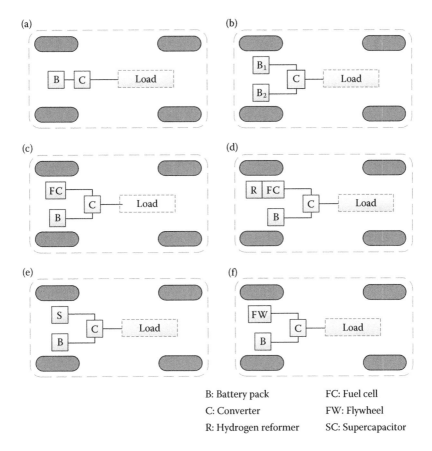

B: Battery pack     FC: Fuel cell

C: Converter     FW: Flywheel

R: Hydrogen reformer     SC: Supercapacitor

FIGURE 15.8 Energy supply source configurations: (a) basic battery-based energy supply, (b) dual-battery energy supply system, (c) fuel cell-based energy supply, (d) fuel cell energy supply with hydrogen reformer onboard, (e) supercapacitor-based energy supply system, and (f) flywheel-based energy supply system.

decelerating or braking. The main component in a simple flywheel is a plane disk spinning around its axis. When the plane disk slows down, the kinetic energy stored in the flywheel can be released. Figure 15.11a shows that a generator can be used to generate electricity from the energy stored in the flywheel, or employed to accelerate the flywheel. Figure 15.11b shows that the flywheel can be connected to the vehicle wheels through a clutch and a gearbox through which the kinetic energy stored in the flywheel can directly drive the electric vehicle.

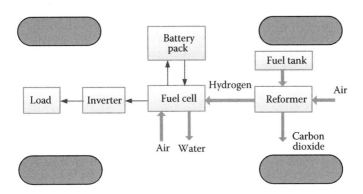

FIGURE 15.9 Fuel cell energy system with a hydrogen reformer.

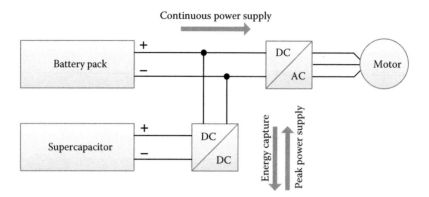

**FIGURE 15.10** Combination of supercapacitor and battery pack as energy source for electric vehicle.

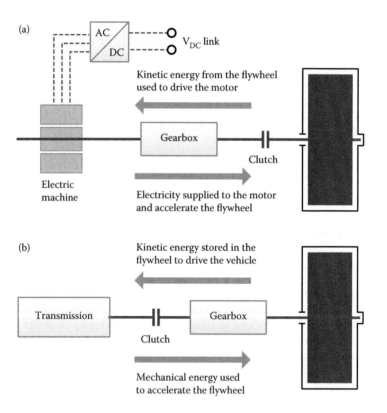

**FIGURE 15.11** Two types of flywheel arrangements used in electric vehicles: (a) mechanical energy from flywheel converted to electric energy and (b) mechanical energy from flywheel propelled the transmission directly.

## 15.3 EV PERFORMANCE

### 15.3.1 POWER DISTRIBUTION FOR ELECTRIC VEHICLES

The power from the powertrain system provides the vehicle's acceleration, enables the vehicle to ascend a gradient, overcomes aerodynamic resistance, and surpasses rolling resistance, as shown in Figure 15.12. Furthermore, additional power is required to overcome the internal transmission and conversion losses, accelerate the rotating components, for instance, wheels and

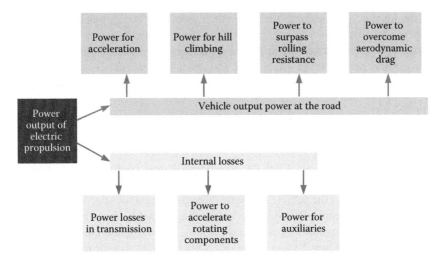

FIGURE 15.12   Power distribution for an electric vehicle.

mechanical transmission, and power other auxiliaries, such as light, wipers, horn, radio, heating, and air conditioning.

As shown in Figure 15.13, the vehicle's longitudinal acceleration can be determined by

$$\frac{dV_x}{dt} = \frac{1}{f_m M}(F_{xf} + F_{xr} - F_r), \tag{15.1}$$

where $M$ is the overall mass of the vehicle, $V_x$ is the vehicle speed, $f_m$ is the mass factor that converts the rotational inertia of rotating components into equivalent translational mass, $F_{xf}$ is the longitudinal force on the vehicle at the front-wheel ground contact, $F_{xr}$ is the longitudinal force on the vehicle at the rear-wheel ground contact, and $F_r$ is the total resistive force.

If the electric vehicle is front-wheel drive, the above equation can be modified into

$$\frac{dV_x}{dt} = \frac{1}{f_m M}(F_{xf} - F_r). \tag{15.2}$$

The total resistive force can be determined by the following equation:

$$F_r = (F_{zf} + F_{zr})\cos\theta \cdot C_r + F_d + Mg\sin\theta, \tag{15.3}$$

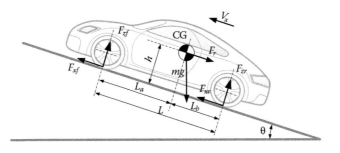

FIGURE 15.13   Vehicle longitudinal dynamics.

where $F_{zf}$ is the force perpendicular to the road on the vehicle at the front-wheel ground contact, $F_{zr}$ is the force perpendicular to the road on the vehicle at the rear-wheel ground contact, $\theta$ is the gradient of the road, $C_r$ is the coefficient of rolling resistance between tires and road surface, and $F_d$ is the aerodynamic air drag acting on the vehicle. The total resistive force can be expressed as a function of the vehicle's speed. The aerodynamic air drag can be calculated by the following equation:

$$F_d = \frac{1}{2}C_d\rho A(V_x - V_W)^2 \operatorname{sgn}(V_x - V_W),\qquad(15.4)$$

where $\rho$ is the mass density of air, $C_d$ is the aerodynamic drag coefficient that characterizes the shape of the vehicle body, $A$ is the effective frontal vehicle cross-sectional area, and $V_W$ is the component of wind speed on the vehicle's moving direction. The wind speed $V_W$ has a positive sign when it is in the same direction as the vehicle's longitudinal speed $V_x$, and a negative sign when it is opposite to the vehicle's longitudinal speed. Let us assume that the aerodynamic air drag acts on the vehicle's center of gravity. When the wind speed $V_W$ is not considered in the model, the aerodynamic air drag can be rewritten as

$$F_d = \frac{1}{2}C_d\rho AV_x^2 \operatorname{sgn}(V_x).\qquad(15.5)$$

Forces perpendicular to the road on the vehicle at the front- and rear-wheel ground contact are given as

$$F_{zf} = \frac{-h(F_d + Mg\cdot\sin\theta + M\dot{V}_x) + L_b\cdot Mg\cdot\cos\theta}{L},\qquad(15.6)$$

$$F_{zr} = \frac{h(F_d + Mg\cdot\sin\theta + M\dot{V}_x) + L_a\cdot Mg\cdot\cos\theta}{L},\qquad(15.7)$$

where $L_a$ is the distance between vertical projection points of the front axle and the vehicle's center of gravity, $L_b$ is the distance between vertical projection points of the rear axle, $h$ is the height of the vehicle's center of gravity, and $L = L_a + L_b$.

## 15.3.2 Vehicle Acceleration

The initial acceleration force is specified to propel the vehicle from full stop to its rated velocity $V_{rv}$, in $t_{rv}$ seconds. At motor's rated speed, the electric vehicle achieves its speed of $V_{rm}$. It is assumed that $V_{rm}$ is less than $V_{rv}$. When the vehicle accelerates from standstill to $V_{rm}$ in the electric motor's constant torque region, the traction force $F_x$ during this stage is obtained by

$$F_{x,rm} = \frac{P_m}{V_{rm}},\qquad(15.8)$$

where $P_m$ is the motor's rated power. In the electric motor's constant power region, the traction force can be estimated by

$$F_x(V) = \frac{P_m}{V}.\qquad(15.9)$$

The vehicle's acceleration $a$ can be defined as

$$a = \frac{dV_x}{dt} = \frac{1}{f_m M}(F_x - F_r).$$ (15.10)

When the vehicle accelerates from standstill to $V_{rm}$, the time spent can be calculated by

$$t_{rm} = f_m M \int_0^{V_{rm}} \frac{dV}{F_{x,rm} - F_r(V)}.$$ (15.11)

The total resistive force $F_r(V)$ can be expressed as a function of vehicle speed. When the vehicle accelerates from $V_{rm}$ to $V_{rv}$, the time spent can be

$$\Delta t = t_{rv} - t_{rm} = f_m M \int_{V_{rm}}^{V_{rv}} \frac{dV}{F_x(V) - F_r(V)}.$$ (15.12)

Thus, the acceleration time for the vehicle to accelerate from standstill to the rated velocity can be obtained by

$$t_{rv} = f_m M \left[ \int_0^{V_{rm}} \frac{dV}{F_{x,rm} - F_r(V)} + \int_{V_{rm}}^{V_{rv}} \frac{dV}{F_x(V) - F_r(V)} \right].$$ (15.13)

The initial acceleration force is specified to propel the vehicle from full stop to its rated velocity $V_{rv}$, in $t_{rv}$ seconds, as shown in Figure 15.14. The traction motor can deliver sufficient force to propel the electric vehicle at the rated speed. The cruising range is related to the battery's capacity.

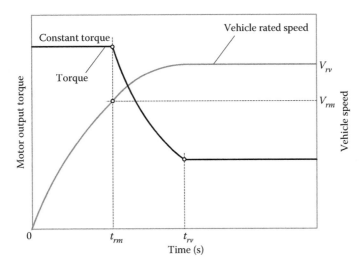

**FIGURE 15.14** Electric vehicle acceleration graphic.

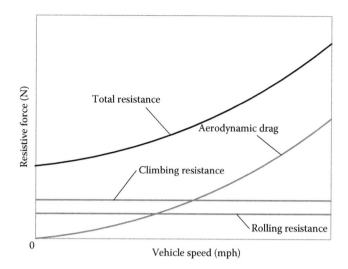

FIGURE 15.15 Resistive force expressed as a function of vehicle speed.

### 15.3.3 Vehicle's Maximum Speed

When the vehicle is cruising at its maximum speed $V_{max}$, the tractive power requirement can be estimated as

$$P_{V\,max} = V_{max}F_r = V_{max}\left[Mg\sin\theta + \frac{1}{2}C_d\rho A(V_{max} - V_W)^2\,\mathrm{sgn}(V_{max} - V_W) + Mg\cos\theta\cdot C_r\right]. \quad (15.14)$$

Aerodynamic drag is significant at high-speed cruising compared to low-speed driving conditions, as shown in Figure 15.15. Normally, $P_m$ will be much greater than the $P_{V\,max}$ value to achieve greater acceleration performance. Otherwise, $P_{V\,max}$ will define the traction motor power rating.

## 15.4  RANGE-EXTENDED ELECTRIC VEHICLE

### 15.4.1 Range-Extended Electric Vehicle Introduction

Owing to the limited capacity and longer charging time of battery technologies, BEVs are inferior to conventional and hybrid vehicles in terms of achieving higher driving range. Therefore, BEVs are suitable for short driving objectives such as urban commutes, and so on. REEVs, however, increase the driving range of the vehicle by incorporating an auxiliary electrical power source to the propulsion system, as shown in Figure 15.16. As shown in Figure 15.17, the range extender can be a small internal combustion engine with a generator. Instead of powering the vehicle directly, the engine in a range extender is acting as an electricity generator to recharge the batteries. The battery capacity for a REEV is designed to satisfy a customer's average daily usage, while the range extender allows the vehicle to maintain an acceptable long drive range. Compared to the conventional ICEVs, fuel consumption and carbon dioxide emission for REEVs are significantly reduced.

### 15.4.2 Range Extenders

The engine-based range extender, as shown in Figure 15.18a, is commonly designed to be extremely compact, lightweight, and low-cost. The engine is controlled to operate in its economic zone with high efficiency. Other types of energy sources can also be used as the range extender. As shown in

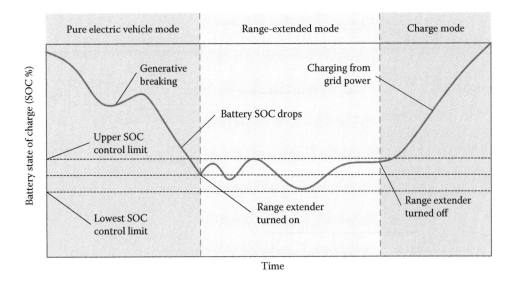

FIGURE 15.16    Range-extended electric vehicle operating modes.

Figure 15.18b, a fuel cell system can also be incorporated as a range extender to provide electric power to the powertrain system when the battery state of charge (SOC) reaches its power limit. Instead of carrying a bulky and heavy battery pack to achieve a long driving range, a fuel cell-based range extender can be used to downsize the battery pack and reduce the cost of the battery.

### 15.4.3    RANGE EXTENDER CONNECTION

Figure 15.19 shows two ways of connecting a range extender. In the first case, the range extended is connected to the battery, as shown in Figure 15.19a. The energy from the range extender flows in two directions: to the battery pack and to the load via a DC–DC boost converter and a DC–AC

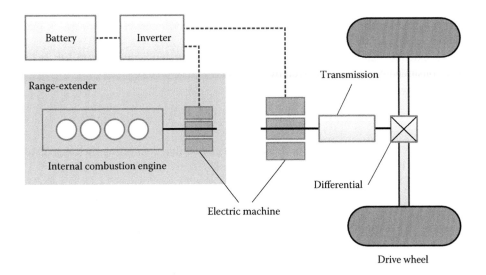

FIGURE 15.17    Powertrain configuration for a range-extended electric vehicle.

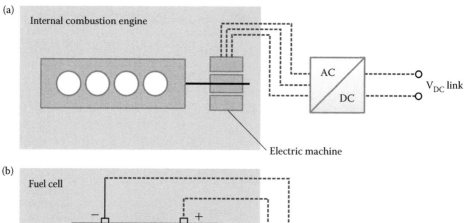

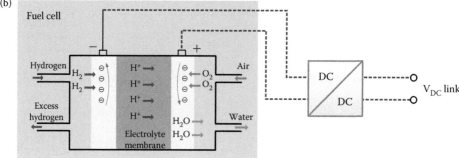

FIGURE 15.18   Two types of range extenders: (a) internal combustion engine-based range extender and (b) fuel cell-based range extender.

inverter. In this case, the total energy $E_{\text{load}}$ flowing from the range extender to the load can be calculated by the following equation:

$$E_{\text{load}} = E_{RE}(\alpha \cdot \eta_{BAT} + \beta)\eta_{DC} \cdot \eta_{AC}, \tag{15.15}$$

where $E_{RE}$ is the energy provided by the range extender, $\alpha$ is the percentage of the energy flowing from the range extender to the battery pack, $\beta$ ($0 < \beta \leq 1$) is the percentage of the energy flowing from the range extender to the boost converter, $\eta_{BAT}$ is the energy conversion efficiency for the battery pack, $\eta_{DC}$ is the energy conversion efficiency for the boost converter, and $\eta_{AC}$ is the energy conversion efficiency for the inverter.

The range extender can also be connected to the DC link, as shown in Figure 15.19b. In this case, the energy consumed by the load is given by

$$E_{\text{load}} = E_{RE}[\alpha \cdot (\eta_{DC})^2 \cdot \eta_{BAT} + \beta] \cdot \eta_{AC}. \tag{15.16}$$

Comparing the two equations above, it can be concluded that connecting the range extender to the DC link has a higher efficiency if the following holds true:

$$\beta > \alpha \cdot \eta_{DC} \cdot \eta_{BAT}. \tag{15.17}$$

If the total energy from the range extender flows to the load, the range extender should be connected to the DC link in order to increase the overall efficiency. It is preferable that the battery pack keeps its SOC during the extended range driving stage. The energy used to charge the battery pack from the range extender would finally move to the load in which process the energy conversion efficiency is reduced.

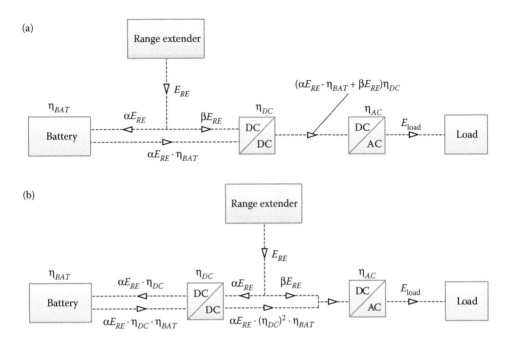

**FIGURE 15.19** Types of range extender connections: (a) connected to the battery and (b) connected to the DC link.

## 15.5　FUEL CELL ELECTRIC VEHICLE

### 15.5.1　Fuel Cell Electric Vehicle Introduction

FCEVs run on electricity generated by the fuel cells onboard. The fuel cells combine oxygen and hydrogen, in which process electricity and water are generated. The fuel cells are much more efficient than most other types of energy converters, such as internal combustion engines and chemical batteries. Furthermore, the by-products of fuel cells are only water and heat, and depending on the fuel source, very small amount of nitrogen dioxide and other emissions. Fuel cell powertrain minimizes noise and enhances the driving comfort. However, there are plenty of challenges for the FCEV to overcome before it gains its foothold in electric mobility. Compared with internal combustion engines and other energy sources, fuels cells are more expensive. Waste water vapor needs to be well managed and would be another concern for foggy and misty places. Hydrogen preparation, storage, transportation, and distribution are all important issues to be properly dealt with.

### 15.5.2　Fuel Cell Introduction

A fuel cell is an electrochemical energy conversion device, which combines hydrogen and oxygen to produce electricity and emits water as the by-product of this reaction. The proton exchange membrane fuel cell (PEMFC) and the alkaline fuel cell are two of the most commonly developed fuel cells for electric vehicle application.

Figure 15.20a shows the mechanism of a PEMFC. The electrolyte is an ~0.1-mm-thick proton-conducting plastic membrane, coated with a platinum catalyst. At the anode, hydrogen gives up its electron to the anode with the help of the catalyst. The electrolyte membrane is designed to allow only hydrogen ions to pass through. At the cathode, the hydrogen ions, oxygen, and electrons are bonded to form water. In this process, electrons flow from the anode to the cathode through the external load. As shown in Figure 15.20b, for alkaline fuel cells, alkaline electrolyte only allows

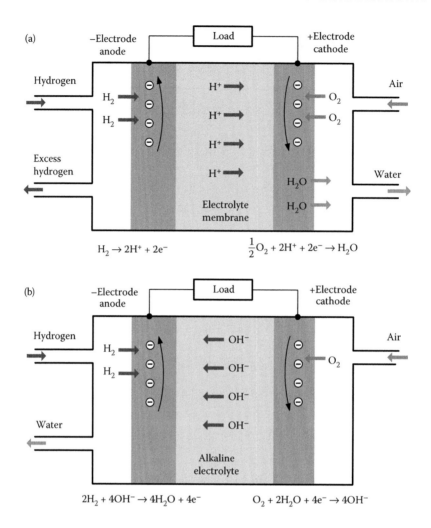

$$H_2 \rightarrow 2H^+ + 2e^-$$

$$\frac{1}{2}O_2 + 2H^+ + 2e^- \rightarrow H_2O$$

$$2H_2 + 4OH^- \rightarrow 4H_2O + 4e^-$$

$$O_2 + 2H_2O + 4e^- \rightarrow 4OH^-$$

FIGURE 15.20   Schematic of fuel cells: (a) proton exchange membrane fuel cell and (b) alkaline fuel cell.

hydroxide to pass through. At the anode, hydrogen combined with hydroxide generates water and electrons; at the cathode, oxygen, water, and electrons are combined to create hydroxide.

### 15.5.3   Fuel Cell Electric Vehicle Powertrain

Figure 15.21 shows a powertrain system for an FCEV. The primary components in the powertrain consist of a fuel tank, a fuel processor, a fuel cell as primary energy source, a battery pack, an electric machine as the traction motor, and so on. The vehicle controller takes command signals from the accelerator pedal and the brake pedal, the speed signal, the fuel cell power signal, and the battery signal, and sends the control signal to the fuel cell system. The power from the fuel cell and the battery pack combine to provide energy for the electric machine, which propels the vehicle through the transmission system.

Figure 15.22 shows that the fuel cell can be combined with other types of energy storage devices to provide power for the powertrain. Energy from the battery pack and the fuel cell can be combined to drive the electric motor, as shown in Figure 15.22a. The fuel cell in this configuration is more like a range extender. Figure 15.22b shows the parallel combination of the supercapacitor, and the

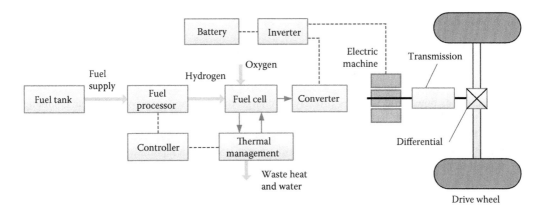

**FIGURE 15.21**   Powertrain system for a fuel cell electric vehicle.

fuel cell can work as the energy storage device for the electric vehicle. As stated earlier, the supercapacitor has high specific power and can assist the fuel cell in providing high power. In regenerative braking, peak power can be absorbed by the supercapacitor. Figure 15.22c and d shows two arrangements of the combination of the fuel cell and the flywheel. The mechanical energy stored in the flywheel can be converted into electrical energy first, and then combined with the energy from the fuel cell, as shown in Figure 15.22c. Figure 15.22d shows that the energy flowing out from the fuel cell can be converted into mechanical energy first and the energy from two energy devices then can be combined mechanically.

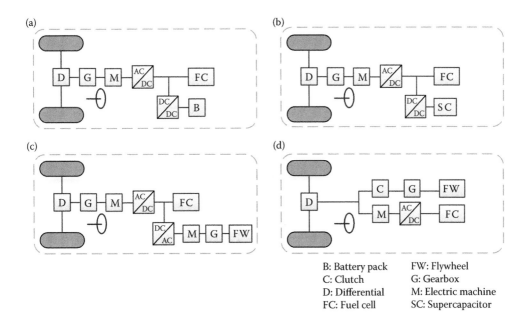

B: Battery pack     FW: Flywheel
C: Clutch           G: Gearbox
D: Differential     M: Electric machine
FC: Fuel cell       SC: Supercapacitor

**FIGURE 15.22**   Fuel cells combined with other energy sources: (a) fuel cell combined with battery pack as energy source, (b) fuel cell combined with supercapacitor, (c) fuel cell combined with flywheel, and (d) another type of combination of fuel cell and flywheel, in which flywheel can provide mechanical energy directly and propel the vehicle through mechanical transmission.

## 15.6   SOLAR ELECTRIC VEHICLE

### 15.6.1   Solar Electric Vehicle Introduction

Allan Freeman in England built the first solar-powered vehicle in 1979. Solar-powered vehicles capture the energy from the sun through photovoltaic cells and convert the solar energy into electricity, which can either power the vehicles directly or charge the batteries. The World Solar Challenge (WSC) is one of the most well-known events for solar-powered vehicles, crossing the Australian continent from Darwin to Adelaide, as shown in Figure 15.23. The WSC is now held every 3 years and attracts teams from all over the world. Similar events are held in Japan, the United States, South Africa, and so on.

### 15.6.2   Solar Electric Vehicle Powertrains

There are several configurations for the solar-powered vehicle's powertrain, as shown in Figure 15.24. Figure 15.24a shows, for the conventional solar-powered powertrain, that the solar energy harnessed from the sun is first converted into electricity by solar panels, and then the electricity directly powers the traction motor, which drives the vehicle forward. The conventional solar-powered powertrain is easy to implement and operate, but has a short drive range, limited acceleration capacity, and low efficiency, and is highly sensitive to weather conditions.

Rather than simply connecting the solar panels to the battery pack, the method of maximum power point tracking, frequently referred to as MPPT, is used to connect the solar panel and battery, in order to maximize the power generation from solar panels at different irradiance levels, as shown in Figure 15.24b. Simply speaking, MPPT is an output variable electronic DC–DC converter, which constantly tracks the output from the solar panel, compares it with the voltage from the battery pack, and prepares the optimized voltage to deliver the maximum power into the battery pack, as shown in Figure 15.25. The power from MPPT can power the electric machine directly, and the power

FIGURE 15.23    Route map of World Solar Challenge held in Australia.

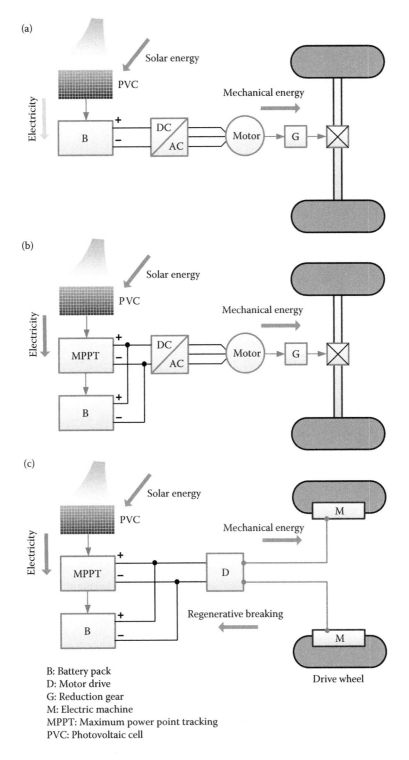

B: Battery pack
D: Motor drive
G: Reduction gear
M: Electric machine
MPPT: Maximum power point tracking
PVC: Photovoltaic cell

**FIGURE 15.24** Solar-powered electric vehicle powertrain configurations: (a) conventional solar-powered powertrain, (b) solar-powered powertrain with maximum power point tracking, and (c) solar-powered in-wheel drive with MPPT and regenerative braking.

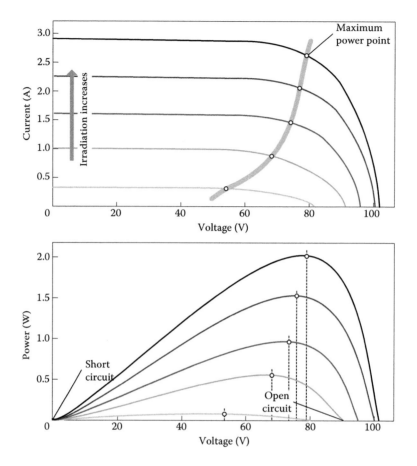

**FIGURE 15.25** Photovoltaic array current–voltage and power–voltage relationships at different irradiance levels.

surplus can be used to charge the battery pack. The energy stored in the battery pack can be used to assist in powering the vehicle when the power obtained from MPPT is low, or when the weather is cloudy. Figure 15.24c shows another powertrain configuration for solar-powered vehicles; in-wheel motors drive the vehicle directly, which improves the efficiency by removing the transmission. The overall efficiency for the powertrain also gets enhanced by gathering the kinetic energy from regenerative braking.

### 15.6.3 SOLAR-POWERED CHARGING STATION

Solar electric vehicle currently is not practical for everyday mobility. Instead of being installed onboard, the solar panels would be mounted at charging stations at home, in parking lots, or elsewhere as a solar power source and used to convert solar energy into electricity, which then charges electric vehicles. Figure 15.26 shows a parking lot being transformed into a solar-powered charging station.

## 15.7 ELECTRIC BICYCLE

### 15.7.1 ELECTRIC BICYCLE INTRODUCTION

Affordable, efficient, and convenient, electric bicycles now play an important role in electric mobility. Battery-powered electric bicycles can be seen in China, Japan, and other places across the world.

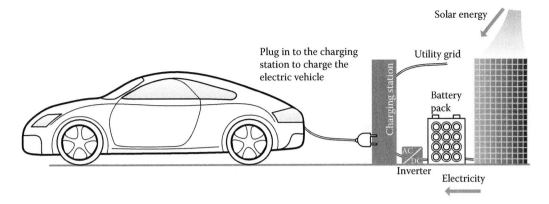

FIGURE 15.26    Solar-powered charging station.

More people choose to replace traditional bikes or even vehicles with electric bicycles for daily short-range commutes, especially in urban areas. Electric-powered vehicles have been supported by many local governments to reduce pollution and traffic congestion in the city areas. Currently, the travel range on a single full charge is acceptable for most electric bicycles. They are mostly driven directly by brushed DC electric machines, which are powered by lead-acid or nickel-cadmium batteries, or propelled by drivers with electric motors as auxiliary energy source.

## 15.7.2   Electric Bicycle Propulsion System

Figure 15.27 shows the scheme of an electric propulsion system for an electric bicycle. The major components for this system include a battery pack, an inverter, a battery management system, an accelerator, an in-wheel motor, and an ECU. The ECU takes the acceleration signal from the driver, motor's rotational speed obtained through Hall speed sensor installed in the motor, and battery information to provide the command signal to the inverter. The energy from the battery pack controlled by a battery management system powers the in-wheel motor via the inverter. Some electric

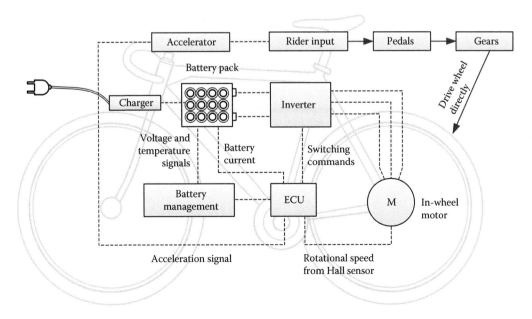

FIGURE 15.27    Electric bicycle propulsion system.

bicycles can capture kinetic energy through regenerative braking system when the rider applies the brake. The electric bicycle can be driven by either an electric machine or a combination of electric machine and the rider.

### 15.7.3 ELECTRIC BICYCLE POWER DISTRIBUTION

Electric bicycles are expected to travel at a low speed, compared with other types of electric motilities. The time used for an electric bicycle to accelerate from standstill to its expected speed is short. Thus, the total power $P_{total}$ provided by the electric propulsion system, or by the rider power input through pedals, or the combination of the two, is used to overcome the power for hill climbing $P_{hc}$, the power for surpassing rolling and bearing resistance $P_f$, and the power for overcoming aerodynamic drag $P_d$, as illustrated in Figure 15.28, which can be calculated using the following equation:

$$P_{total} = P_{hc} + P_f + P_d. \tag{15.18}$$

The power consumed in hill climbing can be calculated by

$$P_{hc} = MgV_g \sin(\theta), \tag{15.19}$$

where $M$ is the total weight, $g$ is the gravitational acceleration, $V_g$ is the ground speed, and $\theta$ is the gradient of the road. The power required to overcome the bearing and tire friction can be measured when the bicycle is traveling at a constant speed with negligible wind speed, which can be estimated by

$$P_f = C_r MgV_g, \tag{15.20}$$

where $C_r$ is the resistance coefficient. The power used to overcome the aerodynamic drag can be calculated by

$$P_d = \frac{1}{2}C_d\rho AV_r^2 V_g, \tag{15.21}$$

where $\rho$ is the mass density of air, $C_d$ is the aerodynamic drag coefficient that characterizes the shape of the bicycle body, $A$ is the effective frontal cross-sectional area, and $V_r$ is the relative speed in air.

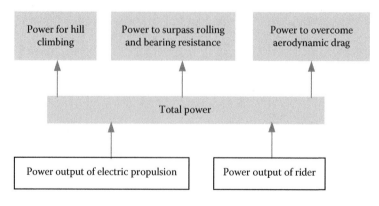

**FIGURE 15.28**   Power distribution for electric bicycle.

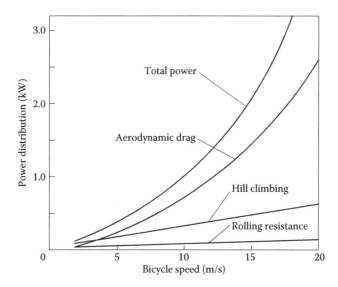

**FIGURE 15.29** Power distribution for electric bicycle at different speeds with 6% road gradeability, and 2 m/s headwind. (Adapted from W. C. Morchin, Battery-powered electric bicycles, *IEEE Northcon/94 Conference Record*, 269–274, October, 1994.)

When the electric bicycle is traveling at speeds less than 3 m/s, the total power is primarily consumed to overcome hill climbing and rolling and bearing resistance, as shown in Figure 15.29. When the road is flat, the total power is mainly used to overcome rolling and bearing resistance. The power to overcome hill climbing and rolling resistance increases almost linearly with the increase of bicycle speed, provided the road gradability is steady. At speeds greater than 3 m/s and on flat road, the majority of the total power is used to overcome the aerodynamic resistance. The power for overcoming aerodynamic resistance increases with the cube of the bicycle speed. When the bicycle is traveling on steep hills, the bicycle can only travel at a low speed and more power is required to cross the slope.

## ASSIGNMENT

1. Survey recent commercial and research electric vehicle programs around the world.
2. Survey various standardizations and regulations related to electric vehicle charging.
3. List and compare energy sources for electric vehicles. Discuss the recent development, advantages, and disadvantages for each type of energy source.
4. Survey the current status of commercial electric bicycles manufacture and sales worldwide. Compare the regulations regarding electric bicycles in China, Japan, and the United States.

## BIBLIOGRAPHY

1. M. Ehsani, Y. Gao, and A. Emadi, *Modern Electric, Hybrid Electric, and Fuel Cell Vehicles: Fundamentals, Theory, and Design*, Second Edition, CRC Press, Boca Raton, FL, 2009.
2. A. Emadi, Transportation 2.0, *IEEE Power & Energy Magazine*, 9(4), 18–29, June 2011.
3. M. Ehsani, K. M. Rahman, and H. A. Toliyat, Propulsion system design of electric and hybrid vehicles, *IEEE Transactions on Industrial Electronics*, 44(1), 19–27, February 1997.
4. Z. Q. Zhu and D. Howe, Electrical machines and drives for electric, hybrid, and fuel cell vehicles, *Proceedings of the IEEE*, 95(4), 746–765, April 2007.
5. C. C. Chan, The state of the art of electric, hybrid, and fuel cell vehicles, *Proceedings of the IEEE*, 95(4), 704–718, April 2007.

6. C. C. Chan and K. T. Chau, *Modern Electric Vehicle Technology*, Oxford University Press, New York, 2001.

7. P. Simon and Y. Gogotsi, Materials for electrochemical capacitors, *Nature Materials*, 7, 845–854, November 2008.

8. J. Larminie and J. Lowry, *Electric Vehicle Technology Explained*, Second Edition, John Wiley & Sons, Chichester, UK, 2012.

9. R. Garcia-Valle and J. A. P. Lopes, *Electric Vehicle Integration into Modern Power Networks*, Springer, New York, 2013.

10. C. Mi, M. A. Masrur, and D. W. Gao, *Hybrid Electric Vehicles: Principles and Applications with Practical Perspectives*, John Wiley & Sons, New York, 2011.

11. M. H. Westbrook, *The Electric Car: Development and Future of Battery, Hybrid and Fuel-Cell Cars*, IET, London, 2001.

12. I. Husain, *Electric and Hybrid Vehicles: Design Fundamentals*, CRC Press, Boca Raton, FL, 2011.

13. J. M. Miller, *Propulsion Systems for Hybrid Vehicles*, IET, London, 2008.

14. R. Hodkinson and J. Fenton, *Lightweight Electric/Hybrid Vehicle Design*, Butterworth-Heinemann, Oxford, UK, 2001.

15. A. E. Fuhs, *Hybrid Vehicles and the Future of Personal Transportation*, CRC Press, Boca Raton, FL, 2009.

16. P. Mulhall, S. M. Lukic, S. G. Wirashingha, Y.-J. Lee, and A. Emadi, Solar-assisted electric auto rickshaw three-wheeler, *IEEE Transactions on Vehicular Technology*, 59(5), 2298–2307, June 2010.

17. A. Muetze and Y. C. Tan, Electric bicycles—A performance evaluation, *IEEE Industry Applications Magazine*, 13(4), 12–21, July 2007.

18. McKinsey & Company, A portfolio of power-trains for Europe: A fact-based analysis. The role of battery electric vehicles, plug-in hybrids and fuel cell electric vehicles, Available at http://www.fch-ju.eu/sites/default/files/documents/Power_trains_for_Europe.pdf, last visited in December 2013.

19. I. Aharon and A. Kuperman, Topological overview of powertrains for battery-powered vehicles with range extenders, *IEEE Transactions on Power Electronics*, 26(3), 868–876, March 2011.

20. W. C. Morchin, Battery-powered electric bicycles, *IEEE Northcon/94 Conference Record*, 269–274, October, 1994.

# 16 Vehicle-to-Grid Interface and Electrical Infrastructure

*Giampaolo Carli, Arash Shafiei, Florence Berthold, and Sheldon S. Williamson*

## CONTENTS

## 16.1   INTRODUCTION

Conventional vehicles (CVs), which use petroleum as the only source of energy, represent a majority of the existing vehicles today. As shortage of petroleum is considered as one of the most critical worldwide issues, costly fuel becomes a major challenge for CV users. Moreover, CVs emit greenhouse gases (GHG), thus making it harder to satisfy stringent environmental regulations. One of the most attractive alternatives includes electric vehicles (EVs) or zero emission vehicles (ZEVs), which only consume electric energy. However, due to the limited energy densities of the current commercially available battery packs, the performance of EVs are restrained as neighborhood vehicles, with limitations of low speed, short autonomy, and heavy battery packs. As a successful example, Canada-based ZENN's commercialized EV has an average speed of 25 mph and 30–40 miles driving range per charge.

Currently, the most promising and practical solution is the hybrid electric vehicles (HEVs). Its propulsion energy is usually from more than two types of energy storage devices or sources, and one of them has to be electric. HEV drive trains are basically divided into series and parallel hybrids. Series hybrids are electric-intensive vehicles, as the electric motor is the only traction source, and the internal combustion engine (ICE) merely works at its maximum efficiency, as an on-board generator, to charge the battery.

Keeping in mind the goals of creating an energy-wise, cost-effective, and overall sustainable society, plug-in hybrid electric vehicles (PHEVs) are recently being widely touted as a viable alternative to both conventional and regular HEVs. PHEVs are equipped with sufficient onboard electric power to support daily driving (an average of 40 miles/day) in an all-electric mode, only using the energy stored in batteries, without consuming a drop of fuel. This, in turn, causes the embedded ICE to use only a minimal amount of fossil fuel to support further driving beyond 40 miles, which further results in reduced GHG emissions.

PHEVs can reduce fuel consumption by charging its battery from the grid. It is, thus, a valid assumption that moving into the future, a large number of PHEV users will most definitely exist, and the overall influence of charging the onboard energy storage system (ESS) cannot be neglected. Related literature firmly states that by the year 2020, the market share of PHEVs will increase to about 25%. Based on these data, the additional electric energy demanded from the distribution grid for 5 million PHEVs would roughly be about 50 GW h/day. Also, the typical charging time would be 7–8 h, which might make it hard to accommodate these additional loads in the load curve without increasing the peak load. Also, the required additional charging energy would have a possible impact on the utility system.

Expanding the electric system the conventional way, with large generating plants located far from the load centers, would require upgrading the transmission and distribution systems too. Besides the high costs, this can take many years before obtaining the right-of-way. Alternatively, smaller power plants based on renewable energy, such as wind energy, is a cost-effective renewable energy, in addition to many utilities. Also, solar energy can be installed in a fraction of that time on the distribution system, which is commonly referred to as "distributed generation (DG)." Photovoltaic (PV) presents a modular characteristic and can be easily deployed in the roof top and facades of residences and buildings. Many corporations are adopting the green approach for distributed energy generation. For instance, *Google* has installed 9 MWh/day of PV on its headquarters, Googleplex, in Mountain View, California. At the moment, it is connected to Mountain View's section of electricity grid. Alternatively, it could be used for charging PHEVs during work hours, being a great perk for environmentally concerned employees. The energy stored in the batteries could also be used for backup during faults. In Canada, the latest projections (2000) indicate that by 2010, renewable DG sources will represent at least 5% of the total energy produced and 20% of cogeneration, from the actual figures of 1% and 4%, respectively. Therefore, from the environment point of view, charging PHEVs with solar power will be the most attractive solution.

This chapter primarily aims at addressing the practical issues for commercialization of current and future PHEVs, and focuses primarily on power electronics-based solutions for both current and future EV technologies. New PHEV power system architectures are discussed in detail. Key EV battery technologies are explained as well as corresponding battery management issues are summarized. Advanced power electronics intensive charging infrastructures for EVs and PHEVs are also discussed in detail.

## 16.2 EV AND PHEV CHARGING INFRASTRUCTURES

### 16.2.1 EV/PHEV Batteries and Charging Regimes

Replacing the conventional internal combustion engines with electric vehicles (EVs) and plug-in hybrid EVs (PHEVs) in a large scale can result in tremendous prosperities for saving our world from the dangerous ever-increasing rate of pollutants. The majority of benefits such as pollution reduction and decrease of oil consumption resulting from moving toward using EVs and PHEVs are mainly based on using batteries as a green source of energy. The chemical nature of batteries makes them to have a highly nonlinear behavior and dependent on many factors such as chemistry, temperature, aging, load profile, and charging algorithm. Besides, to have a specific amount of energy for a reasonable all electric range (AER), tens or hundreds of cells should be connected in series and parallel to make the desirable voltage and current ratings of the battery pack [1,2]. This causes the nonlinear behavior of cells to be amplified in some aspects. Furthermore, there are phenomena that are observed only in battery packs and not in single cells, such as thermal unbalance among the cells in packs.

EV and PHEV battery packs are relatively expensive compared to the price of the whole car, due to high number of cells, chemistry types such as lithium-based, protective circuits, and so on. Accordingly, the life cycle of these battery packs are very important. Therefore, reduction in cost for the final customer can be achieved with increasing the battery pack life cycle, which results in posterior need for replacement of the whole pack. Just to get an idea about the price of the battery packs, a real example from Honda Civic is mentioned here [3]. Recently, there was news about Honda Company regarding battery packs of Honda Civics produced during 2006–2008. Apparently, some of the battery packs in second-generation Honda Civic hybrids, which went into production 5 years ago, are failing prematurely. According to regulations in California, there is a 10 year-, 150,000-mile warranty requirement on the components of the hybrid system. Honda Company has taken some actions to solve the problem; however, some customers are not satisfied and prefer to change the battery packs themselves. The price of these battery packs is about $2000 excluding shipping and installations.

The above-mentioned case shows the importance of the price of battery packs in commercialization of EVs and PHEVs on a large scale. A factor that highly impacts the life cycle of battery packs is the charging algorithm. There are also other factors involved such as the charging time that plays an important role in high attraction to EVs and PHEVs. These topics and all other ones related to this area should be mainly handled with a multilevel control and power system called battery management system (BMS) which takes care of all or some of the aspects affecting batteries in any way. The more accurate and comprehensive the BMS is the more reliable, safer, and faster the charging procedure will be. Designing a high-efficient BMS needs very good understanding of the behavior of single cells according to the variations of different parameters and also mutation of these behaviors in a packed state with a large number of cells.

First, in the following sections, we try to describe and mention some basic definitions and aspects in the field of batteries and based on those come up with some results such as appropriate charging algorithms which improve the life cycle of batteries. In the following section, we do not want to describe very detailed mathematical definitions of different parameters of batteries, which can be used for solving problems and designing purposes, rather, it is intended to give some basic definitions, which help in understanding the later sections for readers who may not be familiar with these topics.

### 16.2.1.1 Battery Parameters

#### 16.2.1.1.1 Battery Capacity

This parameter can be simply assumed as the amount of charge, which can be drawn from a fully charged battery until it gets fully discharged. An important effect in batteries is that the higher the amount of current drawn from a battery, the lower the capacity the battery will have. Hence, theoretically, battery capacity is defined as the amount of current drawn from a battery that completely discharges it in exactly 1 h. For example, a battery capacity of 10 Ah means that if a constant current of 10 A is drawn from the battery, it will get discharged completely after 1 h. However, in practice, battery manufacturers may use other definitions. Usually, a table of different test results is provided, which shows the amount of time the battery runs with different constant current loads and also another table with different constant power loads. In practice, this table provides much more practical information rather standard definitions, because after production, different loads with different characteristics may be connected to the battery. Nevertheless, the amount of time that a battery runs is not predictable exactly, because not all the loads are constant current or constant power loads. Even if they are one of these types, those tables are valid for new batteries and not for aged ones. Therefore, in many design procedures, just rough estimates of battery runtime are calculated. The battery capacity is shown in the literature with letters such as "C" or "Q" or other notations. The main unit for battery capacity is ampere hour (Ah); however, based on the size of the battery, alternative units such as mAh or even mAs in the case of very small batteries are used.

#### 16.2.1.1.2 C-Rate

This parameter is used to show the amount of current used for charging the battery or that of a load, which is drawn from the battery. For example, in the previous case of 10 Ah battery, when it is mentioned to terminate the charging process while the charging current falls below C/10 rate (10 h rate), it means that the charging should be stopped when current becomes less than the amount of current with which the battery is discharged after 10 h; in other words: 10 Ah/10 h = 1 Amp.

#### 16.2.1.1.3 State of Charge

In its simplest form, state of charge (SOC) can be visualized as the percentage of the remaining water to the whole capacity of a water tank. In terms of charge, it means the percentage of charge available from a battery to the whole capacity of the battery. Assuming the battery as a water tank gives a good idea; however, it is very preliminary and not accurate because of some effects in the batteries such as relaxation effect, which will be described in the following sections. Besides, according to aging, the rated capacity of the battery reduces over time, hence, for determining SOC, the rated capacity should be measured or calculated regularly.

#### 16.2.1.1.4 Depth of Discharge

Again using the water tank concept, depth of discharge (DOD) can be assumed as the percentage of water that has been drawn from the water tank to the whole capacity of the tank. In terms of charge, the water can be replaced with electric charge. This parameter is usually used in discharge patterns recommendations. For example, the battery manufacturer may recommend the user not to go over 30% DOD according to lifetime issues.

### 16.2.1.1.5 Energy Density

Energy density can be defined in two ways. One is "volumetric energy density," which is defined as the amount of available energy from a fully charged battery per unit volume (W h/L). The unit "liter" is mainly used for measuring the volume of liquids. Mostly, the batteries have liquid electrolyte, so in such cases, it easily makes sense; however, even for solid-state electrolytes such as lithium polymer batteries, the same unit is usually used. The other way of defining the energy density is "gravimetric energy density," which is usually referred to as "specific energy," and defined as the available energy from a fully charged battery per unit weight (W h/kg). Based on application and based on the importance of the volume or weight, either definition can be used. In the case of EVs and PHEVs, usually weight is a more important factor than volume; hence, mostly specific energy would be seen in the literature for this specific application.

### 16.2.1.1.6 Charging Efficiency

The chemical reactions inside the battery during charge and discharge are not ideal and there are always losses involved. In other words, not all the energy used to charge the battery is available during discharge. Some of this energy is wasted in some types of energy dissipation such as heat energy dissipation. The charging efficiency can be defined as the ratio of available energy from the battery due to a complete discharge to the amount of energy needed to completely charge the battery. This parameter may be mentioned by other names such coulombic efficiency or charge acceptance. The types of losses that reduce coulombic efficiency are mainly losses in charging process due to chemical reactions, such as electrolysis of water or other redoxation reactions in the battery. In general, the coulombic efficiency for a new battery is high, however, reducing as the battery ages.

Hereafter, this chapter will discuss some aspects of batteries in the case of EVs and PHEVs regarding charging battery packs. This will greatly help in designing more efficient and flexible chargers based on battery behavior, which will finally lead to improvement of battery pack life cycle.

### 16.2.1.2 Important Characteristics of Common Battery Chemistries

There exist many types of batteries, which can be found in battery reference books such as [4]; however, a big part of them are just produced in laboratory conditions and still under investigation and not commercialized because of many factors such as nonmaturity, low-energy density, safety, high rate of toxic materials, and price. Hence, a small group of batteries are commercially available and mostly used which mainly are Pb-acid, Ni–Cd, Ni–MH, Li-ion, and Li-polymer. Batteries in the first view can be divided to two big categories, primary and secondary. Primary batteries are simply those which can be used only once, and after a full discharge they cannot be used any more. This is because the chemical reactions happening inside them are irreversible. Secondary batteries, however, can be used many times by recharging. In the case of automotive and traction applications, mostly secondary batteries are of interest, since utilizing primary batteries in these applications seems unreasonable. Here, we will only consider secondary-type batteries and when we are talking about batteries we mean secondary batteries, otherwise stated.

### 16.2.1.2.1 Lead Acid

For over one century, lead-acid (PbA) batteries have been utilized for various applications including traction. Their well-improved structure has led to valve-regulated lead-acid (VRLA) batteries, which can be considered as maintenance-free batteries, which is a desirable characteristic for PHEVs. In terms of efficiency, they have a high efficiency in the range of 95%–99%. The main disadvantage of PbA batteries is their weight; in other words, they have a low specific energy (30–40 W h/kg) compared to their counterparts.

### 16.2.1.2.2  Nickel–Cadmium

Considering low-power applications, nickel–cadmium (Ni–Cd) batteries also benefit from a mature technology, but considering traction applications, their specific energy is low as well. The typical specific energy for this type is 45–60 Wh/kg. They are mainly used where long life and price are of high importance. The main applications, which this type is utilized, are portable devices; however, in cases that high instantaneous currents are necessary, they are desirable. Considering environmental issues, they contain toxic metals [5].

### 16.2.1.2.3  Nickel–Metal Hydride

Comparing to previous types, they have higher specific energy but lower cycle life. In general, for the same size batteries, NiMH batteries can have up to two or three times energy of a Ni–Cd type. The typical value for the specific energy of the present technology NiMH batteries is in the range of 75–100 Wh/kg. This type is widely used in EVs and PHEVs.

### 16.2.1.2.4  Lithium-Ion

This type has noticeably high specific energy, specific power, and great potential for technological improvements providing EVs and PHEVs with perfect performance characteristics such as acceleration. Their specific energy is in the range of 100–250 Wh/kg. Because of their nature, Li-ion batteries can be charged and discharged faster than Pb-acid and Ni-MH batteries, nominating them as a good candidate for EV and PHEV applications. Besides all, Li-ion batteries have an outstanding potential for long life if managed in proper conditions, otherwise, their life can be a disadvantage. One of the main reasons is almost the absence of memory effect in Li-based batteries. A weak point of Li-based batteries is their safety issues. Overcharge of Li-ion batteries should be carefully prevented, as they are highly potential for explosion due to overheating caused by overcharging. They can almost easily absorb extra charge and get exploded. Utilizing advanced battery management systems (BMS) can ensure reliable range of operation of Li-ion batteries even in cases of accidents. Besides, Li-ion batteries have environmentally friendly materials compared to nickel-based batteries.

### 16.2.1.2.5  Lithium-Polymer

Lithium-polymer (Li-Po) batteries have the same energy density as the Li-ion batteries but with lower cost. This specific chemistry is one of the most potential choices for EVs and PHEVs. There have been significant improvements in this technology. Formerly, in the maximum, discharge current of Li-Po batteries was limited to about 1 C rate; however, recent enhancements have led to maximum discharge rates of almost 30 times the 1 C rate, which greatly improves and simplifies the storage part of the EVs and PHEVs in terms of power density, since this can even eliminate the need for ultracapacitors in some cases. Besides, there have been outstanding improvements in charging times. Recent advances in this technology have led to some types which can reach over 90% SOC in a couple of minutes which can significantly increase the attraction toward EVs and PHEVs because of noticeable reduction of charging time. Because this type is a solid-state battery, having solid electrolyte, the materials would not leak out even in the case of an accident. One of the other advantages of this type is that it can be produced in any size or shape that offers flexibility to vehicle manufacturers.

### 16.2.1.3  Basic Requirements of EV/PHEV Batteries

The basic preferred characteristics of PHEV batteries can be summarized as follows [6]:

1. High specific energy which results in higher all electric range (AER) and less recharge cycles required.
2. High specific power, which results in high acceleration characteristics of the PHEV due to high rates of currents available from the battery without causing any permanent damage to the battery pack.

3. High number of charge/discharge cycles available and high safety mechanisms built into the battery because of high power ratings of battery packs.

4. Environmental friendly aspect of the battery, that is, being recyclable and including low amounts of toxic materials.

Cost is also an important concern for commercializing EVs and PHEVs in a large scale.

### 16.2.1.4 EV Battery Charging Methods

Charging in general is the action of putting energy back to the battery in terms of charge or current. Different chemistries need different charging methods. Other factors affecting choosing the charging method are capacity, required time, and so on. The most common techniques are mentioned here.

#### 16.2.1.4.1 Constant Voltage

As it is clear from the name "constant voltage" or CV, a constant voltage is applied to the battery pack. This voltage is a preset value by the manufacturer. Besides, this method is accompanied with a current-limiting circuit most of the time, especially for the beginning periods of charging when the battery can easily accept high rates of current compared to its capacity. The current limitation value mainly depends on the capacity of the battery. Depending on the battery type to be charged, this preset voltage value is chosen. For example, for Li-ion cells, the value of $4.200 \pm 50$ mV is desirable. The accurate set point is necessary, since overvoltage can damage the cell and undervoltage causes partial charge, which will reduce life cycle over time. Therefore, the circuit used for charging, which can be a simple buck, boost, or buck/boost topology depending on the voltage ratio of input and output, should be accompanied with a controller to compensate for source and load changes over time. When the cell reaches the preset voltage value, this causes the battery to be in a standby mode, ready for later use. However, the amount of this idle time should not be very long and should be limited based on the manufacturer's recommendations. This method is usually used for PbA batteries and also for Li-ion batteries while using current limiter to avoid overheating the battery especially in the first stages of the charging process [7].

#### 16.2.1.4.2 Constant Current

Constant current (CC) charging simply means applying a constant current to the battery with a low percentage of current ripples regardless of the battery state-of-charge or temperature. The abbreviation for this method is CC in the literature. This is achieved by varying the voltage applied to the battery using control techniques such as current mode control to keep the current constant. CC technique can be implemented using a "single rate current" or "split rate current." In single rate, only one preset current value is applied to the battery, which is useful in balancing the cells; however, backup circuits must be used to avoid overcharging. In the split rate CC, different rates of current are applied based on time of charge, voltage, or both in different stages of charging. This gives more accurate and balanced charging; besides, circuits should be used to avoid overvoltage of the cells. In some cases, for prolonging dead batteries, CC method with high rates and low duration can be utilized to extend the lifetime of the battery. However, this is a very cautious procedure and should be done carefully. Ni–Cd and Ni-MH batteries are charged using this method. Ni-MH batteries can be easily damaged due to overcharging, so, they should be accurately monitored during charging [8].

#### 16.2.1.4.3 Taper Current

This can be used when the source is a nonregulated DC source. It is usually implemented with a transformer with a high output voltage compared to the battery voltage. A resistance should be used to limit the current flowing to the battery. A diode can also be used to ensure unidirectional power flow to the battery. In this method, the current starts at full rating and gradually decreases as the cell gets charged. As an example, for 24 V 12 A battery, the charging begins with 12 A when the battery voltage is 24 V, then 6 A when the voltage reaches 25 and then 3 A for 26 V and finally 0.5 Amp for 26.5 V. This was just a hypothetical example and the values are not necessarily valid. This technique is only applicable to sealed lead-acid (SLA) batteries. Taper charging has other disadvantages. As mentioned before, this technique uses transformers, which adds to the weight of charger and generates heat.

### 16.2.1.4.4   Pulse Charge

This technique involves using short-time current pulses for charging. By changing the width of pulses, the average of the current can be controlled. Plus charging provides two significant advantages. One is the noticeably reduced charging time and the other one is conditioning effect of this technique, which highly improves the life cycle. The intervals between pulses called rest times play an important role. They provide some time for chemical reactions inside the battery to take place and stabilize. In addition, this method can reduce undesirable chemical reactions that may happen at the electrodes. These reactions can be mentioned such as gas formation and crystal growth, which are the most important reasons of life cycle reduction in batteries.

### 16.2.1.4.5   Reflex Charge

During charging procedure, some gas bubbles appear on the electrodes. This is amplified specially during fast charging. This phenomenon is called "burping." Applying short discharge pulses or negative pulses which can be achieved, for example, by shortcircuiting the battery for very small intervals in a current limited fashion, typically 2–3 times bigger than the charging pulses during the charging rest period resulting in depolarizing the cell will speed up the stabilization process and hence the overall charging process. This technique is called with other names such as "burp charging" or "negative pulse charging." Different control modes of charging along with waveforms and diagrams can be found in [9]. Besides, there are other charging methods such as current interrupt or CI, which will be thoroughly explained in the charging algorithm section.

### 16.2.1.4.6   Float Charge

For some applications when the charging process is complete and the battery is fully charged, the batteries should be maintained at 100% SOC for a long time to be ready for time of use. Uninterruptable power supplies (UPS) are one of such applications. The batteries should always remain fully charged. However, because of self-discharge of batteries, they get discharged over time; for example, they may lose 20% or 30% of their charge per month. To compensate for self-discharge, a constant, which is determined based on the battery chemistry and ambient temperature, is applied. This voltage is called "float voltage." In general, float voltage should be decreased with the increase of temperature. This causes a very low rate of current, for example, C/300 to C/100 rate to the battery, which continuously compensates for the self-discharge rate and also prevents sulfate formation on the plates. This technique is not recommended for Li-ion and Li-Po batteries. Besides, this method is not necessary for EV/PHEVs, which are frequently used every day. In addition, float charging involves a protection circuit, which avoids overcharging. This circuit adjusts the float voltage automatically and interrupts charging at some intervals based on battery voltage and temperature.

### 16.2.1.4.7   Trickle Charge

Mainly, trickle charging is the same as float charging just with small differences. One is the usual absence of protection circuit that avoids overcharging. Hence, it is very important to make sure in the design procedure that the charging current is less than self-discharge rate. If so, they can be left connected to the battery pack for long time.

### 16.2.1.5   Termination Methods

When the charging is in procedure, it is very important when to terminate the charging. This is because of two main reasons. One is to avoid undercharge. That is, to make sure the battery is fully charged, not partially, in order to use the full capacity of the batteries. The other one is to avoid overcharging which is very dangerous especially in the case of high energy density lithium-based EV/PHEV battery packs. If not terminated on time, the overcharging of batteries can lead to over gassing of the cells, especially in liquid electrolyte cells which results in increase in the volume of individual cells that cannot be tolerated in a battery pack which is rigidly packed. Another issue is overheating of the cells especially in lithium-based batteries, which can easily lead to the explosion and firing of the whole pack, since lithium is a very active material and easily combines with oxygen in the air. The only thing needed to begin the combination is enough heat.

Choosing different termination criteria leads to different termination methods. Selecting the type of termination of charging process depends on different factors such as application and the environment that the battery is used. Enlisted below are the different termination methods.

### 16.2.1.5.1 Time

Using time is one of the simplest methods, which is mainly used as a backup for fast charging or normally used for regular charging for specific types of batteries. This method can be cheaply implemented; however, because of diminishing battery capacity over time due to aging, the time should be set for a reduced capacity aged battery to avoid overcharging of old batteries. Therefore, the charger would not work efficiently for new batteries and leads to lifetime reduction.

### 16.2.1.5.2 Voltage

As mentioned before, voltage can be used as a termination factor. The charging process is stopped when the battery voltage reaches a specific value. However, this method has some inaccuracies, since real open-circuit voltage is obtained when the battery is left disconnected for some time after the charging. This is because chemical actions taking place inside the battery need some time to stabilize. Nevertheless, this method is widely used. Besides, this technique is usually used with constant current technique to avoid overheating damage to the battery.

### 16.2.1.5.3 Voltage Drop (dV/dT)

In some chemistries like Ni–Cd, if charged using constant current method, the voltage increases up to the fully charged state point and then the voltage begins to decrease. This is due to oxygen build up inside the battery. This decrease is significant enough, so the negative derivative of the voltage versus time can be measured to be a sign of overcharge. When this parameter becomes positive, it shows that we are passing the fully charged state and the temperature also begins to rise. After this point, the charging method can be switched to trickle or float charge or terminated completely.

### 16.2.1.5.4 Current

In the last stages of charging, if constant voltage method is used, the current begins to decrease as the battery reaches fully charge state. A preset current value such as C/10 rate can be defined, and when the current goes below this value, the charging would be terminated.

### 16.2.1.5.5 Temperature

In general, increase in temperature is a sign of overvoltage. However, using temperature sensors highly adds to the cost of the system. Nevertheless, for some chemistries such as Ni-MH, methods such as voltage drop is not recommended, since the voltage drop after full charge state is not significant to be relied on. In this case, temperature increase is a good sign of overvoltage, and can be used.

### 16.2.1.6 Cell Balancing

For high-power and energy-demanding applications such as EV/PHEVs, numerous cells should be connected in series to provide high voltages and connected in parallel to produce high currents, hence in general, high-power and high-energy rates for traction applications are achieved. This seems great; however, there are disadvantages involved. Single cells produced by different manufacturer's are claimed to be possible to be recharged hundreds of times; nonetheless, while connected in series, the life cycle dramatically declines. This is because of cell imbalances. Just to get an idea about the significance of this effect, the results of a real experiment from [10] is mentioned here. In an experiment, 12 cells were connected in series. Despite claiming life cycles of 400 cycles by the manufacturer, it reduced to only 25–30 cycles in a string. This shows how devastating this effect can be. To deal with this, the reasons of cell imbalance should be known and managed. Batteries are electrochemical devices. Even in the case of a simple resistor, while produced, there is a percentage of error. In the case of batteries, this is magnified. Two different cells produced in the same factory at the same time will have slight difference in their parameters. One of these parameters is capacity difference. In the case of a battery pack, there are different reasons leading to cell imbalance. As mentioned in [11], there are four fundamental factors leading to cell imbalance. They are manufacturing variations, differences in self-discharge rate, differences in cell age, and also charge acceptance variance. Similarly, in [12], cell imbalance is classified as internal sources which include "variations in charge storage volume" and "variations in internal battery impedance" and external sources resulting from "protection circuits" and "thermal differential across the battery pack."

To simply explain what is happening, again we refer to the water tank visualization of cells. Suppose different cells with different capacities are connected in series. It is like assuming different water tanks with different volumes are connected using pipes at the bottom of tanks. If the first tank is supplied with water, the level of water in all the tanks evenly rises. After some time, those tanks with lower capacity get full of water while others are partially filled with water. To completely fill up higher capacity tanks, there is no way other than over filling the lower capacity tanks.

Coming back to the real situation, now it is easy to guess what happens in the case of battery strings. Fully charging the high-capacity cells involves overcharging low-capacity cells. This will lead to excessive gassing and premature dry out of lower capacity cells and at the same time sulfate formation in partially charged cells leads to their life cycle reduction. How to deal with this effect and solve it is the main task of cell equalization circuits and their control algorithms. A point which should be mentioned here is that in the case of EVs, the batteries are usually completely charged up to 100% SOC. Hence, cell balancing is an important issue; however, in PHEVs, batteries are intended be kept in the range of 40%–80% so that they can provide enough energy, while being able to absorb regenerative power at the same time. Cell equalization techniques for series strings fall into three main groups: (1) charging, (2) passive, and (3) active.

It is important to note that, in cell balancing, in general, the SOC is the key point and not voltage itself, although voltage is a good sign of SOC. However, if other techniques can be used that can determine SOC more accurately. As mentioned in [13], cell balancing in a series string really means equalizing the SOC of the cells, which is equivalent to voltage balancing. Voltage is a useful indicator of SOC. Different SOC estimation techniques will be studied later.

1. *Charging:* Charging method is simply continuing charging the cells until they are all balanced to some extent. This implies overcharging the cells in a controlled manner, which leads to the full charge of high-capacity cells. This method is applicable to PbA and nickel-based batteries since they can tolerate somewhat overcharge without significant damage; however, this should be implemented carefully since extra overcharge leads to overheating the cells and finally premature drying of the electrolyte. Despite simplicity and low cost of this method, there are disadvantages such as low efficiency and long times required to obtain cell balance. Experimental results from [14] show that for actual cell equalization of 48 V batteries of a specific chemistry, weeks of time are required. Furthermore, results from [10] show that the extra time needed using this method increases with the square ratio of the number of cells added.

2. *Passive:* In this method, the extra energy in lower capacity cells is dissipated in resistive elements connecting two terminals of the cells. This will provide enough time for higher

capacity cells to get fully charged. This method has also low efficiency because of energy dissipation; nevertheless, it has a higher speed than charging method. Passive technique is also cheap and easy to implement and also the control algorithm can be easily designed.

3. *Active:* Active cell balancing involves using active electrical elements such as transistors, op-amps, and diodes to control the power flow between different cells. This flow can be between groups of cells or single cells. Obviously, extra charge is removed from lower capacity cells and transferred to higher capacity cells. This highly speeds up the charging procedure, since no energy is dissipated. Just small amount of energy is dissipated in the circuitry, which can be minimized using zero voltage or zero current switching techniques if possible.

Consider lithium-ion batteries, which are one of the most attractive candidates for EV/PHEVs. In this chemistry, the voltage should be carefully monitored to be rigorously controlled in the typical range of 4.1–4.3 V/cell since the threshold voltage leading to breaking down the cell is very close to fully charged cell voltage. As mentioned before, lithium batteries cannot tolerate overcharging. Hence, the charging technique is not applicable to them. According to safety issues related to lithium-based batteries, the only reliable cell equalization technique for them is active balancing.

Various types of cell balancing techniques can be found in the literature. Hence, there is a need to categorize them based on a criterion. Based on energy flow, they can be classified into four different groups: (1) dissipative, (2) single cell to pack, (3) pack to single cell, and (4) single cell to single cell. It is easy to imagine the operation of each category based on the name. There are advantages and disadvantages for each group. For instance, dissipative shunting resistor technique is a low-cost technique. Besides, it is easy to control because of simple structure leading to simple implementation [15].

In addition to energy flow criterion for categorizing, cell balancing techniques can be split into three main groups based on the circuit topology: (1) shunting, (2) shuttling, and (3) energy converter. Nondissipative techniques such as PWM-controlled shunting technique have high efficiency but it needs accurate voltage sensing and is somewhat complex to control [16]. Besides, the high number of elements leads to an expensive system. On the one hand, using resonant converters highly increases the efficiency because of very low switching losses, but on the other, it increases the complexity of the control system [17].

Shuttling techniques work based on transferring extra charge of high-capacity cell or cells to an energy storing component such as a capacitor or a group of capacitors and then transferring it to the low-capacity cell or cells [18]. The system would be cheaper using only one high-capacity capacitor; however, because of the existence of only one element for charge transfer, the speed of the equalization is lower compared to when a group of capacitors are used. Utilizing a group of low-capacity cells instead of one high-capacity cell is a good idea, although it increases the complexity of the control system.

Most of the energy converter cell equalization techniques utilize transformers. The achieved isolation from transformers is an advantage; however, they suffer from more costly weight. A model and transfer function of the energy converter cell equalization system is derived in [19] which can be used for control designing purposes.

The abovementioned cell balancing techniques are all summarized and explained along with circuit topologies in [20]. The question that arises here is that how much the cells should be balanced. Should the balancing range be allocated in Volts or milliVolts? As experiments from [13] show that, for PbA batteries, cell-to-cell voltage matching should be in the range of 10 mV which corresponds to SOC to provide reasonable improvement in life cycle. This is an important factor, since, for example, if the voltage matching should be in the range of 1 mV, it means that the sensors should be 10 times more accurate and also the algorithm may need to be improved for this case. This means more cost and complexity. Therefore, there is a tradeoff between expense and life cycle. This parameter should be experimentally verified for different chemistries, environments, and applications.

Since EV/PEHV battery packs do not possess a mature technology and also not many experimental data are available; sometimes contradictory claims may be seen in the literature, one of which is mentioned here. As mentioned before, battery packs used in HEVs are usually controlled to remain in the midrange of SOC. This is in order to keep the battery in a state which has the ability of absorbing enough regenerative current while being able to support enough power during acceleration. If the battery is in 100% SOC, absorbing regenerative current will lead to the overcharge of the battery. Cell overcharge is usually sensed through measuring the cell voltage. Some researchers believe that switched capacitor cell equalization technique (shuttling method) is a suitable candidate for applications with no end of charge state like HEVs. Because there is no need for intelligent control, it can work in both charge and discharge modes [20]. However, some others believe that according to the nearly flat shape of open-circuit terminal voltage of lithium-ion cells in the range of 40%–80%, the suitability of charge shuttling methods for HEV applications is denied because of the negligible voltage deviation of cells [15].

### 16.2.1.7 SOC Estimation

One of the important information needed for safe charging is SOC. Charging algorithms are mainly based on SOC directly or indirectly. Hence, the knowledge of SOC value is a key parameter in accurate charging. Unfortunately, directly measuring SOC is somehow impossible or at least very hard and expensive to implement and in some applications does not make sense, so, mostly SOC is estimated based on other variables or states of the battery. This involves battery models based on which different estimation methods can be utilized or observers can be designed. Precise estimation of SOC is not an easy task, although in usual applications, battery voltage, which is a sign of SOC, can be used. In the case of high-power/high-energy EV/PHEV battery packs, more accurate methods are advisable although being more expensive and complex in implementation. The more accurate the SOC estimation, the better the charging algorithms can be implemented, resulting in life cycle improvement.

As mentioned before, SOC is mainly the ratio of available charge to the rated capacity of the cell. One of the important points in SOC estimation is rated capacity change over time due to aging resulting from degradation of electrolyte, corrosion of plates, and other factors. Dealing with this issue is in the field of analyzing the state of health of the battery and is called "state of health estimation," which is a field of research and is not mentioned here.

Here, we will mention some SOC estimation techniques. One of the simplest methods is to completely *discharge* the battery and measure the SOC. Although simple, it is very time-consuming and does not seem logical to completely discharge a battery just to measure SOC. Knowledge of the SOC is useful for the current situation of the battery, so, if the battery is discharged, the state of the battery has been changed and there is no more use of previous state SOC knowledge. Especially, in the case of EV/PHEV, this method is not applicable. Although this method is not used in battery packs, it may be used periodically after long intervals to calibrate other SOC methods.

Another method is *Ampere Hour Counting*, which measures and calculates the amount of charge entering the battery or leaving it with integrating the current over time. This is one of the most common methods used; however, there are some deficiencies. There are always inaccuracies in sensors. Even very small, because it is being integrated over time, it can sum up to a considerable value leading to significant errors. Besides, even supposing a very accurate current sensor, because this integration is implemented usually by digital circuits and numerical methods, there are always calculation errors involved and again can show up high errors over time. Even if assuming both deficiencies to be solved in some way, there is another reason leading to inaccuracy. Even if the amount of charge entering the battery is exactly calculated, because of coulomb efficiency mentioned before, less amount of charge is available, and is also dependent on discharge rate while leaving the battery. One way to reduce these inaccuracies is to recalibrate the integration process each time a specific known set point such as fully discharged state is reached.

Another method for SOC estimation is *Measurement of Physical Characteristics of Electrolyte*. Obviously, this method is mostly applicable to liquid electrolyte batteries, not solid ones like Li-Po.

In this method, a chemical relationship is used, which depicts the change in important parameters of the electrolyte with the change in SOC. One of these parameters is the density of the acid. There is an almost linear relation between change in acid density and SOC. This method is very well known especially in PbA batteries. The density can be measured directly or indirectly using parameters such as viscosity, conductivity, ion concentration, refractive index, and ultrasonic effect.

As discussed before, the *Open Circuit Voltage* of the batteries can be used as an indicator of SOC. The uncertainty in this method is the fact that batteries under operation need some rest time for their open-circuit voltage to become stable. This time for some cases can be up to hours. However, this method is also widely used. The key point in this method is the linear relation of open-circuit voltage versus SOC in a specific range of SOC. This range and its slope are different in different chemistries, which should be taken into account.

There are other techniques categorized under soft computation techniques such as fuzzy neural network [21] or adaptive neuro-fuzzy modeling [22], which can also be utilized for SOC estimation. Other approaches such as heuristic interpretation of measurement curves mentioned in [23] such as Coup de fouet, linear model, artificial neural network, impedance spectroscopy, internal resistance, and Kalman filters, which are more precise methods but more complicated to implement, can also be utilized.

### 16.2.1.8 Charging Algorithms

Charging algorithm can be defined as the combination of what was mentioned up to here and controlling all or part of the parameters affecting battery performance and life cycle in such a way to achieve charging the battery pack safely, efficiently, and terminating on time. Managing the charging procedure of a high-power battery pack with hundreds of cells involves many issues as mentioned before. Controlling all of these aspects needs efficient and accurate algorithms with reliable safety and backup circuits. The trends toward fast charging, with huge amount of current flowing to the battery pack producing lots of heat, need accurate and reliable supervisory control algorithms to ensure safe charging. Managing this complicated task can be handled with some advanced control topics such as fuzzy logic, supervisory control, and decentralized control. In general, each battery-chemistry needs its own charging algorithm. However, depending on the algorithm, it may be applied to other types also; however, this should be carefully done according to life cycle issues.

For precise battery charging, the charge/discharge profile of the battery provided by the manufacturer may be used. However, the profile is valid for new batteries; hence, it is better to use other techniques such as data-acquisition methods to acquire the charge/discharge profile of the battery. Novel techniques regarding this issue are being introduced in the literature very often [24].

As mentioned before, PbA batteries have mature technology, and infrastructure is already there; however, they have poor life cycles in the order of 300–400 cycles. Huge effort has to be put into research for increasing the life cycle of this chemistry because of its many advantages such as cost and availability. This chemistry has a common algorithm, which includes four different stages or three based on application. In the first stage, a predefined constant current is applied to the battery pack, which charges the cells with a high speed. In this stage, the cell voltages increase gradually because of increase in SOC. This is called *bulk charge* stage. The process is continued until a predefined maximum voltage is reached. These values are recommended by the manufacturer, in the datasheet. In the next stage called *absorption charge* stage, constant voltage is applied to the battery pack. At this stage, the current decreases gradually until it again reaches a predefined C rate value. Now the cells are approximately charged but not equalized because of cell imbalance. At this stage, a relatively higher voltage than constant voltage stage can be applied to the pack to balance all the cells inside the pack. This stage is called *equalization charge* stage. This can be achieved with other techniques as mentioned before, especially during the previous two stages. After some time, the charger switches to float charge mode to keep the battery in a ready state. Depending on the application, this stage can be omitted. This stage is called *float charge* stage. This is illustrated in Figure 16.1.

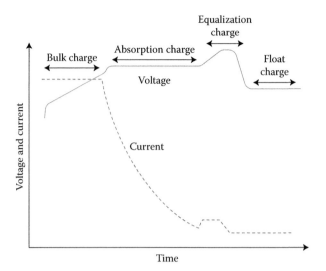

FIGURE 16.1   Charging algorithm of a typical PbA battery.

As the battery ages, its internal characteristics also changes, hence, an adaptive charging algorithm should be used to take into account these changes also. Experiments show that the value of voltage of the third stage should be increased over time to get the same amount of energy as the battery ages [25–27]. The equalization stage is the key part of this algorithm and has a great influence on the life cycle of the battery. As mentioned, the voltage of this stage should be increased but this increases the current and also the heat generated which has negative impact on the life cycle. One way to get the same amount of current with lower heat dissipation is using pulses of current. This technique seems the same as pulse charging, but it is different, since the time intervals are significantly bigger than pulse charge time periods, which are in the range of kHz. This method is called *current interrupt* or CI. This technique has shown significant life cycle improvements [28]. Using this algorithm, the battery can reach 50% of the initial capacity after 500 cycles, which is a significant improvement in life cycle. Although this algorithm is useful, it puts the battery under stress while it reaches the end of life because of permanently increasing the overvoltage value. This algorithm can be implemented in an alternative way. Instead of using this method in each cycle, which puts high stress on the battery, it can be utilized after every 10 cycles. This algorithm is called partial-state-of-recharge cycling (PSOR) [28], which has approximately the same effect with the advantage of lower stress on the battery. This algorithm has been claimed to enable the battery deliver up to 80% of initial capacity at the cycle number of 780, which is a really noticeable improvement in life cycle.

As can be seen, these complicated algorithms cannot be done using simple PI or PID controllers. They need DSP-based controllers to be programmable based on chemistry, state of health of the battery, and other factors. Continuously, different algorithms are being proposed every day and tested for improving life cycle of the batteries. This is a vast research area and developed every day and getting more attention as the EV/PHEVs become more and more popular.

## 16.3   POWER ELECTRONICS FOR EV AND PHEV CHARGING INFRASTRUCTURE

In its simplest incarnation, a charging facility for EVs would merely consist of a unidirectional AC/DC converter-charger connected to the power grid. Power would simply flow on demand from the power grid through a power conditioner into the vehicle battery pack; once the battery is fully charged, the connection to the grid no longer performs any useful work. This simple setup may have been appropriate for small private commercial vehicle fleets, or where electric cars represent a very small fraction of the active road vehicles. However, as society's efforts to electrify our means of transportation intensify, it is clear that a smarter exploitation of the vehicle-to-grid (V2G) inter-action is in order. To the power utility, the bulk of EVs connected to its grid appears as an energy storage agent that is too significant to be left untapped. This view is reinforced by the outcome of several statistical studies [29] that show that more than 90% of all vehicles are parked at all times, and thus potentially connected to the grid. Assuming a 50% EV market penetration, simple calcu-lations show that, the total storage capacity available would be in the order of thousands of GWh. Therefore, the V2G connection should be bidirectional, giving the owner of each vehicle the ability to "sell" back a portion of this stored energy to the utility, presumably at an advantageous rate. The same requirement of bidirectionality also applies when the vehicle is connected to a microgrid pow-ered by a distributed resource. In a grid-connected solar carport, for instance, many vehicles can be charged by PV panels (or by the grid; or by both), depending on load and insolation (time of day, meteorological conditions, time of year, etc.). In the case of overproduction, energy from the panels can be fed back to the grid for a profit, while the EV batteries function to buffer the characteristic solar intermittence. Similarly, the DC/DC converters that condition the power from the solar panels to each charging vehicle should also be bidirectional in order to allow the owner of a plug-in EV (PEV) to exchange a portion of his energy with the operator of the microgrid. A typical PV-powered gird-tied carport architecture is shown in Figure 16.2.

These considerations demonstrate that bidirectionality is a highly desirable feature in any power conditioner utilized in vehicle charging–discharging applications, including interactions to and from the grid, microgrid, or residential loads and renewable energy generators. On this basis, the reader should note that the discussion that follows makes no distinction between vehicle-to-grid and grid-to-vehicle communication, both being classified by the acronym (V2G). Similarly, V2H will designate either the vehicle-to-home or the home-to-vehicle interface.

Other requirements for the optimal charging infrastructure are harder to identify. This is due to pervasive lack of standardization involving battery technology and nominal voltage, safety strategy, connector configuration, communication protocols, location of charger (on-board or off-board), and more. In the following sections, these issues are treated especially with reference to their impact on local power generation and utilization.

### 16.3.1  CHARGING HARDWARE

Like any other means of transportation, EV/PHEVs benefit markedly from minimizing their weight. These vehicles are even more sensitive to that issue, considering the unavoidable presence of heavy battery/ultracapacitor energy packs. The electronic power converters intended for the charging function can be bulky and heavy in their own right, and their deployment onboard seems to make little engineering sense. Yet, at the time of this writing, the great majority of PEVs in North America contains their own power rectifier and connects directly to 120 V or 240 V household plugs. This

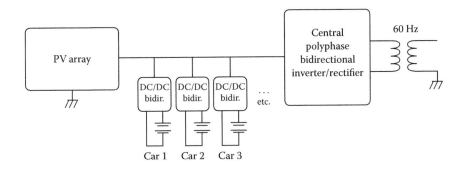

FIGURE 16.2   Typical PV-powered gird-tied carport architecture.

can be explained by two considerations. First, while the household AC voltages are fully standardized, at least within a country, the DC nominal battery voltage for PEVs is definitely not. Different manufacturers have adopted ad hoc energy storage technologies and safety strategies, resulting in strikingly different bus voltages and current requirements. An unsophisticated external converter could then be optimized for only one vehicle brand or model.

Second, some techniques have been developed that do not add significant weight to the vehicle. The critical idea is to utilize the power electronic circuitry that is already onboard in order to perform the rectifying function. This charging circuit is commonly referred to as an "integrated charger"; it makes use of the bidirectional inverter that drives the electric motor as well as the windings of the motor itself. Figure 16.8 shows a well-known example of this concept.

With regard to Figure 16.3, it is important to realize that inductors LS1, LS2, and LS3 are not added magnetic devices, but the actual winding leakage inductances of the electric motor. Thus, the only added components are the two relays K1 and K2, which are activated in order to reconfigure the schematic from a three-phase motor driver, during normal vehicle propulsion operation, to a single-phase boost rectifier, during charging.

The above two considerations are consistent with relatively slow charging strategies. In the first instance, it is because the amount of electric power available in a residential setting does not usually exceed 10 kW at a household plug. In the second instance, it is because the electronics that drive a PEV electric machine are sized for its propulsion needs. Thus, the average charging power must be limited to a level comparable to the motor's rated power, which is of the order of 10–50 kW in smaller cars.

Slow charging strategies are commonly referred to as levels 1 and 2. The former is associated with a connection to a regular AC household plug (120 V, 15 A), while the latter involves power that can be as high as 14.4 kW or 240 V at 60 A, which is also normally available in residential settings. Moreover, these power levels are compatible with the average generating capacity of the microgrids and corresponding distributed resources. Then, it would appear that whether a car is charged through a regular residential wall socket, or a microgrid outlet, the available power levels justify the location of the rectifier onboard the vehicle.

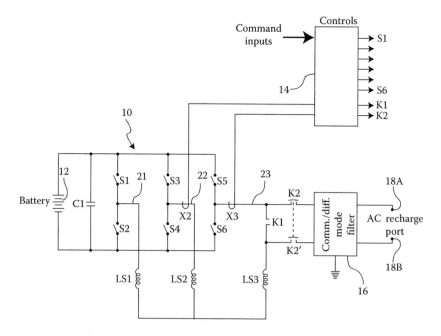

FIGURE 16.3   Integrated charger based on boost converter. (Adapted from Cocconi AG. 1994. Combined motor drive and battery recharge system, US Patent no. 5,341,075.)

However, EV manufacturers are quickly recognizing that long charging periods may be acceptable to consumers only if quick charging is available as well, albeit at higher cost. Two solutions are presently under consideration. The first solution is the so-called battery swapping, whereby a car owner simply drives to a service station and allows an automated system to safely replace a spent battery with a fully charged one. Along the same lines, the battery could be of the redox flow type. In this case, the battery casing is not replaced; rather, it is drained and then filled with fresh liquid electrolyte. In either forms, the obvious drawback is the need for the exact standardization of battery size, chemistry, and capacity.

The second solution consists of allowing direct access to the battery DC terminals, so that a large offboard rectifier can be connected and reenergize the battery pack using powers of the order of up to several hundred kilowatts. This is known as level 3 charging, allowing an electric "fill-up" service stop to last only a few minutes. In this instance, although the battery itself may not need a high level of standardization, it would be subject to extremely high currents at high voltages. This renders the practical implementation of this second solution strongly dependent on needed improvements to battery and ultracapacitor technologies. Furthermore, a public charging station capable of servicing many cars simultaneously would represent a local load of several megawatts as seen by the grid.

Despite these difficulties, it is highly likely that either the battery swapping or the fast-charging strategy will eventually be universally available to complement, or even replace, the onboard charger.

## 16.3.2  Grid-Tied Infrastructure

Assuming fast charging through direct DC connection becomes the method of choice, car owners will have two options. They may still prefer to slow charge their vehicles overnight by plugging to an AC/DC charger (or electric vehicle supply equipment: EVSE), most probably in their homes. This converter will deliver relatively low powers of the order of 5–10 kW because it is limited by the residential connection, as mentioned earlier. However, as further explained in Section 15.5, this method may involve some financial returns. The alternative method will be to use a fast-charging public facility, corresponding to a familiar service gas station that is capable of multimegawatt power transfers. Although the cost per kWh will be high, the owner benefits from charge times in the order of minutes rather than hours.

In both cases, V2G capability, enabled by smart grid technology will become a standard feature with all EVSEs, whether they are public, commercial, semipublic, or private. This will allow the subsistence of a very significant distributed storage resource at the disposal of electric utilities. More specifically, the EV fleet will be optimally positioned to become a significant provider of some ancillary services and play a role in offering dispatch-able peak power. These services to the electricity supplier will be analyzed separately.

### 16.3.2.1  EVs as "Peakers"

A peaker is a small but nimble generating units that can supply the grid with a relatively fast response. Historically, natural gas turbines or small hydroelectric plants were the devices of choice for this task. They are active for only a few hours every day and therefore provide only limited energy. Thus, a substantial fleet of EVs can carry out this task as a highly distributed resource without significantly depleting their batteries. Unfortunately, as long as peak power is not considered as a "service," the utility operator will compensate the car owner solely for the energy sold, albeit at a higher peak-demand rate [31]. This may not constitute a strong enough incentive to the car owner who has to consider other factors, such as the additional battery and power electronic wear and tear for his vehicle. Nevertheless, future adjustments in energy market models are under study to address this among other issues.

### 16.3.2.2   EVs as Spinning and Nonspinning Reserves

Two of the most lucrative ancillary services are the spinning and nonspinning reserves. The former consists of generators that are on line, but normally run at very low capacity. In the case of a disruption, such as a failure in baseload generation or transmission, these generators are commanded to provide the missing power. They must be able to ramp up in less than 10 min and provide power for as long as 1 h or more. Nonspinning reserves are not on line and are required to ramp up to full power within 30 min. Because this is a service, the utility company will pay for the availability of the power as well as its amount. In fact, this service is paid even when no power is ever delivered. An EV owner can provide this service naturally and be reimbursed starting at the time he plugs his vehicle to the grid even if the battery is never discharged. Also, it must be noted that plug-in hybrid electric vehicles (PHEV) have smaller battery capacity than all-EVs, but contain an ICE that can be started on a V2G command to generate electricity and function as a spinning reserve as well.

### 16.3.2.3   EVs as Voltage/Frequency Regulation Agents

An ancillary service that is even better tailored for EVs is regulation. It consists of delivering or absorbing limited amounts of energy on demand and in real time. Normally, the request is automated in order to match exactly the instantaneous power generation with the instantaneous load. Failure to do so results in dangerous shifts in line with frequency and voltage. The dispatched amount of energy has short duration—in the order of few minutes—but it is requested relatively frequently. Therefore, this is a continuous service. It is important to underline that the amount of energy involved is relatively small and changes direction quite rapidly and regularly, implying minimal EV battery discharge for any reasonably short time interval. The near instantaneous response time and the distributed nature of the EV fleet explains why regulation is probably the most competitive application for V2G from the point of view of the utility operators.

### 16.3.2.4   EVs as Reactive Power Providers

Most electronic topologies used for the inverter/rectifier function in the interface of the EV to the grid are fully capable of shaping the line current to have low distortion and varying amounts of phase shift with respect to the AC line voltage. This implies that reactive power can be injected into the grid on demand and in real time [32]. Furthermore, since reactive power translates in no net DC currents, this service can be provided without any added stress to the EV battery.

## 16.4   VEHICLE-TO-GRID AND VEHICLE-TO-HOME CONCEPTS

The advantages described in the preceding sections are not presently exploitable due to a general lack of the required hardware infrastructure, as well as the thorny transition to new business models that include the V2G concept. The roadmap toward achieving this goal will probably consists of the following several milestones.

1. The first milestone is rather rudimentary as it does not yet require bidirectional converters. It will consist of a simple owner-selectable option afforded by the vehicle battery management system (BMS) user interface that allows the grid to schedule when to activate and deactivate charging. In return, the owner pays lower per-kWh rates. Communication between the grid operator and the BMS can be done through existing cell phone technology, requiring no additional infrastructure or hardware.
2. The straightforward "grid-friendly" charging time-window strategy described above will evolve to include more sophisticated algorithms. For instance, the grid might broadcast any updates to the current per-kWh cost and let the vehicles BMS choose whether to activate charging. Some ancillary services, such as regulation "down" could become feasible, while regulation "up" will be limited by the lack of reverse power flow capability of the EVSE at this stage. The use of aggregators will also become widespread. Aggregators are

intermediate communication and power distribution nodes between a group of vehicles, located in proximity to each other, and the grid. This allows the grid to macromanage a single installment of several vehicles, corresponding to significant power level blocks with somewhat predictable behavior, akin to the other distributed energy resources. Furthermore, because the aggregator's consumption will be in the MW range, it will allow purchases of power on the wholesale market, reducing the cost for each participant vehicle.

3. Eventually, bidirectionality will become a standard feature for all EVSEs. However, this capability will not be immediately harnessed to achieve controlled reverse power flow to the grid. Rather, the EV battery will, most likely, initially service the surrounding premises, probably the owner's home. This scenario, called V2H, will probably precede the full implementation of V2G [33] because it effectively bypasses several large infrastructure and technical issues needed for V2G, while achieving many of the same results. Through pricing incentives, an EV parked in the residential premises and connected on the customer's side of the meter can be exploited to absorb energy from the grid during times of low demand, and transfer it to the household appliances, during times of high demand. This will indirectly shrink the power peaking for the grid while reducing the electrical bill to the user. It will also reduce overall transmission losses over the V2G strategy, because line current will flow only in one direction, from grid to vehicle, and will then be consumed locally.

   Moreover, if the household is geared with renewable source generators, the vehicle can immediately serve as storage and, during blackouts, as backup power. Although one can find some similarities between the concepts of V2H and V2G, there are important distinctions. In practical terms, these differences stem from the fact that V2H cannot take advantage of the high predictability deriving from statistical averages afforded by very high numbers of vehicles available for V2G operations. Simply stated, the real benefits of V2H are not easily estimated because they are dependent on many exceedingly uncertain variables. Some of these are: the number of available vehicles, commute schedule, time duration and distance, PEV energy storage capacity, presence and quantity of quasi-predictable local generation (e.g., solar panels), presence and quantity of unpredictable local generation (e.g., wind power), residence-specific energy consumption profile, and presence of additional storage. Despite the fact that these issues will require complicated management algorithms in order to optimize the use of V2H, some benefits such as emergency backup are available immediately with a relatively minor upgrades to the residential infrastructure. These upgrades consist mainly in the installation of a transfer switch to disconnect the residence from the grid during backup operation, and expand the design of the power converter to detect islanding conditions. Furthermore, the EVSE must be capable of controlling output current into the line when connected to the grid, but reverting to controlling output voltage when acting as a backup generator.

4. Full V2G implemented with automated options for V2H. The connection will be metered and could also include any locally generated renewable energy management.

### 16.4.1 Grid Upgrade

The electric transmission and distribution networks in most industrialized nations must consider changes and upgrades in order to fully benefit from the introduction of EVs as distributed resources. First, we must consider the extent by which the current production capacity will have to be expanded. Various studies [34] have suggested that once the typical charging profile for an EV is scrutinized and hopefully optimized—charging mostly at night—the installation of new generation will be unnecessary or minimal at most. In fact, it will have the effect of diminishing reliance on more expensive load-following plants, since the overall 24-h demand curve will average closer to the baseload. Therefore, the main effort should be in effectively introducing intelligence into the grid. The hardware and communication standards for implementing such intelligence are still under study. Wideband digital interface can take the form of PLC (power line communication) or utilize separate communication channels that have some market penetration already. In either case, the EV will most likely be treated as any other managed load by this smart grid, with the exception of a sophisticated onboard metering device that will have to be reconciled with the utility's pricing model. Currently, the two major obstacles to the utilization of EVs as distributed resources are the lack of bidirectionality in the power converters and the lack of recognized standards, both software protocols and hardware, for the smart grid function. Of the two, the former is by far the easiest to implement, given the well-established characterization of suitable power electronic topologies.

### 16.4.1.1 Renewable and Other Intermittent Resource Market Penetration

Owing to recent well-known trends, renewable resources are increasingly prominent in the complex energy market mosaic. As long as their penetration level is low, they can be easily handled by the current infrastructure, but at present incremental rates, this will not be the case in the future. The intermittent nature of solar and wind generation will require a far more flexible compensation mechanism than what is available now. Because of this, large battery banks that act as buffers between the generator and the grid invariably accompany today's renewable energy installations. Wind power, in particular, is not only intermittent put has no day-average predictability, as winds can differ hour-to-hour as easily as night as during the day, adding an extra amount of irregularity to an already varying load. This suggests that EVs will be called to perform not only the more manageable regulation task, but also aid in providing peak power. As noted earlier, this may not find the approval of the EV owner unless the pricing model is modified. Nevertheless, it is reasonable to ask whether a large EV contracted fleet could perform this task on a national (United States) level. Studies have shown [35] that the answer is yes. With an overconfident 50% estimation for the market penetration of wind energy and 70 million EVs available, peak power can be provided at the expense of approximately 7 kWh of battery energy per day or about 10%–20% of an average PEV reserve.

### 16.4.1.2 Dedicated Charging Infrastructure from Renewable Resources

The traditional microgrid often relies on diesel generators as a single source of energy. Even in this case, any load fluctuations are quite difficult to negotiate, relying solely on the intrinsically slow ramp up speeds of the generator itself. The new trend toward integrating renewable resources into microgrids greatly amplifies this problem due to their notorious intermittent nature. However, the dedicated generation from renewables for the explicit purpose of EV charging is gaining more credibility as a means to eliminate transmission losses and greatly reduces the overall carbon footprint associated with EVs. Such installation would fall into two categories: (1) small installations with or without a grid tie, (2) large installation with grid tie. Small installations can be somewhat arbitrarily defined at less than a total of 250 kW of peak production. This would be sufficient to slow charge about 20 vehicles and would certainly require local external storage in order to buffer the peaks and valleys in local energy production. This is more evident in the case of islanded installations; if any energy is produced in excess, it cannot be sent back to the grid, so it will need a long-term storage capability. Large installation with a grid tie can inject or draw power to and from the grid as a means to equalize the grid during overproduction and to draw from the line. However, depending on the number of vehicles connected, which can be accurately predicted with statistical methods, some of the EV resource can be utilized to minimize the size of the external storage. Nonetheless, it appears that EVs can alleviate the inherent issues associated with local renewable production for the dedicated purpose of EV charging, but cannot totally eliminate them.

## 16.5 POWER ELECTRONICS FOR PEV CHARGING

The PEV charging process will be enabled by sophisticated power electronic circuits found in the EVSE. Such equipment will be optimally designed depending on the different possible sites and types of power connection. We will begin by looking at EVSE connected to the main power grid and then analyze dual-sourced systems such as grid-tied renewable energy installations dedicated to EV charging. A short discussion on basic safety compliance strategy follows.

### 16.5.1 SAFETY CONSIDERATIONS

For offboard chargers, there are only a few important safety needs that affect significantly the power converter design. These are (1) isolation of the battery pack with respect to chassis and the grid terminals; (2) ground fault interrupters (GFI) to detect any dangerous leakage current from either the grid or the battery circuit; (3) connector interface; (4) software. A typical EVSE and related connections are shown in Figure 16.4.

Two GFIs detect any breakdown or current leakage on either side of the isolation barrier in order to ensure complete protection to the user and disconnect the high-power circuit immediately in case of fault. The battery pack is fully isolated from the chassis since it cannot be grounded properly during charging without heavily oversizing the connector cable. In fact, some existing safety recommendations require that an active breakdown test be performed on the battery pack prior to every charging cycle. At the time of writing, the de facto standard for level 3 DC charging is the CHAdeMO standard developed by the Tokyo Electric Power Company. Although competing standards may eventually overtake it in popularity, the description the CHAdeMO connector demonstrates the safety concerns involved. The connector itself will have mechanical means to lock itself onto the car receptacle in order to prevent accidental removal when energized. It will not only carry the power leads, but also communication wires that include a CAN bus digital interface as well as several optically isolated analog lines for critical commands such as on/off and start/stop. Every analog signal sent by the EV to the charger (or vice versa) is received and acknowledged through the analog lines. This analog interface is sturdier than a digital one and less susceptible to electromagnetic interference. The CAN bus is activated only when more complex information is exchanged. Prior to the start-charge command, the EVSE communicates its parameters to the EV (maximum output voltage and currents, error flag convention, etc.), and the PEV communicates its parameters to the EVSE (target voltage, battery capacity, thermal limits, etc.), and a compatibility check is performed. During charging, the EV continuously updates the EVSE with its instantaneous current request (every 100 ms or so) and all accompanying status flags. Once charging is finished, the operator can safely unlock the connector and drive away.

As can be seen, the presence of safety devices, such as the GFIs as well as a sturdy method of analog and digital communications, renders the charging process extremely safe, leaving the power electronic designer of the EVSE with the relatively simple task of ensuring only the isolation barrier between the grid voltage and the EV floating battery. In fact, the utilization of an isolation transformer can actually simplify some designs due to the added voltage amplification capability afforded by the transformer's turns ratio. This could prove very beneficial if much higher battery voltages become necessary in order to increase storage capacity.

### 16.5.2 GRID-TIED RESIDENTIAL SYSTEMS

As noted earlier, only levels 1 and 2 are feasible within the confines of a residential setting. This can be accomplished through integrated chargers when available or an external EVSE. In the latter case, the most obvious circuit configuration is a single-phase bidirectional rectifier/inverter powered by a 240 V AC/60 A circuit that is readily available from the distribution transformer. The DC-link voltage is then processed by a bidirectional DC/DC converter that performs the isolation function. This simple topology shown in Figure 16.5 can be called the canonical topology, as will be repeated, with minor changes, for most grid-tied systems irrespective of power rating.

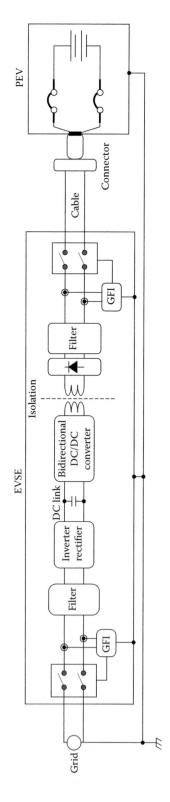

FIGURE 16.4 Typical EVSE safety configuration.

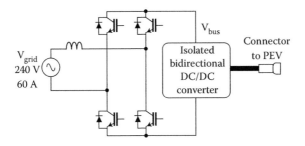

FIGURE 16.5    Canonical single-phase EVSE configuration.

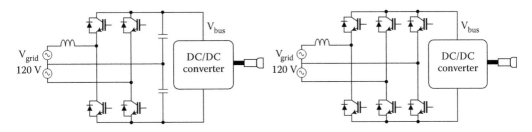

FIGURE 16.6    Split-phase sourced EVSE configurations.

In North America, the 240 V from the residential distribution transformer is in the form of a split 120 V supply, suggesting small modifications to the canonical topology. Figure 16.6 shows two possibilities.

The two topologies in the figure are similar, but the one on the right has better voltage utilization and is better equipped to counter unbalanced loads on the split supply [36]. For the DC/DC converter, many bidirectional isolated circuit topologies have been proposed [37]. Typical circuits are shown in Figure 16.7.

When the two controlled bridges are independently driven in phase-shift modulation (PSM), these are generally referred to as dual active bridge (DAB) topologies. In their simplest operation mode, when power needs to be transferred from the left-side circuit to the right-side circuit, for instance, the right-side IGBT switches are left undriven, leaving their antiparallel diodes in the form of a regular diode bridge. Under these circumstances, the topology becomes identical to a regular PSM converter, which is simple to operate, but not very flexible in terms of voltage gain. However, when both bridges are modulated, power transfer can be accomplished in both directions and with great variability ranges on the input and output voltages. In addition, zero voltage switching (ZVS) can be assured for all switches for reduced switching loss and generated electrical noise (EMI). Other topologies [38,39] based on the DAB have been proposed with purported additional benefits, such as better switch utilization, extended ZVS operating range, and more flexible voltage amplification.

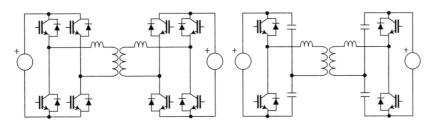

FIGURE 16.7    Typical isolated bidirectional buck–boost DC/DC converter topologies.

### 16.5.3 GRID-TIED PUBLIC SYSTEMS

A public parking/charging installation would deliver only level 2 power, given the relatively long plug-in times. Because there are several parking locations in close proximity, the power configuration used for residential use may not be optimal. Rather, a single transformer can be installed at the grid, delivering isolated power to all vehicles in the facility. This way, cheaper and more efficient nonisolated DC/DC converters can be used without violating safety rules. Figure 16.8 illustrates this configuration for each charging station. For the whole installation, the architectures shown in Figure 16.9 are possible.

In the centralized architecture [40], a single, large polyphase, 50/60 Hz step-down transformer connects to the grid, providing isolation for the whole facility. A large bidirectional rectifier that produces a single high-voltage DC bus follows this. Each parking station uses inexpensive high-efficiency, nonisolated DC/DC converters to process this bus voltage into the appropriate charging

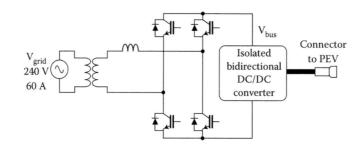

**FIGURE 16.8**  Configuration with isolation at the grid.

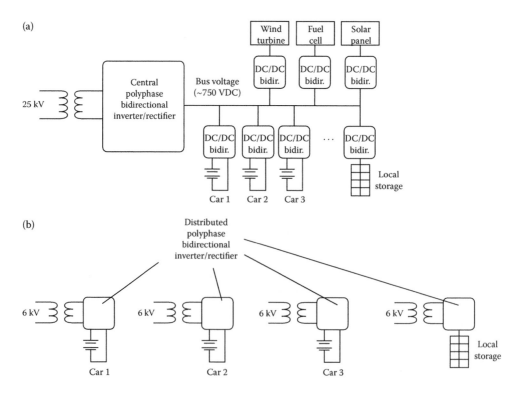

**FIGURE 16.9**  Centralized architecture (a); distributed architecture (b).

current for the individual EVs. Because isolation is either desirable or required, especially on PV panels depending on local electrical codes, additional storage, or generating resources, such as wind turbines and fuel cells, it can also benefit from simpler interface to the DC bus. Moreover, the single transformer connection guarantees that no DC current is injected into the grid, doing away with complicated active techniques to achieve the same purpose.

The advantages just noted for centralized configuration are somewhat offset by the following drawbacks: (1) the need for a bulky and usually inefficient line-frequency transformer, (2) an expensive high-power polyphase inverter/rectifier, (3) single-fault vulnerability in the transformer and central inverter rectifier, (4) lack of voltage amplification in each nonisolated DC/DC converter (otherwise afforded by the turns ratio of high-frequency transformer in isolated topologies).

In a level 3 (fast-charging) public facility, other technical challenges must be considered. For instance, with battery pack-rated voltages in the range of 200–600 V, the overall currents required for fast charging will be of the order of thousands of amps. The current must necessarily flow through cables and especially connectors, causing local thermal issues and loss of efficiency due to ohmic loss. In addition, the charging stations will appear as a concentrated load to the grid, so that any power transients produced by the stations are very likely to cause local sags or surges.

The first issue can be partially countered by brute force methods such as the development of advanced submilliohm connectors and minimizing cable lengths by placing the grid step-down transformer in physical proximity of the vehicle. It is obvious that any intervening power conditioning electronic circuit should be added only when absolutely necessary. This immediately suggests that the architecture of the charging station should be distributed rather than central. As can be seen from Figure 16.10, a distributed architecture could potentially reduce the number of processors from grid to battery from two to one. To be fair, this single stage may not be feasible when managing large input–output voltage ranges, especially if buck–boost operation is required (see discussion on the Z-converter). Nevertheless, if an additional DC/DC stage should prove necessary, it will be easily integrated locally with the inverter for improved efficiency. Furthermore, a central processor, besides constituting a single point of failure as already noted, would have to be rated for the full service station power, which could be of the order of a megawatt. On the contrary, a distributed architecture benefits from repeated circuitry (economies of scale), redundancy for higher reliability, and the possibility of power conditioning in physical proximity to the vehicles, reducing ohmic loss.

The issue of power line quality deterioration caused by the service station operating transients has only been studied for specific geographic locations [41], but possible voltage fluctuation of up to 10% have been reported depending on the length of the feeding high-voltage transmission line. The obvious and perhaps sole approach to mitigate this problem is the integration of flywheel, battery, or ultracapacitor banks into the charging station. This storage will smoothen out the load transients by delivering local power when needed and storing power during periods of lower demand. Moreover,

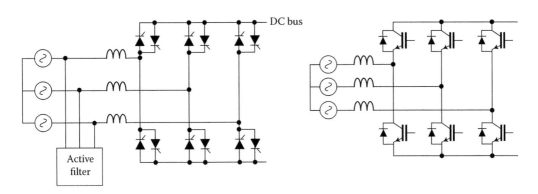

FIGURE 16.10   Thyristor bridge and active line filter (left); IGBT bridge (right).

it will average out the draw from the grid so that the distribution equipment can be rated at much lower peak powers (as much as 40%) [42].

The task of discriminating between the various available electronic topologies is made easier when considering the sheer power handled by fast chargers; to wit, up to 250 kW. Obviously, a good candidate must be very efficient, with inherently low noise, with low component count, and capable of high-frequency operation in order to control physical size [43]. For the inverter/rectifier section, we must also add the requirement that no significant harmonic content should be present in the line current. In order to obtain input currents that are sinusoidal and free of ripple noise, several methods of increasing complexity exist.

One method uses a three-phase thyristor bridge. The devices are very rugged and efficient in terms of conduction loss and have enough controllability to roughly regulate the DC bus [15]. In order to remove unwanted current harmonics, an active filter is added. This filter is based on IGBT devices, but only processes a small portion of the total power. A second method uses a fully controlled IGBT bridge in order to achieve excellent input current shaping for extremely low input current distortion and well regulated, ripple-free DC bus voltage.

Moreover, fewer components and much higher switching frequencies can be achieved resulting in smaller magnetic components. However, IGBTs have switching losses and more significant conduction losses than thyristors. Yet other techniques, although less sophisticated, have the potential of realizing the required low current distortion limit without the addition of an active filter. The uncontrolled 12-pulse rectifier shown in Figure 16.11 (left) can certainly do this, albeit with the addition of significant inductive filtering. Because the output DC bus will not be regulated, the subsequent DC/DC converter design cannot be optimized. Using thyristors can achieve regulation of the bus and possibly still achieve the required input current shaping. It is important to note that of the four topologies mentioned here, only those in Figure 16.10 are bidirectional and, therefore the only choice if V2G is to be implemented. For the final DC/DC converter, all common basic topologies, that is, boost, buck–boost, buck, Cuk, SEPIC, and ZETA can be used, as long as they are rendered bidirectional, by replacing the diode with a transistor device. In this case, these topologies function differently, depending on the direction of power flow (see Figure 16.12).

Different design requirements might suggest different topologies [40], but some of these are objectively more difficult to justify. For instance, using the buck–boost/buck–boost (bottom left in Figure 16.12) produces a voltage inversion from positive to negative that may be undesirable. It also places higher electrical stress on the switches; it requires a more sophisticated design for the inductor and draws pulsed current from the battery. Similarly, the ZETA/SEPIC topology has a higher part count, including a capacitive, rather than inductive energy-transferring element. However, as long as the DC bus is guaranteed to exceed the battery voltage—a requirement that is assured by

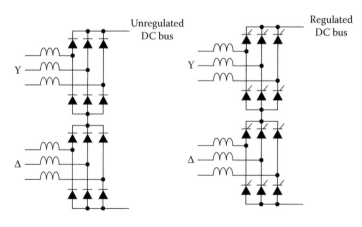

FIGURE 16.11   12-Pulse rectifier circuits.

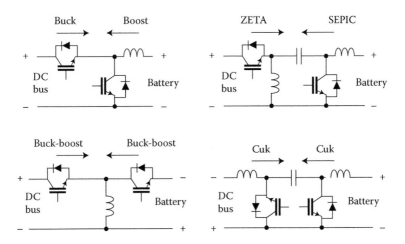

**FIGURE 16.12**   Basic bidirectional nonisolated topologies.

the use of the controlled bridge discussed earlier—the buck/boost topology (top left in the figure) is quite attractive. Furthermore, this topology is readily modified in order to divide the task of handling a very large power flow among paralleled modules [41].

This is shown in Figure 16.13; the amount of converted power can be split among $n$ identical sections and the battery ripple current greatly reduced by the well-known technique of phase-shift interleaving. Using this circuit with $n = 3$ and a switching frequency of 2 kHz, for a typical 125 kW application, efficiencies as high as 98.5% have been reported.

### 16.5.4   GRID-TIED SYSTEMS WITH LOCAL RENEWABLE ENERGY PRODUCTION

As noted earlier, when relatively large energy production from intermittent sources is to be tied to the grid, a statistically predictable EV presence could serve the purpose of minimizing onsite dedicated storage. This would be the case for municipal carports powered by wind and/or solar generation and where the vehicles must be able to interact intelligently with both locally generated and grid distributed power at the same time. The possible scenario described in Figure 16.9a may not be ideal when the renewable resource is meant to generate the dominant share of EV charging energy. Rather, by realizing the advantages of the distributed configuration, as in Figure 16.9b, one stage of conversion can be eliminated so long as a conversion topology with wide input–output voltage range capability can be found.

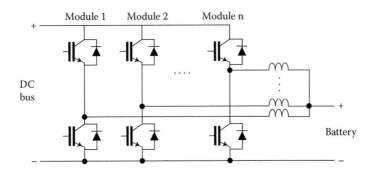

**FIGURE 16.13**   Interleaved modular approach for the DC/DC converter.

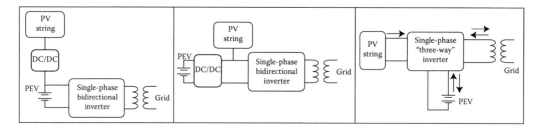

**FIGURE 16.14**  Possible configurations for solar carport.

Figure 16.14 shows some possible configurations for one of the several charging stations in a solar carport. The architecture depicted on the left has the disadvantage of inserting a DC/DC converter into the main intended power flow, from PV to battery. Moreover, the power drawn from a single-phase connection is pulsed at twice the line frequency. This pulsating power takes the form of an undesirably high ripple current into the battery. The configuration shown in the middle of Figure 16.14 removes the ripple issue, but adds an additional conversion stage between the grid and the battery. The configuration on the right requires a converter that is capable of bidirectional flow between the EV and the grid, as well as steering of PV power to either the EV or the grid in a controlled fashion. Furthermore, this should ideally be achieved by a single conversion stage for all power flow paths and with wide voltage range capability. A good candidate for this task is the Z-loaded inverter/rectifier topology shown in Figure 16.15.

The operating characteristics of the Z-loaded converter have been described extensively in the literature [44–46]. The most salient feature of this conversion topology is its controllability through two distinct modulation modes within the same switching cycle, designated by duty cycle $D$ and "shoot-through" duty cycle, $D_o$. The gating patterns shown in Figure 16.15 describe the meaning of $D$ and $D_o$. As can be seen, during period $D_o$, all four switches are closed simultaneously, causing the inductors to charge and ultimately boost the voltage across the capacitor, the battery, and the grid terminals. Thus, $D_o$ can be understood as the duty cycle associated with operation akin to that of a current-sourced inverter. However, during period $D$, the bridge operates in a manner similar to that of a voltage-sourced inverter, which is essentially a buck. Therefore, with the appropriate utilization of $D$ and $D_o$, both buck and boost operations can be achieved, so that the battery voltage can be either higher or lower than the peak of the line voltage. This allows a wide line and battery voltage range. Most significantly, due to the double modulation, both the grid and the battery current can be controlled precisely in amplitude and shape (sinusoidal for the line current and rippleless DC for the battery). The MPPT function for the PV string can then be achieved by managing the simple addition of these two power flows.

The topology shown in Figure 16.15 must be modified in order to achieve isolation of the battery pack. Therefore, the DAB converter shown in Figure 16.7 (right) can be integrated resulting in the

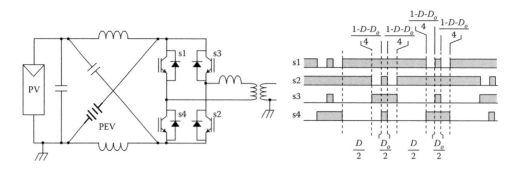

**FIGURE 16.15**  Z-loaded rectifier (left); gating pattern (right).

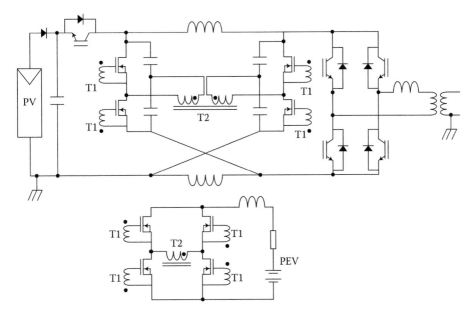

FIGURE 16.16    Z-converter application to single-phase, grid-tied PV charging station.

detailed schematic of Figure 16.16. The apparent complexity of the isolation stage is deceptive; in fact, this is a simple bidirectional converter that uses a small and inexpensive high-frequency transformer and that runs in open loop at full duty cycle and where all eight switches are driven by the same signal. In addition, since the duty cycle is always 100%, ZVS is assured, resulting in efficient operation executed by relatively small devices.

With the inclusion of the isolated DC/DC converter, the need for the 50/60 Hz isolation transformer may be called into question. In North America, the grounding of one side of the PV panel has traditionally been the required norm. Although the National Electric Code has allowed recent conditional exceptions to this safety regulation, utility companies have resisted this change, mainly because a direct connection to the AC/DC bridge converter can inject dangerous levels of DC current into the distribution transformer. However, should this constraint become less binding in North America, as it is currently in Europe, other circuits could be proposed that could prove more reliable and efficient. Many so-called transformerless topologies have been proposed [47, 48], and Figure 16.17 depicts a simplified schematic for one such possibility.

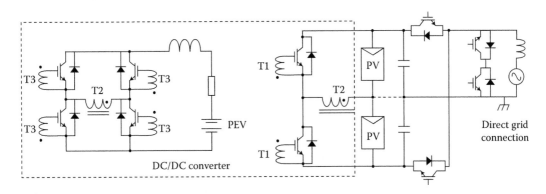

FIGURE 16.17    Transformerless topology.

In this case, the DC/DC conversion and the rectifier/inverter section are controlled separately, rendering the control strategy much simpler. However, the DC/DC converter is now governed by a feedback loop, meaning that it no longer takes advantage of the low switching loss normally associated with 100% duty cycle operation. With allowances from the regulatory safety agencies, the PV panels can be floating as long as the circuit has additional protection afforded by GFIs and that it produce no leakage currents to ground during normal operation. The last requirement is attained only if the topology guarantees very little common mode voltage on the PV panels during normal operation (note that this cannot be achieved with the Z-converter). Nevertheless, the midpoint can still be grounded, as indicated by the dashed line in the figure, but at the expense of performance.

Whichever architecture is chosen, it is clear that the energy transfer cannot be controlled to fully satisfy any arbitrary current demands of the PV, the grid, and the EV/PHEV battery simultaneously. In fact, many renewable resources are themselves subject to maximum power point tracking (MPPT) control so that the simple power balance in Equation 16.1 must be satisfied:

$$P_{MPPT} = P_{PEV} + P_G \tag{16.1}$$

Here, MPPT is the power draw requested by the distributed resource. It has to equal the sum of the power absorbed by the grid and the EV battery ($P_{PEV}$ and $P_G$, respectively). Since PMPPT is determined by external factors, such as clouding in the case of PV, either PPEV or PG can be controlled independently, but not both. Which of these is controlled will depend heavily on how the EV owner decides to utilize his vehicle storage resource. Thus, in installations where charging power comes primarily from intermittent sources, the need for a significant presence of additional storage on the premises will be diminished, but not eliminated.

## 16.6  EV BATTERY CHARGING SPECIFICATIONS AND SAFETY ISSUES

The major concern is the present limit of EV batteries' practical drive range of 100–120 km on a single charge. This is coupled with a pervasive shortage of charging stations that, even when available, cannot generate enough power for fast charging (<15 min). Even in the likely event that fast charging will be available soon, it is conceivable that the necessary high-power availability will come with an additional price to the user. It appears that EVs will continue to benefit from slower charging strategies. For instance, standard level 2 (SAE J1772) charging bounds the available power to less than 14.4 kW so that a fully depleted 35 kWh battery pack will require 2½ h to achieve full SOC. In realistic circumstances, this naturally implies that the charging station will also double as a parking facility, where the vehicle is expected to reside for relatively long time periods. As will be seen through the topics covered in this section, where PV generation is convenient, power levels compatible with both levels 1 and 2 are possible, making solar carports ideal for EV/PHEV charging applications. Different battery chemistries and their typical charging time from a residential outlet, for 40 miles energy usage, are summarized in Table 16.1.

Even though it is the auto industry's preliminary battery, PbA batteries are out of favor for EV applications, because of its low energy density. In comparison, the nickel-metal hydride (Ni-MH) battery is favored more, because of its higher energy density, shorter charging time, and long life cycle, but it presents an immature recycling system. The lithium-ion (Li-ion) battery chemistry is considered as a definite future trend, but compared to the other two candidates, it has lower durability, which is an issue that needs to be focussed upon while charging.

In terms of charge efficiency PbA has a high efficiency in the range of 95%–99%. However, Li-ion batteries can be charged and discharged faster than PbA and Ni-MH batteries, making them a good candidate for EV and PHEV applications. Furthermore, lithium-polymer (Li-Po) batteries have the same energy density as the Li-ion batteries but with lower cost. Formerly, the maximum discharge current of Li-Po batteries was limited to about 1 C rate; however, recent enhancements

**TABLE 16.1**

**Typical Charging Time and Energy Densities of Popular Battery Candidates for EVs**

| Battery | PbA | Ni-MH | Li-ion | Unit |
|---|---|---|---|---|
| Charging time | 8–10 | 6–14 | 5–7 | h |
| Energy density | 60 | 80 | 180 | Wh/kg |

have led to maximum discharge rates of almost 30 times the 1 C rate, which greatly improves and simplifies the storage part of the EVs and PHEVs in terms of power density, since this can even eliminate the need of ultracapacitors. Besides, there have been outstanding improvements in charging times. Recent advances in this technology have led to some types which can reach over 90% SOC in a couple of minutes, which can significantly increase the application toward EVs/PHEVs.

One of the major concerns for Li-ion and Li-polymer PHEV batteries is that the battery lifetime is only limited to a specific number of charge/discharge cycles. Too many charging/discharging routines disrupt the total life cycle of the battery pack. With reference to life cycle, the battery can suffer significant degradation in its capacity, depending on the charging level and time (level 2 or 3). Furthermore, the internal resistance also increases with each charging cycle. Also, according to the chemistry and the quality of the cells, a battery typically loses about 20% of its initial capacity after about 200–2000 full cycles, also known as the 100% SOC cycles. The life cycle can be greatly increased by reducing SOC usage, by avoiding complete discharges of the pack between recharging or full charging. Consequently, a significant increase is obtained in the total energy delivered, whereby the battery lasts longer. In addition, overcharging or overdischarging the pack also drastically reduces the battery lifetime.

### 16.6.1 EV CHARGING LEVELS, SPECIFICATIONS, AND SAFETY

EV/PHEV batteries are typically charged by a DC power supply that is normally derived from an AC source. Many earlier EVs utilized offboard chargers for this purpose, in order to avoid adding weight to the vehicle. With technological advances in the field of power electronics and in consideration of the added convenience to the user, most EVs today use an onboard rectifier. In fact, today's EV/PHEV manufacturer's loosely follow SAE J1771 recommendations that require an AC electrical connection and power capacity not to exceed 14.4 kW. Moreover, common wisdom may suggest that this strategy has the supplemental benefit of simplifying future public infrastructure. To this point, the California Air Resources Board (CARB), through their report "*ZEV Infrastructure: A Report on Infrastructure for Zero Emission Vehicles,*" invokes economic factors to suggest that EVs be minimally equipped with an onboard rectifier. Table 16.2 shows the standard charging levels and

**TABLE 16.2**

**Summary of Charging Levels for EVs/PHEVs**

**Standard EV charging levels**

| | |
|---|---|
| Level 1 | "Level 1" EV charging employs cord and plug connected portable EV supply equipment (EVSE) that can be transported with an EV. This equipment is used specifically for EV charging and shall be rated at 120 V AC and 15 A, and shall be compatible with the most commonly available grounded electrical outlet (NEMA 5–15R). |
| Level 2 | EV charging employs permanently wired EVSE that is operated at a fixed location. This equipment is used specifically for EV charging and is rated at less than or equal to 240 V AC, less than or equal to 60 A, and less than or equal to 14.4 kW. |
| Level 3 | EV charging employs permanently wired EVSE that is operated at a fixed location. This equipment is used specifically for EV charging and is rated at greater than 14.4 kW. |

their significance. While levels 1 and 2 are in line with the above considerations, SAE J1772 also includes provision for the so-called fast charging (level 3), which allows for the transfer of much higher power levels to EV/PHEV batteries.

Currently, fast charging is not commonly used, mainly because commercially available battery technologies do not allow for excessive charging currents. Furthermore, at this time, such quick charging schemes can only be regarded as a marketing gimmick, because of the required massive power demand.

### 16.6.2 EV/PHEV Battery Charging Voltage Levels

To charge a 35-kWh battery in 10 min, it requires 250 kW of power—five times as much as the average office building consumes at its peak. Notwithstanding these important impediments for the development of public level 3 infrastructure, the market drive for the possibility of quick refuelling stops at charging stations (as is now done with gas stations) makes such development very likely in the future. This is especially true in view of continuing efforts to improve battery chemistry and ultracapacitor specifications to allow higher currents. The advent of fast charging will most certainly presuppose a return to offboard rectifiers, so that direct access to the battery pack will become a necessity. It is, thus, reasonable to predict that future EVs/PHEVs will have both AC and DC plugs for battery charging. Such an assumption is critical for the purpose of renewable energy-based DC charging of EV batteries.

Furthermore, battery technology for EVs/PHEVs is still being developed, and has not yet reached maturity, let alone any kind of standardization. Each manufacturer is free to choose different chemistries and system designs; hence, it is currently impossible to establish a common specification. For charging solutions, the most problematic parameter is the voltage level, given that in some cases, a transformer and corresponding inverter must be added, to allow for voltage flexibility. Even with addition of a transformer, it is impossible to correctly quantify the performance of the power processor, if the battery voltage is allowed to possess a very wide range. Therefore, in wishful expectation that such voltage be standardized at some point in the future, a reasonable assumption must be made in the short term.

Nominal battery voltages for various EV/PHEV models suggest that a range between 275 and 400 V represents a reasonable assumption, although some neighborhood vehicles still use PbA batteries, and have much lower nominal voltages.

### 16.6.3 Charging Safety Issues

Vehicles are intrinsically nonstationary, and consequently, cannot be grounded through a permanent safety conductor. As the battery pack voltages are quite unsafe, they must be isolated from the chassis under operating conditions, as described or assumed by many standards such as UL 2202, UL 2231, ISO 6469, J1772, and J1766. However, this clause is not strictly necessary during charging, because the charger's conductive coupling can be used to force both the chassis and the floated battery to safety ground. This could possibly relieve the charger power system from providing galvanic isolation to the battery, but also entails that the grounding conductor be oversized, while imposing strict safety regulation on all external wiring, connectors, and interlocks. Moreover, SAE J1772 calls for ground fault detectors (or equivalent means) to ascertain the isolation integrity of the battery pack, making it impractical to assume that the battery pack could be anything else than fully floating at all times. The immediate consequence is that the charger must be designed to provide galvanic isolation from the AC line.

## 16.7 IMPACT OF EV CHARGING AND V2G POWER FLOW ON THE GRID

The demand of energy will only escalate, according to energy predictions. More than half of the auto industry market will consist of EVs and/or PHEVs. This implies an additional energy charge in the evenings, when the cars reach home. On the one hand, reducing $CO_2$ pollution and, on the other, increasing polluting power plants is not a good energy balance equation for the future. However, by introducing vehicle-to-grid (V2G) power flow and AC grid charging, the system energy balance can be altered. The bidirectional EV battery charger can not only charge the car battery, but also use the energy from the battery to supply household energy consumption.

The average home-to-work vehicle trip length is about 12 miles (19.4 km), which gives a gap to use the energy contained in the EV battery. The EV can supply the home loads, instead of charging its battery directly, when the driver reaches home. The charge of the battery will be shifted during the night. At this time, the EV battery works as a buffer or an ESS. In addition, to increase energy production by renewable energy sources, the home power supply (AC grid) will include local renewable energy sources, such as wind turbines and solar power. Using V2G technology is obviously beneficial from the standpoint of peak shaving, as is evident from the description above. However, some critical issues need to be taken into account, in order to achieve V2G power flow safely. Some of the key technical aspects of interconnectivity between an EV Li-ion battery pack, the AC grid, and renewable wind/solar power infrastructures are enlisted below.

### 16.7.1 LINE STABILITY ISSUES

As aforementioned, the AC grid could perform the function of "storing" excess energy produced by the renewable energy resource of the carport, thus eliminating the need for local storage (batteries and flywheels). In fact, the scenario is somewhat more complicated than that. It is well known that intermittent generators, such as solar and wind installations, can potentially cause problems to the grid. In fact, should these generators become very widespread, or be connected to remote locations on the grid, the energy they produce may exceed the available load. Simple energy conservation theory dictates that such a condition is untenable and must be remedied either by storing the energy for later use or by decreasing power generation, thus underutilizing the generator's capacity. Even when produced energy is not in excess, the flow of current back toward the power substation can cause the local point of common coupling to experience a voltage boost that can be severe when line impedance is significant.

Furthermore, and especially in the case of PV sources, the instantaneous level of generated power can experience rapid variations (up to 15% per second, due to clouding) that cannot be compensated in real time by the grid, thus causing voltage flicker. These serious issues concerning distributed generation are presently the object of intense study and mobilization by public and private parties. Thus, it is important to consider the following points: (a) present grid penetration levels of distributed systems, in general, and PVs, in particular, are very low, less than 2%–3% in North America. Several studies have established that flicker and voltage boost are significant only when the penetration levels at the power substation are more than 5% and 15%, respectively, in most cases; (b) in the case of a carport, the load is intrinsically a storage type EV/PHEV batteries. In other words, especially for large carports, there is a statistically high probability that the storage and the load are one and the same. Scenarios involving an empty carport and overproduction occurring simultaneously can be deemed so rare that decreasing production in such instances would be fairly acceptable. In fact, grid-connected EV/PHEV batteries could be beneficial to the utility company for the reasons mentioned in the previous section.

### 16.7.2 INVERTER DISTORTION AND DC CURRENT INJECTION

As the renewable energy resources generate DC voltage, a DC/AC inverter is needed in order to connect to the AC grid. Several regulatory agencies, such as UL and IEC, have imposed common specifications for inverter performance. Minimally, the inverter must produce low-harmonic distortion currents (<5%) and near-unity power factors (displacement PF compensation is not yet allowed in most cases). Furthermore, the inverter is not allowed to inject a DC current component into the grid, as this could cause distribution transformers to saturate. Commonly followed standards IEEE

929, IEC 61727, and EN 61000 specify between 0.5% and 1% of rated output current as a maximum, while some national European standards add a 5 mA absolute requirement in addition to this minimum percentage.

### 16.7.3 LOCAL DISTRIBUTION CONFIGURATION

The solutions considered in the previous sections assume the use of 1-phase connections. The main appeal of the 3-phase system is the enhanced power capacity, the fact that this power is delivered and absorbed without any line-frequency components, and the elimination of third-harmonic currents in the distribution and grounding wiring. These features are quite attractive and include the advantage of potentially eliminating undesirable pulsating charging current to the EV battery. However, the single-phase configuration allows for less expensive and simpler distributed inverters. Especially, considering the fact that PV can only provide 5 kW of power to the inverter, and that intervening DC/DC converters need to be added to eliminate any pulsating current, it appears that a 1-phase system is much better adapted to EV charging application using PV resources.

## 16.8 RENEWABLE ENERGY AND EV/PHEV BLEND IN A SMART GRID

When defining the technical goals for a distributed power converter system to be used in a PV/wind-powered, grid-tied carport, suitable compromises must be made, in order to contain costs while providing acceptable performance. Essentially, the main design objective is dictated by the fact that the carport will be a public or semi-public structure. Hence, it is crucial that the system is robust, reliable, and offers high availability. It has already been ascertained that both the PV/wind resource and the power conversion system must be distributed, providing flexibility and redundancy, while choosing topologies that are characterized by low component count and stress levels, in order to ensure a high reliability.

Another important consideration is that renewable resources are not easily harnessed. This simple fact makes energy conservation through high conversion efficiency a priority. As mentioned above, the number of conversion stages should be minimized, at least for the most utilized power flow paths. In the case of renewable charging, the power will flow most frequently from the PV/wind to the EV battery and from the AC grid to the EV battery. Therefore, given a choice, the flow paths should be optimized in line with the following priorities: in decreasing order of importance, renewable source to battery, grid to battery, renewable resource to grid, and battery to grid [49]. A schematic, showing possible power flow paths is depicted in Figure 16.18 specifically for a PV/AC grid interface. A proposed structure with PV, AC grid-connected home, and an EV battery system is shown in Figure 16.19.

The configuration of Figure 16.19 has unique advantages. First, the PV power can be directly injected to the battery pack and not to the grid, then from the grid to the load, which reduces the overall losses. Second, this configuration provides great flexibility for different power flows. Depending on the amount of available power from PV panels and required power by the battery pack, different modes of operation may occur. If available power from PV is more than the required

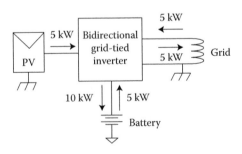

FIGURE 16.18 Possible power flow paths with specified power level.

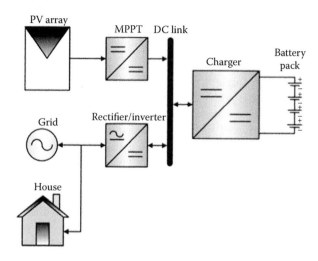

FIGURE 16.19 EV, AC grid-connected home, and PV interconnection.

power for charging the battery pack, the remaining can be injected to the grid (mode 1). If PV power is not enough, grid will be involved to supply the remaining power (mode 2). If there is no power available from the PV panels, battery pack can be charged solely by the grid (mode 3). If the battery pack is not connected to the system, PV power can be injected to the grid to reduce the electricity bill of the house (mode 4). If needed, the battery pack can be discharged to the grid during specified times performing a configuration supporting V2G mode (mode 5). Even in the case of blackouts, the EV battery pack can perform as an energy source, and supply the power to the house for some time, depending on the battery pack capacity (mode 6).

### 16.8.1 VEHICLE-TO-GRID: TEST CASE

The EV/PHEV battery does not completely solve the problem during peak production. However, it can be used as an external storage system, which can help the AC grid to supply the home and decrease peak production. As most renewable energy sources are intermittent, ESSs as well as advanced controls will have to be well developed. EV/PHEV batteries can have an important role to play in such an infrastructure as a storage system as well as an energy producer. A typical house powered from an AC grid, renewable wind energy, and PV, connected to a plug-in PHEV is as shown in Figure 16.20. This is the concept of net-zero energy consumption power system.

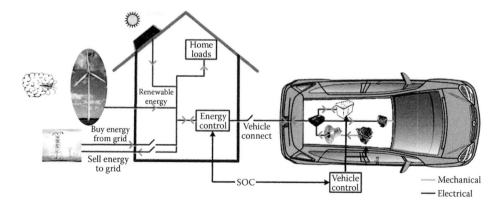

FIGURE 16.20 PHEV connected to a net-zero energy consumption home.

Considering most vehicles are at home between 8.00 pm and 7.00 am and the fact that most people use their vehicle between 8.00 am and 9.00 am and from 4.00 pm to 5.00 pm, 11 h are available to use the vehicle as a storage/production system (depending on battery state-of-charge, SOC). During this time, the EV/PHEV can be charged or discharged, depending on user needs. Thus, the charge can be shifted during the night, when the demand on the grid is low. In addition, in most cases, one-way home-to-work does not exceed 25 km. For example, the Chevy Volt® has an electric range of 38 miles (61 km). This means that, the car battery will have at least 11 km range available at the end of the day [50]. The overall system can be divided into two parts: one when the vehicle is connected to home and the other when the vehicle is disconnected. In the first case, renewable energy sources supply the household load directly. The energy difference supplies either the grid or the vehicle battery. However, when renewable energy sources are not enough, the grid, or EV battery, or both, help to supply the household.

Typically, to solve the cost-based power flow problem, a dynamic programming-based optimization function needs to be devised. The objective of the optimization is to minimize the cost of energy, including the AC grid electricity cost, which has usually two prices: one for the offpeak hours and one for peak hours. Fuel prices as well as cost of renewable energy sources also need to be taken into account. The proposed optimization problem has three constraints: battery capacity, ICE size, and grid capacity. In a typical PHEV, the battery SOC range is set between 30% (minimum SOC) and 90% (maximum SOC). The AC grid size is characterized by the power of the home outlet, which is considered on an average 3.0 kVA [50]. To help formulate the optimization algorithm, production and consumption profiles have to be predicted as well as the vehicle drive cycle. Computing at each time the 8-h energy, which is available, usually assists the predictions for the optimization problem.

A typical PHEV and AC grid optimization problem control uses approximately 30% of renewable energy, 40% of AC grid energy, and 30% of ICE energy, to supply the vehicle trip as well as household energy loads. A practical test case developed at Concordia University, in Montreal, Canada, based on the system of Figure 16.20, is considered as an example here. Practical test results are shown in Figure 16.21.

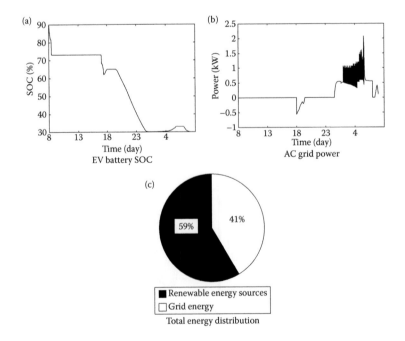

FIGURE 16.21 Test case for a typical summer day (initial SOC = 90%).

The test starts with a fully charged battery (90% SOC). The results show that the PHEV battery helps supply home loads and sells energy to the grid, keeping the optimization-based cost functions in order.

## QUESTIONS

16.1 Enlist the various ancillary services that could potentially be utilized from electric and PHEVs. Explain each of these services in brief, giving suitable examples.

16.2 Explain the power conversion stages in a typical grid-tied EV/PHEV charging system. Highlight the key role played by power electronic converters in the various conversion options.

16.3 Compare and contrast EV/PHEV centralized and decentralized public charging infrastructures.

16.4 Highlight the operational characteristics of various nonisolated DC/DC bidirectional converter options, to connect EVs/PHEVs to the high-voltage DC bus. Explain the advantage of each individual DC/DC converter topology from an overall performance standpoint.

16.5 What is the advantage of using an interleaved DC/DC converter in the context of designing an efficient, high-performance EV/PHEV battery charging infrastructure? Support your answers with a conceivable high-capacity EV battery charging scenario.

16.6 Explain the futuristic role of PV panels for EV/PHEV charging. Highlight the various EV/PV/grid interface options/scenarios. Highlight the comparative features between each interface scenario.

16.7 Explain the concept of renewable energy interface with an EV/PHEV and the smart grid. Using the simulation example highlighted in the last section of this chapter, consider that your home is powered by a suitable sized PV panel. Model and simulate the best EV charge/discharge scenario (including vehicle-to-home, V2H power flow), keeping in mind the EV battery state-of-charge (SOC), AC grid power, and the total energy distribution between the AC grid and PV source. In your test-case, ensure that maximum penetration of PV is obtained, without sacrificing EV battery charge/discharge efficiency and maintaining PV as well as AC grid efficiency.

## REFERENCES

1. Ehsani M, Gao Y, Emadi A. 2010. *Modern Electric, Hybrid Electric and Fuel Cell Vehicles, Fundamentals, Theory and Design*, 2nd edn., CRC Press, New York.
2. Husain I. 2005. *Electric and Hybrid Vehicles, Design Fundamentals*, CRC Press, New York.
3. IEEE spectrum, Software Fix Extends Failing Batteries in 2006–2008 Honda Civic Hybrids: Is Cost Acceptable? http://spectrum.ieee.org/riskfactor/green-tech/advanced-cars/software-fix-extends-failing-batteries-in-2006-2008-honda-civic-hybrids-is-cost-acceptable. Accessed 23 August 2011.
4. Crompton TR. 2000. *Battery Reference Book*, 3rd edn., Newnes, Oxford, UK.
5. Buchmann I. 2001. *Batteries in a Portable World: A Handbook on Rechargeable Batteries for Non-Engineers*, 2nd edn., Cadex Electronics Inc, Richmond, BC, Canada.
6. Dhameja S. 2002. *Electric Vehicle Battery Systems*, Newnes, Boston, MA.
7. Chen LR. 2008. Design of duty-varied voltage pulse charger for improving Li-ion battery-charging response. *IEEE Transactions on Industrial Electronics*, 56(2), 480–487.
8. Park SY, Miwa H, Clark BT, Ditzler D, Malone G, D'souza NS, Lai JS. 2008. A universal battery charging algorithm for Ni–Cd, Ni–MH, SLA, and Li-ion for wide range voltage in portable applications, in *Proceedings of IEEE Power Electronics Specialists Conference*, Rhodes, Greece, pp. 4689–4694.
9. Hua CC, Lin MY. 2000. A study of charging control of lead acid battery for electric vehicles, in *Proceedings of IEEE International Symposium on Industrial Electronics*, Cholula, Puebla, Mexico, vol. 1, pp. 135–140.
10. West S, Krein PT. 2000. Equalization of valve-regulated lead acid batteries: Issues and life tests, in *Proceedings of IEEE International Telecommunications Energy Conference*, Phoenix, AZ, pp. 439–446.

11. Brost RD. 1998. Performance of valve-regulated lead acid batteries in EV1 extended series strings, in *Proceedings of IEEE Battery Conference on Applications and Advances*, Long Beach, CA, pp. 25–29.

12. Bentley WF. 1997. Cell balancing considerations for lithium-ion battery systems, in *Proceedings of IEEE Battery Conference on Applications and Advances*, Long Beach, CA, pp. 223–226.

13. Krein PT, Balog RS. 2002. Life extension through charge equalization of lead-acid batteries, in *Proceedings of IEEE International Telecommunications Energy Conference*, Montreal, QC, Canada, pp. 516–523.

14. Lohner A, Karden E, DeDoncker RW. 1997. Charge equalizing and lifetime increasing with a new charging method for VRLA batteries, in *Proceedings of IEEE International Telecommunications Energy Conference*, Melbourne, Australia, pp. 407–411.

15. Moore SW, Schneider PJ. 2001. A review of cell equalization methods for lithium ion and lithium polymer battery systems, in *Proceedings of SAE 2001 World Congress*, Detroit, MI.

16. Nishijima K, Sakamoto H, Harada K. 2000. A PWM controlled simple and high performance battery balancing system, in *Proceedings of IEEE 31st Annual Power Electronics Specialists Conference*, vol. 1, Galway, Ireland, pp. 517–520.

17. Isaacson MJ, Hoolandsworth RP, Giampaoli PJ. 2000. Advanced lithium ion battery charger, in *Proceedings of IEEE Battery Conference on Applications and Advances*, Long Beach, CA, pp. 193–198.

18. Pascual C, Krein PT. 1997. Switched capacitor system for automatic series battery equalization, in *Proceedings of 12th Annual Applied Power Electronics Conference and Exposition*, Atlanta, GA, vol. 2, pp. 848–854.

19. Hung ST, Hopkins DC, Mosling CR. 1993. Extension of battery life via charge equalization control, *IEEE Transactions on Industrial Electronics*, 40(1), 96–104.

20. Cao J, Schofield N, Emadi A. 2008. Battery balancing methods: a comprehensive review, in *Proceedings of IEEE Vehicle Power and Propulsion Conference*, Harbin, China, pp. 1–6.

21. Lee YS, Wang WY, Kuo TY. 2008. Soft computing for battery state-of-charge (BSOC) estimation in battery string systems, *IEEE Transactions on Industrial Electronics*, 55(1), 229–239.

22. Shen WX, Chan CC, Lo EWC, Chau KT. 2002. Adaptive neuro-fuzzy modeling of battery residual capacity for electric vehicles, *IEEE Transactions on Industrial Electronics*, 49(3), 677–684.

23. Piller S, Perrin M, Jossen A. 2001. Methods for state-of-charge determination and their applications, *Journal of Power Sources*, 96(1), 113–120.

24. Ullah Z, Burford B, Dillip S. 1996. Fast intelligent battery charging: neural-fuzzy approach, *IEEE Aerospace and Electronics Systems Magazine*, 11(6), 26–34.

25. Atlung S, Zachau-Christiansen B. 1994. Failure mode of the negative plate in recombinant lead/acid batteries, *Journal of Power Sources*, 52(2), 201–209.

26. Feder DO, Jones WEM. 1996. Gas evolution, dryout, and lifetime of VRLA cells an attempt to clarify fifteen years of confusion and misunderstanding, in *Proceedings of IEEE International Telecommunications Energy Conference*, Boston, MA, pp. 184–192.

27. Jones WEM, Feder DO. 1996. Behavior of VRLA cells on long term float. II. The effects of temperature, voltage and catalysis on gas evolution and consequent water loss, in *Proceedings of IEEE International Telecommunications Energy Conference*, Boston, MA, pp. 358–366.

28. Nelson RF, Sexton ED, Olson JB, Keyser M, Pesaran A. 2000. Search for an optimized cyclic charging algorithm for valve-regulated lead–acid batteries, *Journal of Power Sources*, 88(1), 44–52.

29. Kempton W, Tomic J, Brooks A, Lipman T, Davis. 2001. Vehicle-to-grid power: Battery, hybrid, and fuel cell vehicles as resources for distributed electric power in California, UCD-ITS-RR-01-03.

30. Cocconi AG. 1994. Combined motor drive and battery recharge system, US Patent no. 5,341,075.

31. Kempton W, Kubo T. 2000. Electric-drive vehicles for peak power in Japan, *Energy Policy*, 28(1), 9–18.

32. Kisacikoglu1 MC, Ozpineci B, Tolbert LM. 2010. Examination of a PHEV bidirectional charger system for V2G reactive power compensation, *IEEE Applied Power Electronics Conference*, Palm Springs, California.

33. Tuttle DP, Baldick R, The evolution of plug-in electric vehicle–grid interactions, *IEEE Trans. on Smart Grid*, 3(1), 500–505.

34. Jenkins SD, Rossmaier JR, Ferdowsi M. 2008. Utilization and effect of plug-in hybrid electric vehicles, in *The United States Power Grid, Vehicle Power and Propulsion Conference,* Arlington, TX.

35. Kempton W, Tomic J. 2005. Vehicle-to-grid power implementation: From stabilizing the grid to supporting large-scale renewable energy, *Journal of Power Sources*, 144(1), 280–294.

36. Wang J, Peng FZ, Anderson J, Joseph A, Buffenbarger R. 2004. Low cost fuel cell converter system for residential power generation, *IEEE Transactions on Power Electronics*, 19(5), 1315–1322.

37. Han S, Divan D. 2008. Bi-directional DC/DC converters for plug-in hybrid electric vehicle (PHEV) applications, in *Applied Power Electronics Conference and Exposition,* Austin, TX, pp. 784–789.

38. Peng FZ, Li H, Su G-J, Lawler JS. 2004. A new ZVS bidirectional DC–DC converter for fuel cell and battery application, *IEEE Transactions on Power Electronics*, 19(1), 54–65.
39. Xiao H, Guo L, Xie L. 2007. A new ZVS bidirectional DC-DC converter with phase-shift plus PWM control scheme, in *Applied Power Electronics Conference,* Anaheim, CA, pp. 943–948.
40. Du Y, Zhou X, Bai S, Lukic S, Huang A. 2010. Review of non-isolated bi-directional DC-DC converters for plug-in hybrid electric vehicle charge station application at municipal parking decks, in *Applied Power Electronics Conference and Exposition,* Palm Springs, CA, pp. 1145–1151.
41. Aggeler D, Canales F, Zelaya H, Parra DL, Coccia A, Butcher N, Apeldoorn O. 2010. Ultra-fast DC-charge infrastructures for EV-mobility and future smart grids, *Innovative Smart Grid Technologies Conference Europe,* Gothenburg, Sweden.
42. Bai S, Du Y, Lukic S. 2010. Optimum design of an EV/PHEV charging station with DC bus and storage system, *Energy Conversion Congress and Exposition,* Atlanta, GA, pp. 1178–1184.
43. Buso S, Malesani L, Mattavelli P, Veronese R. 1998. Design and fully digital control of parallel active filters for thyristor rectifiers to comply with IEC-1000-3-2 standards, *IEEE Transactions on Industry Applications*, 34(3), 508–517.
44. Peng Z. 2003. Z-source inverter, *IEEE Transactions on Industry Applications*, 39(2), 504–510.
45. Peng FZ, Shen M, Holland K. 2007. Application of Z-source inverter for traction drive of fuel cell—battery hybrid electric vehicles, *IEEE Transactions on Power Electronics*, 22(3), 1054–1061.
46. Carli G, Williamson S. 2009. On the elimination of pulsed output current in Z-loaded chargers/rectifiers, *Proceedings of IEEE Applied Power Electronics Conference and Exposition*, Washington, DC.
47. González R, López J, Sanchis P, Marroyo L. 2007. Transformer-less inverter for single-phase photovoltaic systems, *IEEE Transactions on Power Electronics*, 22(2), 693–697.
48. Kerekes T, Teodorescu R, Borup U. 2007. Transformer-less photovoltaic inverters connected to the grid, *Proceedings of IEEE Applied Power Electronics Conference and Exposition,* Anaheim, CA, pp. 1733–1737.
49. Carli, G, Williamson, S. 2013. Technical considerations on power conversion for electric and plug-in hybrid electric vehicle battery charging in photovoltaic installations, *IEEE Transactions on Power Electronics*, 28(12), 5784–5792.
50. Berthold, F, Blunier, B, Bouquain, D, Williamson, S, Miraoui, A. 2012. Offline and online optimization of plug-in hybrid electric vehicle energy usage (home-to-vehicle and vehicle-to-home), *Proceedings of IEEE Transportation Electrification Conference and Exposition*, Dearborn, MI.

# 17 Energy Management and Optimization

*Ilse Cervantes*

## CONTENTS

## 17.1  INTRODUCTION

In contrast to conventional vehicles, hybrid and electric vehicles (HEV and EV, respectively) are more efficient devices in terms of energy utilization mainly due to three characteristics.

First, the average tank-to-wheels efficiency (i.e., chemical-to-mechanical energy conversion efficiency) is higher. That is, even in the case of HEV, where an internal combustion engine (ICE) is used, the utilization of electric power from an energy-storage system (ESS) leads to an increment of the total average efficiency due to the high ESS efficiency.

The second characteristic is the regenerative braking. Charging the ESS during the braking process opens the possibility of recycling mechanical energy that in conventional vehicles is dropped, further increasing the efficiency of the vehicle.

Finally, the third characteristic is the controlled power split among the energy sources. By manipulating such power split, a decision can be made on how much fuel is consumed. The more frequent use of the ESS, the more efficient vehicle operation and the less energy dropped.

One of the main subjects of this chapter is to provide methods and theoretical foundation about how to perform this power split; in other words, about how to design an energy management strategy (EMS). Most of the existing EMS is essentially focused on reducing the fuel consumption. The relevance of this subject is particularly important once the design stage of the vehicle is accomplished, as any fuel saving or operation cost reduction related with fuel provision is mainly provided by the power-split decision. That is, given the mechanic characteristics of a vehicle (i.e., geometry design, air resistance, tire friction, etc.), we wonder whether what a *good* decision of power split is, for a given driving cycle.* I have used the word good instead of "the best" to emphasize that many of the EMS in the literature are not always, what mathematicians call, optimal. We will revisit this subject later in Section 17.3, where we will establish the necessary and sufficient conditions for optimality.

EMS can be categorized as (i) optimization based and (ii) heuristic- or rule-based. An advantage of the optimization-based strategies is that their solution can be called "minimal" (i.e., the fuel consumption is minimal). Moreover, they have the theoretical support of formal results that allow their application in a variety of vehicles; however, they have the disadvantage that the solution may be a hard find and may involve too much computation resources for real-time applications. Depending on the problem formulation, the optimization can be performed only by taking into account the instantaneous power demand (i.e., no knowledge of the future demand is required) or can be optimal for the entire vehicle trajectory.† Solving the optimization problem for the entire vehicle trajectory requires the knowledge of the driving cycle but instead, terminal conditions on the ESS state of charge (SOC) can be taken into account, allowing more convenient ESS operation.

Heuristic approaches exploit the experience or *a priori* knowledge of the designer. In general, they are susceptible of being applied in real time because their complexity is usually linked to the design stage of the strategy, rather than to their operation. This classification belongs to those based on fuzzy logic [1–3], neural networks (NNs) [4,5], and frequency (i.e., based on the power demand frequency) [6–8]; most of them require somewhere a piecewise continuous description of the power split that depends on the designer experience or criterium (switching mode conditions). Many of the research performed on this topic is focused on providing evidence of functioning as well as on providing theoretical support to the design criteria.

Once the decision is made about what energy source will be used and in what proportion, the next step is to ensure that the power split is performed as expected. To this end, electronic control units (ECUs) are used. In the automotive industry, ECU is a generic term to name controlling and monitoring devices. In general, an ECU is expected to receive electrical signals from various sensors that measure the state of the engine/motor/battery, and so on. From these signals, the controller generates electrical signals to the actuators that determine the fuel delivery and the electrical power

---

* A driving cycle is a typical velocity profile of the vehicle for a given road condition, for example, urban, semiurban/extra urban, and highway.
† A trajectory is the time evolution of system states for a given initial condition.

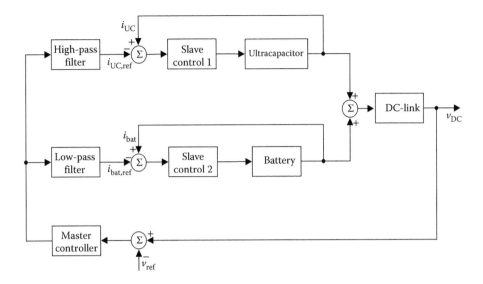

**FIGURE 17.1**   Cascade control structure for an EV. The EMS acts as a current-reference generator to perform power split, while the slave controllers track the desired current provided by the voltage control (master). The EMS is constituted of high- and low-pass filters.

delivered by the ESS. The EMS acts as a power-reference generator, while the ECU is the tracking control. It is worth noting that the tracking control in the ECU may also have a cascade structure (e.g., the voltage and current loop in a converter controller, see Figure 17.1).

The design of the ECU involves both the mathematical design of the controller and the electronic and actuator setup. Since this subject is too vast for a single chapter, we will focus only on the general aspects of the control design, referring the interested reader to References 9–11 for issues regarding the control implementation and its electronic design.

### 17.1.1   ENERGY MANAGEMENT PROBLEMS: EXISTING SOLUTIONS

As stated above, there exist two different trends in the EMS design, one is based on optimization and the other is based on heuristic rules. In the following, we will perform a brief review of some of the most relevant approaches. The intention is not to be vast or complete, but only to visit the different ideas proposed to reduce fuel consumption and to discuss how they work.

As the first step to introduce this description, observe that EMS for HEVs and EVs are focused on reducing the fuel consumption; therefore, such an objective can be accomplished in EVs only if a fuel cell (FC) is used to supply part of the power demand; these vehicles are also called fuel cell HEVs (FCHEVs) or fuel cell vehicles. For HEVs or FCHEV, there exist similar ideas that can be used for the design of EMS. For example, in HEVs or FCHEVs, the engine/FC can be forced to operate in a region where they are more efficient to improve the overall efficiency of the vehicle. Such regions can be defined *a priori* and can be taken into account using either (i) heuristic rules using piecewise continuous description of the power split or (ii) optimization in a restricted set. Other considerations in the formulation of the energy management problem are: battery endurance, FC longevity, FC humidity, and healthiness of the ESS, among the most important.

Among the optimization-based strategies, it is worthy discussing the following:

*Strategies that consider ESS operation restrictions.* For the case of FCHEV and HEV, the fuel consumption minimization problem has a trivial solution, that is, complete electrical operation. However, if the idea is that the ICE or FC provides the main power of the vehicle, and the ESS provides transient power demands or to be active in certain torque or velocity conditions, then the fuel consumption minimization problem can be formulated in a congruent manner using restrictions on FC/engine power and ESS SOC, as well as using what is called an equivalent fuel consumption of the electrical motor (see Reference 12). The equivalent consumption strategy allows to take into account both the fuel consumption and the energy consumption of the batteries in the minimization objective function (OF), and it is used to reduce ESS energy consumption.* For these kinds of strategies, usually, the inclusion of some FC power/current restrictions has to be considered for a suitable operation of the cell. For example, high-current demands placed on the FC lead to flooding while low-current demands lead to FC membrane dryness, and such behavior can be prevented using a minimum and maximum power demand restrictions. The set where all the restrictions are satisfied is called *feasible set* and the solution must belong to it. Examples of this kind of EMS can be found in References 13–15.

*Strategies that consider FC/engine efficiency.* In this case, the restrictions of the optimization problem are fixed according to a maximum FC/engine efficiency [16]. For the case of FC, the efficiency is dependent on the current load, being high at low-current load. In these conditions, current losses are related to the required energy to initialize the oxidation–reduction reaction (open-circuit zone). The strategy may also take into account restrictions on the rate of change of the FC-delivered power that ensures that the FC does not flood nor dry. Such restrictions are also important because FC dynamics limit the rate of delivered power, and fast power demands cannot be satisfied by the FC. Although convenient in terms of fuel economy (FE), operation of the cell at the maximum-efficiency cell zone, over-dimensions the size of the cell and its power generation is not completely exploited [17].

*Strategies that consider battery SOH or longevity.* Usually, the battery SOC is the only dynamic state taken into account when designing an optimization-based strategy, since voltage can be inferred from a static (algebraic) relationship of the battery SOC. However, the SOH can also be considered. To explain what an SOH is, assume that the battery operates under constant conditions. In this situation, a battery can withstand a certain amount of energy that can be used to compute an equivalent number of charge/discharge cycles. The SOH constitutes a quantification of how close from the end of life the battery is, taking into account the idealized scenario described above. The optimization problem can now be subjected to additional constraints of the battery, involving the restriction of a minimum SOH allowed [18]. The end of life is usually considered to be reached when the cell or battery delivers only 80% of its rated capacity. Nickel–metal hydride (NiMH) batteries typically have a life cycle of 500 cycles, while Ni–Cd batteries can have a life of more than 1000 cycles.

*Strategies based on analytical solutions.* In this approach, the objective is to formulate an optimization-based strategy capable to be solved offline. Such a scenario is satisfied only for special OFs and restrictions (typically linear). In this case, the real-time implementation may be guaranteed for both the restricted and unrestricted case. As the solution is analytic, analysis of the effect of ESS efficiency or parameter sensitivity can be performed [19].

*Online and offline strategies.* Depending on the OF, the optimization problem can be formulated as static (i.e., the OF is instantaneous considering only the current state of the fuel consumption) or dynamic (i.e., the OF depends on the integral of the instantaneous fuel consumption or involves derivatives). In this chapter, we will study both approaches.

---

* This topic is revisited in Section 17.2.4.

Among the heuristic-based EMS, the following can be mentioned:

*Piecewise continuous description of the power split.* This is probably the simplest heuristic strategy. In this case, the power split is directly assigned with the use of a piecewise continuous function, which is basically inspired in the designer criterion. Such function may depend on the ESS SOC or/and SOH, FC power, and so on. The definition of this function is directly related to the performance of the strategy and therefore, it is the subject for many researches.

*Frequency-based power split.* This is a particular case of the kind of strategies above; however, in this case, the power split is defined by the frequency characteristics of the power demand. As stated before, FC cannot fulfill rapid power demands; this is also the case of small engines; therefore, the ESS is used. The frequency partition is the result of the analysis of the frequency response of the power sources. Typically, super capacitors (SCs) are used for high-frequency power changes, batteries are used for medium-power demands, and FC/engines are used for low-frequency demands. The SOC of the ESS must be taken into account in this partition, since for example, depleted ESS cannot satisfy the power demand and the frequency-based partition is pointless.

*Fuzzy.* The first step in the design of this strategy is to perform what is called the fuzzification. That is, based on the knowledge or the criterion of the designer, the fuzzy concepts or qualifications are translated to sets. Once these sets are defined, the next step is to assign membership functions to every concept; these membership functions may overlap, which may assign a true value to a point belonging to two or more membership functions. By applying fuzzy rules (fuzzy logic), it is inferred what action or strategy applies (which can be more than one). The objective is to define a set decisions based on, for example, the SOC of the ESS (e.g., (i) High SOC: The FC/engine works around its optimal running zone; (ii) low SOC: The FC/engine operates higher than its optimal running point). The defuzzification procedure will assign a deterministic value of the power share, for implementation purposes. The definition of membership functions can be modified to obtain different results of the EMS. For example, in Reference 3, such functions are manipulated to approach an offline strategy based on optimization. The authors show that this strategy displays a close behavior to the offline-optimized strategy even if no predictive action of the driving cycle is available.

*NN.* The neural network approach has the advantage that it can deal with uncertain and unstructured systems. It departs from a black-box model in which a number of inputs are connected to a layer of cells. A cell is basically constituted of a low-order dynamical system, which in turn, can be interconnected to more cells. By manipulating the parameters of the cell, the number of layers, and the interconnections, the output of the NN can be manipulated (training). This manipulation can be performed either by trial and error (based on the designer experience), or using a feedback of the deviation error from a desired response. In the latter case, for some NNs, proof of convergence can be given, and such characteristics make them very valuable for control scientists.[*] For EMS design, the NN may be used (a) to learn from an optimal strategy to emulate it or (b) to learn about the driving cycle to use it in an optimal strategy, among other uses.

## 17.1.2 Chapter Organization

In the rest of this chapter, we will focus on formalizing the EMS design problem and some of the existing solutions as well as on discussing some control design criteria for ECUs. The subjects in this chapter are mainly oriented to undergraduate students, but there are also some topics for

---

[*] In this case, the strategy is no longer heuristic.

postgraduate-level students. Basic concepts of optimization, functional analysis, and dynamic systems are required. In some cases, references are given to the inexpert reader.

Section 17.2 is devoted to formalizing the problem formulation of the EMS design. Particularities of the optimization and heuristic-based EMS are also discussed. In Section 17.3, we discuss two of the most important strategies for solving optimal problems, that is, dynamic programming (DP) and variational calculus (VC). The main differences among them are discussed, and, in some cases, necessary and sufficient conditions for optimality are given. In Section 17.4, the illustration of optimal EMS is performed for the case of an FCHEV.

Finally, Section 17.5 is devoted to discuss some generalities of the control design and to give insight into the design of some of the most important control strategies nowadays; that is, robust and optimal control. The interested reader in control design strategies may skip Sections 17.2 through 17.4, and begin with Section 17.5 without detriment.

The design of electronic control circuits is not covered in this chapter. Numerical methods for optimization as well as the illustration of intelligent EMS are not given in this chapter.

## 17.2 ENERGY MANAGEMENT PROBLEM FORMULATION

In Section 17.1.1, some generalities of the EMS design problem formulation have been introduced. In this section, we will gain precision in this formulation since, as it will be clear later, the well poseness of the problem is crucial to establish the existence of the solution and its method of computation. To this end, the mathematical models of every component are required. As the particularities of ESS and ICEs have been introduced in Chapters 3, 7, and 8, the component description in the next section is mainly focused on hydrogen proton exchange membrane FC (PEMFC), which constitutes one of the most appropriate FC technologies for vehicles nowadays.

### 17.2.1 PEMFC DESCRIPTION

An FC is an electrochemical device that consumes hydrogen and air (oxygen) and produces heat, water, and electrical energy. An FC has a similar electrochemical principle that batteries, but unlike them, the fuel and the oxidant (hydrogen and air, respectively) are externally provided. This characteristic is very convenient for FCHEV, since the fuel tank can be quickly filled, similar to ICE vehicles. The conversion of chemical energy into electric energy occurs without combustion, and therefore, FCs are silent, do not operate at high temperatures, and are more important; their efficiency is not limited by the Carnot cycle; such characteristics make them more efficient than any ICE. The oxidation–reduction reaction that generates the electricity is performed without explosion by the use of a membrane electrode assembly (MEA), which besides of preventing gas mixing is also an ion conductor. During the reaction, the hydrogen is decomposed into protons (positively charged particles) and electrons (negatively charged). Protons pass through the polymer to the cathode, whereas the electrons travel along an external electric circuit to create the electrical current. Once this ion conduction is performed, water molecules are formed. It is known that humidity modifies the membrane conductivity, the drier it is, the less conductivity capacity it has; however, flooding is also undesirable that may block the reaction. The higher the extracted current, the more the water produced.

FCs can be considered as variable-voltage, current-dependent DC sources. A polarization curve is a function relating steady-state voltage and current of an FC (i.e., $V_{FC} = g(I_{FC})$).[*] In general, a polarization curve is composed of three regions (see Figure 17.2). The first region (I) is at low currents and high voltages, where current losses are related with the required energy to initialize the reaction (open-circuit zone). The second region (II) is where current losses are mainly attributed to

---

[*] Sometimes, when chemical crossover occurs, such a relation may not be a function.

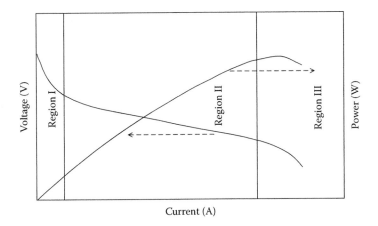

FIGURE 17.2 Typical polarization curve of a PEMFC.

the Ohmic resistance of the electrolyte resistivity and the external resistance of electrodes and connections (linear region), and the third region (III) is located at high currents and low voltages, where
the power extraction is limited by the mass transfer rate.

The FC power can be computed using the voltage and current of the FC, as shown in Figure 17.2.
Such curves are dependent on the $H_2$ pressure that modifies the voltage of the FC and therefore the
total extracted power. A normal operation of the FC avoids Regions I and III; therefore, the current
(power) demands to the FC must be limited.

The hydrogen consumption ($\dot{m}_{H_2}$) of the FC is proportional to the current delivered by the FC
($I_{FC}$); that is,

$$\dot{m}_{H_2}(t) = -\frac{\bar{N} M_{H_2} I_{FC}(t)}{2\bar{F}} \tag{17.1}$$

where $\bar{N}$ is the number of cells in the stack, $M_{H_2}$ is the molar mass of $H_2$ and $\bar{F}$ is the Faraday constant, and $I_{FC}(t)$ is the current provided by the FC. This relationship will be used later when solving
EMS for FCHEVs.

### 17.2.2 COMPUTATION OF ELECTRIC POWER DEMAND: FCHEV

As stated in the earlier chapters, mechanical power demanded by the vehicle depends on its dynamics; that is, it depends not only on parameters such as rolling resistance, gravitational acceleration,
frontal area, road inclination, and so on, but it also depends on the efficiency of the movement
transmission given by the mechanical design of the vehicle. For the case of FCHEV, this mechanical
power demand can be transformed into electric power demand for a given power train. Such demand
is larger than the mechanical power due to unavoidable energy losses present in the electrical motor
and its driver.

Let $P_{mech}(t)$ be the net mechanical power demanded; hence, the corresponding electrical power
demand ($P_{elec}$) is given by

$$P_{mech}(t) = \eta_{md} P_{elec}(t) \tag{17.2}$$

where $\eta_{md}$ is the overall efficiency of the motor and its drive. Let us denote the current and power
demanded by the vehicle $I_{load}$ and $P_{load}$ respectively, while $I_{bat}$, $P_{bat}$ and $I_{FC}$, $P_{FC}$ are the current and
power from the battery and from the FC, respectively, as illustrated in Figure 17.3.

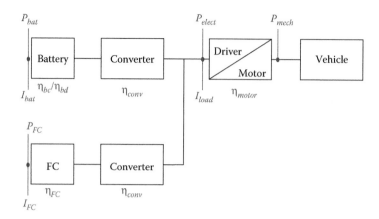

FIGURE 17.3 Schematic diagram of the electric power demand in an FCHEV.

To satisfy the current demand of the load, the following must be satisfied:

$$P_{load} = P_{bat}\eta_{eff,bat} + P_{FC}\eta_{eff,FC} \tag{17.3}$$

where $\eta_{eff,bat}$ is the overall electric efficiency of the battery with

$$\eta_{source}\,\eta_{eff,bat} = \begin{cases} \eta_{bc}\,\eta_{conv} & \text{Battery charging} \\ \eta_{bd}\,\eta_{conv} & \text{Battery discharging} \end{cases}$$

where $\eta_{conv}$ is the converter efficiency and $\eta_{bc}$ and $\eta_{bd}$ are the internal battery efficiency during charging and discharging processes, respectively. $\eta_{eff,FC} = \eta_{FC}\eta_{conv}$ is the overall electric efficiency of the FC and $\eta_{FC}$ is the internal efficiency of the FC. Typical efficiencies of the source are [20]: $\eta_{bc} = 1$, $\eta_{bd} = 0.95$, $\eta_{SC} = 1$, and $\eta_{FC} = 0.5$.[*]

Notice that $P_{mech}(t)$ changes with time because it depends on the driving cycle; that is, on the vehicle velocity profile. In other words, the driver and the road conditions fix at every time, a time varying power demand that must be satisfied by the power train.

### 17.2.3 Problem Formulation: Heuristic Approach

Heuristic EMS cannot be mathematically formulated in a unique form and depends on the type of heuristic approach used. In the case of piecewise continuous EMS, the problem formulation is mostly performed using words.

#### 17.2.3.1 Piecewise Continuous EMS: PHEV Case

Let us consider the case of a plug-in HEV (PHEV). The vehicle is provided with an engine, an electric motor, and a set of batteries. Assume that the vehicle is also provided with a variable transmission so that the engine can be operated at a desired (optimal) speed.

The set of operation rules depends on the battery SOC, the motor power, and the engine optimal operation point, while the the outputs are the battery, engine, and motor states. An example of the said-heuristic rules is given in Table 17.1 [21], where $P_{load}$ is the power demand, $P_{eng,opt}$ is the

---

[*] Higher FC efficiencies are reported in the literature but usually, they do not take into account the power losses related with the auxiliary systems and the balance of plant (BoP) of the FC. The BoP is the power plant of the FC, plus auxiliary monitoring and control devices.

TABLE 17.1

**Operation Modes of the PHEV**

| Conditions | Motor State | Engine State | Battery State |
|---|---|---|---|
| $SOC \leq SOC_{min}, P_{load} < P_{eng,opt}$ | REG | ON at $P_{eng,opt}$ | Charging |
| $SOC \leq SOC_{min}, P_{load} \geq P_{eng,opt}$ | OFF | ON | — |
| $SOC > SOC_{min}, P_{load} < P_{motor,max} > 0$ | ON | OFF | Discharging |
| $SOC > SOC_{min}, P_{load} \geq P_{motor,max} > 0$ | ON | ON | Discharging |
| $SOC < SOC_{max}, P_{load} < 0$ | REG | OFF | Charging |

power provided by the engine at the optimal speed, $P_{motor,max}$ is the maximum power delivered by the electric motor, and $SOC_{min}$ is the minimum allowed level of the SOC, while REG stands for the regeneration mode [21].

### 17.2.3.2 Frequency-Based EMS: EV Case

To apply this kind of strategy in EVs, it is necessary that the ESS is composed of both batteries and SCs. The main objective is to decide the power split based on the frequency analysis of the driving cycle, exploiting the high- power density of SCs and the high-energy density of the batteries. The effectiveness of this EMS is crucially dependent on the variability of the driving cycle; therefore, frequency-based EMS is not a good option for highway-driving cycles.

The first step is to design low- and high-pass filters according to the designer criterion and the ESS dynamics. To this end, the current demand under a given driving cycle is analyzed using, for example, fast Fourier transform (FFT) (see Figure 17.4).[*] The adequate filter cutoff frequencies in accordance with the ESS dynamics can be readily chosen using Figure 17.4. The filters are used to

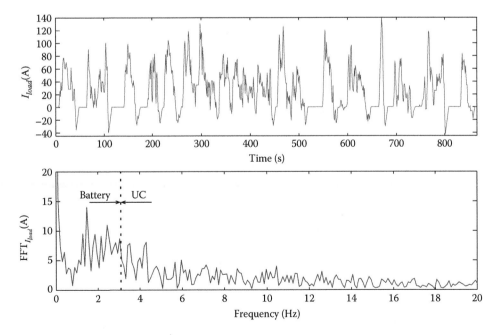

FIGURE 17.4 Frequency analysis of the current demand of a driving cycle. The cutoff frequency of the high- and low-pass filters can be readily chosen from the frequency analysis of the current demand and the ESS dynamics.

---

[*] Wavelet transform can also be used (see References 5 and 28).

separate the actual frequency content of the power demand. The output of the high-pass filter is used to create the reference of the SC current controller, while the output of the low-pass filter is used to create the reference of the battery current controller, as shown in Figure 17.1.

### 17.2.4 PROBLEM FORMULATION: OPTIMIZATION APPROACH

As it was stated earlier, the objective of the optimization-based EMS is to minimize a function of fuel consumption ($\dot{m}_{Fuel}(t)$). This problem can be formulated using different, the so-called *OFs*. Some options are

$$J_1 = K\dot{m}_{Fuel}(t) \tag{17.4}$$

$$J_2 = K(\dot{m}_{Fuel}(t))^2 \tag{17.5}$$

$$J_3 = K\int_{t_0}^{t} \dot{m}_{Fuel}(\sigma)d\sigma \tag{17.6}$$

$$J_4 = K_1\int_{t_0}^{T} \dot{m}_{Fuel}(\sigma)d\sigma + K_2[SOC(T) - SOC^*] \tag{17.7}$$

$$J_5 = \int_{t_0}^{T} [K_1\dot{m}_{Fuel}^2(\sigma) + K_2(SOC(\sigma) - SOC^*)^2]\,d\sigma \tag{17.8}$$

where $K$, $K_1$, and $K_2$ are suitable constants and $t_0$ and $T$ are, respectively, the initial and final time of the driving cycle. OFs (17.4) and (17.5) do not take into account the past or future fuel consumption; if the problems were formulated using such equations, the minimization problem would be instantaneous.

The optimization of this class of OF can be performed using standard optimization tools, for example, those given in Reference 22. However, OFs (17.6) through (17.8) are quite different, because the OF is dynamic and in any of them, the objective would be to minimize the *trajectories* of the fuel consumption from time $t_0$ to $T$. Observe that to solve the optimization problem, such set of trajectories must not be empty. The existence of such set of trajectories depends on the driving cycle and the initial conditions of the power sources.

The term $K_2[SOC(T) - SOC^*]^2$ in Equation 17.8 is known as equivalent fuel consumption of the ESS and it is used to reduce ESS utilization. The use of this term shifts the optimal point by an amount proportional to the difference of the ESS SOC and the reference value ($SOC^*$); this shift usually leads to an increment of fuel consumption. Observe that Equation 17.8 takes into account the battery SOC deviation from a desired value $SOC^*$ along the entire system trajectory.

The term $K_2[SOC(T) - SOC^*]$ in Equation 17.7 is called *terminal cost* since it only affects the OF at the final time $T$. It is used to penalize ESS SOC deviation from a desired value at the end of the driving cycle.[*]

The minimization of Equations 17.6 through 17.8, must be performed using either VC or DP that are relatively more complex optimization tools; however, observe that some unconstrained problems are trivial to solve, for example, $\min_{\dot{m}_{Fuel}} J_i$ for $i = 1, 2, 3, \ldots$ has a trivial solution at $\dot{m}_{Fuel}(t) = 0$. So, to formulate the problem in a congruent and useful manner, it is necessary to include restrictions; such restrictions can be dynamic or static, as shown in the following sections.

---

[*] Terminal restrictions are also helpful to this end as we will see later.

Depending on the knowledge of the driving cycle, the choice of the OF can be made accordingly. If no information is available, OF (17.4) and (17.5) are more convenient options, while for OFs given by Equations 17.6 through 17.8, predictive knowledge of the driving cycle or accurate characterization of the driving cycle is required.

As a final worthy observation, let the fuel consumption of the vehicle using only the engine or FC be $\dot{\tilde{m}}_{Fuel}$ the FE is defined as

$$FE \triangleq \frac{[\dot{\tilde{m}}_{Fuel}(t) - \dot{m}_{Fuel}(t)]}{\dot{\tilde{m}}_{Fuel}(t)} \times 100\%$$

hence, the minimization of the fuel consumption has the same solution as the maximization of the FE.

### 17.2.4.1  Instantaneous Optimization Problem Formulation: FCHEV Case

Let us consider the case of an FCHEV; since the fuel consumption is proportional to the current demanded from the FC, the optimization problem can be formulated as

$$\min_{I_{FC}} J_2 = \min_{I_{FC}} K(\dot{m}_{H_2}(t))^2 = \min_{I_{FC}} K\left\{\frac{-\bar{N}M_{H_2}I_{FC}(t)}{2\bar{F}}\right\}^2 \qquad (17.9)$$

subjected to

$$P_{load} = P_{ESS}\eta_{eff,ESS} + P_{FC}\eta_{eff,FC} \qquad (17.10)$$

$$\dot{SOC}(t) = -I_{ESS}(t)/C \text{ with } C > 0 \qquad (17.11)$$

$$SOC_{min} \leq SOC(t) \leq SOC_{max} \text{ with } SOC_{min} > 0, SOC_{max} \leq 1 \qquad (17.12)$$

$$P_{FCmin} \leq P_{FC}(t) \leq P_{FCmax} \text{ with } P_{FCmin}, P_{FCmax} > 0 \qquad (17.13)$$

where $\eta_{eff,ESS}$ and $\eta_{eff,FC}$ are the overall electrical efficiency of the ESS and FC, respectively. Note that the restrictions must be satisfied for every time $t \geq t_0$; therefore, they are also terminal restrictions (i.e., restrictions to be satisfied at the final time of the cycle, T).

Furthermore, observe that Equation 17.12 constitutes restrictions over ESS SOC, such that the ESS cannot be depleted beyond $SOC_{min}$ or charged beyond $SOC_{max}$, while restriction (17.11) relates the battery current with the battery SOC and constitutes a dynamical restriction given by the battery inner dynamics.

Restriction (17.13) constitutes the limits of the operating power range for the FC. Observe that restriction (17.10) is actually a power conservation equation and it states that the load power demand must be satisfied by the ESS and the FC. Under the worst conditions, the FC design should be able to satisfy the maximum power demand of the vehicle; otherwise, such restriction cannot be satisfied. Note that once the FC power is fixed, the ESS power is fixed for a given power load. That is, there is only one design variable for optimization, and the problem can be formulated accordingly, and the most natural choice is $I_{FC}$ as in Equation 17.14.

It is worth noting that the solution of the optimization-based EMS crucially depends on the existence of a nonempty feasible set. It is expected that as the battery is depleted, such set becomes narrower, leading to a point where no optimization can be performed and the engine or the FC must provide all the power. That is, the existence of emergency modes (heuristic rules) is unavoidable for extreme conditions.

## 17.2.4.2 Dynamic Optimization Problem Formulation: FCHEV Case

In this case, the OF can be chosen as any of functions (17.6) through (17.8). In this case, let us set it as

$$\min_{I_{FC}} \int_{t_0}^{T} \dot{m}_{Fuel}^2(\sigma)d\sigma = \min_{I_{FC}} \int_{t_0}^{T} \left\{ \frac{-\bar{N}M_{H_2}I_{FC}(\sigma)}{2\bar{F}} \right\}^2 d\sigma \tag{17.14}$$

subjected to restrictions (17.10) through (17.13) and the terminal restriction

$$SOC(T) \geq SOC(t_0) \tag{17.15}$$

The terminal restriction (17.15) is used here to ensure that the battery is replenished at least to the SOC level as it was at the beginning of the driving cycle.

The satisfaction of terminal restrictions in the dynamical optimization problem (DOP) substantially depends on the initial conditions. Such restrictions are only satisfied if the initial conditions belong to, what is called, the *reachable set*. That is, the set of initial conditions from which the terminal conditions are reachable for the set of all possible power-sharing actions.

The existence of the reachable set requires that the dynamic system is controllable. The set of control actions (power-sharing decisions) such that the terminal restrictions are satisfied are called *admissible control inputs*.

Observe that the existence of a reachable set depends not only on the power-sharing decision but also on the driving cycle. This point will be illustrated in Section 17.4. If the initial conditions do not belong to the reachable set, the solution of the DOP does not exist; therefore, it is necessary to ensure that such condition is satisfied, although in most of works in the literature, it is implicitly assumed.

## 17.3 SOLUTION OF THE DOP

In Section 17.2.4, it has been stated that two different optimization problems can be formulated, those are instantaneous and dynamic. The instantaneous optimization problem (IOP) has been extensively discussed in the literature and the existence of a global or local optimum is closely related to the type of OFs or/and the type of restrictions. To clarify this statement, let us consider, for example, the theorem of classical Lagrange multipliers for equality constraints. The theorem states that it is required (necessary condition for optimality) that the Jacobian matrix associated to the equality constraint has full row rank at the minimum; such condition is called qualification (constraint qualification). This condition is independent of the OF; therefore, if the constraint holds, the theorem holds for the same constraints and any differentiable OF. Constraint qualifications may constitute necessary and/or sufficient conditions for the existence of an optimum. There are numerous research in the literature focused on determining the weakest qualifications and also on extending the results to general problems.

Among the most important optimization results is the one due to Karush, Khun, and Tucker (KKT) [23] that establishes first- and second-order conditions (necessary and sufficient) for convex* OF and restrictions. First-order KKT conditions have also shown to be qualifications for pseudo-convex [24], invex [25], type I [26], strong-pseudo-quasi-type I-univex [27], weak-pseudo-quasi-type I-univex [27], OFs, and constraints, among others.

If an OF is strictly convex, its minimum is unique, and therefore global. However, in general, global optimization addresses the computation and characterization of global solutions to nonconvex continuous problems, including differential–algebraic, nonfactorable, and mixed-integer problems.

On the other hand, the solution of the DOP can be derived using the basic static optimization tools mentioned above; however, such derivation is not direct and in most cases, constitutes a very complicated matter. There exist two main trends in the literature to solve this class of problems. The DP and VC methods. In both cases, the objective is to transform dynamic optimization problems into instantaneous (static) ones. In the following sections, both methodologies are described; in every case, when restrictions are considered, it is assumed the existence of a nonempty feasible set, otherwise, any methodology of the solution is pointless.

### 17.3.1 RESULTS OF CALCULUS OF VARIATIONS

Calculus of variations has a long list of results. In this section, we will focus only on Pontryagin's principle, which states necessary conditions for optimality for the so-called optimal control problem (OCP) (also known as Mayer problem). Such problem can be stated as follows:

Let $M(t,x,u)$ be a continuous scalar function $C^1$ in† $(x,u)$ with $x$ and $u$, $n$-, and $m$-dimensional vectors, respectively. Let the OF be

$$\bar{J}(x,u) = G(x(T)) + \int_{t_0}^{T} M(t,x(t),u(t))\, dt \tag{17.16}$$

restricted to

$$\dot{x} = f(t,x,u) \tag{17.17}$$

$$x(t_0) = \bar{x}(t_0) \tag{17.18}$$

$$\phi(T,x(T)) = 0 \tag{17.19}$$

with $t_0$ and $T$ the initial and final time and $G(x(T))$ is the terminal cost. Restrictions (17.18) and (17.19) state that the initial and final states are fixed, and they are also known as *end restrictions*. Restriction (17.19) is also called terminal restriction.

*DOP1: Find u that minimizes Equation 17.16 restricted to Equations 17.17 through 17.19.*

In the following, when no confusion may arise, the time argument of states and inputs will be omitted for brevity; also, the following notations are introduced. Let $x'$ stand for the transpose of $x$, $x_j$ with $j = 1\ldots n$ stands for the $j$th entry of $x$ and the operator $\nabla_x \triangleq (\partial/\partial x)$.

To establish necessary conditions for optimality, let us define the Hamiltonian,

$$H(t,x(t),u(t),\lambda(t)) = M(t,x(t),u(t)) + \lambda' f(t,x(t),u(t)) \tag{17.20}$$

---

* A scalar continuous function $f(x)$ is convex in the convex set $\Omega$ if for any two points $x_1$ and $x_2$ in $\Omega$ and $0 \le \sigma \le 1$, $f$ satisfies $f(\sigma x_1 + (1 - \sigma)x_2) \le \sigma f(x_1) + (1 - \sigma)f(x_2)$. The function is strictly convex if $f(\sigma x_1 + (1 - \sigma)x_2) < \sigma f(x_1) + (1 - \sigma)f(x_2)$.
† The class $C^1$ contains all the differentiable functions whose first derivative is continuous. They are also called continuously differentiable.

**Theorem 17.1  (Pontryagin's Principle)**

Necessary conditions for optimality of the DOP1 are

$$\dot{\lambda}(t) = -\nabla_x' f(t,x,u)\lambda(t) - \nabla_x' M(t,x,u) \tag{17.21}$$

$$\nabla_u' H(t,x,u,\lambda) = \nabla_u' f(t,x,u)\lambda(t) + \nabla_u' M(t,x,u) = 0 \tag{17.22}$$

$$\lambda(T) = [\nabla_x \phi(t,x) + \nabla_x G(x)]\,|_{t=T} \tag{17.23}$$

Conditions (17.21) through (17.23) are called transversality conditions or boundary conditions and Equation 17.21 is called the adjoint equation.

Observe that in Theorem 17.1, $\phi(T,x(T)) = 0$ must be satisfied; moreover, $\lambda_j(t_0) = 0$ whenever $x_j(t_0)$ is not specified. The result in Theorem 17.1 cannot be applied if inequality constraints on the states and control variable exist. For example, with a set of restrictions

$$R(t,x(t),u(t)) \le 0 \tag{17.24}$$

where $R(t,x(t),u(t))$ is a $p$-dimensional vector $C^1$ in $(x,u)$ and $\le$ stands in this case, for element-wise inequality. For these cases, Theorem 17.2 is introduced.

*DOP2: Find u that minimizes Equation 17.16 restricted to Equations 17.17 through 17.19, and 17.24.*

To propose a solution to DOP2 and to introduce Theorem 17.2, let us consider the modified Hamiltonian,

$$\bar{H}(t,x,u,\lambda(t),v(t)) = M(t,x,u) + \lambda' f(t,x,u) + v' R(t,x,u) \tag{17.25}$$

with

$$v_j = \begin{cases} > 0, & R_j(t,x,u) = 0 \\ = 0, & R_j(t,x,u) < 0 \end{cases} \tag{17.26}$$

Note that in the case when $R_j(t,x,u) > 0$, no solution exists because restrictions are not satisfied. The following result is in order.

**Theorem 17.2**

Necessary conditions for optimality of the DOP2 are

$$\dot{\lambda}(t) = -\nabla_x' f(t,x,u)\lambda(t) - \nabla_x' M(t,x,u) - \nabla_x' R(t,x,u)v(t) \tag{17.27}$$

$$\nabla_u \bar{H} = \lambda(t)' \nabla_u f(t,x,u) + \nabla_u M(t,x,u) + v' \nabla_u R(t,x,u) = 0 \tag{17.28}$$

$$\lambda'(T) = [\nabla_x \phi(t,x) + \nabla_x G(x)]\,|_{t=T} \tag{17.29}$$

Observe that as in Theorem 17.1, $\phi(T,x(T)) = 0$ must be satisfied, and $\lambda_j(t_0) = 0$ whenever $x_j(t_0)$ is not specified. At this point, it is natural to ask whether the conditions of Theorems 17.1 and 17.2 are also sufficient. In general, sufficient conditions for optimality requires the introduction of DP concepts (see Reference 28). However, for the case of the optimal problem with free terminal point, that is, the problem where the initial and final times are fixed but there are no conditions on the final state, necessary and sufficient conditions are quite simple.

**Theorem 17.3**

Necessary and sufficient conditions for the ODP1 with free terminal point are

$$\dot{\lambda}(t) = -\nabla'_x f(t,x,u)\lambda(t) - \nabla'_x M(t,x,u) \tag{17.30}$$

$$\nabla'_u H(t,x,u,\lambda) = \nabla'_u f(t,x,u)\lambda(t) + \nabla'_u M(t,x,u) = 0 \tag{17.31}$$

$$\lambda'(T) = \nabla_x G(x) \mid_{t=T} \tag{17.32}$$

Moreover, if in addition, $M(t,x,u)$ is strictly convex in $(x,u)$, the optimal control $u(t)$ is global (unique).

The proofs of Theorems 17.1 through 17.3 are highly involved and they will not be presented here. However, the verification of Theorem 17.2 is presented in Section 17.4.

### 17.3.2  DYNAMIC PROGRAMMING

Pontryagin's principle can provide a nice (but complicated) geometric interpretation of optimality; however, as stated before, if one is looking for sufficient conditions for optimality, DP tools are needed. To understand the connections between DP and Pontryagin's principle, one must first review the optimality existence conditions. That is, from Theorem 17.1, we have established necessary conditions for optimality; in particular, for the existence of extremals. However, such extremals are not necessary to be a minimum (i.e., the minimum is necessarily an extremal but not vice versa). DP tools depart from the existence of such extremal and this constitutes the general principle of DP. To introduce DP, let us define the following:

**Definition 17.1**

Let $V(t,x) \triangleq \inf_u J(t,x,u)$, where *inf* stands for *infimum*, that is, the greatest lower bound of the scalar function $J(t,x,u)$. $V(t,x)$ is known as Bellman function or optimal return function.
    Moreover, we are focused on the class of DOP that satisfies the following properties:

*Property 1.* For all $T, x$, there exists a continuous control function $u$ that satisfies $V(t,x) = J(t,x,u)$.
*Property 2.* For all $t$, $J(t,x,u)$ is a nondecreasing function of time evaluated along any trajectory corresponding to an admissible control input of the initial state $x(t_0)$ (see Figure 17.5).
*Property 3.* $J(t,x,u)$ evaluated along any optimal trajectory is constant.

Considering the DOP1, let us introduce the associated Hamilton–Jacobi–Bellman (HJB) equation

$$-\frac{dV(t,x)}{dt} = \min_u [\nabla_t V(t,x) + \nabla'_x V(t,x)f(t,x,u)] \tag{17.33}$$

which can be alternatively written as

$$-\frac{dV(t,x)}{dt} = \min_u [M(t,x,u) + \nabla'_x V(t,x)f(t,x,u)] \tag{17.34}$$

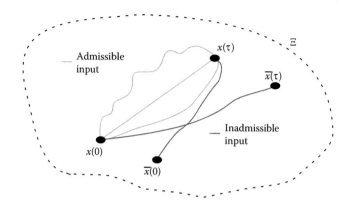

**FIGURE 17.5** Admissible and unadmissible inputs for the optimization problem. The admissible inputs are the set of the control inputs $u$ such that the terminal conditions (both initial and end) are satisfied.

Observe that for the terminal cost to be satisfied, Bellman function must also satisfy

$$V(T,x) = G(x) \tag{17.35}$$

It can be shown that the HJB equation is a sufficient condition for optimality for an ODP1 satisfying properties 1–3. In this case, there exists a unique smooth optimal control for every initial time and initial condition such that V satisfies the HJB equation (17.33) with (17.35). That is, by solving the partial differential equation (17.34) with boundary condition (17.35), it is possible to find the optimal control. Unfortunately, solving such equation is a very complicated matter (depends on the nonlinearity $f(t,x,u)$), and in just a few cases, it can be computed explicitly.

Let us assume that OF (17.16) and restriction (17.17) admit the following expressions, respectively:

$$\bar{J}(x,u) = x(T)'Sx(T) + \int_{t_0}^{T} [x(t)'F(t)x(t) + u(t)'N(t)u(t)]\, dt \tag{17.36}$$

$$\dot{x} = A(t)x + B(t)u \tag{17.37}$$

Since $V(x(T),T) = x(T)'Sx(T)$, the Bellman function must to be quadratic (i.e., $V(t,x) = x'P(t)x$); therefore, it is reasonable that the optimal control must have the form $u(t) = K(t)x(t)$ (otherwise the Bellman function cannot be quadratic); hence, the HJB equation takes the form

$$\dot{P} = P(t)B(t)N^{-1}(t)B'(t)P(t) - P(t)A(t) - A'(t)P(t) - F(t) \tag{17.38}$$

$$P(T) = S \tag{17.39}$$

$$K(t) = -N(t)^{-1}B(t)'P(t) \tag{17.40}$$

where the optimal gain is given by Equation 17.40. The matrix differential equation (17.38) is known as the Riccati differential equation. Observe that for the DOP2, that is, the problem with restricted control and states, the solution is also Equations 17.38 through 17.40 in the feasible region, as long as the restrictions are not active.[*]

---

[*] A restriction $R(t,x,u) \leq 0$ is called active in the feasible set, if $R(t,x,u) = 0$.

## 17.4  OPTIMAL EMSS FOR FCHEV

The objective of this section is to illustrate some of the EMS designs introduced before. Two cases of optimization-based EMS are chosen: the instantaneous and the dynamic problem. Such election allows us to analyze and contrast the complexity of both problems as well as the effect of the parameters and the current demand. In this section, two driving cycles are used; these are: the Economic Commission for Europe Driving Cycle (ECE) also called ECE-15 and the urban driving cycle (City II). See Figure 17.6.

### 17.4.1  STATIC OF

In this case, the optimization problem is given by Equation 17.9 subjected to Equations 17.10 through 17.13. Let us consider the case where the ESS is given by a battery bank; therefore, $I_{ESS} = I_{bat}$. Since the problem is instantaneous, the feasible region can be easily illustrated, by plotting the FC and battery restrictions. To illustrate this point, let us consider the polarization curve in Section 17.2.1. Using this curve, it is possible to express the FC voltage in Region II as a linear function of the FC current (see Figure 17.2); that is,

$$V_{FC} = mI_{FC} + b \tag{17.41}$$

where $V_{FC}$ is the voltage of the FC. Notice that $m < 0$ and $b > 0$ (see Figure 17.2). Substituting Equation 17.41 in restriction (17.13) leads to

$$P_{FCmin} - mI_{FC}^2 - bI_{FC} \leq 0 \tag{17.42}$$

$$-P_{FCmax} + mI_{FC}^2 + bI_{FC} \leq 0 \tag{17.43}$$

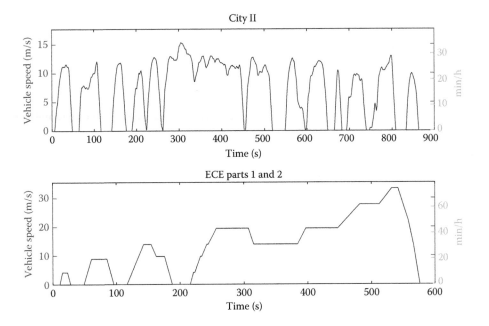

FIGURE 17.6  City II and ECE driving cycles.

Let us define

$$\alpha = \frac{P_{bat}(t)\eta_{eff,bat}}{P_{load}} \qquad (17.44)$$

that is, $\alpha$ is the proportion of the power demand that is provided by the battery bank. Note that $0 \le \alpha \le 1$; therefore, restrictions (17.10) through (17.13) can be written as

$$\frac{P_{load}(t)(1-\alpha)}{\eta_{eff,FC}} = P_{FC}$$

$$P_{FCmin} - \frac{P_{load}(1-\alpha)}{\eta_{eff,FC}} \le 0 \qquad (17.45)$$

$$-P_{FCmax} + \frac{P_{load}(1-\alpha)}{\eta_{eff,FC}} \le 0 \qquad (17.46)$$

$$\frac{-C\eta_{eff,bat}V_{bat}\dot{SOC}(t)}{P_{load}} = \alpha \qquad (17.47)$$

Using the expressions above, it is now possible to visually identify the feasible region in terms of the current load (shaded region in Figure 17.7). Restrictions (17.45) and (17.46) constitute curves that approach $\alpha = 1$ as $P_{load}$ increases and they are denoted with $P_{FCmin}$ and $P_{FCmax}$, respectively. Note that as the restriction of the battery is dynamic, the curves A and B correspond to curves with initial conditions $SOC = SOC_{min}$ and $SOC = SOC_{max}$ respectively, but in an actual application, battery $SOC$ will be varying according to the power demand displaying a curve between curves A and B (dashed line in Figure 17.7). Observe that reductions of fuel consumption are obtained as $\alpha$ increases. Therefore, since the solution must belong to the feasible region, the solution necessary resides at the boundary of the feasible set and for low-power loads, this solution is the curve of minimum FC power. This fact agrees with intuition, since to minimize the fuel consumption, the use of the FC must be minimized. This condition rapidly changes as the current demand increases. In this case, if the battery is sufficiently charged, the solution will move from the curve of minimum FC power to the maximum FC power along the line $\alpha = 1$; otherwise, the solution will lay on the actual $SOC$ trajectory (dashed line in Figure 17.7) and the value of $\alpha$ will depend on the current demand.

Figure 17.7 is also valuable as an auxiliary tool to design the power sources of an FCHEV. To clarify this point, first observe that given a driving cycle, the maximum power demand ($P_{load,max}$) is fixed for a given vehicle and driving cycle. The maximum FC power (the maximum capacity of the FC) can be fixed at the intersection of $P_{load,max} = P_{load}$ and $\alpha = 1$, to maximize the feasible region, and therefore the optimization events. Let us denote the power demand at such intersection as $P_{load,C}$ (see point C in Figure 17.7); then a final worthy comment is in order: power demands greater than $P_{load,C}$ are not reachable for any value of $\alpha$. That is, the implicit controllability assumption from which we have departed to solve the optimization problem is not satisfied for $P_{load} > P_{load,C}$.

The time evolution of the system under the instantaneous optimal problem is displayed in Figure 17.8 for the initial condition $SOC(t_0) = 1$ and for the driving cycle City II. Observe that the fuel minimization problem can be solved due to the battery utilization; therefore, in such optimal strategy, the battery will be depleted to its minimum (allowed) level. Moreover, observe that the case when the battery is charged by the FC is never an optimal solution since more fuel

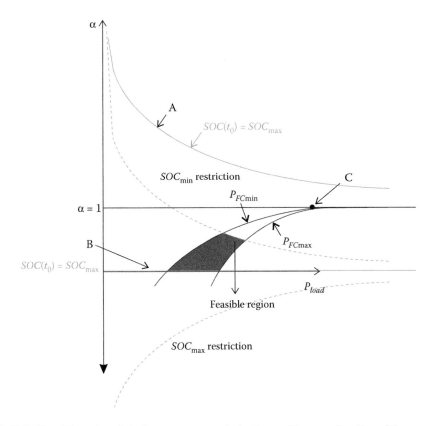

**FIGURE 17.7** Feasible region of the instantaneous optimization problem as a function of the current load.

is consumed by this charging process if the FCs were used to provide all the power demand.[*] To penalize the battery utilization, the equivalent fuel consumption term as in OF (17.8) can be used (i.e., $J = K\dot{m}_{Fuel}^2(\sigma) + K_2(SOC(\sigma) - SOC^*)^2$). Such strategy does not avoid battery depletion (to its minimum allowable level) but the rate of discharging is smaller than in the optimization strategy above, as a penalization of the battery utilization is performed. However, the use of the equivalent fuel consumption has the disadvantage of having increased fuel consumption, as can be seen in Figure 17.9, where the FE as a function of the gain $K_2$ is displayed.

### 17.4.2 INTEGRAL OF

Let us consider the DOP given in Section 17.2.4.2, given by Equation 17.14, under restrictions (17.10) through (17.13). The optimization problem has an integral OF defined from time $t_0$ to $T$. Here, VC methods introduced in Section 17.3.1 will be used to solve the optimization problem. To this end, let us consider, as in Section 17.4.1, $\alpha(t) = u(t)$ as the proportion of the total power demand provided by the battery. Since the FC is operating in the Ohmic region $I_{FC}(t) \propto P_{FC}(t)$, hence, the objective function (17.14) becomes

$$\min_\alpha \int_{t_0}^T a^2 \bar{K}^2 P_{FC}^2(\sigma) d\sigma = \min_\alpha \int_{t_0}^T a^2 \bar{K}^2 \frac{(1-\alpha)^2 P_{load}^2(\sigma)}{\eta_{eff,FC}^2} d\sigma = \min_u \int_{t_0}^T a^2 \bar{K}^2 \frac{(1-u)^2 P_{load}^2(\sigma)}{\eta_{eff,FC}^2} d\sigma$$

$$(17.48)$$

---

[*] This is due to energy drop given by the converters' efficiency.

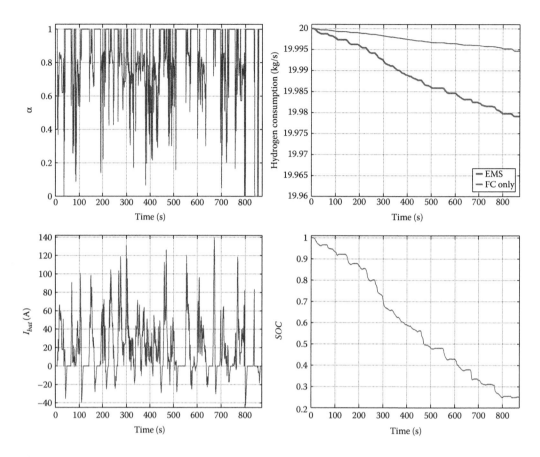

**FIGURE 17.8** Time evolution of an optimal EMS for a vehicle under the City II driving cycle.

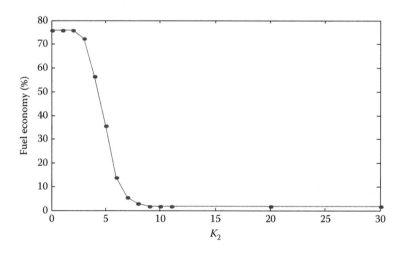

**FIGURE 17.9** FE as a function of the gain of the equivalent fuel consumption term.

where $\bar{K} = \{-\bar{N}M_{H_2}/2\bar{F}\}$ and the dynamic of the battery SOC is given by Equation 17.47. Therefore, the functions in Theorem 17.2 are

$$f(t,x,u) = B(t)u \tag{17.49}$$

$$M(t, x(t), u(t)) = \frac{a^2 \bar{K}^2 P_{load}^2 (1-u)^2}{\eta_{eff,FC}^2} \tag{17.50}$$

$$G(x(T)) = 0 \tag{17.51}$$

$$\phi(T,x(T)) = 0 \tag{17.52}$$

$$R(t,x,u) = \begin{bmatrix} P_{FCmin} - \dfrac{P_{load}(1-u)}{\eta_{eff,FC}} \\ -P_{FCmax} + \dfrac{P_{load}(1-u)}{\eta_{eff,FC}} \\ SOC_{min} - x \\ x - SOC_{max} \\ -u \\ u - 1 \end{bmatrix} \tag{17.53}$$

Note that $u(t) = \alpha(t)$ and $x(t) = SOC(t)$ have been used, and $B(t) \triangleq (-P_{load}(t)/CV_{bat}\eta_{eff,bat})$. Since the OF and the dynamical system do not depend on the state but only on the input $u$, hence

$$\nabla_x f(t,x,u) = 0 \tag{17.54}$$

$$\nabla_u f(t,x,u) = B(t) \tag{17.55}$$

$$\nabla_x M(t,x,u) = 0 \tag{17.56}$$

$$\nabla_u M(t,x,u) = -\frac{2a^2 \bar{K}^2 P_{load}^2(t)(1-u)}{\eta_{eff,FC}^2} \tag{17.57}$$

$$\nabla_x' R(t,x,u) = (0 \quad 0 \quad -1 \quad 1 \quad 0 \quad 0) \tag{17.58}$$

$$\nabla_u' R(t,x,u) = (\rho(t,u) \quad -\rho(t,u) \quad 0 \quad 0 \quad -1 \quad 1) \tag{17.59}$$

with $\rho(t,u) = \dfrac{P_{load}(t)}{\eta_{eff,FC}}$. According to Theorem 17.2, necessary conditions for the optimum are

$$\dot{\lambda}(t) = \nu_3 - \nu_4 \tag{17.60}$$

$$\lambda(T) = 0 \tag{17.61}$$

and from Equation 17.28

$$\lambda B(t) + \Xi(t,u) + \nu_1 \rho(t,u) - \nu_2 \rho(t,u) - \nu_5 + \nu_6 = 0 \tag{17.62}$$

with $\Xi(t,u) = -\dfrac{2a^2 \bar{K}^2 P_{load}^2 (1-u)}{\eta_{eff,FC}^2}$ and

$$\nu_j = \begin{cases} > 0, & R_j(t,x,u) = 0 \\ = 0, & R_j(t,x,u) < 0 \end{cases}$$

Observe from Equation 17.53 that $\nu_1$ and $\nu_2$ cannot be positive at the same time.[*] The same is true for $\nu_3$ and $\nu_4$[†] as well as for $\nu_5$ and $\nu_6$.[‡] That means that Equation 17.60 can be rewritten as

$$\dot{\lambda} = \begin{cases} > 0, & \text{if } SOC = SOC_{min} \\ = 0, & \text{if } SOC_{min} < SOC < SOC_{max} \\ < 0, & \text{if } SOC = SOC_{max} \end{cases}$$

that is, $\dot{\lambda} = \nu_3$ when $SOC = SOC_{min}$ and $\dot{\lambda} = -\nu_4$ when $SOC = SOC_{max}$. Let $\theta(t,u) = \lambda B(t) + \Xi(t,u) + \nu_1 \rho - \nu_2 \rho$; then Equation 17.62 can be rewritten as

$$\theta(t,u) > 0 \quad \text{if } u = 0 \tag{17.63}$$

$$\theta(t,u) = 0 \quad \text{if } 0 < u < 1 \tag{17.64}$$

$$\theta(t,u) < 0 \quad \text{if } u = 1 \tag{17.65}$$

Since the inequality constraints (17.63) through (17.65) depend only on the control variable $u$, the minimum will always require the control variable $u(t)$ (i.e., $\alpha$) to be at the boundary of the feasible region. That is, the minimum is given by one of expressions (17.63) through (17.65). To find the solution, first observe that the cases of Equations 17.63 through 17.65 are trivial because the solution is directly $u = 0$ and $u = 1$, respectively. However, for Equation 17.64, we must analyze all the boundaries; these are

*Case (i).* $SOC_{min} < SOC(t) < SOC_{max}$ and $P_{FCmin} = P_{FC}(t)$,
*Case (ii).* $SOC_{min} < SOC(t) < SOC_{max}$ and $P_{FCmax} = P_{FC}(t)$,
*Case (iii).* $SOC(t) = SOC_{min}$ and $P_{FCmin} = P_{FC}(t)$,
*Case (iv).* $SOC_{max} = SOC(t)$ and $P_{FCmin} = P_{FC}(t)$,
*Case (v).* $SOC_{min} = SOC(t)$ and $P_{FCmax} = P_{FC}(t)$, and finally
*Case (vi).* $SOC_{max} = SOC(t)$ and $P_{FCmax} = P_{FC}(t)$.

### 17.4.2.1 Case (i)

For the first case, we know that restriction (17.64) becomes

$$-\frac{2a^2 \bar{K}^2 P_{load}^2 (1-u)}{\eta_{eff,FC}^2} + \nu_1 \frac{P_{load}(t)}{\eta_{eff,FC}} = 0$$

---

[*] Otherwise, the FC power could be minimum and maximum at the same time.
[†] Otherwise, the battery SOC could be minimum and maximum at the same time.
[‡] That would imply that $0 = u = 1$!

which leads to

$$u = 1 - \frac{v_1 \eta_{eff,FC}}{2a^2 \bar{K}^2 P_{load}}$$

(17.66)

### 17.4.2.2  Case (ii)

For $P_{FC} = P_{FC,max}$ and $SOC_{max} < SOC(t) < SOC_{min}$ restriction (17.64) becomes

$$u = 1 + \frac{v_2 \eta_{eff,FC}}{2a^2 \bar{K}^2 P_{load}}$$

(17.67)

Note that since $v_1 > 0$, $v_2 > 0$, $m < 0$, and $b > 0$, the curve of function (17.67) is always above the curve of function (17.66), as can be observed in Figure 17.10, where both curves are displayed. Hence, the solution of the optimization problem for $SOC_{max} < SOC(t) < SOC_{min}$, as in the case of the example of Section 17.4.1, is along the curve of the minimum FC power $P_{FC,min}$, since $P_{FC,max}$ restriction leads to $\alpha > 1$ when power is demanded by the load. Note that the solution of the optimization problem does not depend on the battery size. However, this condition changes as the battery SOC level takes the maximum or the minimum value allowed, as it is shown below.

### 17.4.2.3  Case (iii)

For $SOC(t) = SOC_{min}$ and $P_{FCmin} = P_{FC}(t)$, we know that restriction (17.64) becomes

$$\lambda B(t) - \frac{2a^2 \bar{K}^2 P_{load}^2 (1-u)}{\eta_{eff,FC}^2} + v_1 \frac{P_{load}(t)}{\eta_{eff,FC}} = 0$$

with $\dot{\lambda} = v_3$ that leads to

$$u = 1 - \frac{v_1 \eta_{eff,FC} \eta_{eff,bat} CV_{bat} - \lambda \eta_{eff,FC}^2}{2 P_{load} \eta_{eff,bat} Ca^2 \bar{K}^2 V_{bat}}$$

(17.68)

### 17.4.2.4  Case (iv)

If $SOC(t) = SOC_{max}$ and $P_{FCmin} = P_{FC}(t)$, then $\dot{\lambda} = -v_4$ and restriction (17.64) is also given by Equation 17.68.

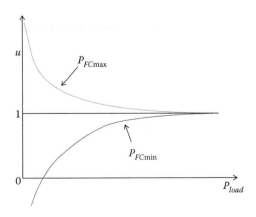

**FIGURE 17.10**  Restrictions associated with the minimum and maximum FC power as a function of power demand, for a dynamic optimization problem and $SOC_{min} < SOC(t) < SOC_{max}$.

### 17.4.2.5 Case (v)

If $SOC(t) = SOC_{min}$ and $P_{FCmax} = P_{FC}(t)$, then $\dot{\lambda} = v_3$ and restriction (17.64) becomes

$$u = 1 + \frac{v_2 \eta_{eff,FC} \eta_{eff,bat} CV_{bat} + \lambda \eta_{eff,FC}^2}{2 P_{load} \eta_{eff,bat} Ca^2 \overline{K}^2 V_{bat}} \tag{17.69}$$

### 17.4.2.6 Case (vi)

If $SOC(t) = SOC_{max}$ and $P_{FCmax} = P_{FC}(t)$, then $\dot{\lambda} = -v_4$ and restriction (17.64) can also be rewritten as Equation 17.69.

By comparing the control functions of Cases (iii) and (v), it is possible to observe that the solution of the optimization problem will depend on $\lambda$ (given by a an integrator system) and the final condition of this variable given by Equation 17.29. Note that in such a situation, the optimum strategy depends on the battery size. Finally, recall that since we have transformed the dynamic optimization problem into a static one by virtue of Theorem 17.2, the total fuel consumption must be computed by the integration of the system along the entire driving cycle.

## 17.5 CONTROL SYSTEMS

As stated before, once the decision has been made about what energy source will be used and in what proportion (i.e., once the energy management strategy is defined), the next step is to design and implement a controller to ensure that the power split is performed as expected. Such a task is performed by the ECU.

ECU design involves two aspects, the theoretical foundation of the controller and its electromechanical implementation. Regarding the last aspect, it is worth mentioning that there are several factors limiting control performance and stability in a real control implementation; some of them are sample time, sensor noise, sensor resolution as well as the actuator dynamics, and resolution.

Passing from theory to practice in the electronic control design can be painful if no attention is being paid to the explicit and implicit assumptions made about the system under study. Assuming a continuous system when slow[*] sensors or processors are used may lead to unacceptable performance or instability. Therefore, minimum sample time (including processing time) requirements must be explicitly computed before starting any control implementation and, system identification plays a fundamental role in this task. In general, to preserve the information of a signal, it is necessary to sample at least twice the maximum frequency of the signal, and such a principle is known as Nyquist minimal sample rate.

The resolution of the sensor and the actuator is also important because the conversion of analog signals (as found in nature) into digital signals (as computers process) is sampled in both frequency and magnitude. Coarse sensor resolution may induce persistent perturbations that may prevent algorithms to converge; in these extreme cases, the sampling principle cannot be satisfied. The sampling principle establishes that all the energy (related with the amplitude and the information that a signal contains) is contained in a countable number of samples taken at a fast-enough rate.

A final worthy observation is that, actuator and sensor limitations become particularly important when dealing with unstable systems. It has been brilliantly stated in Reference 29 that unstable systems are (i) quantifiable and more difficult to control, (ii) their operation is always critical and susceptible of failure, and (iii) the kind of stability that can be achieved is always local (i.e., the stable point/region can only be reached from a bounded set) and the design of control for such kind of systems must be performed carefully.

---

[*] Slow compared to system dynamics.

### 17.5.1 Control Design

Current control designs follow two main trends: time based and frequency based. The frequency-based techniques are suitable for linear systems and they are mainly the result of the effort of control scientists of the first part of the twentieth century. An advantage of frequency domain analysis is that the design techniques are, in general, simple and intuitive.

Proportional–integral (PI), proportional–integral–derivative (PID), and proportional–derivative (PD) controllers are some of the most common controllers derived from this design trend. PI and PD controllers can be seen as particular cases of PID control and the matter of choosing appropriate tuning parameters has been kept busy to control engineers for a long time.

Integral action gives to PID controller the capability to reject uncertainty and compensate steady-state bias; the proportional and derivative actions provide stabilization properties, while derivative action, which can be seen as a predictive action, has the property of providing overshoot and undershoot damping for improved performance. For appropiate control parameters, PID has displayed relatively close performance or stabilization capabilities to adaptive and robust controllers; however, how to obtain such performance using a systematic procedure for an arbitrary application has not been solved yet.

Since a lot of research has been done on this topic, there exist several reviews and compilations in the literature. In Reference 30, a comprehensive review of tuning procedures, patents, software, and commercial hardware modules is presented for PID controllers (see also Reference 31). In spite of the simplicity of frequency-based control design techniques, they are limited in practice because the derived stability results are valid only locally (the nature always displays some kind of nonlinearity).

However, the time-based control design provides the advantage that the results can be applied to either linear or nonlinear systems; however, they have an increased design complexity as nonlinear systems may display incredibly rich behavior that rarely admits closed-form solutions. Unlike linear control, nonlinear control lacks general methods that may provide a unified treatment to approach a wide class of problems. In this chapter, we will review some of the most common control techniques, namely robust control, linear quadratic regulator (LQR), and time-optimal control.

### 17.5.2 Stability Notions and Tools

Before discussing the controllers, some notions and stability tools will be introduced. In general, the stability definitions are developed only for autonomous systems,[*] since closed-loop systems can be written in this form. Also, without loss of generality, the stability is studied about the origin.

**Definition 17.2**

The equilibrium point $x_{eq} = 0$ of the system

$$\dot{x} = f(t, x) \tag{17.70}$$

(i.e., $f(t, x_{eq}) = 0$) is stable if for all $\delta > 0$, there exists an $\varepsilon > 0$ such that whenever $|x(t_0)| < \delta$, then $|x(t)| < \varepsilon \ \forall t \geq t_0$.

Note that the definition above is more related with a trajectory confinement that is approaching to a point. The stability in Definition 17.2 is also known as stability in the sense of Lyapunov.

**Definition 17.3**

The origin of system (17.70) is asymptotically stable if (i) the origin is stable and (ii) if all $|x(t_0)| < \delta$, then $|x(t)| \to 0$ as $t \to \infty$ (i.e., the origin is attractive).

---

[*] That is, no perturbed systems of the form $\dot{x} = f(t, x)$.

It is said that an equilibrium point is unstable if it is not stable. Observe that asymptotic stability implies stability in the sense of Lyapunov but the converse is not true. In general, although otherwise stipulated, we concern about the asymptotic stability of an isolated equilibrium point.

**Definition 17.4**

The set

$$\Sigma = \left\{ \beta \in R^{nv} \mid \sum_{i=1}^{nv} \beta_i = 1, \beta_i \geq 0 \right\}$$

(17.71)

is called simplex

**Definition 17.5**

The matrix $A$ is called polytopic if

$$A = \left\{ A(\beta) \mid \sum_{i=1}^{nv} A_i \beta_i = A, \beta_i \geq 0 \right\}$$

(17.72)

and $nv$ is called the number of vertexes

**Definition 17.6**

The linear system

$$\dot{x} = A(\beta)x$$

(17.73)

is robustly stable if it is asymptotically stable for all $\beta_i$ in the simplex.

### 17.5.3   ROBUST CONTROL DESIGN

In this section, we will introduce a very simple result of a robust control design that can be seen as a *minimax* stability problem formulated in the time domain, analogous to what $H_\infty$ does in the frequency domain. The result is easy to verify and admits a simple and natural description of the uncertainty. The strategy is actually an early result of robust control that uses polytopic matrices where the vertices of the polytope are the extremal values of the matrix, given the uncertainty bounds. That is, if the parameters of a matrix describing a linear dynamical system are uncertain, or subjected to sudden changes, but their entries have known upper and lower bounds, a convex combination of matrices or a polytopic matrix can be obtained. By proving stability of the vertexes, the stability of all possible linear descriptions within the simplex is ensured: that is, the main idea of robust control. In contrast to the traditionally linear description of a system, not only small variations to the parameters are valid, but also large variations.

Following the ideas above, the control problem can be formulated as

*Robust control problem.* Consider the linear system

$$\dot{x} = \bar{A}(\beta)x + \bar{B}(\beta)u$$

(17.74)

Find a control gain $K$ for $u = -Kx$ such that the origin is robustly stable.

It is known, since the system is linear, that Equation 17.74 can be stabilized using standard eigenvalues assignment [32]; this result is summarized in the following theorem.

## Theorem 17.4

The linear system (17.74) is robustly stable if there exists a symmetric positive definite matrix $P$ such that

$$PA(\beta) + A'(\beta)P < 0 \tag{17.75}$$

with $A(\beta) = \bar{A}(\beta) - \bar{B}(\beta)K$ and the matrix inequality meaning that the left-hand side of the inequality is a negative definite matrix.[*]

Instead of attempting to assign the closed-loop eigenvalues for every matrix in $A(\beta)$, the fact that $A(\beta)$ is a convex combination of the matrices can be exploited. Since this matrix is a convex function of $\beta$, we, instead, may opt just to assign the eigenvalues of their vertexes by observing that

$$\lambda_{\max}\{PA(\beta) + A'(\beta)P\} \leq \lambda_{\max,i}\{PA(\beta_i) + A'(\beta_i)P\} \tag{17.76}$$

for $i = 1,2\ldots nv-$. Condition (17.76) states that the maximum eigenvalue of the polytopic matrix $A(\beta)$ is upper bounded by the maximum eigenvalues of their vertexes. Transforming condition (17.75) to one, only the vertices are involved. In Reference 33, it is shown that the polytopic system (17.73) is robustly stable if a symmetric positive definite matrix $P$ satisfies

$$PA(\beta_i) + A'(\beta_i)P < 0 \quad i = 1,2\ldots nv \tag{17.77}$$

At this point, we wonder whether a single matrix $P$ exists for a given polytopic matrix. The following result states sufficient and necessary conditions for its existence, which constitutes a *minimax* problem [34]. Let $h_i(P) = \lambda_{\max,i}\{PA(\beta_i) + A'(\beta_i)P\}$, for $i = 1,2,\ldots nv$; hence $h_i$ is a convex and continuous function.

## Theorem 17.5

Assume that for the linear system (17.73), if only one vertex is known to be asymptotically stable, then there exists a symmetric positive definite matrix P such that Equation 17.77 is satisfied if [34][†]

$$\min_{\|P\|\leq 1}[\max_i h_i(P)] < 0| \tag{17.78}$$

*for* $i = 1,2\ldots nv$.

For the case of system (17.74), the computation of the matrix $P$ or (equivalently) the matrix gain $K$ can be performed using numerical mathematical methods, solving the system of linear matrix inequalities (LMIs) (see Reference 35).

More relaxed conditions can be derived using what is called homogeneous Lyapunov functions [36], or polynomial–parameter-dependent Lyapunov functions [37]. Such extensions use more

---

[*] A negative definite matrix B satisfies that $x'Bx < 0$ for all $x \neq 0$. Descartes' rule can be used to verify this condition.
[†] That is, if and only if.

parameters, making the inequality matrix much easier to solve. Moreover, the robust stability of nonlinear systems can be studied using input–output properties of the system along with state space and Lyapunov techniques, as shown in Reference 38.

### 17.5.4  OPTIMAL CONTROL DESIGN

#### 17.5.4.1  Linear Quadratic Regulator

The results for optimization-based EMS in Section 17.3.2 can be used to derive the optimal control known as LQR. To see this clearly, first observe that the objective function (17.36) can be seen as a performance index, where the matrix $F(t)$ is a time-varying, weight matrix penalizing deviations of the state from the reference (in this case, the origin), while matrix $N(t)$ has the same effect on the control actions.[*]

Let us assume that $S = 0$, then the solution of the Riccati differential equation (17.38) can be used to construct a linear feedback that minimizes the desired performance index over the time interval $[t_0, T]$. Since in general, the solution of the optimization problem is required for large final times (e.g., infinite time horizon), it is more convenient to use an infinite horizon performance index. Such optimization problem has a very simple solution and it constitutes the steady state of Equation 17.38. By letting $t \rightarrow \infty$ in Equation 17.38, the Riccati algebraic equation is derived, that is,

$$P(t)B(t)N^{-1}(t)B'(t)P(t) - P(t)A(t) - A'(t)P(t) - F(t) = 0 \qquad (17.79)$$

$$K(t) = -N(t)^{-1}B(t)'P(t) \qquad (17.80)$$

with $u = Kx$, which can be computed offline.

#### 17.5.4.2  Minimum Time Controllers

The solution of the ODP1, as shown in Theorem 17.1, can also be used to compute a kind of controllers that is optimum in time: that is, a kind of controllers that brings the system to a final state in a minimum time. Such control is also called bang-bang, because it leads to fast control actions. In this case, if $G(x(T)) = 0$ and $M(t,x(t),u(t)) = 1$, then Equation 17.16 leads to

$$\bar{J}(x,u) = t_0 - T \qquad (17.81)$$

which from Theorem 17.1, the solution can be computed by solving the boundary-value problem

$$\begin{aligned} \dot{x} &= f(t,x,u) \\ \dot{\lambda}(t) &= -\nabla'_x f(t,x,u)\lambda(t) \\ \nabla'_u f(t,x,u)\lambda(t) &= 0 \\ \lambda'(T) &= \nabla_x \phi(t,x) \mid_{t=T} \end{aligned} \qquad (17.82)$$

### EXERCISES

17.1  Implement heuristic EMS given in Section 17.2.3.1 in a PHEV using Table 17.1.

17.2  What is the effect of the battery size in the fuel consumption for the optimization problem in Section 17.4.1?

---

[*] Observe that when the origin is reached, the corresponding control action is zero. For other state references, this may not be true and the corresponding control action can be computed offline from the dynamic equation.

17.3 Introduce the terminal constraint (17.15) to the problem of Section 17.4.2. (i) Compute the solution. (ii) What is the expression of the adjoint equation?

17.4 Consider the problem of Section 17.4.1. (i) How regeneration can be taken into account?

## ACKNOWLEDGMENTS

The author acknowledges financial support from CONACYT-SENER Grant No. 152484. The author also would like to thank Irwin A. Diaz Diaz and Josefa Morales Morales for the technical support in the artwork preparation of this chapter.

## REFERENCES

1. Blunier B., Simoies M.G. and Miraoui A. Fuzzy logic controller development of a hybrid fuel cell-battery auxiliary power unit for remote applications. In *2010 9th IEEE/IAS International Conference on Industry Applications (INDUSCON)*, Sao Paulo, Brazil, pp. 1–6, 2010.
2. Ferreira A.A., Pomilio J.A., Spiazzi G. and de Araujo Silva L. Energy management fuzzy logic supervisory for electric vehicle power supplies system. *IEEE Transactions on Power Electronics,* 23(1):107–115, 2008.
3. Ravey A., Blunier B. and Miraoui A. Control strategies for fuel-cell-based hybrid electric vehicles: From offline to online and experimental results. *IEEE Transactions on Vehicular Technology*, 61(6):2452–2457, 2012.
4. Murphey Y.L., Park J., Chen Z., Kuang M.L., Masrur M.A. and Phillips A.M. Intelligent hybrid vehicle power control—Part I: Machine learning of optimal vehicle power. *IEEE Transactions on Vehicular Technology*, 61(8):3519–3530, 2012.
5. Murphey Y.L., Park, J., Kiliaris L., Kuang M.L., Masrur M.A., Phillips A.M. and Wang Q. Intelligent hybrid vehicle power control—Part II: Online intelligent energy management. *IEEE Transactions on Vehicular Technology*, 62(1):69–79, 2013.
6. Chunting C., Masrur A. and Daniszewki D. Wavelet-transform-based power management of hybrid vehicles with multiple on-board energy sources including fuel cell, battery and ultracapacitor. *Journal of Power Sources*, 185(2), 1533–1543, 2008.
7. Florescu A., Bacha S., Munteanu I. and Bratcu A.I. Frequency-separation-based energy management control strategy of power flows within electric vehicles using ultracapacitors. In *IECON 2012—38th Annual Conference on IEEE Industrial Electronics Society*, pp. 2957–2964. IEEE Press, Quebec, QC, Canada, 2012.
8. Erdinc O., Vural B. and Uzunoglu M. A wavelet-fuzzy logic based energy management strategy for a fuel cell/battery/ultra-capacitor hybrid vehicular power system. *Journal of Power Sources*, 194(1), 369–380, 2009.
9. Archambeault B.R. *PCB Design for Real-World EMI Control.* Kluwer Academic Publishers, Norwell, Massachusetts, USA, 2002.
10. Ott H.W. *Electromagnetic Compatibility Engineering.* John Wiley and Sons, Inc., Hoboken, NJ, USA, 2009.
11. Montrose M.I. *Printed Circuit Board Design Techniques for EMC Compliance: A Handbook for Designers.* IEEE Press Series on Electronics Technology, New York, NY, USA, 2000.
12. Paganelli G., Delprat S., Guerra T.-M., Rimaux J. and Santin J. J. Equivalent consumption minimization strategy for parallel hybrid powertrains. In *IEEE 55th Vehicular Technology Conference, 2002. VTC Spring 2002*, Vol. 4, pp. 2076–2081. IEEE Press, Birmingham Al, 2002.
13. Lin C.C., Peng H., Grizzle J.W. and Jun M.K. Power management strategy for a parallel hybrid electric truck. *IEEE Transactions on Control Systems Technology*, 11:839–849, 2003.
14. Bo Geng, Mills J.K. and Sun D. Energy management control of microturbine-powered plug-in hybrid electric vehicles using the telemetry equivalent consumption minimization strategy. *IEEE Transactions on Vehicular Technology*, 60(9):4238–4248, 2011.
15. Bernard J., Delprat S., Guerra T.M. and Buchi F.N. Fuel efficient power management strategy for fuel cell hybrid powertrains. *Control Engineering Practice*, 18:408–417, 2010.
16. Feroldi D., Serra M. and Riera J. Energy management strategies based on efficiency map for fuel cell hybrid vehicles. *Journal of Power Sources*, 190(2):387–401, 2009.
17. Feroldi D., Serra M. and Riera J. Design and analysis of fuel-cell hybrid systems oriented to automotive applications. *IEEE Transactions on Vehicular Technology*, 58(9):4720–4729, 2009.

18. Bashash S., Moura S.J., Forman J.C. and Fathy H.K. Plug-in hybrid electric vehicle charge pattern optimization for energy cost and battery longevity. *Journal of Power Sources*, 196:541–549, 2011.

19. Tazelaar E., Veenhuizen B., van den Bosch P. and Grimminck M. Analytical solution of the energy management for fuel cell hybrid propulsion systems. *IEEE Transactions on Vehicular Technology*, 61(5):1986–1998, 2012.

20. Bernard J., Delprat S., Bchi F.N., and Guerra M.T. Fuel-cell hybrid powertrain: Toward minimization of hydrogen consumption. *IEEE Transactions on Vehicular Technology*, 58(7):3168–3176, 2009.

21. Banvait H., Anwar S. and Chen Y. A rule-based energy management strategy for plug-in hybrid electric vehicle (phev). In *American Control Conference, 2009. ACC '09*, St. Louis, MO, pp. 3938–3943, 2009.

22. Luenberger D.G. *Optimization by Vector Space Methods.* John Wiley and Sons, New York, 1969.

23. Kuhn H.W. and Tucker A.W. Nonlinear programming. In *Proceedings of the 2nd Berkeley Symposium on Mathematical Statistics and Probability,* University of California, Berkeley, CA, 1951.

24. Mangasarin O.L. *Nonlinear Programming.* SI AM. Classics in Applied Mathematics, Philadelphia, PA, 1994.

25. Hanson M.A. On sufficiency of the Kuhn–Tucker conditions. *Journal of Mathematical Analysis and Applications*, 80:545–550, 1981.

26. Hanson M.A. and Mond B. Necessary and sufficient conditions; Kuhn–Tucker conditions; invexity, type I and type II functions; duality; converse duality. *Mathematical Programming*, 37(1):51–58, 1987.

27. Sashi K.M. and Giorgi G. *Non Convex Optimization and Its Applications. Invexity and Optimization.* Springer Verlag, 2008.

28. Fleming W.H. and Rishel R.W. *Deterministic and Stochastic Optimal Control.* Springer Verlag, New York, NY, 1975.

29. Stein G. Respect the unstable. *IEEE Control Systems*, 23(4):12–25, 2003.

30. Ang K.H., Chong G., and Li Y. PID control system analysis, design, and technology. *IEEE Transactions on Control Systems Technology*, 13(4):559–576, 2005.

31. Li Y., Ang K.H. and Chong G.C.Y. PID control system analysis and design. *IEEE Control Systems*, 26(1):32–41, 2006.

32. Chen C.-T. *Linear System Theory and Design.* Oxford University Press, Inc., New York, NY, USA, 1999.

33. Path V.N. and Jeyakumar V. Stability, stabilization and duality for linear time-varying systems. *Optimization: A Journal of Mathematical Programming and Operations Research*, 59:447–460, 2010.

34. Horisberger H. and Belanger P.R. Regulators for linear, time invariant plants with uncertain parameters. *IEEE Transactions on Automatic Control*, 21(5), 705–708, 1976.

35. Gahinet P., Nemirobskii A., Laub A.J. and Chilali M. *LMI Control Toolbox.* Mathworks, 1994.

36. Chesi G., Garulli A., Tesi A. and Vicino A. Homogeneous Lyapunov functions for systems with structured uncertainties. *Automatica*, 39:1027–1035, 2003.

37. Chesi G., Garulli A., Tesi A. and Vicino A. Robust stability of time-varying polytopic systems via parameter dependent homogeneous Lyapunov functions. *Automatica*, 43:309–316, 2007.

38. Freeman R. and Kokotovic P.V. *Robust Nonlinear Control Design: State-Space and Lyapunov Techniques.* Modern Birkhuser Classics, Ann Arbor, MI, 2008.

# Index

Printed and bound by CPI Group (UK) Ltd, Croydon, CR0 4YY

18/10/2024

01776231-0005